Handbook of
CHIRAL
CHEMICALS

Handbook of
CHIRAL
CHEMICALS

edited by

David J. Ager

NSC Technologies
Mount Prospect, Illinois

MARCEL DEKKER, INC.   NEW YORK · BASEL

ISBN: 0-8247-1058-4

This book is printed on acid-free paper.

Headquarters
Marcel Dekker, Inc.
270 Madison Avenue, New York, NY 10016
tel: 212-696-9000; fax: 212-685-4540

Eastern Hemisphere Distribution
Marcel Dekker AG
Hutgasse 4, Postfach 812, CH-4001 Basel, Switzerland
tel: 44-61-261-8482; fax: 44-61-261-8896

World Wide Web
http://www.dekker.com

The publisher offers discounts on this book when ordered in bulk quantities. For more information, write to Special Sales/Professional Marketing at the headquarters address above.

Current printing (last digit):
10 9 8 7 6 5 4 3 2 1

PRINTED IN THE UNITED STATES OF AMERICA

Preface

The purpose of this book is to highlight the problems associated with the production of chiral compounds on a commercial scale. With the movement by pharmaceutical companies to develop single enantiomers as drug candidates, the focus has turned to problems associated with this subclass of organic synthesis. The major classes of natural products are also discussed since the stereogenic center can be derived from nature through the use of "chiral pool" starting materials.

Despite the explosion of asymmetric methods over the past 20 years, very few can be performed at scale due to limitations in cost, thermodynamics, or equipment. The major reactions that have been used are covered in this volume. Resolution, whether chemical or enzymatic, still holds a key position. This is highlighted by a short discussion of the best-selling compounds of 1996. Many are obtained either by resolution or by fermentation methods.

The most mature chemical method is asymmetric reductions and hydrogenations. This is highlighted by chapters on the uses of new ligands for hydrogenation and hydride-reducing agents. Although we have made considerable advances in this area, the general catalyst is still elusive. The struggle goes on to identify the ultimate hydrogenation catalyst; for example, the use of enzymes and biological systems for the production of chiral compounds continues to increase at an almost explosive rate. Now that we have learned to manipulate nature's catalysts, this area will continue to grow and become more important.

The chapter on amino acid derivatives is the result of a considerable amount of research on the new methods for the preparation of unnatural amino acids and derivatives at scale. Their findings carry over into other classes of compounds, but the principles are highlighted exclusively within this field.

The chapters are grouped by topic. The first three are an introduction and discussion of the requirements of sourcing chiral intermediates. Another chapter presents an overview of the current large-volume chiral compounds and how they are synthesized.

The next three chapters discuss how the key subclasses of the chiral pool are obtained. The amino acid chapter is specific to the chiral pool materials as there are more examples of amino acid syntheses contained within other chapters.

The next eight chapters cover methods that can be used to introduce or control stereogenic centers. In some cases, such as asymmetric hydrogenations, the approach is well established and has been employed for the large-scale synthesis of a number of commercially important compounds. In other cases, such as pericyclic reactions, the potential exists, but has not yet been used. One chapter covers enzymatic methods, an area that seems to be becoming more important as we learn how to manipulate enzymes by allowing them to catalyze new reactions or take new substrates. The rush to market for pharmaceutical companies is forcing the chemical development time to be minimized. This is leading to large-scale usage of chiral auxiliaries.

The chapter on resolutions has a number of examples as illustrations showing that this methodology is still important to obtain chiral compounds. Although, ultimately, it may not be the most cost-effective method, it can provide material in a rapid manner, and can usually be scaled up. The introduction of large-scale chromatographic techniques, as well as the availability of a large number of enzymes that can be used to perform reactions on only one enantiomer, will ensure that this approach remains a useful tool in the future.

The remaining chapters discuss various examples and topics to augment other chapters and provide a perspective of the different methods available.

I would like to thank all the authors who contributed to this book and who have worked on it with me for the past few years. I would especially like to thank my colleagues at NSC Technologies for writing a number of the chapters and for having supplied numerous suggestions and ideas. Not only have they developed new methodology, but they have also proceeded to use it at scale within a very short timeframe. They continue to inspire me, as do many others working in the arena of asymmetric synthetic methodology.

David J. Ager

Contents

Preface   *i*

Contributors   *ix*

1. Introduction   1
   David J. Ager

   1.1. Chirality   1
   1.2. Chiral Pool   2
   1.3. Chiral Reagents   2
   1.4. Chiral Catalysts   2
   1.5. Stoichiometric Reagents   4
   1.6. Resolution   5
   1.7. Synthesis at Scale   5
   1.8. Analysis   8
   1.9. Summary   8
   References   8

2. Sourcing Chiral Compounds for the Pharmaceutical Industry   11
   Graham J. Tucker

   2.1. Introduction   11
   2.2. Consideration of Sources   12
   2.3. Major, Medium, and Minor Players   28
   2.4. Technology   28
   2.5. Available Chiral Compounds   32
   2.6. Conclusion   32

3. Synthesis of Large-Volume Products   33
   David J. Ager

   3.1. Introduction   33
   3.2. Pharmaceuticals   33
   3.3. Food Ingredients   40
   3.4. Agricultural Products   43
   3.5. Summary   45
        References   45

4. Synthesis of Phenylalanine by Fermentation and
   Chemoenzymatic Methods   49
   Ian G. Fotheringham

   4.1. Introduction   49
   4.2. L-Phenylalanine Overproducing Microorganisms   50
   4.3. Biotransformation Routes to L-Phenylalanine   57
   4.4. Resolution-Based L-Phenylalanine Synthesis   60
   4.5. Conclusions   63
        References   64

5. Carbohydrates in Synthesis   69
   David J. Ager

   5.1. Introduction   69
   5.2. Disaccharides   70
   5.3. Monosaccharides and Related Compounds   72
   5.4. Glyceraldehyde Derivatives   75
   5.5. Hydroxy Acids   76
   5.6. Summary   79
        References   79

6. Terpenes: Expansion of the Chiral Pool   83
   Weiguo Liu

   6.1. Introduction   83
   6.2. Isolation   84
   6.3. Monoterpenes   84
   6.4. Reactions of Monoterpenes   92
   6.5. Summary   99
        References   100

7. **Substitution Reactions**   103
   David J. Ager

    7.1.   Introduction   103
    7.2. S_N2 Reactions   103
    7.3.   Epoxide Openings   105
    7.4.   Cyclic Sulfate Reactions   107
    7.5.   Iodolactonizations   108
    7.6.   Allylic Substitutions   108
    7.7.   Summary   110
          References   110

8. **Resolutions at Large Scale: Case Studies**   115
   Weiguo Liu

    8.1.   Introduction   115
    8.2.   Chemical Resolution   116
    8.3.   Enzymatic Resolutions   128
    8.4.   Summary   138
          References   138

9. **Transition Metal Catalyzed Hydrogenations, Isomerizations, and Other Reactions**   143
   Scott A. Laneman

    9.1.   Introduction   143
    9.2.   Homogeneous Catalysts   147
    9.3.   Asymmetric Heterogeneous Catalysts Implemented in Industry   165
    9.4.   Asymmetric Hydrogen Transfer   168
    9.5.   Hydroformylation   169
    9.6.   Hydrosilylation   170
    9.7.   Asymmetric Cyclopropanations   171
    9.8.   Conclusions   171
          References   173

10. **Pericyclic Reactions**   177
    Michael B. East

    10.1. Introduction   177
    10.2. The Diels-Alder Reaction   177
    10.3. Claisen-Type Rearrangements   187

10.4 The Ene Reaction   190
10.5. Dipolar Cycloadditions   191
10.6. [2,3]-Sigmatropic Rearrangements   192
10.7. Other Pericyclic Reactions   194
10.8. Summary   195
        References   196

11. Asymmetric Reduction of Prochiral Ketones Catalyzed
      by Oxazaborolidines   211
      Michel Bulliard

11.1. Introduction   211
11.2. Stoichiometric Reactions   211
11.3. The Catalytic Approach   212
11.4. Industrial Application in the Synthesis of
        Pharmaceuticals   220
11.5. Conclusion   224
        References   224

12. Asymmetric Oxidations   227
      David J. Ager and David R. Allen

12.1. Introduction   227
12.2. Sharpless Epoxidation   227
12.3. Asymmetric Dihydroxylation   232
12.4. Jacobsen Epoxidation   236
12.5. Halohydroxylations   238
12.6. Enzymatic Methods   239
12.7. Summary   239
        References   239

13. Biotransformations: "Green" Processes for the Synthesis
      of Chiral Fine Chemicals   245
      David P. Pantaleone

13.1. Introduction   245
13.2. Biocatalyst Classifications   246
13.3. Metabolic Pathway Engineering   272
13.4. Screening for Biocatalysts   276
13.5. Summary   278
        References   278

14. **Industrial Applications of Chiral Auxiliaries**   **287**
David R. Schaad

    14.1.   Introduction   287
    14.2.   Chiral Auxiliary Structures in Pharmaceuticals   289
    14.3.   Application of Chiral Auxiliaries in Industry   290
    14.4.   Potential Applications of Chiral Auxiliaries   293
    14.5.   Conclusions   298
           References   298

15. **Synthesis of Unnatural Amino Acids: Expansion of the Chiral Pool**   **301**
David J. Ager, David R. Allen, Michael B. East, Ian G. Fotheringham, Scott A. Laneman, Weiguo Liu, David P. Pantaleone, David R. Schaad, and Paul P. Taylor

    15.1.   Introduction   301
    15.2.   The Choice of Approach   302
    15.3.   Small-Scale Approaches   304
    15.4.   Intermediate-Scale Approaches   308
    15.5.   Large-Scale Methods   310
    15.6.   Summary   315
           References   315

16. **Synthesis of L-Aspartic Acid**   **317**
Paul P. Taylor

    16.1.   Introduction   317
    16.2.   Commercial Production   317
    16.3.   General Properties of Aspartase   319
    16.4.   Biocatalyst Development   319
    16.5.   Future Perspectives   323
           References   326

17. **Synthesis of Homochiral Compounds: A Small Company's Role**   **329**
Basil J. Wakefield

    17.1.   Introduction   329
    17.2.   Classical Resolution   329
    17.3.   The Chiral Pool   331
    17.4.   Enzyme-Catalyzed Kinetic Resolution   333
           References   336

18. Asymmetric Catalysis: Development and Applications
    of the DuPHOS Ligands   **339**
    Mark J. Burk

    18.1. Introduction   339
    18.2. Chiral Ligands   340
    18.3. Asymmetric Catalytic Hydrogenation Reactions   343
    18.4. Commercial Development and Application of
           Asymmetric Catalysis   356
    18.5. Summary   358
           References   358

Index   *361*

Contributors

DAVID J. AGER, PH.D.   Fellow, NSC Technologies, Mount Prospect, Illinois

DAVID R. ALLEN, B.S.   Senior Process Chemist, NSC Technologies, Mount Prospect, Illinois

MICHEL BULLIARD, PH.D.   Research and Development Manager, PPG-SIPSY, Avrille, France

MARK J. BURK, PH.D.   Head, ChiroTech Technology Limited, Chiroscience PlC., Cambridge, United Kingdom

MICHAEL B. EAST, PH.D.   Senior Research Scientist, NSC Technologies, Mount Prospect, Illinois

IAN G. FOTHERINGHAM, PH.D.   NSC Technologies, Mount Prospect, Illinois

SCOTT A. LANEMAN, PH.D.   Senior Research Scientist, NSC Technologies, Mount Prospect, Illinois

WEIGUO LIU, PH.D.   Senior Research Scientist, NSC Technologies, Mount Prospect, Illinois

DAVID P. PANTALEONE, PH.D.   Group Leader, Protein Biochemistry, NSC Technologies, Mount Prospect, Illinois

DAVID R. SCHAAD, PH.D.   Senior Research Scientist, Department of Chemical Process Development, NSC Technologies, Mount Prospect, Illinois

PAUL P. TAYLOR, PH.D.   NSC Technologies, Mount Prospect, Illinois

GRAHAM J. TUCKER, B.Sc.   The R-S Directory, Kenley Chemicals, Kenley, Surrey, United Kingdom

BASIL J. WAKEFIELD, PH.D., D.Sc.   Chemicals Director, Ultrafine Chemicals, UFC Pharma, Manchester, United Kingdom

1

Introduction

DAVID J. AGER
NSC Technologies, Mount Prospect, Illinois

This book discusses various aspects of chiral fine chemicals, including their synthesis and uses at scale. There is an increasing awareness of the importance of chirality in biological molecules, as the two enantiomers can sometimes have different effects. [1–4].

In many respects, chiral compounds have been regarded as special entities within the fine chemical community. As we will see, the possession of chirality does not, in many respects, make the compound significantly more expensive to obtain. Methods for the preparation of optically active compounds have been known for well over 100 years (many based on biological processes). The basic chemistry to a substrate on which an asymmetric transformation is then performed can offer more challenges in terms of chemistry and cost optimization than the "exalted" asymmetric step.

1.1. CHIRALITY

The presence of a stereogenic center within a molecule can give rise to chirality. Unless a chemist performs an asymmetric synthesis, equal amounts of the two antipodes will be produced. To separate these, or to perform an asymmetric synthesis, a chiral agent has to be employed. This can increase the degree of complexity in obtaining a chiral compound in a pure form. However, nature has been kind and does provide some chiral compounds in relatively large amounts. Chirality does provide an additional problem that is sometimes not appreciated by those who work outside of the field: analysis of the final compound is often not a trivial undertaking.

1.2. CHIRAL POOL

Nature has provided a wide variety of chiral materials, some in great abundance. The functionality ranges from amino acids to carbohydrates to terpenes (Chapters 4–6). All of these classes of compounds are discussed in this book. Despite the breadth of functionality available from natural sources, very few compounds are available in optically pure form at large scale. Thus, incorporation of a ''chiral pool'' material into a synthesis can result in a multistep sequence. However, with the advent of synthetic methods that can be used at scale, new compounds are being added to the chiral pool, although they are only available in bulk by synthesis. When a chiral pool material is available at large scale, it is usually inexpensive. An example is provided by L-aspartic acid (Chapter 16), where the chiral material can be cheaper than the racemate (see also Chapter 15).

How some of these chiral pool materials have been incorporated into synthesis of biologically active compounds is illustrated in this book. In addition, chiral pool materials are often incorporated, albeit in derivatized form, into chiral reagents and ligands that allow for the transfer of chirality from a natural source into the desired target molecule.

1.3. CHIRAL REAGENTS

Chiral reagents allow for the transfer of chirality from the reagent to the prochiral substrate. Almost all of these reactions involve the conversion of an sp^2 carbon to an sp^3 center. For example, reductions of carbonyl compounds (Chapter 11), asymmetric hydrogenations (Chapter 9), and asymmetric oxidations of alkenes (Chapter 12) are all of this type. The reagents can be catalytic for the transformation they bring about, or stoichiometric. The former is usually preferred because it allows for chiral multiplication during the reaction—the original stereogenic center gives rise to many product stereocenters. This allows for the cost of an expensive catalyst to be spread over a large number of product molecules.

1.4. CHIRAL CATALYSTS

Considerable resources are being expended in the quest for new asymmetric catalysts for a wide variety of reactions (Chapter 9). In many cases, these catalysts are based on transition metals, where the ligands provide the chiral environment. However, as our understanding of biotransformations increases, coupled with our ability to produce mutant enzymes at scale, biocatalysts are beginning to become key components of our asymmetric synthetic tool box (Chapters 13 and 15).

1.4.1. Chemical Catalysts

The development of transition metal catalysts for the asymmetric reduction of functionalized alkenes allowed synthetic chemists to perform reactions with a stereochemical fidelity approaching that of nature (Chapter 9). We now have a number of reactions at our disposal that can be performed with chemical catalysts, and the number continues to grow. However, there are still problems associated with this approach because many catalysts have specific substrates requirements, often involving just one alkene isomer of the substrate. The chiral multiplication associated with use of a chiral catalyst often makes for attractive economical advantages. However, the discovery and development of a chemical catalyst to perform a specific transformation is often tedious, time consuming, and expensive. There are many reports of chiral ligands in the literature, for example, to perform asymmetric hydrogenation, yet very few have been used at scale (Chapter 9). This highlights the problem that there are few catalysts that can be considered general. As previously mentioned, the preparation of the substrate is often the expensive part of a sequence, especially with catalysts that have high turnover numbers and can be recycled.

1.4.2. Biological Catalysts

Biological catalysts have been used for asymmetric transformations in specific cases for a considerable period of time, excluding the chiral pool materials. However, until recently, the emphasis has been on resolutions with enzymes rather than asymmetric transformations (Chapter 13). With our increasing ability to produce mutant enzymes that have different or broad-spectrum activities compared with the wild types, the development of biological catalysts is poised for major development. In addition to high stereospecificities, an organism can be persuaded to perform more than one step in the overall reaction sequence, and may even make the substrate (Chapter 15).

Unlike the design of a chemical catalyst, which has to be semi-empirical in nature and is therefore very difficult to apply to a completely different transformation, screening for an enzyme that performs a similar reaction is relatively straightforward and often gives the necessary lead for the development of a potent biological catalyst. The use of molecular biology, site-specific mutagenesis, and enzymology all contribute to the development of such a catalyst. This approach is often ignored because these methods are outside of traditional chemical methodologies.

There are a large number of reports of abzymes, or catalytic antibodies, in the literature [5–10]. Although catalysis has been observed in a large number of examples, the problems associated with the production of large amounts of abzymes, compounded by the low turnover numbers often observed, makes this technology only a laboratory curiosity. The increasing use of mutant enzymes without isolation from the host organisms makes this latter approach economically more attractive.

1.5. STOICHIOMETRIC REAGENTS

To understand a specific transformation, chemists have often developed asymmetric synthetic methods in a logical, stepwise manner. Invariably, the mechanism of the reaction and the factors that control the stereochemical outcome of a transformation are paramount in the design of an efficient catalyst for use at scale with a wide variety of substrates. There are some noticeable exceptions to this approach, such as the empirical approach used to develop asymmetric hydrogenation catalysts (Chapter 9). In other instances, an empirical approach provided sufficient insight to allow for the development of useful chiral catalysts, such as the empirical rule for the oxidation of alkenes [11,12] that led to the asymmetric hydroxylation catalysts [13,14]. The first generation of asymmetric reagents are often chiral templates or stoichiometric reagents. These are then superseded by chiral auxiliaries, if the substrate has to be modified, or chiral catalysts in the case of external reagents.

1.5.1. Chiral Auxiliaries

This class of compounds modifies the substrate molecule to introduce a stereogenic center that will influence the outcome of a reaction to provide an asymmetric synthesis. The auxiliary has to be put onto the substrate and removed. Although this involves two steps, concurrent protection of sensitive functionality can also take place, so that one inefficient sequence (protection and deprotection) is traded for another (auxiliary introduction and removal). A large number of asymmetric transformations have been performed with chiral auxiliaries, providing a wealth of literature. Thus, a precedence for most reactions is available, providing for a large degree of certainty that a specific reaction, even with a new substrate, will work (see Chapters 14 and 15). Due to the curtailed timelines for the development of new pharmaceutical products, coupled with the decrease in the costs of many auxiliaries, this approach is now being used at larger scale.

Although an auxiliary is recovered intact after the asymmetric transformation and has the potential to be recycled, there are often problems associated with the practical implementation of this concept.

1.5.2. Chiral Templates

Chiral templates can be considered a subclass of chiral auxiliaries. Unlike auxiliaries that have the potential to be recycled, the stereogenic center of a template is destroyed during its removal. Although this usually results in the formation of simple by-products that are easy to remove, the cost of the template's stereogenic center is transferred to the product molecule. The development of a template is usually the first step in understanding a specific transformation, and the knowledge gained is used to develop an auxiliary or catalyst system.

1.6. RESOLUTION

The separation of enantiomers through the formation of derivative diastereoisomers and the subsequent separation of these by physical means has been practiced at large scale for many years (Chapter 8). In addition, the racemic mixture can be reacted with a chiral reagent, where the rates of reaction are very different for the two enantiomers, allowing for a resolution. This approach is applicable to chemical agents, such as the Sharpless epoxidation procedure [15–17], and biological agents, such as enzymes (see Chapters 8, 12, and 13). Unless a meso-substrate is used, or the wrong isomer can be converted back to the racemate in situ, to provide a dynamic resolution, the "off-isomer" can present an economic problem. It either has to be disposed of—this results in a maximum overall yield of 50% from the racemic substrate—or epimerized to allow for recycle through the resolution sequence. The latter approach often involves additional steps in a sequence that can prove to be costly. In either case, recovery of the resolving agent also has to be considered. As the development of robust, general, asymmetric methods to a class of compounds becomes available, it is becoming apparent that the more traditional resolution approaches are not economically viable. However, there is still a place for resolutions (see Chapters 8 and 15).

1.7. SYNTHESIS AT SCALE

There are many problems associated with conducting asymmetric synthesis at scale. Many asymmetric transformations reported in the literature use the technique of low temperature to allow differentiation of the two possible diastereomeric reaction pathways. In some cases, the temperature requirements to see good asymmetric induction can be as low as $-100°C$. To obtain this temperature in a reactor is not only costly in terms of cooling, but also presents problems associated with materials of construction and the removal of heat associated with the exotherm of the reaction itself. It is comforting to see that many asymmetric catalytic reactions do not require the use of low temperature. However, the small number of "robust" reactions often leads development chemists to resort to a few tried and tested approaches, namely, chiral pool synthesis, use of a chiral auxiliary, or resolution. In addition, the scope and limitations associated with the use of a chiral catalyst often result in a less-than-optimal sequence either because the catalyst does not work well on the necessary substrate, or the preparation of that substrate is long and costly. Thus, the availability of a number of different approaches helps to minimize these problems (Chapter 15).

A short overview of the synthesis of some of the large-scale and monetary value chiral products is given in Chapter 3. This illustrates the relative importance of some of the approaches discussed within this book, especially the power of biological approaches.

1.7.1. Reactions That Are Amenable to Scale

When reactions that are "robust" are considered, only a relatively small number are available. Each of these reaction types are discussed within this book, although some do appear under the chiral pool materials that allowed for the development of this class of asymmetric reagent. Such an example is the use of terpenes, which have allowed for the development of chiral boranes (Chapter 6).

1.7.1.1. Biological Methods

This class of reagents holds the most promise for rapid development in the near future as most reactions are asymmetric. The problems being overcome are the tight substrate specificity of many enzymes and the need for cofactor regeneration. Systems are now being developed for asymmetric synthesis rather than resolution approaches. Some of these reactions are discussed in Chapter 13.

1.7.1.2. Transition Metal Catalyzed Oxidations

The development of simple systems that allow for the asymmetric oxidation of allyl alcohols and simple alkenes to epoxides or 1,2-diols has had a great impact on synthetic methodology as it allows for the introduction of functionality with concurrent formation of one or two stereogenic centers. This functionality can then be used for subsequent reactions that usually fall into the substitution reaction class. Because these transition metal catalysts do not require the use of low temperatures to ensure high degrees of induction, they can be considered robust. However, the sometimes low catalyst turnover numbers and the synthesis of the substrate can still be crucial economic factors. Aspects of asymmetric oxidations are discussed in Chapter 12.

1.7.1.3. Transition Metal Catalyzed Reductions

As mentioned elsewhere (Chapter 9), the development of transition metal catalysts that allowed for high enantioselectivity in reduction reactions showed that chemists could achieve comparable yields and enantiomeric excesses (ee's) to enzymes. A large number of transition metal catalysts and ligands are now available. A number of reactions that use an asymmetric hydrogenation for the key chiral step have been scaled up for commercial production (Chapters 9 and 18).

1.7.1.4. Transition Metal Catalyzed Isomerizations

These reactions are closely related to asymmetric hydrogenations, especially as similar catalysts are used. A number of these reactions are used at scale (Chapter 9).

1.7.1.5. Reductions

The reduction of a carbonyl group to an alcohol has been achieved at a laboratory scale with a large number of reagents, many stoichiometric with a ligand derived from the chiral pool. The development has culminated in boron-based reagents that perform this transformation with high efficiency (Chapter 11).

1.7.1.5.1. Hydroborations.

In addition to being useful reagents for the reductions of carbonyl compounds, boron-based reagents can also be used for the conversion of an alkene to a wide variety of functionalized alkanes. Because the majority of these reagents carry a terpene substituent, they are discussed under these chiral pool materials (Chapter 6).

1.7.1.6. Pericyclic Reactions

Many of these reactions are stereospecific and, because they have to be run at temperatures higher than ambient, are very robust. It is somewhat surprising that there are very few examples of pericyclic reactions being run at scale, especially in light of our understanding of the factors that control the stereochemical course of the reaction, either through the use of a chiral auxiliary or catalyst (Chapter 10).

1.7.1.7. Substitution Reactions (S_N2)

This heading has been used to describe the conversion of one stereogenic center to another. Of course, this means that the substrate stereogenic center has had to be obtained by one of the reaction types outlined above, from the chiral pool, or by resolution. Reactions that fall into this category include epoxide and cyclic sulfate openings, and iodolactonizations (Chapter 7).

I.8. ANALYSIS

The analysis of chiral compounds to determine their optical purity is still not a trivial task. The analysis method has to differentiate between the two antipodes, and, thus, has to involve a chiral agent. However, the development of chiral chromatography, especially high-performance liquid chromatography, has done a significant amount to relieve this problem. The purpose of this book is to discuss large-scale synthetic reactions, but the development of chiral analytical methods may not have been a trivial undertaking in many examples.

I.9. SUMMARY

The development of optically active biological agents such as pharmaceuticals has led to the increase in large-scale chiral synthesis. The chirality may be derived from the chiral pool or a chiral agent such as an auxiliary, template, reagent, or catalyst. There are, however, relatively few general asymmetric methods that can be used at scale.

REFERENCES

1. Blashke, G., Kraft, H. P., Fickenscher, K., Kohler, F. *Arzniem-Forsch/Drug Res.* 1979, *29*, 10.
2. Blashke, G., Kraft, H. P., Fickenscher, K., Kohler, F. *Arzniem-Forsch/Drug Res.* 1979, *29*, 1140.
3. Powell, J. R., Ambre, J. J., Ruo, T. I. in *Drug Stereochemistry*; Wainer, I. W., Drayer, D. E., Eds. Marcel Dekker: New York, 1988, p. 245.
4. Ariëns, E. J., Soudijn, W., Timmermas, P. B. M. W. M. *Stereochemistry and Biological Activity of Drugs*; Blackwell Scientific: Palo Alto, 1983.
5. Lerner, R. A., Benkovic, S. J., Schultz, P. G. *Science* 1991, *252*, 659.
6. Blackburn, G. M., Kang, A. S., Kingsbury, G. A., Burton, D. R. *Biochem. J.* 1989, *262*, 381.
7. Schultz, P. G., Lerner, R. A. *Acc. Chem. Res.* 1993, *26*, 391.
8. Hilvert, D. *Acc. Chem. Res.* 1993, *26*, 552.
9. Stewart, J. D., Liotta, L. J., Benkovic, S. J. *Acc. Chem. Res.* 1993, *26*, 396.
10. Stewart, J. D., Benkovic, S. J. *Chem. Soc. Rev.* 1993, *22*, 213.
11. Cha, J. K., Christ, W. J., Kishi, Y. *Tetrahedron Lett.* 1983, *24*, 3943.
12. Cha, J. K., Christ W. J., Kishi, Y. *Tetrahedron* 1984, *40*, 2247.
13. Sharpless, K. B., Behrens, C. H., Katsuki, T., Lee, A. W. M., Martin, V. S., Takatani, M., Viti, S. M., Walker, F. J., Woodard, S. S. *Pure Appl. Chem.* 1983, *55*, 589.
14. Sharpless, K. B., Verhoeven, T. R. *Aldrichimica Acta* 1979, *12*, 63.

15.   Brown, J. M. *Chem. Ind. (London)* 1988, 612.
16.   Gao, Y., Hanson, R. M., Klunder, J. M., Ko, S. Y., Masamune, H., Sharpless, K. B. *J. Am. Chem. Soc.* 1987, *109*, 5765.
17.   Ager, D. J., East, M. B. *Asymmetric Synthetic Methodology*; CRC Press: Boca Raton, 1995.

2

Sourcing Chiral Compounds for the Pharmaceutical Industry

GRAHAM J. TUCKER
The R-S Directory, Kenley Chemicals, Kenley, Surrey, England

2.1. INTRODUCTION

In this chapter, some generalizations have been attempted, an exercise that is always open to disagreement. It will no doubt be possible to cite exceptions, particularly in the field of chiral compounds. The focus is on sourcing for drug candidates in development.

Sourcing a chiral raw material or intermediate is different from sourcing other fine chemicals. There is a perception in some quarters that chirality offers participation in a "sunrise" industry, somehow to be regarded similarly to biotechnology. As such, there are more "start-up" companies to consider, some backed by venture capital, compared with the situation with other fine chemicals.

Another different factor in sourcing chiral materials is that, in many cases, the ways to potentially obtain them can be diverse and often more novel than for racemates. The overwhelming majority of racemic intermediates is still made through the use of organic, often classical, chemical processes with conventional physical methodology used to isolate and purify the product. However, there are some exceptions in which a biological process is used to make an achiral material such as acrylamide. Nonetheless, nature is decidedly not evenhanded, and biological-based processes with prochiral substrates often yield chiral materials. Such processes join asymmetric chemical synthesis as candidate technologies to access a specific enantiomer. A further possibility to obtain a single antipode is separation of a racemic compound through the less conventional physical technique of chiral chromatography. All of these methods are in a state of rapid evolution,

11

fueled by the increase in number of chiral compounds required for pharmaceutical development.

How important is chirality? Based on the percentage of chiral compounds entering phase I development in the last year, the importance of chirality varies greatly between companies. In one case, all phase I candidates were chiral, in another, just 5%. In the recent past, some discovery groups were deliberately avoiding chiral molecules to circumvent perceived complications.

2.2. CONSIDERATION OF SOURCES

Suppliers can be classified as chiral center creators (CCCs), of which there are relatively few, or chiral center elaborators (CCEs), of which there are many.

CCCs may obtain chiral compounds by classical resolution, kinetic resolution using chemical or enzymatic methods, biocatalysis (enzyme systems, whole cells, or cell isolates), fermentation (from growing whole microorganisms), and stereoselective chemistry (e.g., asymmetric reduction, low-temperature reactions, use of chiral auxiliaries). CCCs may also be CCEs by capitalizing on a key raw material position and "going downstream." Along with companies manufacturing chiral molecules primarily for other purposes, such as amino acid producers, these will be the key sources for the asymmetric center.

CCCs may be companies that specialize in chiral compounds, such as Celgene, Chiroscience, and Oxford Asymmetry, or may have developed into a chiral raw material supplier to an industry other than its original main customer base [e.g., Takasago, a flavor manufacturer, who developed a catalytic route to (−)-menthol, and used related technology to make beta-lactam intermediates].

CCEs elaborate an available chiral raw material to a further chiral molecule, for example, an amino acid to an amino alcohol and then to an oxazolidinone auxiliary. Thus, there are at least nine suppliers of (S)-phenylalaninol and six of (S)-4-benzyl-2-oxazolidinone, although not all of these might be considered commercial in terms of potential for competitive production at scale. Often, only a few compounds are made by CCEs, just in response to market demand. Many fine chemical intermediate manufacturers can be CCEs, although, to be viable sources in the longer term, they must possess expertise or patented/proprietary technology covering the utilization of the chiral raw material. There will also be cases of advantageous access to the required chiral raw material or, perhaps, the process will involve a hazardous reactant that a particular plant is equipped to handle. Although offering a chiral product, these companies' role as competitive suppliers arises from factors other than particular expertise in chirality itself (Table 1). One might suspect that some fine chemical producers mention chiral compounds not because they have anything special to offer, but merely because they want to participate in a field with perceived growth potential.

TABLE I   Chiral Compounds Available at Scale[a]

Compound	Formula	Supplier
(3R,4R)-4-Acetoxy-3-[(R)-t-butyldimethylsilyloxy)ethyl]-2-azetidinone	$C_{13}H_{25}NO_4Si$	Kaneka, Nippon Soda, Takasago
N-Acetyl-(4S)-benzyl-2-oxazolidinone	$C_{12}H_{13}NO_3$	NSC Technologies
(R)-O-Acetylmandelic acid	$C_{10}H_{10}O_4$	Yamakawa
(S)-O-Acetylmandelic acid	$C_{10}H_{10}O_4$	Yamakawa
(S)-(−)-Acetyl-3-mercapto-2-methylpropionic acid	$C_6H_{10}O_3S$	DSM Andeno, Kaneka, Sumitomo Seika
N-Acetyl-D-3-(2-naphthyl)alanine	$C_{15}H_{15}NO_3$	Rexim/Degussa, Synthetech
(R)-Acetylthio-2-methylpropionyl chloride	$C_6H_9O_2ClS$	Kaneka
D-Alaninamide HCl	$C_3H_9ClN_2O$	Tanabe, Synthetech
D-Alanine	$C_3H_7NO_2$	Ajinomoto, Kaneka, Nippon, Kayaku, Recordati, Rexim/Degussa, Tanabe, Toray
(R)-(−)-2-Aminobutanol	$C_4H_{11}NO$	Themis
(S)-(+)-2-Aminobutanol	$C_4H_{11}NO$	Themis
(1R,2S)-(−)-erythro-2-Amino-1,2-diphenylethanol	$C_{14}H_{15}NO_2$	Yamakawa
(1S,2R)-(+)-erythro-2-Amino-1,2-diphenylethanol	$C_{14}H_{15}NO_2$	Yamakawa
(−)-cis-(1S,2R)-Aminoindan-2-ol	$C_9H_{11}NO$	Chirex
(R)-(−)-2-Amino-3-methyl-1-butanol (D-Valinol)	C_5H_9NO	Boehringer, Newport, Rexim/Degussa
(S)-(+)-2-Amino-3-methyl-1-butanol (L-Valinol)	$C_5H_{13}NO$	Boehringer, Newport, Rexim/Degussa
L-(+)-2-Amino-2-methyl-3-(3,4-dimethoxyphenyl)propionitrile	$C_{12}H_{16}N_2O_2$	Alfa Chemicals Italiana
(S)-(+)-2-Amino-3-methyl-1-pentanol (L-(+)-Isoleucinol)	$C_6H_{15}NO$	Boehringer, Rexim/Degussa
(R)-(−)-2-Amino-4-methyl-1-pentanol (D-(−)-Leucinol)	$C_6H_{15}NO$	Rexim/Degussa
(S)-(+)-2-Amino-4-methyl-1-pentanol (L-(+)-Leucinol)	$C_6H_{15}NO$	Boehringer, Rexim/Degussa
[R-(R.R)]-2-Amino-1-[4-(methylthio)phenyl]-1,3-propanediol	$C_{10}H_{15}NO_2S$	Zambon
[S-(R,R)]-2-Amino-1-[4-(methylthio)phenyl]-1,3-propanediol	$C_{10}H_{15}NO_2S$	Zambon
(R)-(−)-2-Amino-2-phenylbutyric acid	$C_{10}H_{13}NO_2$	Sipsy
(S)-(+)-2-Amino-2-phenylbutyric acid	$C_{10}H_{13}NO_2$	Sipsy

(Table continues)

TABLE 1   Continued

Compound	Formula	Supplier
(R)-(−)-2-Amino-1-propanol (D-Alaninol)	C_3H_9NO	Boehringer, Newport
(S)-(+)-2-Amino-1-propanol (L-Alaninol)	C_3H_9NO	Boehringer, Newport, Synthetech
D-Arginine	$C_6H_{14}N_4O_2$	Ajinomoto, Rexim/Degussa, Tanabe
D-Aspartic acid	$C_4H_7NO_4$	Ajinomoto, Rexim/Degussa, Tanabe
D-(−)-α-Azidophenylacetic acid	$C_8H_7N_3O_2$	Dynamit Nobel
D-(−)-α-Azidophenylacetyl chloride	$C_8H_6ClN_3O$	Dynamit Nobel
(1S,2R)-(+)-cis-2-Benzamidocyclohexanecarboxylic acid	$C_{14}H_{17}NO_3$	Yamakawa
(1R,2S)-(−)-cis-2-Benzamidocyclohexanecarboxylic acid	$C_{14}H_{17}NO_3$	Yamakawa
(R)-(−)-1-Benzoyl-2-t-butyl-3-methyl-4-imidazolidinone	$C_{15}H_{20}N_2O_2$	Rexim/Degussa
(S)-(+)-1-Benzoyl-2-t-butyl-3-methyl-4-imidazolidinone	$C_{15}H_{20}N_2O_2$	Rexim/Degussa
(R)-(−)-Benzylglycidyl ether	$C_{10}H_{12}O_2$	Daiso
(S)-(+)-Benzylglycidyl ether	$C_{10}H_{12}O_2$	Daiso
(1S,2R)-(+)-cis-N-Benzyl-2-(hydroxymethyl)cyclohexylamine	$C_{14}H_{21}NO$	Yamakawa
(1R,2S)-(−)-cis-N-Benzyl-2-(hydroxymethyl)cyclohexylamine	$C_{14}H_{21}NO$	Yamakawa
(R)-N-Benzyl-3-hydroxypyrrolidine	$C_{11}H_{15}NO$	Kaneka, Toray
(S)-N-Benzyl-3-hydroxypyrrolidine	$C_{11}H_{15}NO$	Kaneka, Toray
Benzyl (R)-(−)-mandelate	$C_{15}H_{14}O_3$	Yamakawa
Benzyl (S)-(+)-mandelate	$C_{15}H_{14}O_3$	Yamakawa
(R)-(+)-N-Benzyl-α-methylbenzylamine	$C_{15}H_{17}N$	Yamakawa
(S)-(−)-N-Benzyl-α-methylbenzylamine	$C_{15}H_{17}N$	Yamakawa
(R)-(+)-4-Benzyl-2-oxazolidinone	$C_{10}H_{11}NO_2$	Boehringer, Nagase, Newport, Rexim/ Degussa, Urquima
(S)-(−)-4-Benzyl-2-oxazolidinone	$C_{10}H_{11}NO_2$	Boehringer, Nagase, Newport, NSC Technologies, Rexim/Degussa
Benzyl (S)-1,2,3,4-tetrahydroisoquinoline-3-carboxylate p-toluenesulphonate	$C_{24}H_{25}NO_5S$	Fine Organics, NSC Technologies, Tanabe

Benzyl (R)-2-tosyloxypropionate	$C_{17}H_{18}O_5S$	Daicel, Tanabe
(R)-(+)-1,1-Bi-2-naphthol	$C_{20}H_{14}O_2$	Finetech, Mitsubishi, Oxford, Strem
(S)-(−)-1,1-Bi-2-naphthol	$C_{20}H_{14}O_2$	Finetech, Mitsubishi, Oxford, Strem
(R)-(−)-1,1′-Binaphthyl-2,2-diyl hydrogen phosphate	$C_{20}H_{13}O_4P$	Finetech
(S)-(+)-1,1′-Binaphthyl-2,2-diyl hydrogen phosphate	$C_{20}H_{13}O_4P$	Finetech
(4R,5R)-Bis[bis(3′,5′-dimethyl-4′-methoxyphenyl)phosphinomethyl]-2,2-dimethyl-1,3-dioxolane	$C_{42}H_{56}O_6P_2$	Toyotama
(4S,5S)-Bis[bis(3′,5′-dimethyl-4′-methoxyphenyl)phosphinomethyl]-2,2-dimethyl-1,3-dioxolane	$C_{42}H_{56}O_6P_2$	Toyotama
(R)-(+)-2,2′-Bis(diphenylphosphino)-1,1′-binaphthyl	$C_{44}H_{32}P_2$	Finetech
(S)-(−)-2,2′-Bis(diphenylphosphino)-1,1′-binaphthyl	$C_{44}H_{32}P_2$	Finetech
(2R,3R)-(+)-2,3-Bis(diphenylphosphino)butane	$C_{28}H_{28}P_2$	Finetech
N-α-BOC-D-3-(3-Benzothienyl)alanine	$C_{16}H_{19}NO_4S$	Synthetech
N-α-BOC-L-3-(3-Benzothienyl)alanine	$C_{16}H_{19}NO_4S$	Synthetech
N-α-BOC-D-3-(4-Biphenyl)alanine	$C_{20}H_{23}NO_4$	Synthetech
N-α-BOC-L-3-(4-Biphenyl)alanine	$C_{20}H_{23}NO_4$	Synthetech
(R)-(+)-1-BOC-2-t-Butyl-3-methyl-4-imidazolidinone	$C_{13}H_{24}N_2O_3$	Rexim/Degussa
(S)-(−)-1-BOC-2-t-Butyl-3-methyl-4-imidazolidinone	$C_{13}H_{24}N_2O_3$	Rexim/Degussa
N-α-BOC-D-3-(4-Chlorophenyl)alanine	$C_{14}H_{18}ClNO_4$	Rexim/Degussa, Synthetech
N-α-BOC-L-3-(4-Chlorophenyl)alanine	$C_{14}H_{18}ClNO_4$	Synthetech
N-α-BOC-D-3,3-Diphenylalanine	$C_{20}H_{23}NO_4$	Synthetech
N-α-BOC-L-3,3-Diphenylalanine	$C_{20}H_{23}NO_4$	Synthetech
N-α-BOC-L-Octahydroindole-2-carboxylic acid	$C_{14}H_{23}NO_4$	Synthetech
N-BOC-D-Ornithine	$C_{10}H_{20}N_2O_4$	Tanabe
N-α-BOC-D-3-(4-Pentafluorophenyl)alanine	$C_{14}H_{14}F_5NO_4$	Synthetech
N-α-BOC-L-3-(4-Pentafluorophenyl)alanine	$C_{14}H_{14}F_5NO_4$	Synthetech
N-α-BOC-D-3-(2-Pyridyl)alanine	$C_{13}H_{18}N_2O_4$	Synthetech
N-α-BOC-L-3-(2-Pyridyl)alanine	$C_{13}H_{18}N_2O_4$	Synthetech
N-α-BOC-(3S,4S)-Statine	$C_{13}H_{25}NO_5$	Synthetech
(2R)-Bornane-10,2-sultam	$C_{10}H_{17}NO_2S$	Newport

(Table continues)

TABLE 1 Continued

Compound	Formula	Supplier
(2S)-Bornane-10,2-sultam	$C_{10}H_{17}NO_2S$	Newport
(R)-(+)-2-Bromobutyric acid	$C_4H_7BrO_2$	Linz
(S)-(−)-2-Bromobutyric acid	$C_4H_7BrO_2$	Linz
(+)-3-Bromocamphor	$C_{10}H_{15}BrO$	Calaire
(+)-3-Bromo-8-camphorsulphonic acid ammonium salt	$C_{10}H_{18}BrNO_4S$	Calaire
(−)-3-Bromo-8-camphorsulphonic acid ammonium salt	$C_{10}H_{18}BrNO_4S$	Calaire
(R)-(+)-2-Bromopropionic acid	$C_3H_5BrO_2$	Linz
(S)-(−)-2-Bromopropionic acid	$C_3H_5BrO_2$	Linz, Zeneca
(R)-(−)-1,3-Butanediol	$C_4H_{10}O_2$	Boehringer, Daicel, Tanabe
(2R,3R)-(−)-2,3-Butanediol	$C_4H_{10}O_2$	Urquima
(R)-(−)-3-Butene-2-ol	C_4H_8O	Boehringer, Chiroscience
(S)-(+)-3-Butene-2-ol	C_4H_8O	Boehringer, Chiroscience
(S)-(−)-3-t-Butylamino-1,2-propanediol	$C_7H_{17}NO_2$	DSM Andeno
Butyl (S)-(−)-2-chloropropionate	$C_7H_{13}ClO_2$	Daicel, Tanabe
(3S,4S)-3-(R)-(t-Butyldimethylsilyloxy)ethyl)-4-[(R)-carboxyethyl]-2-azetidinone	$C_{14}H_{27}NO_4Si$	Kaneka, Takasago
(2R,3R)-3-Butylglycidol	$C_7H_{13}O_2$	Sipsy
(2S,3S)-3-Butylglycidol	$C_7H_{13}O_2$	Sipsy
(S)-(−)-Butyl lactate	$C_7H_{14}O_3$	Boehringer
(S)-4-t-Butyl-2-oxazolidinone	$C_7H_{13}NO_2$	Rexim/Degussa
(S)-3-t-Butyl-2,5-piperazindione	$C_8H_{14}N_2O_2$	Rexim/Degussa
(S)-3-Butyn-2-ol	C_4H_6O	DSM Andeno
(+)-Camphoric acid	$C_{10}H_{16}O_4$	China Camphor
(+)-Camphor-10-sulphonic acid	$C_{10}H_{16}O_4S$	Calaire, China Camphor
(−)-Camphor-10-sulphonic acid	$C_{10}H_{16}O_4S$	Calaire
(+)-Camphor-10-sulphonyl chloride	$C_{10}H_{15}ClO_3S$	Calaire
(+)-(Camphorsulphonyl)oxaziridine	$C_{10}H_{15}NO_3S$	Newport

(−)-(Camphorsulphonyl)oxaziridine	$C_{10}H_{15}NO_3S$	Newport
N-α-CBZ-D-3-(1-Naphthyl)alaninol	$C_{21}H_{21}NO_3$	Synthetech
N-α-CBZ-D-3-(2-Naphthyl)alaninol	$C_{21}H_{21}NO_3$	Synthetech
(R)-2-Chlorobutyric acid	$C_4H_7ClO_2$	Kaneka
(R)-(+)-4-Chloro-3-hydroxybutyronitrile	C_4H_6ClNO	Daiso
(S)-(−)-4-Chloro-3-hydroxybutyronitrile	C_4H_6ClNO	Daiso
(R)-3-Chlorolactic acid	$C_3H_5ClO_3$	Kaneka
1-[(S)-3-Chloro-2-methylpropionyl]-L-proline	$C_9H_{14}ClNO_3$	Kaneka
D-3-(4-Chlorophenyl)alanine	$C_9H_{10}ClNO_2$	Degussa, Synthetech
L-3-(4-Chlorophenyl)alanine	$C_9H_{10}ClNO_2$	Synthetech
(R)-(3-Chlorophenyl)-1,2-ethanediol	$C_8H_9ClO_2$	Chirex
(S)-(3-Chlorophenyl)-1,2-ethanediol	$C_8H_9ClO_2$	Chirex
(R)-2-(4-Chlorophenyl)-3-phenylpropionic acid	$C_{15}H_{13}ClO_2$	Sumitomo
(S)-2-(4-Chlorophenyl)-3-phenylpropionic acid	$C_{15}H_{13}ClO_2$	Sumitomo
(R)-(−)-3-Chloro-1,2-propanediol	$C_3H_7ClO_2$	Kaneka
(R)-(+)-2-Chloropropionic acid	$C_3H_5ClO_2$	Linz, Marks
(S)-(−)-2-Chloropropionic acid	$C_3H_5ClO_2$	BASF, Linz, Marks, Zeneca
(R)-3-Chlorostyrene oxide	C_8H_7ClO	Kaneka, Chirex, Sipsy
(S)-3-Chlorostyrene oxide	C_8H_7ClO	Kaneka, Chirex, Sipsy
(R)-(−)-Citramalic acid	$C_5H_8O_5$	Lonza
(S)-(+)-Citramalic acid	$C_5H_8O_5$	Lonza
D-Citrulline	$C_6H_{13}N_3O_3$	Rexim/Degussa
D-Cyclohexylalanine	$C_9H_{16}NO_2$	Rexim/Degussa, NSC Technologies
L-Cyclohexylalanine	$C_9H_{16}NO_2$	Rexim/Degussa, NSC Technologies
(S)-Cyclohexylalaninol	$C_9H_{18}NO$	NSC Technologies
D-Cyclohexylglycine	$C_8H_{14}NO_2$	Rexim/Degussa
L-Cyclohexylglycine	$C_8H_{14}NO_2$	Rexim/Degussa
D-Cysteine	$C_3H_7NO_2S$	Ajinomoto, Nippon Rikagakuyakuhin, Tanabe
D-Cystine	$C_6H_{12}N_2O_4S_2$	Nippon Rikagakuyakuhin

(Table continues)

TABLE I Continued

Compound	Formula	Supplier
(1R,2R)-(−)-1,2-Diaminocyclohexane	$C_6H_{14}N_2$	Chirex, Toray
(1S,2S)-(+)-1,2-Diaminocyclohexane	$C_6H_{14}N_2$	Chirex, Toray
(S)-(+)-2,6-Diamino-1-hexanol (L-(+)-Lysinol)	$C_8H_{16}N_2O$	Boehringer
(R)-1,2-Diaminopropane (as salt)	$C_3H_{10}N_2$	Toray
(S)-1,2-Diaminopropane (as salt)	$C_3H_{10}N_2$	Toray
Dibenzoyl-D-(+)-tartaric acid	$C_{18}H_{14}O_8$	Elso Vegyi, Toray, Uetikon
Dibenzoyl-L-(−)-tartaric acid	$C_{18}H_{14}O_8$	Elso Vegyi, Knoll, Toray, Uetikon
(R)-2,2-Dibenzyl-2-hydroxy-1-methylethylamine	$C_{17}H_{21}NO$	Sumitomo
(S)-2,2-Dibenzyl-2-hydroxy-1-methylethylamine	$C_{17}H_{21}NO$	Sumitomo
(R)-N,N-Dibenzylphenylalaninol	$C_{23}H_{25}NO$	NSC Technologies
(S)-N,N-Dibenzylphenylalaninol	$C_{23}H_{25}NO$	NSC Technologies
(+)-[(8,8-Dichlorocamphoryl)sulphonyl]oxaziridine	$C_{10}H_{13}Cl_2NO_3S$	Newport
(−)-[(8,8-Dichlorocamphoryl)sulphonyl]oxaziridine	$C_{10}H_{13}Cl_2NO_3S$	Newport
L-3-(3,4-Dichlorophenyl)alanine	$C_9H_9Cl_2NO_2$	Synthetech
cis-(1S,2R)-1,2-Dihydro-3-bromocatechol	$C_6H_7BrO_2$	Zeneca
cis-1,2-Dihydrocatechol	$C_6H_8O_2$	Zeneca
cis-(1S,2R)-1,2-Dihydro-3-chlorocatechol	$C_6H_7ClO_2$	Zeneca
cis-(1S,2R)-1,2-Dihydro-3-fluorocatechol	$C_6H_7FO_2$	Zeneca
(2S,3S)-(+)-2,3-Dihydro-3-hydroxy-2-(4-methoxyphenyl)-1,5-benzothiazepin-4(5H)-one	$C_{16}H_{15}NO_3S$	Delmar, Zambon
cis-(1S,2R)-1,2-Dihydro-3-methylcatechol	$C_7H_7F_3O_2$	Zeneca
cis-(1S,2R)-1,2-Dihydro-3-methylcatechol	$C_7H_{10}O_2$	Zeneca
D-(−)-α-Dihydrophenylglycine	$C_8H_{11}NO_2$	Deretil, DSM Andeno
D-(−)-α-Dihydrophenylglycine chloride hydrochloride	$C_8H_{11}Cl_2NO$	DSM Andeno
D-(−)-α-Dihydrophenylglycine Dane salt methyl sodium	$C_{13}H_{16}NO_4$	Deretil, DSM Andeno
D-(3,4-Dihydroxy)-α-phenylglycine	$C_8H_9NO_4$	Kaneka

Compound	Formula	Supplier
(+)-Diisopinocampheylchloroborane	$C_{20}H_{34}BCl$	Callery
(−)-Diisopinocampheylchloroborane	$C_{20}H_{34}BCl$	Callery
Diisopropyl-D-(+)-tartrate	$C_{10}H_{18}O_6$	Toray, Uetikon
Diisopropyl-L-(−)-tartrate	$C_{10}H_{18}O_6$	Uetikon
(R)-3,3-Dimethyl-2-aminobutane	$C_6H_{15}N$	Celgene
(S)-3,3-Dimethyl-2-aminobutane	$C_6H_{15}N$	Celgene
(3R-cis)-3,6-Dimethyl-1,4-dioxane-2,5-dione (R-Lactide)	$C_6H_8O_4$	Boehringer
(3S-cis)-3,6-Dimethyl-1,4-dioxane-2,5-dione (S-Lactide)	$C_6H_8O_4$	Boehringer
(R)-(−)-2,2-Dimethyl-1,3-dioxolane-4-methanol	$C_6H_{12}O_3$	Chemi S.p.A.
(S)-(+)-2,2-Dimethyl-1,3-dioxolane-4-methanol	$C_6H_{12}O_3$	Chemi S.p.A., Inalco
(R)-(−)-2,2-Dimethyl-1,3-dioxolan-4-ylmethyl tosylate	$C_{13}H_{18}O_5S$	Chemi S.p.A.
(S)-(+)-2,2-Dimethyl-1,3-dioxolan-4-ylmethyl tosylate	$C_{13}H_{18}O_5S$	Chemi S.p.A.
Dimethyl-D-(−)-tartrate	$C_6H_{10}O_6$	Uetikon
Dimethyl-L-(+)-tartrate	$C_6H_{10}O_6$	Uetikon
(−)-2,2-Dimethyl-α,α,α',α'-tetraphenyl-1,3-dioxolane-3,4-dimethanol	$C_{31}H_{30}O_4$	Urquima
(R)-2,2-Diphenyl-2-hydroxy-1-methylethylamine	$C_{15}H_{17}NO$	Sumitomo
(S)-2,2-Diphenyl-2-hydroxy-1-methylethylamine	$C_{15}H_{17}NO$	Sumitomo
(R)-(+)-Diphenylprolinol	$C_{17}H_{19}NO$	Boehringer, Sipsy
(S)-(−)-Diphenylprolinol	$C_{17}H_{19}NO$	Boehringer, Rexim/Degussa, Sipsy, Urquima
(R)-2,3-Diphenylpropionic acid	$C_{15}H_{14}O_2$	Sumitomo
(S)-2,3-Diphenylpropionic acid	$C_{15}H_{14}O_2$	Sumitomo
(R)-Diphenylvalinol	$C_{17}H_{21}NO$	Newport
(S)-Diphenylvalinol	$C_{17}H_{21}NO$	Newport
Di-p-toluoyl-D-tartaric acid	$C_{20}H_{18}O_8$	Elso Vegyi, Toray, Uetikon
Di-p-toluoyl-L-tartaric acid	$C_{20}H_{18}O_8$	Elso Vegyi, Linz, Toray, Uetikon
(R)-(−)-Epichlorhydrin	C_3H_5ClO	Daiso, Kaneka, Nagase
(S)-(+)-Epichlorhydrin	C_3H_5ClO	Daiso, Nagase
N-(1-(S)-Ethoxycarbonyl-3-phenylpropyl)-L-alanine	$C_{15}H_{21}NO_4$	Daicel, DSM Andeno, Kaneka, Tanabe
N-(1-(S)-Ethoxycarbonyl-3-phenylpropyl)-L-alanyl-N-carboxyanhydride	$C_{16}H_{19}NO_5$	DMS Andeno, Kaneka
Ethyl (S)-(−)-2-chloropropionate	$C_5H_9ClO_2$	Daicel, Tanabe

(Table continues)

TABLE I Continued

Compound	Formula	Supplier
Ethyl (R)-2-hydroxy-4-phenylbutyrate	$C_{12}H_{16}O_3$	Daicel, DSM Andeno, Kaneka, Recordati, Tanabe
Ethyl (R)-(−)-mandelate	$C_{10}H_{12}O_3$	Yamakawa
Ethyl (S)-(+)-mandelate	$C_{10}H_{12}O_3$	Yamakawa
Ethyl (R)-(+)-2-tosyloxypropionate	$C_{12}H_{16}O_5S$	Boehringer
Ethyl (S)-(−)-2-tosyloxypropionate	$C_{12}H_{16}O_5S$	Boehringer
D-3-(4-Fluorophenyl)alanine	$C_9H_{10}FNO_2$	Synthetech
L-3-(4-Fluorophenyl)alanine	$C_9H_{10}FNO_2$	Synthetech
N-α-FMOC-D-3-(4-Biphenyl)alanine	$C_{30}H_{25}NO_4$	Synthetech
N-α-FMOC-L-3-(4-Biphenyl)alanine	$C_{30}H_{25}NO_4$	Synthetech
N-α-FMOC-L-Octahydroindole-2-carboxylic acid	$C_{24}H_{25}NO_4$	Synthetech
(R)-(−)-Formylmandeloyl chloride	$C_9H_7ClO_3$	DSM Andeno, Toray
(S)-(+)-Formylmandeloyl chloride	$C_9H_7ClO_3$	Toray
D-Glutamic acid	$C_5H_9NO_4$	Ajinomoto, Kaneka, Rexim/Degussa, Tanabe
D-Glutamine	$C_5H_{10}N_2O_3$	Ajinomoto
(R)-(−)-Glycerol-1-tosylate	$C_{10}H_{14}O_5$	Chemi S.p.A.
(S)-(+)-Glycerol-1-tosylate	$C_{10}H_{14}O_5S$	Chemi S.p.A.
(R)-(+)-Glycidol	$C_3H_6O_2$	Sipsy
(S)-(−)-Glycidol	$C_3H_6O_2$	Sipsy
(R)-(−)-Glycidyl-3-nosylate	$C_9H_9NO_6S$	Daiso, Sipsy
(S)-(+)-Glycidyl-3-nosylate	$C_9H_9NO_6S$	Daiso, Sipsy
(R)-(−)-Glycidyl tosylate	$C_{10}H_{12}O_4S$	Daiso, Lonza, Sipsy
(S)-(+)-Glycidyl tosylate	$C_{10}H_{12}O_4S$	Daiso, Sipsy
D-Histidine	$C_6H_9N_3O_2$	Ajinomoto, Rexim/Degussa, Tanabe
L-Homocysteine	$C_4H_9NO_2S$	Tanabe
D-Homocysteinethiolactone HCl	C_4H_8ClNOS	Rexim/Degussa

Compound	Formula	Supplier
L-Homocysteinethiolactone HCl	C_4H_8ClNOS	Rexim/Degussa
D-Homophenylalanine	$C_{10}H_{13}NO_2$	Recordati, Synthetech
L-Homophenylalanine	$C_{10}H_{13}NO_2$	Daicel, Rexim/Degussa, Synthetech, Tanabe
L-Homoserine	$C_4H_9NO_3$	Tanabe
(S)-(−)-3-Hydroxybutyrolactone	$C_4H_6O_3$	Kaneka
L-5-Hydroxylysine	$C_6H_{14}N_2O_3$	Rexim/Degussa
(R)-2-Hydroxy-4-phenylbutyric acid	$C_{10}H_{12}O_3$	Kaneka
D-(−)-α-p-Hydroxyphenylglycine	$C_8H_9NO_3$	Alfa Chemicals Italiana, Deretil, DSM Andeno, Kaneka, Nippon Kayaku, Recordati
D-(−)-α-p-Hydroxyphenylglycine chloride HCl	$C_8H_9Cl_2NO_2$	Nippon Kayaku
D-(−)-α-p-Hydroxyphenylglycine Dane salt ethyl potassium	$C_{14}H_{16}NO_5K$	Nippon Kayaku
D-(−)-α-p-Hydroxyphenylglycine Dane salt methyl sodium	$C_{13}H_{14}NO_5Na$	Nippon Kayaku
D-(−)-α-p-Hydroxyphenylglycine Dane salt (potassium methyl)	$C_{13}H_{14}KNO_5$	Alfa Chemicals Italiana, Deretil, DSM Andeno, Kaneka, Recordati
D-cis-4-Hydroxyproline	$C_5H_9NO_3$	Rexim/Degussa
L-(+)-Hydroxyprolinol	$C_5H_{11}NO_2$	Boehringer, Rexim/Degussa
(R)-3-Hydroxypyrrolidine	C_4H_9NO	Rexim/Degussa, Toray
(S)-3-Hydroxypyrrolidine	C_4H_9NO	Rexim/Degussa, Toray
(R)-3-Hydroxypyrrolidine hydrochloride	$C_4H_{10}ClNO$	Kaneka, Rexim/Degussa
(S)-3-Hydroxypyrrolidine hydrochloride	$C_4H_{10}ClNO$	Kaneka, Rexim/Degussa
(S)-(−)-Indoline-2-carboxylic acid	$C_9H_9NO_2$	DSM Andeno
D-3-(4-Iodophenyl)alanine	$C_9H_{10}INO_2$	Synthetech
L-3-(4-Iodophenyl)alanine	$C_9H_{10}INO_2$	Synthetech
Isobutyl (S)-(−)-2-chloropropionate	$C_7H_{13}ClO_2$	BASF, Daicel, Tanabe
D-allo-Isoleucine	$C_6H_{13}NO_2$	Rexim/Degussa
Isoleucinol see 2-Amino-3-methyl-1-pentanol		
2,3-O-Isopropylidene-(R)-glyceraldehyde	$C_6H_{10}O_3$	Chemi S.p.A.
(R)-(+)-N -Isopropyl-α-methylbenzylamine	$C_{11}H_{17}N$	Yamakawa

(Table continues)

TABLE 1   Continued

Compound	Formula	Supplier
(S)-(+)-N-Isopropyl-α-methylbenzylamine	$C_{11}H_{17}N$	Yamakawa
(R)-(+)-4-Isopropyl-2-oxazolidinone	$C_6H_{11}NO_2$	Boehringer, Newport, Rexim, Degussa
(S)-(−)-4-Isopropyl-2-oxazolidinone	$C_6H_{11}NO_2$	Boehringer, Newport, Rexim/Degussa, Urquima
(R)-(−)-3-Isopropyl-2,5-piperazindione	$C_7H_{12}N_2O_2$	Rexim/Degussa
(S)-(+)-3-Isopropyl-2,5-piperazindione	$C_7H_{12}N_2O_2$	Rexim/Degussa
(S)-(−)-Isoserine	$C_3H_7NO_3$	Rexim/Degussa
(R)-(+)-Lactamide	$C_3H_7NO_2$	Boehringer
(S)-(−)-Lactamide	$C_3H_7NO_2$	Boehringer
Lactide see cis-Dimethyl-1,4-dioxane-2,5-dione		
D-Leucine	$C_6H_{13}NO_2$	Ajinomoto, Nippon Rikagakuyakuhin, Rexim/Degussa, Tanabe
L-tert-Leucine [L-tert-Butylglycine; 2-Amino-3,3-dimethylbutyric acid]	$C_6H_{13}NO_2$	Chiroscience, NSC Technologies, Rexim/Degussa
Leucinol see 2-Amino-4-methyl-1-pentanol		
L-tert-Leucinol	$C_6H_{15}NO$	Rexim/Degussa
D-Lysine HCl	$C_6H_{15}ClN_2O_2$	Ajinomoto, Kaneka, Rexim/Degussa, Tanabe
L-(+)-Lysinol, see 2,6-Diamino-1-hexanol		
(R)-(−)-Mandelic acid	$C_8H_8O_3$	Nippon Chemical, Nippon Kayaku, Nitto, Uetikon,Yamakawa, Zeeland
(S)-(+)-Mandelic acid	$C_8H_8O_3$	Nippon Kayaku, Norse, Uetikon, Yamakawa, Zeeland
(S)-1-Mercaptoglycerol	$C_3H_8O_2S$	Kaneka
D-Methionine	$C_5H_{11}NO_2S$	Ajinomoto, Kaneka, Rexim/Degussa, Tanabe
L-Methioninol	$C_5H_{13}NOS$	Rexim/Degussa
2-Aminotetralin derivatives.	$C_{10}H_{15}NO$	Celgene

(R)-α-Methylbenzylamine	$C_8H_{11}N$	Celgene, Dynamit Nobel, Yamakawa, Zeeland
(S)-α-Methylbenzylamine	$C_8H_{11}N$	Celgene, Dynamit Nobel, Yamakawa
(4S)-2-Methyl-4-benzyloxazole	$C_{11}H_{13}NO$	NSC Technologies
(R)-α-Methyl-4-chlorobenzylamine	$C_8H_{10}ClN$	Celgene, Yamakawa
(S)-α-Methyl-4-chlorobenzylamine	$C_8H_{10}ClN$	Celgene, Yamakawa
Methyl (S)-(−)-2-chloropropionate	$C_4H_7ClO_2$	Daicel, Tanabe
(R)-(−)-Methyl glycidyl ether	$C_4H_8O_2$	Daiso, Sipsy
(S)-(+)-Methyl glycidyl ether	$C_4H_8O_2$	Daiso, Sipsy
(R)-(−)-Methyl 3-hydroxybutyrate	$C_5H_{10}O_3$	Kaneka, NSC Technologies
(S)-(+) Methyl 3-hydroxybutyrate	$C_5H_{10}O_3$	Kaneka
(R)-(−)-Methyl-β-hydroxyisobutyrate	$C_5H_{10}O_3$	Kaneka
(S)-(+) Methyl-β-hydroxyisobutyrate	$C_5H_{10}O_3$	Kaneka
Methyl (R)-(−)-3-hydroxypentanoate	$C_6H_{12}O_3$	Kaneka
Methyl (S)-(+)-3-hydroxypentanoate	$C_6H_{12}O_3$	Kaneka
(R)-(+)-Methyl α,β-isopropylideneglycerate	$C_7H_{12}O_4$	Chemi S.p.A.
(R)-(+)-Methyl lactate	$C_4H_8O_3$	Boehringer, Daicel, Tanabe
(S)-(−)-Methyl lactate	$C_4H_8O_3$	Boehringer
Methyl (R)-(−)-mandelate	$C_9H_{10}O_3$	Yamakawa
Methyl (S)-(+)-mandelate	$C_9H_{10}O_3$	Yamakawa
(R)-α-Methyl-2-methoxybenzylamine	$C_9H_{13}NO$	Celgene, Sumitomo
(S)-α-Methyl-2-methoxybenzylamine	$C_9H_{13}NO$	Celgene, Sumitomo
(R)-α-Methyl-3-methoxybenzylamine	$C_9H_{13}NO$	Celgene, Sumitomo, Yamakawa
(S)-α-Methyl-3-methoxybenzylamine	$C_9H_{13}NO$	Celgene, Sumitomo, Yamakawa
(R)-α-Methyl-4-methoxybenzylamine	$C_9H_{13}NO$	Celgene, Sumitomo
(S)-α-Methyl-4-methoxybenzylamine	$C_9H_{13}NO$	Celgene, Sumitomo
(R)-α-Methyl-4-methylbenzylamine	$C_9H_{13}N$	Celgene, Yamakawa
(S)-α-Methyl-4-methylbenzylamine	$C_9H_{13}N$	Celgene, Yamakawa
(R)-α-Methyl-4-nitrobenzylamine HCl	$C_8H_{11}ClN_2O_2$	EMS-Dottikon, Yamakawa
(S)-α-Methyl-4-nitrobenzylamine HCl	$C_8H_{11}ClN_2O_2$	EMS-Dottikon, Yamakawa

(Table continues)

TABLE 1   Continued

Compound	Formula	Supplier
(R)-Methyloxazaborolidine	$C_{18}H_{20}BNO$	Callery
(S)-Methyloxazaborolidine	$C_{18}H_{20}BNO$	Callery
(4R,5S)-4-Methyl-5-phenyl-2-oxazolidinone	$C_{10}H_{11}NO_2$	Urquima
(R)-(−)-1-Methyl-3-phenylpropylamine	$C_{10}H_{15}N$	Toray
(S)-(+)-1-Methyl-3-phenylpropylamine	$C_{10}H_{15}N$	Toray
(R)-(−)-2-Methylpiperazine	$C_5H_{12}N_2$	Toray, Yamakawa
(S)-(+)-2-Methylpiperazine	$C_5H_{12}N_2$	Toray, Yamakawa
Methyl (R)-2-tosyloxypropionate	$C_{11}H_{11}O_5S$	Daicel
(R)-(+)-4-Methyl-4-(trichloromethyl)-2-oxetanone	$C_5H_5Cl_3O_2$	Lonza
(S)-(−)-4-Methyl-4-(trichloromethyl)-2-oxetanone	$C_5H_5Cl_3O_2$	Lonza
D-3-(1-Naphthyl)alanine	$C_{13}H_{13}NO_2$	Rexim/Degussa, Synthetech
L-3-(1-Naphthyl)alanine	$C_{13}H_{13}NO_2$	Synthetech
D-3-(2-Naphthyl)alanine	$C_{13}H_{13}NO_2$	Rexim/Degussa, Synthetech
L-3-(2-Naphthyl)alanine	$C_{13}H_{13}NO_2$	Synthetech
(R)-(+)-1-(1-Naphthyl)ethylamine	$C_{12}H_{13}N$	Yamakawa
(S)-(−)-1-(1-Naphthyl)ethylamine	$C_{12}H_{13}N$	Yamakawa
L-Neopentylglycine	$C_7H_{15}NO_2$	Rexim/Degussa
D-3-(4-Nitrophenyl)alanine	$C_9H_{10}N_2O_4$	Synthetech
L-3-(4-Nitrophenyl)alanine	$C_9H_{10}N_2O_4$	EMS-Dottikon, Synthetech
(1R,2R,4R)-(+)-endo-Norbornenol [(1R,2R,4R)-Bicyclo[2.2.1]hept-5-en-2-ol]	$C_7H_{10}O$	Chiroscience
(1S,2R,4R)-(−)-endo-Norborneol [(1S,2R,4R)-Bicyclo[2.2.1]heptan-2-ol]	$C_7H_{12}O$	Chiroscience
D-Ornithine hydrochloride	$C_5H_{13}ClN_2O_2$	Kaneka
(S)-1,2-Pentanediol	$C_5H_{12}O_2$	Kaneka
α-Phenetylamine See α-Methylbenzylamine		

Compound	Formula	Suppliers
D-Phenylalanine	$C_9H_{11}NO_2$	Ajinomoto Rexim/Degussa, Kaneka, NSC Technologies, Recordati, Tanabe
(R)-(+)-Phenylalaninol	$C_9H_{13}NO$	Boehringer, Newport, NSC Technologies, Rexim/Degussa, Urquima
(S)-(−)-Phenylalaninol	$C_9H_{13}NO$	Boehringer, Newport, NSC Technologies, Rexim/Degussa, Synthetech, Tanabe
(S)-(+)-Phenylethane-1,2-diol	$C_8H_{10}O_2$	Boehringer, Kaneka, Chirex
(R)-(+)-1-Phenylethanol	$C_8H_{10}O$	Linz
(S)-(−)-1-Phenylethanol	$C_8H_{10}O$	Linz
α-Phenylethylamine, see α-Methylbenzylamine		
(2R,3R)-(+)-3-Phenylglycidol	$C_9H_{10}O$	Sipsy
D-(−)-α-Phenylglycine	$C_8H_9NO_2$	Alfa Chemicals Italiana, Deretil, DSM Andeno, Kaneka, Nippon Kayaku, Recordati, Themis. Zhejiang
L-α-(+)-Phenylglycine	$C_8H_9NO_2$	Nippon Kayaku
D-(−)-α-Phenylglycine chloride hydrochloride	$C_8H_9Cl_2NO$	Alfa Chemicals Italiana, Deretil, DSM Andeno, Kaneka, Nippon Kayaku
D-(−)-α-Phenylglycine Dane salt (potassium, ethyl)	$C_{14}H_{16}KNO_4$	Deretil, DSM Andeno, Themis
D-(−)-α-Phenylglycine methyl ester HCl	$C_9H_{12}ClNO_2$	DSM Andeno
(R)-(−)-α-Phenylglycinol	$C_8H_{11}NO$	Boehringer, Newport, Synthetech
(S)-(+)-α-Phenylglycinol	$C_8H_{11}NO$	Newport, Rexim/Degussa
(4R)-Phenyl-2-oxazolidinone	$C_9H_9NO_2$	Urquima
(4S)-Phenyl-2-oxazolidinone	$C_9H_9NO_2$	Rexim/Degussa, Urquima
(S)-(−)-1-Phenyl-1,3-propanediol	$C_9H_{12}O_2$	Kaneka
(R)-(+)-1-Phenylpropylamine	$C_9H_{13}N$	Celgene, Yamakawa
(S)-(−)-1-Phenylpropylamine	$C_9H_{13}N$	Celgene, Yamakawa
D-Proline	$C_5H_9NO_2$	Ajinomoto, Rexim/Degussa, Tanabe
(S)Prolinol	$C_5H_{11}NO$	Boehringer, Newport, Rexim/Degussa
(R)-(−)-Propane-1,2-diol	$C_3H_8O_2$	Boehringer, Daicel, Tanabe
(S)-(+)-Propane-1,2-diol	$C_3H_8O_2$	Boehringer

(Table continues)

TABLE I Continued

Compound	Formula	Supplier
(R)-(+)-Propyleneoxide	C_3H_6O	Boehringer
(S)-(−)-Propyleneoxide	C_3H_6O	Boehringer
D-3-(3-Pyridyl)alanine	$C_8H_{10}N_2O_2$	Rexim/Degussa, Synthetech
L-3-(3-Pyridyl)alanine	$C_8H_{10}N_2O_2$	Synthetech
D-Pyroglutamic acid	$C_5H_7NO_3$	EMS-Dottikon
D-Serine	$C_3H_7NO_3$	Nippon Rikagakuyakuhin, Recordati, Rexim/Degussa, Tanabe
(R)-(+)-Styrene oxide	C_8H_8O	Boehringer, Kaneka
(S)-(−)-Styrene oxide	C_8H_8O	Boehringer, Kaneka, Chirex
D-(−)-Tartaric acid	$C_4H_6O_6$	Norse, Toray
(S)-(−)-Tetrahydro-2-furoic acid	$C_5H_8O_3$	Yamakawa
1,2,3,4-Tetrahydroisoquinoline-3(S)-carboxylic acid	$C_{10}H_{11}NO_2$	Fine Organics, Nippon Steel, NSC Technologies
1,2,3,4-Tetrahydroisoquinoline-3(R)-carboxylic acid	$C_{10}H_{11}NO_2$	Fine Organics, NSC Technologies, Tanabe
D-Thiazolidine-4-carboxylic acid	$C_4H_7NO_2S$	Nippon Rikagakuyakuhin
L-Thiazolidine-4-carboxylic acid	$C_4H_7NO_2S$	Nippon Rikagakuyakuhin, Rexim/Degussa
D-3-(4-Thiazolyl)alanine	$C_6H_8N_2O_2S$	Synthetech
L-3-(4-Thiazolyl)alanine	$C_6H_8N_2O_2S$	Synthetech
D-3-(2-Thienyl)alanine	$C_7H_9NO_2S$	Synthetech
L-3-(2-Thienyl)alanine	$C_7H_9NO_2S$	Synthetech
D-2-Thienylglycine	$C_6H_7NO_2S$	Kaneka
D-allo-Threonine	$C_4H_9NO_3$	Rexim/Degussa
D-Threonine	$C_4H_9NO_3$	Ajinomoto, Tanabe
L-Threoninol	$C_4H_{11}NO_2$	Rexim/Degussa
(R)-(−)-4-(Trichloromethyl)-2-oxetanone	$C_4H_3Cl_3O_2$	Lonza
(S)-(+)-4-(Trichloromethyl)-2-oxetanone	$C_4H_3Cl_3O_2$	Lonza

(R)-Tritylglycidol	$C_{22}H_{20}O_2$	Sipsy
(S)-Tritylglycidol	$C_{22}H_{20}O_2$	Sipsy
D-Tryptophan	$C_{11}H_{12}N_2O_2$	Ajinomoto, Rexim/Degussa, Tanabe
D-Tyrosine	$C_9H_{11}NO_3$	Ajinomoto, Rexim/Degussa, Tanabe
D-Valine	$C_5H_{11}NO_2$	Ajinomoto, DSM Special Products, Kaneka, Recordati, Rexim/Degussa, Tanabe

Valinol, see 2-Amino-3-methyl-1-butanol

[a] The compounds have been listed in catalogues as available, although some of them may only be potentially available.

2.3.   MAJOR, MEDIUM, AND MINOR PLAYERS

Majors, such as DSM and Kaneka, have often developed a technology base from experience gained in making, for example, antibiotic side chains and angiotensin-converting enzyme (ACE) inhibitor intermediates. For most major suppliers, chiral compounds are only a part of their overall activity.

Medium and minor players will typically have an exploitable intellectual property base from which they are developing or seeking to establish themselves; chiral compounds are often the only products offered. However, as potential commercial suppliers at scale, they will be dependent on subcontracted production, including alliances with larger companies with available production capacity or significant capital investment. Examples are Synthon Corporation, which makes C4 synthons from carbohydrate raw materials, and Oxford Asymmetry, which exploits S. Davies' work and patents and is in alliance with Cambrex. Customers will need to assess the adequacy of financial arrangements and/or the suitability of the chosen production partner.

Major and smaller players will both very likely use patented or proprietary technology to access a target. Majors will generally be less willing to grant licenses. Issues can then arise with respect to the problem of single sourcing—if a compound happens to only be accessible by one company's proprietary technology, it may prove difficult to enable a second source to use this technology.

How is a single sourcing situation to be avoided in these instances? The compound may be available from other companies using a different route, but the impurity profile may be different. This will be increasingly important the further along the synthesis route the compound is used. For late-stage intermediates with critical impurity profile criteria, it will be best if all suppliers use the same route and preferably the same process. This could be another reason to introduce the chiral center early and to develop in-house processes for later synthesis stages that can then be subcontracted, generally under a confidential disclosure agreement (CDA). The CDA is embedded in the culture of many pharmaceutical companies and is, perhaps, overused. Unless truly valuable intellectual property is being disclosed, it may be better to allow the subcontractor carte blanche to use the process freely. If the particular intermediate can be offered to the market at large, it will be easier to have it treated as a raw material rather than an intermediate requiring a DMF, etc.

2.4.   TECHNOLOGY

Understanding the technology potentially applicable to the required chiral compound is important, as is knowledge of suppliers' particular areas of strength and specialization. The best potential partners can be identified and the subsequent dialogue can be more fruitful. Company culture varies; in many cases, the sourcing is undertaken by development chemists, whereas in others it is a purchasing/procurement function. In the latter instance, it is important that there has been technical input. To source effectively there must be some technical knowledge. Equally, suppliers' representatives or agents should also have a grasp of technical issues.

If the desired compound can be obtained starting from an available chiral raw material, there should usually be few problems in sourcing, even at high volume, or in establishing sufficient price competition. Any CCE can be a possible source, providing they can handle the chemistry. Similarly, if the compound is accessible, for example, from a currently unavailable amino acid, sources for the latter usually could be relatively easily developed, given the extensive pool of technology for amino acid synthesis and production. Companies that supply these compounds include Ajinomoto, DSM Andeno, Kaneka, NSC Technologies, Rexim Degussa, and Tanabe.

Classical resolution by crystallization, the oldest method, is usually the development chemist's first approach to obtain a single enantiomer. It is attractive given the wealth of expertise that has been accumulated, much of it in amino acid production.

First, there is a 5–10% chance that the racemate is a conglomerate; in these cases, crystallization is a simple way to the desired product and will often prove the most economic.

Then, there are a number of acidic and basic resolving agents available, although some of the synthetic agents can be expensive. Classical resolution can be variable in application, ranging from difficult and expensive to straightforward. Consider the case of naproxen, the majority of which is still produced by a process involving resolution, despite the potentially attractive possibility of an asymmetric hydrogenation process. Furthermore, in the case of L-dopa, resolution of a conglomerate intermediate results in a process that is competitive to the asymmetric hydrogenation methodology.

Kinetic resolution methods, including biological-based processes, offer another alternative. Again, particularly for amino acids, considerable expertise has been established.

A further option is provided by preparative high-performance liquid chromatography (HPLC) using a chiral stationary phase (CSP). This option is increasingly used in early development as a quick and convenient way to obtain both enantiomers for biological tests. It obviates the need to develop an asymmetric synthesis or other methods relatively early in a drug's development. Such work can often prove fruitless—one can recall instances at conferences when it was only possible to describe elegant chemistry because the target had been dropped from development! Chemical development time is at a premium in any successful pharmaceutical company; preparative HPLC can free this up, enabling resources to be deployed on process improvements for candidate compounds that are further into development and with a larger chance of ultimate success.

Nevertheless, it is probably fair to say that the average process development chemist's immediate reaction to a process with a chromatography step is to seek to eliminate it. In addition, unless one is dealing with a very highly active low-volume final drug, a separation on a single column will generally not be viable. However, consider that chromatography has been successfully used on a large industrial scale since the 1960s. UOP's SORBEX Simulated Moving Bed (SMB) technology has been used to isolate isomers from mixtures and to obtain D-fructose (a chiral molecule) from corn syrup. There is a growing belief that application of SMB to separate enantiomers on a production scale is imminent. Costs decrease dramatically with increasing volumes; however, at present, a major contributor to any costing is the CSP. Its contribution does depend on assumptions as to its life, which are usually unduly pessimistic—a single polysaccharide preparative HPLC column has been run in Japan for more than 7 years.

Currently available CSPs were developed for conventional HPLC analytical applications; the wider the diversity of chemical structures that a particular phase could resolve, the better. For a production-scale SMB resolution, which will be for one specific compound, such a property is unnecessary. Furthermore, the physical characteristics (particle size, etc.) required for SMB are different from those needed for HPLC. It is reasonable to expect that new or modified and perhaps tailored CSPs will be developed that will be significantly cheaper than current products. Had such CSPs been available in the 60s, it is likely that antibiotic side chains would have been produced using SMB instead of by classical resolution.

Any resolution method will be favored if the unwanted enantiomer can be racemized and recycled; SMB also avoids resolution agent recycle.

Among stereospecific chemical methods, asymmetric hydrogenation, hydroformylation, hydrocyanation, and isomerization offer some of the potentially most economic routes. As long as the substrate is relatively accessible and the turnover number is high, low costs are achievable. Development times are usually not unduly excessive. Catalyst availability is becoming less of an issue as some patent holders are now prepared to license out for specific applications. Thus, there is a wider choice of potential sources, such as Bayer, Chiroscience, and NSC Technologies.

Asymmetric epoxidation (Jacobsen) and dihydroxylation (Sharpless) are other potentially viable approach to epoxides, diols, and aminodiols.

Numerous other asymmetric syntheses are available, each of which may turn out to be cost effective. Even chemistry using a relatively costly chiral auxiliary or an expensive low-temperature ($-78°C$) reaction may prove to be the best route in a particular case.

Fermentation is often cited as one of the leading methods to chiral compounds, both simple (e.g., monosodium glutamate) and complex (e.g., penicillin). Certainly, it offers low production costs, and large tonnage of chiral materials are made by fermentation. However, its importance is limited to making compounds that were isolated from a fermentation media or to instances in which a large volume requirement is already firmly established. For such high-volume products, the large investment necessitated by long process development times (3 years in the case of L-phenylalanine for use in aspartame) can be justified by prospective economies in production. However, fermentation cannot even be considered as a method to make a novel intermediate that initially has been otherwise derived; the development times are simply too long and costs excessive.

Biocatalysis is increasingly becoming an option, providing the required chiral compound is an analog of a molecule that has already been established as accessible through this methodology, and assuming the necessary substrate is available; if this is not so, then similar cost/time constraints will apply.

There is pressure to reduce both development times and costs throughout the industry. It is usually instinctive to attempt to obtain a given chiral compound, or indeed any other intermediate, at the lowest cost. On occasion, the potentially cheaper processes will require longer development times. There have also been cases in which a drug's development has been slowed by the delivery delay for a particular intermediate. Current efforts to compress the development time to 5–7 years will increase this possibility. Suppose a compound has potential sales of $500m p.a., the active ingredient cost represents 10% of sales value, and a key chiral intermediate contributes 70% of the drug cost, that is, $35m p.a. If the cost of this intermediate can be reduced by 20%, the savings are $7m p.a. However, if obtaining the lower price introduces a 3-month delay in development, one has lost sales of $125m and profits of $112m—16 years' worth of drug cost saving. Cheapest is not always least expensive!

Of course, this is not to say that costs should not be minimized and potential complications, such as impurity profiles, etc., avoided by working with potentially commercial sources as early in the development process as possible. The problem is that only approximately 10% of phase I candidates will eventually be launched, and, furthermore, estimates of ultimate commercial demand will usually only become available toward the end of phase II. Suppliers may incur significant development costs before having any idea of the potential return, if, indeed, there is to be one at all. Obviously, this situation is not unique to chiral compounds and has lead to customers working with a selected number of preferred suppliers who gain access to a continuing number of opportunities. Speed of response will be one very important factor in making this selection.

2.5. AVAILABLE CHIRAL COMPOUNDS

If the chiral pool (natural L-amino acids, terpenes, etc.) is excluded, most compounds that could be regarded as commercially available, or potentially so, are D-amino acids, elaborations of D- and L-amino acids, antibiotic side chains, and intermediates for well-established existing drugs. This can be seen from the list in Table 1 (extracted from the R-S Directory, 2nd edition). However, many more compounds are available in smaller quantities from specialist companies, as well as the established research chemical houses that provide such a vital service to discovery chemists. "Commercially available" will have different connotations depending on the activity of the final drug—a low-dosage peptide might only require tens of kilograms of a particular unnatural amino acid to satisfy all of the annual consumption of the final drug.

2.6. CONCLUSION

In sourcing enantiomerically pure starting materials, one thing is certain: the scientific efforts of biochemists and chemists and collaborations between academia and industry, with growing interest in the business, are all leading to an increase in the availability of chiral compounds. The job is becoming progressively easier.

3

Synthesis of Large-Volume Products

DAVID J. AGER

NSC Technologies, Mount Prospect, Illinois

3.1. INTRODUCTION

The purpose of this chapter is to discuss the syntheses of top-selling chiral compounds so that a perspective may be obtained about the merits of the various asymmetric approaches discussed elsewhere in this book. Of course, the majority of these drugs are mature, which means that a considerable amount of time and money has been expended to reduce costs and optimize the syntheses. In addition, asymmetric synthesis is a new approach, and, although the potential may exist today to use an asymmetric oxidation, this was not a serious option 20 years ago. In addition to pharmaceuticals, some large-volume products in the food and agricultural areas are also discussed.

3.2. PHARMACEUTICALS

The top-ten-selling single enantiomer drugs are given in Table 1. Because we are only concerned with small molecules, peptide products such as insulin and erythropoietin have not been included as they are outside of the realms of chemical approaches.

Although each of the compounds is treated separately, there are a number of similarities between many of them. In addition, a number derive their stereogenic centers from starting materials that are members of the chiral pool, such as amino acids. The number of products on this list that have the stereogenic centers formed in a fermentation process is considerable. Each of them will be discussed in turn.

33

TABLE I   Top-Ten-Selling Single Enantiomer Drugs in 1996[a]

Drug	Sales ($MM)
Enalapril	2,925
Simvastatin	2,800
Pravastatin	2,420
Amoxicillin	1,885
Lisinopril	1,699
Diltiazem	1,575
Captopril	1,373
Sertraline	1,340
Lovastatin	1,255
Cefaclor	1,157

[a] The data were provided by Technology Catalysts International and is gratefully acknowledged.

3.2.1. Enalapril

The synthesis of the angiotensin-converting enzyme (ACE) inhibitor enalapril (1) incorporates the natural amino acids L-alanine and L-proline [1,2]. Although the compound can be thought of as a derivative of homophenylalanine, this part of the compound is prepared by a reductive amination with the keto ester (2), while the amino acids are coupled through an N-carboxy anhydride 3 (Scheme 1) [3–7].

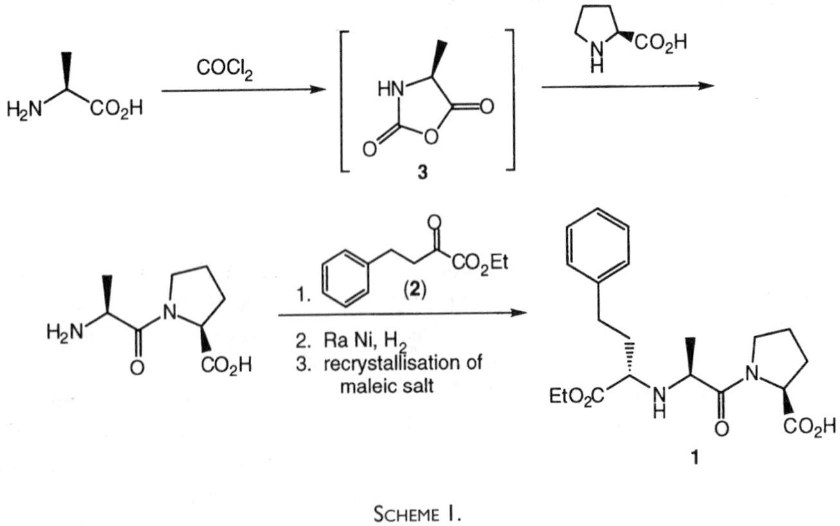

SCHEME 1.

The two chiral amino acids are readily available from fermentation methods.

3.2.2. Simvastatin

This high-value 3-hydroxy-3-methylglutaryl coenzyme A (HMG-CoA) reductase inhibitor (4) is obtained from a fermentation. It is very similar in structure to lovastatin (see Section 3.2.9). A number of routes to various portions of the compound have been reported [3]. However, fermentation is used to provide the core structure and the dimethylbutyric acid side chain is enforced by coupling [8–10], although chemical methylation to form the quaternary dimethylated center has been achieved on synthetic intermediates [11,12].

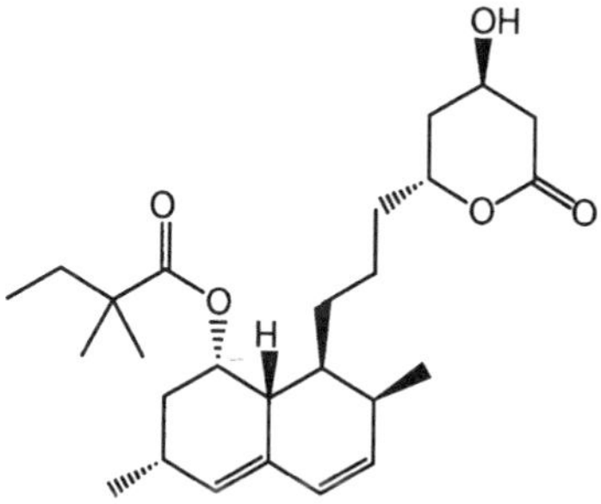

4

3.2.3. Pravastatin

Pravastatin (5) is another HMG-CoA reductase for the inhibition of cholesterol biosynthesis. It is closely related to lovastatin and simvastatin structurally (see Sections 3.2.9 and 3.2.2). It is produced by a two-step sequence: ML-236B (6) is prepared by fermentation of *Penicillium citrinum* [13], then hydroxylated enzymatically to produce (5) [13–15].

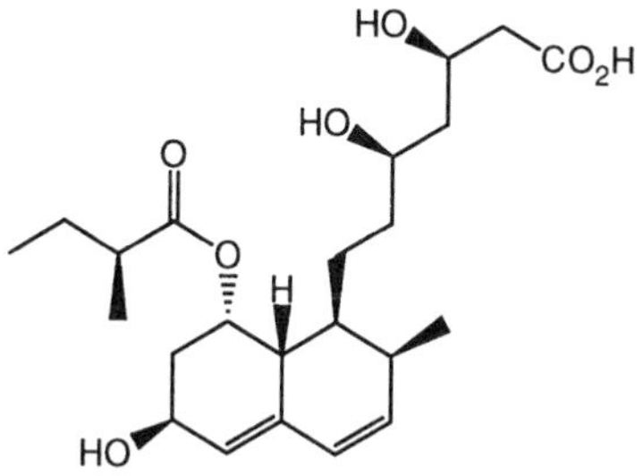

5

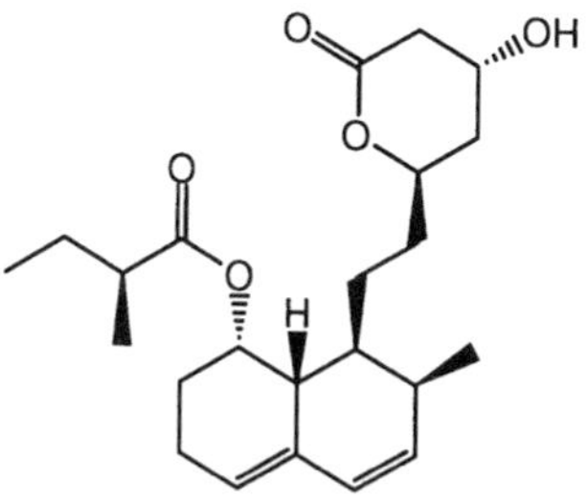

6

3.2.4. Amoxicillin

Amoxicillin (7) is a semisynthetic penicillin antibiotic. The penicillin portion is derived from fermentation of either penicillin-V or -G, and then the side chain is removed chemically to afford 6-aminopenicillanic acid (6-APA) [3,16]. The D-p-hydroxyphenylglycine is then attached as the new side chain—chemical and enzymatic methods are available to achieve this [17–21]. This amino acid is obtained by a classical resolution or by enzymatic hydrolysis of a hydantoin (Chapter 8) [22–26].

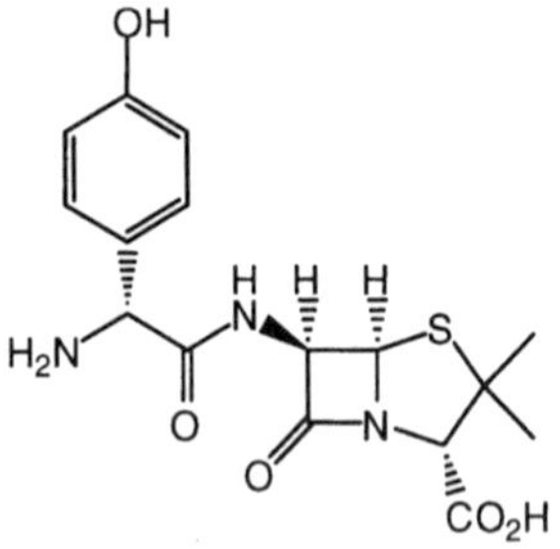

7

3.2.5. Lisinopril

Lisinopril (8) is another ACE inhibitor [1]. Its synthesis is analogous to enalapril, with a protected lysine taking the place of alanine [5,27].

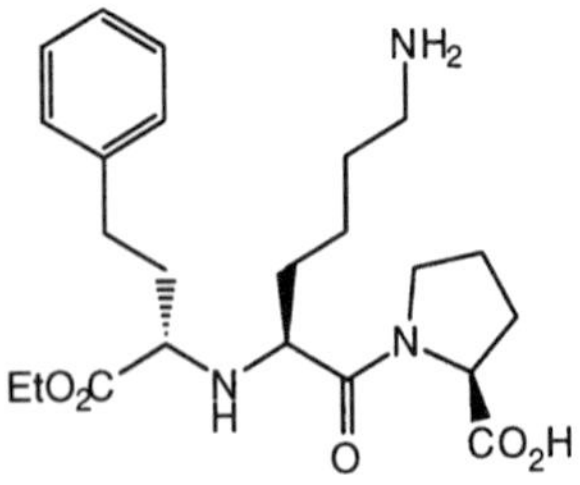

8

3.2.6. Diltiazem

Diltiazem (9) is an antihypertensive drug. It has prompted a number of studies to find an asymmetric Darzen reaction that can be performed at scale, but, to date, none has been successful. This compound can be resolved by crystallization with *R*-phenylethylamine (Chapter 8) [28,29].

An alternative approach is to perform a resolution of the epoxy ester (10), which can be accomplished with a lipase from *Serratia marcescens* (Scheme 2) [30,31]. This approach allows for resolution of an early intermediate that can then be used to prepare just the desired diastereoisomer. Alternatives are to resolve precursors to the epoxy functionality [32].

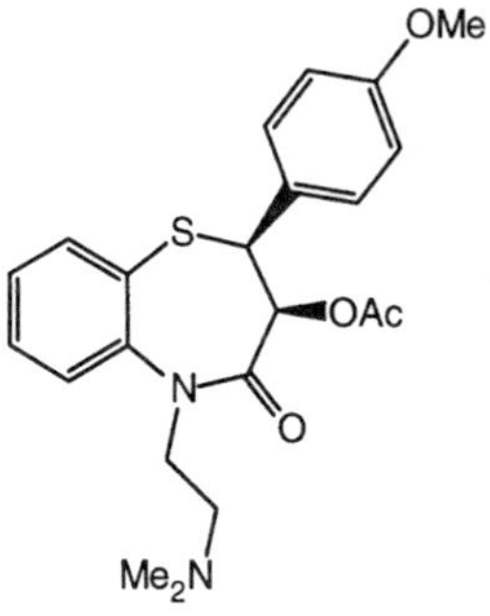

9

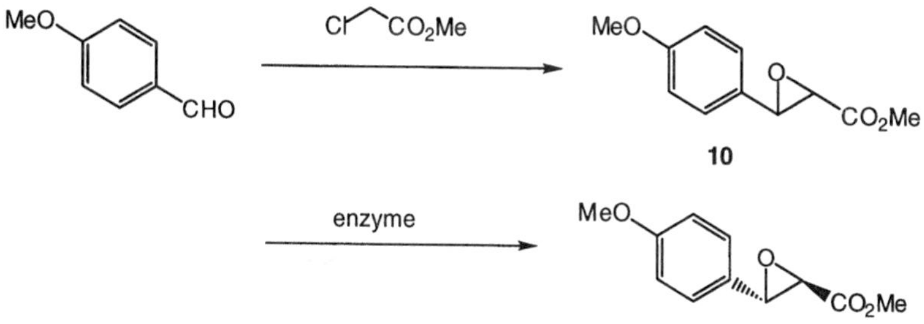

SCHEME 2.

Although the Jacobsen epoxidation can be used to make the chiral epoxide, the number of steps to the precursor alleviates any advantages associated with this approach [33].

3.2.7. Captopril

Captopril (11) is another ACE inhibitor used for the reduction of blood pressure [34,35]. The traditional approach has to been to resolve 3-acyl-2-methylpropionic acids with 1,2-diphenylethylamine, *N*-isopropylphenylethanol, dehydroabiety-

lamine, cinchonidine, or 2-aminobutan-1-ol [3,36–38]. The coupling with L-proline then completes the synthesis (Scheme 3) [36,39]. Various routes have been
proposed in which the sulphur is introduced by a substitution reaction after a
resolution has been used to establish the correct stereochemistry adjacent to the
carboxylic acid group [40].

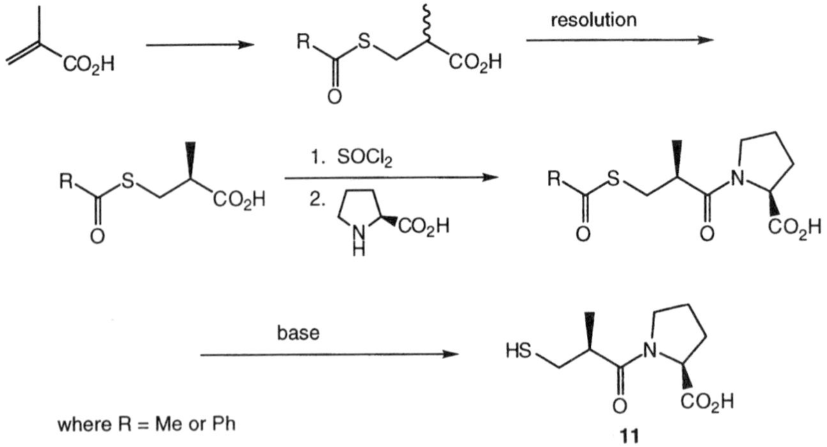

SCHEME 3.

An alternative approach, which has been claimed to be more economical,
is the use of an enzyme to form the stereogenic center (Scheme 4) [36,41,42].

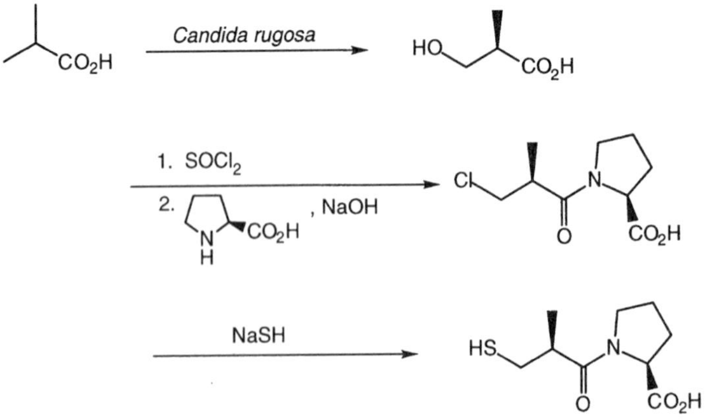

SCHEME 4.

3.2.8.   Sertraline

Sertraline (12) is an antidepressant that inhibits the uptake of serotonin in the central nervous system. One methodology that can be used relies on an asymmetric reduction (Chapter 11) [43–45]. The alternative is a resolution approach [46,47].

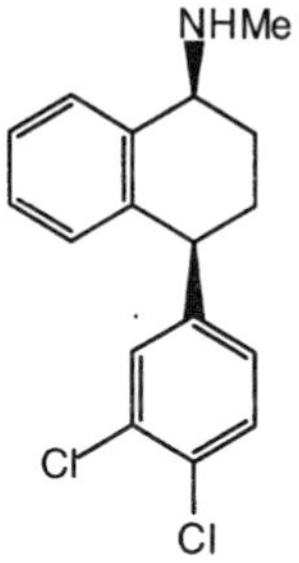

12

3.2.9.   Lovastatin

Lovastatin (13) is a cholesterol-reducing drug that was isolated from *Aspergillus terreus* [48–51]. It is still obtained by fermentation [52], and with the current advances in molecular biology [53,54], chemical approaches will not be able to compete in a cost-effective manner [55–57]. The use of lipases allows for the manipulation of the butyric acid side chain (see Sections 3.2.2 and 3.2.3) [58].

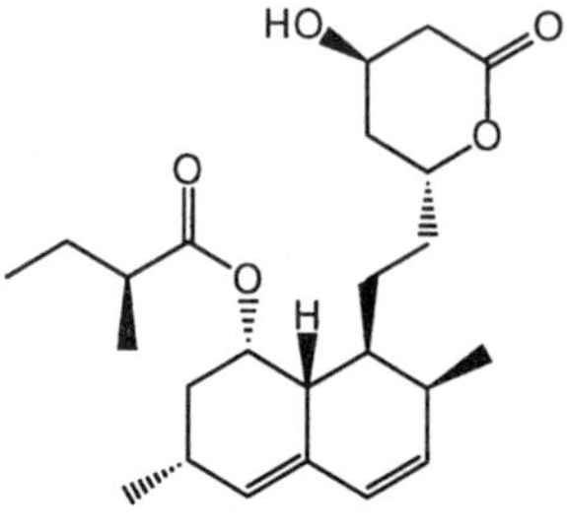

13

3.2.10.   Cefaclor

Cefaclor (14) is a cephalosporin antibiotic. As with penicillins, the side chain is attached through an amide bond to the readily available 7-aminocephalosporanic acid nucleus—obtained by fermentation—and simple chemical transformations [59]. The side chain amino acid is D-phenylglycine, the synthesis of which is discussed in Chapter 8. The use of enzymes to form the amide bond has many advantages over the alternative chemical approaches that usually involve reactive intermediates [17,18,60].

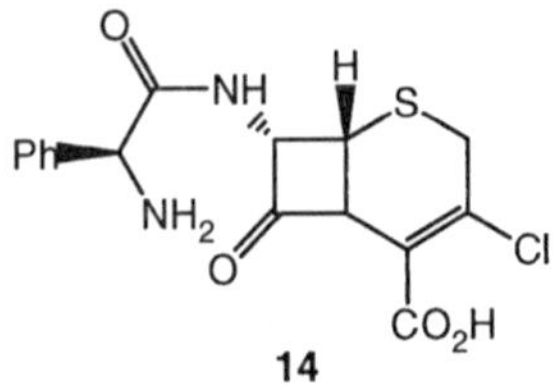

14

3.3. FOOD INGREDIENTS

3.3.1. Menthol

Although this terpene is available from natural sources (Chapter 6), asymmetric synthesis by the Takasago method now accounts for a substantial portion of the market. This synthesis is discussed in detail in Chapter 9.

3.3.2. Aspartame

Aspartame (L-aspartyl-L-phenylalanine methyl ester; APM) (15) is currently the most widely used nonnutritive sweetener worldwide [61].

3.3.2.1. Chemical Synthesis

The chemical methods of industrial significance involve the dehydration of aspartic acid to form an acid anhydride, which is then coupled with the phenylalanine or its methyl ester to give the desired product. The two major processes are known as the Z- and F-processes (Schemes 5 and 6), named after the protecting group used on the aspartyl moiety [62–65]. Both of these processes produce some β-coupled products together with the desired α-aspartame (15), but the selective crystallization removes the undesired isomers. However, since the amino acid raw materials are expensive, they must be recovered from the by-products and waste streams for recycle.

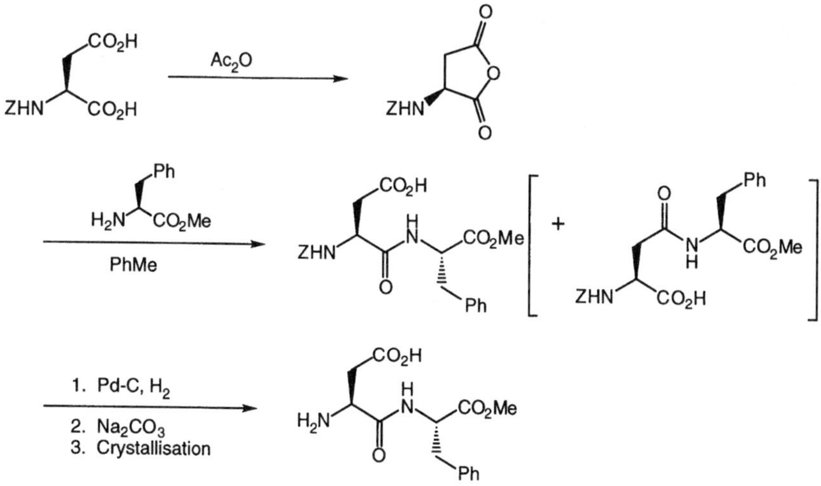

SCHEME 5.

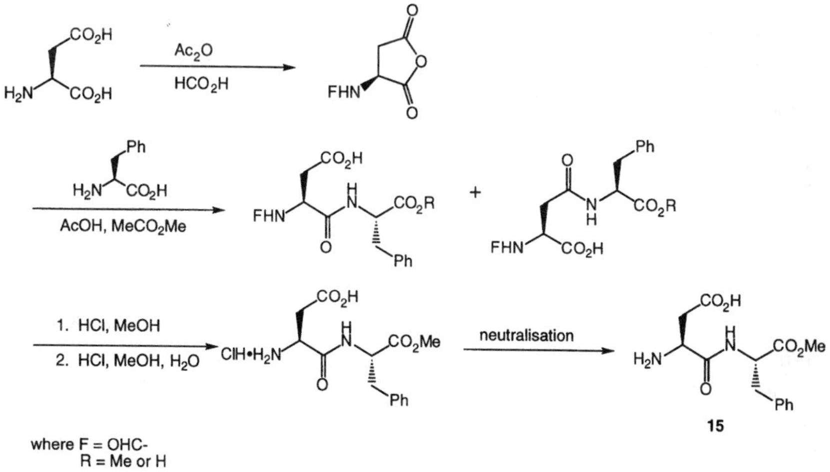

where F = OHC–
      R = Me or H

SCHEME 6.

3.3.2.2.   Enzymatic Synthesis

A protease can be used to catalyze the synthesis of a peptide bond (Scheme 7). When the stoichiometry of the reactions is such that two moles of phenylalanine methyl ester (PM) are used with one mole of Z-aspartic acid, the Z-APM·PM product precipitates and shifts the equilibrium to >95%onversion [66]. This is the basis of the commercial TOSOH process operated by Holland Sweetener that utilizes thermolysin [67]. One significant variation has been the use of racemic PM instead of the L-isomer. Since the enzyme will only recognize the L-PM isomer to form the peptide bond, the unreacted D-PM isomer forms a salt and then, after acidification, the D-PM can be chemically racemized and recycled.

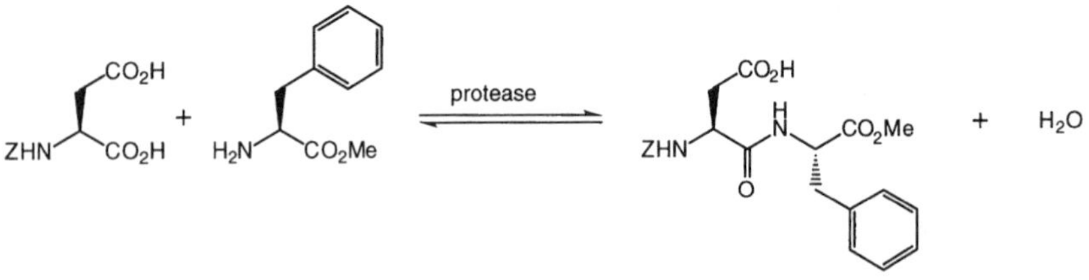

SCHEME 7.

3.3.4. Carnitine

The food additive carnitine (16) is worthy of mention because it has a very close structural relationship to some beta blockers. It is also called vitamin B_T. There are many potential routes to this compound, including an asymmetric hydrogenation method [68,69]. There are two major approaches; one relies on an asymmetric microbial oxidation (Scheme 8) [70].

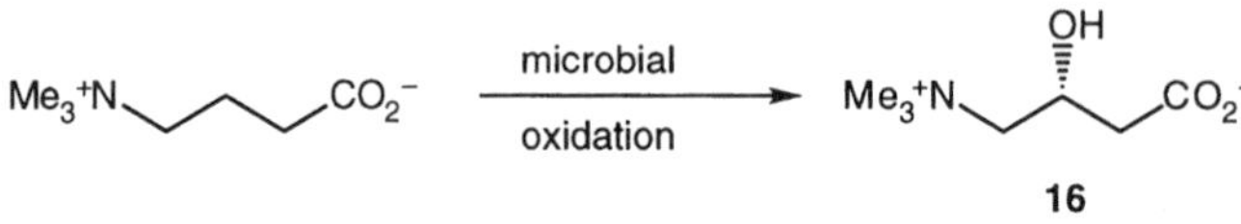

SCHEME 8.

The other approach starts from epichlorohydrin and involves a classical resolution of the amide. The "wrong" amide is also converted to the desired isomer by a dehydration–rehydration sequence that uses hydrolases (Scheme 9) [71].

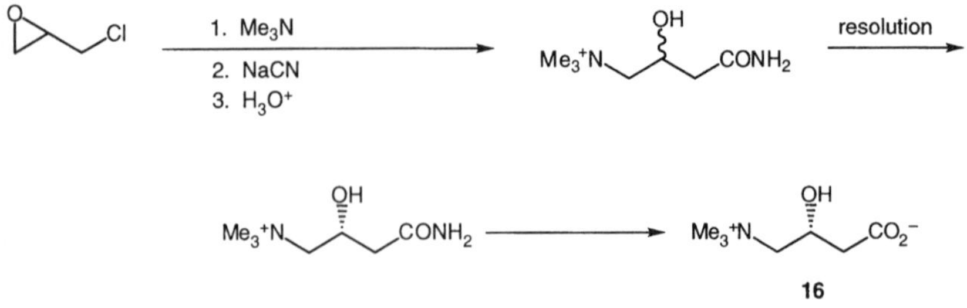

SCHEME 9.

3.4. AGRICULTURAL PRODUCTS

The primary driving force with these compounds is cost. This is reflected in the relatively small number of chiral compounds that are sold at scale.

3.4.1. Metolachlor

The synthesis of metolachlor (17) is described in detail elsewhere (Chapter 9). This compound is sold as an enantioenriched compound (~80% ee) rather than the pure enantiomer due to economic constraints.

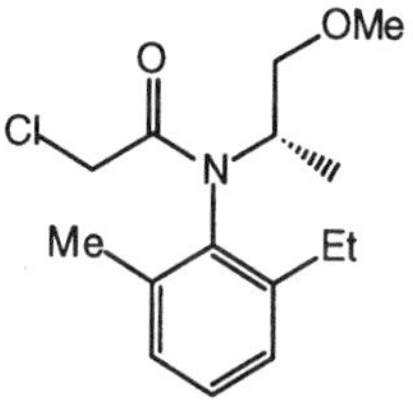

17

3.4.2. Phenoxypropionic Acid Herbicides

These compounds contain a stereogenic center that can be derived from 2-chloropropionic acid and include fluazifop-butyl (18) and flamprop (19) [72].

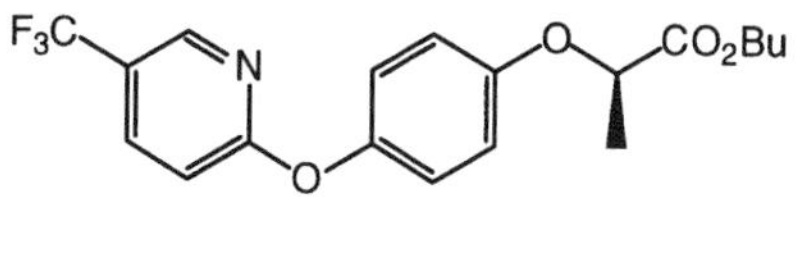

18

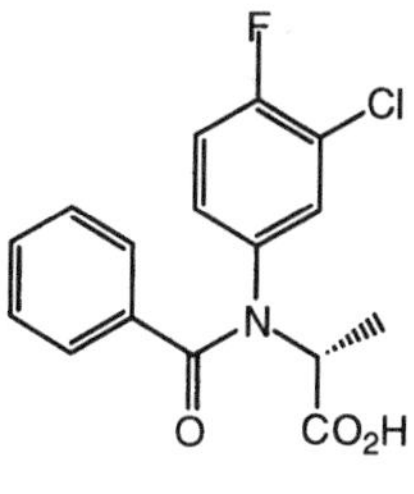

19

A number of enzymatic resolutions have been investigated to the chiral halo ester, including esterification, transesterification, and aminolysis [3,73–75].

Zeneca (ICI) have commercialized the approach that relies on a dehalogenase (Scheme 10) [3,72,75,76].

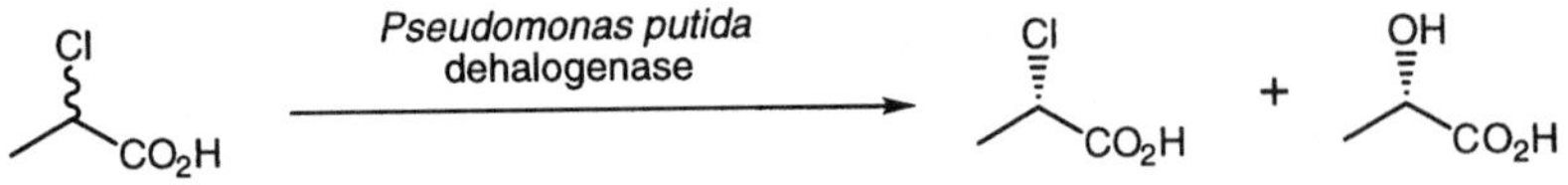

SCHEME 10.

An alternative approach relies a lipase method, but excess substrate has to be present to ensure good ee (Scheme 11) [77].

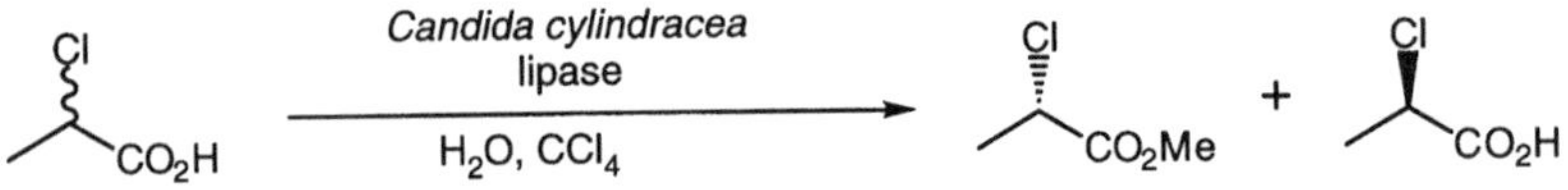

SCHEME 11.

3.4.3. Pyrethroids

These commercially important insecticides usually contain a cyclopropyl unit that is *cis*-substituted and a cyanohydrin derivative. They are usually sold as a mixture of isomers. However, asymmetric routes have been developed, especially as these compounds are related to chrysanthemic esters (Chapter 9) [78]. The pyrethroids can be resolved through salt formation or by enzymatic hydrolysis [79].

Although the methodology does not apply to others, one of the members of this family, deltamethrin (20), is made at scale by a selective crystallization. In the presence of a catalytic amount of base, only one of the isomers crystallizes from isopropanol (Scheme 12) [80].

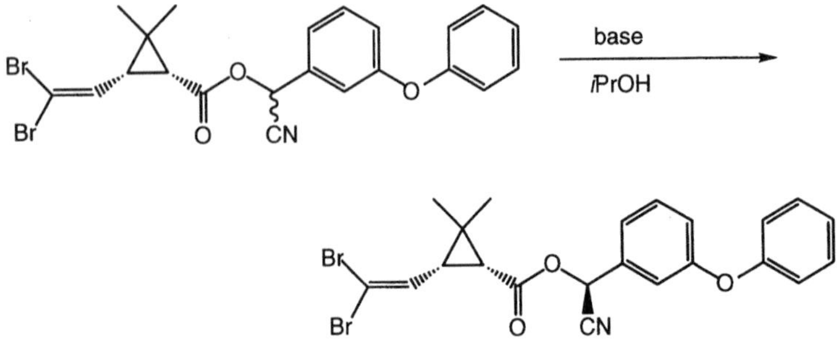

SCHEME 12.

3.5. SUMMARY

Only a small number of examples of chiral compounds that are produced at high volume have been given in this chapter. As the vast majority are mature drugs, the methods of approach reflect the time of their development—resolutions and fermentation methods abound. In other cases, the stereogenic centers are obtained from the chiral pool, as with the "prils." Now that asymmetric reactions are becoming more robust these should have an impact on drugs that are in development or nearing launch. The human immunodeficiency virus protease inhibitors, although not appearing on the 1996 best-seller list, have required a significant amount of process work to bring them to market, and the limitations of many asymmetric reactions have been found. The power of biological approaches must not be underestimated, and this is reflected in some of the chapters in this book. Nature provides us with some very powerful asymmetric catalysts in the form of enzymes. With modern tools, these enzymes can be manipulated to perform a wide variety of reactions on a wide range of substrates.

REFERENCES

1. Harris, E. E., Patchett, A. A., Tristam, E. W., Wyvratt, M. J. *US Patent* 1983, 4374829.
2. Patchett, A. A., Harris, E. Tristam, E. W., Wyvratt, M. J., Wu, M. T. Taub, D., Peterson, E. R., Ikeler, T. J., ten Broeke, J., Payne, L. G., Ondeyka, D. L., Thorsett, E. D., Greenlee, W. J. Lohr, N. S. Hoffsommer, R. D., Joshua, H., Ruyle, W. V., Rothrock, J. W., Aster, S. D., Maycock, A. L., Robinson, F. M., Hirschmann, R., Sweet, C. S., Ulm, E. H., Gross, D. M., Vassil, T. C., Stone, C. A. *Nature* 1980, *288,* 280.
3. Crosby, J. in *Chirality in Industry: The Commercial Manufacture and Applications of Optically Active Compounds*; Collins, A. N., Sheldrake, G. N., Crosby, J., Eds.; Wiley: Chichester, 1992; p. 1.
4. Sheldon, R. A. *Speciality Chem.* 1990, *Feb.,* 31.
5. Blacklock, T. J., Shuman, R. F., Butcher, J. W., Shearin, W. E., Budavari, J., Grenda, V. J. J. *Org. Chem.* 1988, *53,* 836.
6. Greenlee, W. J., Allibone, P. L., Perlow, D. S., Patchett, A. A., Ulm, E. H., Vassil, T. C. *J. Med. Chem.* 1985, *28,* 434.
7. Wyvratt, M. J., Tristam, E. W., Ikeler, T. J., Lohr, N. S., Joshua, H., Springer, J. P., Arison, B. H., Patchett, A. A. *J. Org. Chem.* 1984, *49,* 2816.
8. Conder, M. J., Cianciosi, S. J., Cover, W. H., Dabora, R. L., Pisk, E. T., Stieber, R. W., Tehlewitz, B., Tewalt, G. L. *US Patent* 1993, 5223415.
9. Conder, M. J., Daborah, R. L., Lein, J., Tewalt, G. L. *UK Pat. Appl.* 1992, GB 2255974 A1.
10. Dabora, R. L., Tewalt, G. L. *US Patent* 1992, 5159104.

11. Kubela, R., Radhakrishnan, J. *US Patent* 1995, 5393893.
12. Askin, D., Verhoeven, T. R., Liu, T. M. H., Shinkai, I. *J. Org. Chem.* 1991, 56, 4919.
13. Serizawa, N. *Biotechnol. Ann. Rev.* 1996, 2, 373.
14. Hosobuchi, M., Kurosawa, K., Yoshida, H. *Biotechnol. Bioeng.* 1993, 42, 815.
15. Terahara, A., Tanaka, M. *Ger. Patent* 1981, DE 3123499.
16. Gupta, N., Eisberg, N. *Performance Chem.* 1991, Aug.–Sept. 7, 19.
17. Boesten, W. H. J., Van Dooren, T. J., Smeets, J. C. M. *World Patent* 1996, WO 9623897 A1.
18. Clausen, K., Dekkers, R. M. *World Patent* 1996, WO 9602663 A1.
19. Kaasgaard, S. G., Veitland, U. *World Patent* 1992, WO 9201061 A1.
20. Grossman, J. H., Hardcastle, G. A. *US Patent* 1976, 3980637.
21. Nayler, J. H. C., Long, A. A. W., Smith, H., Taylor, T., Ward, N. *J. Chem. Soc. (C)* 1971, 1920.
22. Williams, R. M. *Synthesis of Optically Active α-Amino Acids*; Pergamon Press: Oxford, 1989.
23. Sheldon, R. A., Hulshof, L. A., Bruggink, A., Leusen, F. J. J., van der Haest, A. D., Wijnberg, H. *Chim. Oggi* 1991, 9, 23.
24. Yamada, H., Takahashi, S., Kii, Y., Kumagai, H. *J. Ferment. Technol.* 1978, 56, 484.
25. Yamada, H., Takahashi, S., Yoneka, K., Amagasaki, H. *Ger. Patent* 1991, 2757980.
26. Drauz, K., Kottenhahan, M., Makryaleas, K., Klenk, H., Bernd, M. *Angew. Chem., Int. Ed. Engl.* 1991, 30, 712.
27. Wu, M. T., Douglas, A. W., Ondeyka, D. L., Payne, L. G., Ikeler, T. J., Joshua, H., Patchett, A. A. *J. Pharm. Sci.* 1985, 74, 352.
28. Bayley, C. R., Vaidya, N. A. in *Chirality in Industry: The Commercial Manufacture and Applications of Optically Active Compounds*; Collins, A. N., Sheldrake, G. N., Crosby, J., Eds.; Wiley: Chichester, 1992; p. 69.
29. Crosby, J. in *Chirality in Industry II: Developments in the Commercial Manufacture and Applications of Optically Active Compounds*; Collins, A. N., Sheldrake, G. N., Crosby, J., Eds.; Wiley: Chichester, 1997; p. 1.
30. Matsumae, H., Furui, M., Shibatani, T. *J. Ferment. Bioeng.* 1993, 75, 93.
31. Keurentjes, J. T. F., Voermans, F. J. M. in *Chirality in Industry II: Developments in the Commercial Manufacture and Applications of Optically Active Compounds*; Collins, A. N., Sheldrake, G. N., Crosby, J., Eds.; Wiley: Chichester, 1997, p. 157.
32. Kanerva, L. T., Sundholm, O. *J. Chem. Soc., Perkin Trans.* 1993, 1, 1385.
33. Jacobsen, E. N., Deng, L., Furukawa, Y., Martinez, L. E. *Tetrahedron* 1994, 50, 4323.
34. Rubin, B., Antonaccio, M. J. in *Pharmacol. Antihypertens. Drugs*; Scriabine, A., Ed.; Raven: New York, 1980; p. 21.
35. Ondetti, M. A., Cushman, D. W. *Ger. Patent* 1977, 2703838.
36. Ohashi, T., Hasegawa, J. in *Chirality in Industry II: Developments in the Commercial Manufacture and Applications of Optically Active Compounds*; Collins, A. N., Sheldrake, G. N., Crosby, J., Eds.; Wiley: Chichester, 1997, p. 269.

37. Sawayama, T., Tsukamoto, M., Sasagawa, T., Naruto, S., Matsumoto, J., Uno, H. *Chem. Pharm. Bull.* 1989, *37,* 1382.
38. Bruggink, A. in *Chirality in Industry II: Developments in the Commercial Manufacture and Applications of Optically Active Compounds*; Collins, A. N., Sheldrake, G. N., Crosby, J., Eds.; Wiley: Chichester, 1997; p. 81.
39. Hen, A. C. *US Patent* 1995, 5387697.
40. Nam, D. H., Lee, C. S., Ryu, D. D. Y. *J. Pharm. Sci.* 1984, *73,* 1843.
41. Hasegawa, J. *US Patent* 1984, 4981794.
42. Shimazaki, M., Hasegawa, J., Kan, K., Nomura, K., Nose, Y., Kondo, H., Ohashi, T., Watanabe, K. *Chem. Pharm. Bull.* 1982, *30,* 3139.
43. Quallich, G. J., Woodall, T. M. *Tetrahedron* 1992, *48,* 10239.
44. Caille, J.-C., Bulliard, M., Laboue, B. in *Chirality in Industry II: Developments in the Commercial Manufacture and Applications of Optically Active Compounds*; Collins, A. N., Sheldrake, G. N., Crosby, J., Eds.; Wiley: Chichester, 1997, p. 391.
45. Quallich, G. J. *World Patent* 1995, WO 9515299.
46. Williams, M., Quallich, G. *Chem. Ind. (London)* 1990, 315.
47. Quallich, G. J., Williams, M. T., Friedmann, R. C. *J. Org. Chem.* 1990, *55,* 4971.
48. Albers-Schonberg, G., Monaghan, R. L., Alberts, A. W., Hoffman, C. H. *US Patent* 1982, 4342767.
49. Moore, R. N., Bigam, G., Chan, J. K., Hogg, A. M., Nakashima, T. T., Vederas, J. C. *J. Am. Chem. Soc.* 1985, *107,* 3694.
50. Witter, D. J., Vederas, J. C. *J. Org. Chem.* 1996, *61,* 2613.
51. Yoshida, Y., Witter, D. J., Liu, Y., Vederas, J. C. *J. Am. Chem. Soc.* 1994, *116,* 2693.
52. Gbewonyo, K., Hunt, G., Buckland, B. *Bioprocess Eng.* 1992, *8,* 1.
53. Vinci, V. A., Conder, M. J., Mcada, P. C., Reeves, C. D., Rambosek, J., Davis, C. R. *World Patent* 1995, WO 9512661 A1.
54. Dahiya, J. S. *Eur. Pat. Appl.* 1993, EP 556699 A1.
55. Hiyama, T. *Pure Appl. Chem.* 1996, *68,* 609.
56. Beck, G., Jendralla, H., Kesseler, K. *Synthesis* 1995, 1014.
57. Rosen, T., Heathcock, C. H. *Tetrahedron* 1986, *42,* 4909.
58. Carta, G., Conder, M. J., Gainer, J. L., Stieberg, R. W., Vinci, V. A., Weber, T. W. *World Patent* 1994, WO 9426920 A1.
59. Chauvette, R. R. *Ger. Patent* 1974, 2408698.
60. Gardner, J. P. *Eur. Pat. Appl.* 1993, EP 567323 A2.
61. Ager, D. J., Pantaleone, D. P., Henderson, S. A., Katritzky, A. R., Prakash, I., Walters, D. E. *Angew. Chem., Int. Ed. Engl.* 1998, in press.
62. Ariyoshi, Y., Nagano, M., Sato, N., Shimizu, A., Kirimura, J. *US Patent* 1974, 3786039.
63. Hill, J. B., Gelman, Y. *US Patent* 1990, 4946988.
64. Boesten, W. H. J. *US Patent* 1975, 3879372.
65. Hill, J. B., Gelman, Y., Dryden, H. L., Erickson, R., Hsu, K., Johnson, M. R. *US Patent* 1991, 5053532.
66. Isowa, Y., Ohmori, M., Mori, K., Ichikawa, T., Nonaka, Y., Kihara, K., Oyama, K., Satoh, H., Nishimura, S. *US Patent* 1979, 4165311.

67. Oyama, K. in *Chirality in Industry*; Collins, A. N., Sheldrake, G. N., Crosby, J., Eds.; Wiley: Chichester, 1992; p. 237.

68. Kitamura, M., Ohkuma, T., Takaya, H., Noyori, R. *Tetrahedron Lett.* 1988, *29*, 1555.

69. Genet, J. P., Pinel, C., Ratovelomanana-Vidal, V., Mallart, S., Pfister, X., Bischoff, L., De Andrade, M. C. C., Darses, S., Galopin, C., Laffitte, J. A. *Tetrahedron: Asymmetry* 1994, *5*, 675.

70. Kulla, H. G. *Chimia* 1991, *45*, 81.

71. Jung, H., Jung, K., Kieber, H. P. *Eur. Patent* 1989, 320460.

72. Holt, R. A. in *Chirality in Industry II: Developments in the Commercial Manufacture and Applications of Optically Active Compounds*; Collins, A. N., Sheldrake, G. N., Crosby, J., Eds.: Wiley: Chichester, 1997; p. 113.

73. Bodnár, J., Gubicza, L., Szabó, L.-P. *J. Mol. Catal.* 1990, *61*, 353.

74. Brieva, R., Rebolledo, F., Gotor, G. *J. Chem. Soc., Chem. Commun.* 1990, 1386.

75. Taylor, S. C. in *Chirality in Industry II: Developments in the Commercial Manufacture and Applications of Optically Active Compounds*; Collins, A. N., Sheldrake, G. N., Crosby, J., Eds.; Wiley: Chichester, 1997; p. 207.

76. Taylor, S. C. *US Patent* 1988, 4758518.

77. Dahod, S. K., Siuta-Mangano, P. *Biotechnol. Bioeng.* 1987, *30*, 995.

78. Elliot, M. *Pestic, Sci.* 1989, *27*, 337.

79. Schneider, M., Engel, N., Boensmann, H. *Angew. Chem., Int. Ed. Engl.* 1984, *23*, 52.

80. Sheldon, R. A. *Chirotechnology: Industrial Synthesis of Optically Active Compounds*; Marcel Dekker: New York, 1993.

4

Synthesis of Phenylalanine by Fermentation and Chemoenzymatic Methods

IAN G. FOTHERINGHAM
NSC Technologies, Mount Prospect, Illinois

4.1. INTRODUCTION

The amino acid phenylalanine is currently manufactured worldwide at an annual scale exceeding 12,000 metric tons. The vast majority of this material is L-phenylalanine synthesized specifically for incorporation into the dipeptide sweetener aspartame. Production of L-phenylalanine grew rapidly during the 1980s in parallel with the demand for aspartame and has grown steadily in recent years. L-phenylalanine is also used in food and medical applications, and demand is increasing in pharmaceutical development as a chiral intermediate or as a precursor of chiral auxiliaries such as benzyloxazolidinone. In contrast, only a few metric tons of D-phenylalanine are used annually, again in pharmaceutical drug development, although this application is increasing. Over the past two decades, a number of manufacturers have developed chemoenzymatic and purely biological routes for L-phenylalanine manufacture (Table 1). Chemoenzymatic syntheses have relied on resolution technology, applied to chemically prepared racemic phenylalanine derivatives or D,L-5-monosubstituted hydantoins and biotransformation routes from chemically prepared achiral precursors of L-phenylalanine. Purely biological routes have concentrated on the direct synthesis of L-phenylalanine through microbial fermentation means. Today, the largest and most commercially successful synthetic routes for L-phenylalanine production use large-scale fermentation of bacterial strains that overproduce L-phenylalanine. Such microorganisms have been isolated through a combination of classical mutagenesis selec-

49

TABLE I Commercial Processes for L-Phenylalanine Synthesis

Company	L-Phenylalanine processes developed
Ajinomoto	Microbial fermentation
Biotechnica	Microbial fermentation
Degussa/Rexim	Enzymatic resolution (aminoacylase)
DSM	Enzymatic resolution (amino acid amidase)
Genex	Biotransformation (phenylalanine ammonia lyase)
Kyowa Hakko	Microbial fermentation
Miwon	Microbial fermentation
Nutrasweet Co.	Microbial fermentation, biotransformation (aminotransferase)
PEI	Biotransformation (aminotransferase)
Tanabe Seiyaku	Enzymatic resolution, biotransformation (aminoacylase, aminotransferase)

tion procedures and molecular genetic manipulation, the latter being derived from a detailed understanding of the molecular biology of the pathways involved in bacterial L-phenylalanine biosynthesis. Conversely, the present commercial production of D-phenylalanine relies solely on chemoenzymatic resolution.

4.2. L-PHENYLALANINE OVERPRODUCING MICROORGANISMS

4.2.1. Common Aromatic and L-Phenylalanine Biosynthetic Pathways

Efforts to develop organisms that overproduce L-phenylalanine have been vigorously pursued by the Nutrasweet Company, Ajinomoto, Kyowa Hakko Kogyo, and others. The focus has centered on bacterial strains that have previously demonstrated the ability to overproduce other amino acids. Such organisms include principally the coryneform bacteria, *Brevibacterium flavum* [1,2] and *Corynebacterium glutamicum* [3,4] used in L-glutamic acid production. In addition, *Escherichia coli* has been extensively studied for L-phenylalanine manufacture due to the detailed characterization of the molecular genetics and biochemistry of its aromatic amino acid pathways and its amenability to recombinant DNA methodology [5]. The biochemical pathway that results in the synthesis of L-phenylalanine from chorismate is identical in each of these organisms (Figure 1), as well as the common aromatic pathway to chorismate. In each case, the L-phenylalanine biosynthetic pathway comprises three enzymatic steps from chorismic acid, the product of the common aromatic pathway, where the precursors are phosphoenolpyruvate (PEP) and erythrose-4-phosphate derived, respectively, from the glycolytic and pentose phosphate pathways of sugar metabolism. Although the intermediate compounds for both pathways are identical in each of the organisms, there are differences in the organization and regulation of the genes and enzymes involved [2,6,7]. Nevertheless, the principal points of pathway regulation are very similar in each of the three bacteria [4,6,8]. Conversely, tyrosine biosynthesis, which is also carried out in three biosynthetic steps from chorismate, proceeds through different intermediates in *E. coli* compared with the coryneform organisms [6,8].

Regulation of phenylalanine biosynthesis occurs both in the common aromatic pathway and in the terminal phenylalanine pathway. The vast majority of efforts to deregulate phenylalanine biosynthesis have focused on two specific rate-limiting enzymatic steps. These are the steps of the aromatic and the phenylalanine pathways carried out, respectively, by the enzymes 3-deoxy-D-arabino-heptulosonate-7-phosphate (DAHP) synthase and prephenate dehydratase. Classical mutagenesis approaches using toxic amino acid analogs and the molecular cloning of the genes encoding these enzymes have led to very significant increases in the capability of host strains to overproduce L-phenylalanine. This has resulted from the elimination of the regulatory mechanisms controlling enzyme

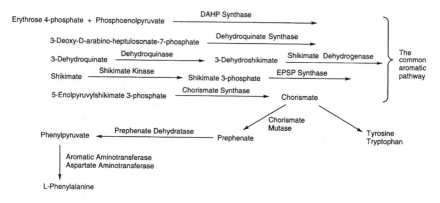

FIGURE 1 The common aromatic and L-phenylalanine pathways.

synthesis and specific activity. The cellular mechanisms that govern the activity of these particular enzymes are complex and illustrate many of the sophisticated means by which bacteria control gene expression and enzyme activity. Chorismate mutase, the first step in the phenylalanine specific pathway and the shikimate kinase activity of the common aromatic pathway, are subject to a lesser degree of regulation and have also been characterized in detail [9,10].

4.2.2. Classical Mutagenesis and Selection

Classical methods of strain improvement have been widely applied in the development of phenylalanine-overproducing organisms [5,11,12]. Tyrosine auxotrophs have frequently been used in efforts to increase phenylalanine production through mutagenesis. These strains often already overproduce phenylalanine due to the overlapping nature of tyrosine and phenylalanine biosynthetic regulation [3,13,14]. Limiting tyrosine availability leads to partial genetic and allosteric deregulation of common biosynthetic steps [15]. Such strains have been subjected to a variety of mutagenesis procedures to further increase the overall titer and the efficiency of phenylalanine production. In general, this has involved the identification of mutants that display resistance to toxic analogs of phenylalanine or tyrosine, such as β-2-thienyl-D,L-alanine and p-fluoro-D,L-phenylalanine [3,5,16]. Such mutants can be readily identified on selective plates where the analog is present in the growth medium. The mutations responsible for the phenylalanine overproduction have frequently been located in the genes encoding the enzymatic activities DAHP synthase, chorismate mutase, and prephenate dehydratase. In turn, this has prompted molecular genetic approaches to further increase phenylalanine production through the isolation and in vitro manipulation of these genes (vide infra).

4.2.3. Deregulation of DAHP Synthase Activity

The activity of DAHP synthase commits carbon from intermediary metabolism to the common aromatic pathway, thus converting equimolar amounts of PEP and E4P to DAHP [17]. In E. coli, there are three isoenzymes of DAHP synthase of comparable catalytic activity encoded by the genes aroF, aroG, and aroH [18]. Enzyme activity is regulated, respectively, by the aromatic amino acids tyrosine, phenylalanine, and tryptophan [15,19–22]. In each case, regulation is mediated both by repression of gene transcription and by allosteric feedback inhibition of the enzyme, although to different degrees.

The aroF gene lies in an operon with tyrA, which encodes the bifunctional protein chorismate mutase/prephenate dehydrogenase. Both genes are regulated by the TyrR repressor protein complexed with tyrosine. The aroF gene product accounts for 80% of the total DAHP synthase activity in wild-type E. coli cells.

The *aroG* gene is repressed by the TyrR repressor protein complexed with phenylalanine and tryptophan. Repression of *aroH* is mediated by tryptophan and the TrpR repressor protein [19]. The gene products of *aroF* and *aroG* are almost completely feedback inhibited, respectively, by low concentrations of tyrosine or phenylalanine, whereas the *aroH* gene product is subject to a maximum of 40% feedback inhibition by tryptophan [23]. In *Brevibacterium flavum*, DAHP synthase forms a bifunctional enzyme complex with chorismate mutase that is feedback inhibited by tyrosine and phenylalanine synergistically but not by tryptophan [24,25]. Similarly, in *Corynebacterium glutamicum*, DAHP synthase is inhibited most significantly by phenylalanine and tyrosine acting in concert [26], but, unlike *B. flavum*, reportedly does not show tyrosine-mediated repression of transcription [3,24].

Many examples of analog-resistant mutants of these organisms display reduced sensitivity of DAHP synthase to feedback inhibition [3,5,7,24,27,28]. In *E. coli*, the genes that encode the DAHP synthase isoenzymes have been characterized and sequenced [6,29,30]. The mechanism of feedback inhibition of the *aroF*, *aroG*, and *aroH* isoenzymes has been studied in considerable detail [30], and variants of the *aroF* gene on plasmid vectors have been used to increase phenylalanine overproduction. Simple replacement of transcriptional control sequences with powerful constitutive or inducible promoter regions and the use of high copy number plasmids has readily enabled overproduction of the enzyme [31,32]. Reduction of tyrosine-mediated feedback inhibition has been described through the use of resistance to the aromatic amino acid analogs β-2-thienyl-D,L-alanine and *p*-fluoro-D,L-phenylalanine [31,33].

4.2.4. Deregulation of Chorismate Mutase and Prephenate Dehydratase Activity

The three enzymatic steps by which chorismate is converted to phenylalanine appear to be identical between *C. glutamicum, B. flavum*, and *E. coli*, although only in the latter case have detailed reports appeared on the characterization of the genes involved. In each case, the principal regulatory step is the one catalyzed by prephenate dehydratase. In *E. coli*, chorismate is converted first to prephenate and then to phenyl pyruvate by the action of a bifunctional enzyme, chorismate mutase/prephenate dehydratase (CMPD) encoded by the *pheA* gene [6]. The final step, in which phenyl pyruvate is converted to L-phenylalanine, is carried out predominantly by the aromatic aminotransferase encoded by *tyrB* [34]. However, both the aspartate aminotransferae, encoded by *aspC* and the branched chain aminotransferase encoded by *ilvE*, can efficiently catalyze this reaction [35]. Phenylalanine biosynthesis is regulated by control of CMPD through phenylalanine-mediated attenuation of *pheA* transcription [6] and by feedback inhibition of the prephenate dehydratase and chorismate mutase activities of the enzyme. Inhibition is most pronounced on the prephenate dehydratase activity, with almost total inhibition observed at micromolar phenylalanine concentrations [36,37]. In contrast, chorismate mutase activity is maximally inhibited by only 40% [37]. In *B. flavum*, prephenate dehydratase and chorismate mutase are encoded by distinct genes. Prephenate dehydratase is again the principal point of regulation with the enzyme subject to feedback inhibition, but not transcriptional repression, by phenylalanine [38,39]. Chorismate mutase, which in this organism forms a bifunctional complex with DAHP synthase, is inhibited by phenylalanine and tyrosine up to a maximum of 65%, but this is significantly diminished by the presence of very low levels of tryptophan [8]. Expression of chorismate mutase is repressed by tyrosine [27,40]. The final step is carried out by at least one transaminase [41]. Similarly, in *C. glutamicum*, the activities are encoded in distinct genes, with prephenate dehydratase again being more strongly feedback inhibited by phenylalanine [3,26,42]. The only transcriptional repression reported is that of chorismate mutase by phenylalanine. The *C. glutamicum* genes encoding prephenate dehydratase and chorismate mutase have been isolated and cloned from analog-resistant mutants of *C. glutamicum* [7,43]. These have been used, along with the cloned DAHP synthase gene, to augment L-phenylalanine biosynthesis in overproducing strains of *C. glutamicum* [43].

Many publications and patents have described successful efforts to reduce and eliminate regulation of prephenate dehydratase activity in phenylalanine-overproducing organisms [5,9,32,33]. As with DAHP synthase, the majority of the reported efforts have focused on the *E. coli* enzyme encoded by the *pheA* gene that is transcribed convergently with the tyrosine operon on the *E. coli* chromosome [6]. The detailed characterization of the *pheA* regulatory region has facilitated expression of the gene in a variety of transcriptional configurations, leading to elevated expression of CMPD and a number of mutations in *pheA* that affect phenylalanine-mediated feedback inhibition have been described [9,31,44]. Most mutations have been identified through resistance to phenylalanine analogs such as β-2-thienylalanine, but there are examples of feedback-resistant mutations arising through insertional mutagenesis and gene truncation [31,44]. In particular, two regions of the enzyme have been shown to reduce feedback inhibition to different degrees. Mutations at position Trp338 in the peptide sequence desensitize the enzyme to levels of phenylalanine in the 2–5 mM range, but are insufficient to confer resistance to higher concentrations of L-phenylalanine [31,44]. Mutations in the region of residues 304–310 confer almost total resistance to feedback inhibition at L-phenylalanine concentrations of at least 200 mM [9]. Feedback inhibition profiles of four such variants (JN305–JN308) are shown in Figure 2, in comparison to the profile of wild-type enzyme (JN302). It is not clear if the mechanism of resistance is similar in either case, but the difference is significant to commercial application as overproducing organisms readily achieve extracellular concentrations of L-phenylalanine greater than 200 mM.

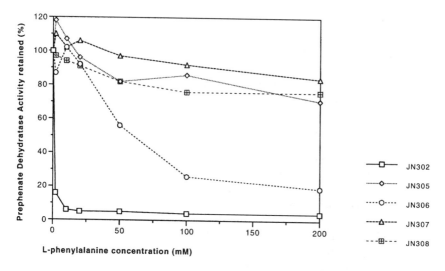

FIGURE 2 Feedback inhibition profiles of prephenate dehydratase variants in *E-coli* K12.

4.2.5. Precursor Supply

The rate-limiting steps in the common aromatic and phenylalanine biosynthetic pathways are obvious targets in the development of phenylalanine overproducing organisms. However, the detailed biochemical and genetic characterization of *E. coli* has enabled additional areas of its metabolic function to be specifically manipulated to determine their effect on aromatic pathway throughput. Besides efforts to eliminate additional lesser points of aromatic pathway regulation, attempts have been made to enhance phenylalanine production by increasing the supply of aromatic pathway precursors and by facilitating exodus of L-phenylalanine from the cell. The precursors of the common aromatic pathway D-erythrose 4-phosphate and phosphoenolpyruvate are the respective products of the pentose phosphate and glycolytic pathways. Precursor supply in aromatic amino acid biosynthesis has been reviewed [45,46]. Theoretical analyses of the pathway and the cellular roles of these metabolites suggest that the production of aromatic compounds is likely to be limited by PEP availability [47,48] since phosphoenolpyruvate is involved in a number of cellular processes, including the generation of metabolic energy through the citric acid cycle [49], and the transport of glucose into the cell by the phosphotransferase system [50]. Strategies to reduce the drain of PEP by these processes have included mutation of sugar transport systems to reduce PEP-dependent glucose transport [51] and modulation of the activities of pyruvate kinase, PEP synthase, and PEP carboxylase that regulate PEP flux to pyruvate, oxaloacetate, and the citric acid cycle [45,49,52]. Similarly, the availability of E4P has been increased by altering the levels of transketolase, the enzyme responsible for E4P biosynthesis [53]. In general, these efforts have been successful to direct additional flux of PEP of E4P into the aromatic pathway, although their effect has not proven to be as predictable as the deregulation of rate-limiting pathway steps, and their overall impact on L-phenylalanine overproduction has not been well characterized.

4.2.6. Secretion of Phenylalanine from the Cell

Exodus of phenylalanine is of manifest importance in the large-scale production of phenylalanine because the amino acid is typically recovered only from the extracellular medium and washed cells. Because lysis of cells to recover additional phenylalanine is not generally practical, L-phenylalanine remaining within the cells after washing is usually lost. Since cell biomass is extremely high in large-scale fermentations, this can represent up to 5% of total phenylalanine produced. Additionally, since L-phenylalanine is known to regulate its own biosynthesis at multiple points through feedback inhibition, attenuation, and TyrR-mediated repression, methods to reduce intracellular concentrations of phenylalanine, through reduced uptake or increased exodus throughout the fermentation, have been sought. Studies have addressed the means by which L-phenylalanine is both taken up and excreted by bacterial cells, and whether this can be altered to increase efflux. In overproducing strains, it is likely that most phenylalanine leaves the cell by passive diffusion, but specific uptake and exodus pathways also exist. In E. coli, L-phenylalanine is taken up by at least two permeases [54], one specific for phenylalanine encoded by pheP, and the other encoded by aroP, a general aromatic amino acid permease responsible for the transport of tyrosine and tryptophan in addition to phenylalanine. The genes encoding the permeases have been cloned and sequenced, and E. coli mutants deficient in either system have been characterized [55,56]. The effect of an aroP mutation has been evaluated in L-phenylalanine overproduction [5]. In addition, specific export systems have been identified that, although not completely characterized, have been successfully used to increase exodus of L-phenylalanine from strains of E. coli [56,57]. In one such system, the Cin invertase of bacteriophage P1 has been shown to induce a metastable phenylalanine hypersecreting phenotype on a wild type strain of E. coli. The phenotype can be stabilized by transient introduction of the invertase on a temperature-sensitive plasmid replicon and is sufficient to establish L-phenylalanine overproduction in the absence of any alterations to the normal biosynthetic regulation [57].

4.2.7. Summary

The incremental gains made from the various levels at which phenylalanine biosynthesis has been addressed, has led to the high-efficiency production strains currently in use today. Almost all large-scale L-phenylalanine manufacturing processes in present operation are fermentations using bacterial strains such as those previously described in which classical strain development and/or molecular genetics have been extensively applied to bring about phenylalanine overproduction. The low substrate costs and the economics of scale associated with this approach have resulted in significant economic advantages. However, a number of alternate biotransformation and chemoenzymatic resolution strategies have been extensively developed and have also demonstrated highly efficient synthesis of L-phenylalanine.

4.3. BIOTRANSFORMATION ROUTES TO L-PHENYLALANINE

4.3.1. L-Phenylalanine Ammonia Lyase Process

The use of phenylalanine ammonia lyase (PAL) in various overproducing strains of the yeast *Rhodoturula* was developed by Tanabe Seiyaku Co. and by Genex Corporation in the 1980s to produce phenylalanine in the reaction (Figure 3). In this process, fermented cells of yeast strains *R. glutinis, R. rubra*, or *R. graminis* were recovered and washed before bioreaction under batch or immobilized conditions with *trans*-cinnamic acid and ammonia [58–61]. The elevated pH and high concentration of ammonia used under the process conditions enabled the normally catabolic PAL reaction to favor formation of L-phenylalanine [62,63]. Significant development work was conducted to determine the appropriate conditions for optimal growth and maximal induction of the PAL enzyme [64], and the appropriate process conditions for retention of PAL activity [65]. The principal drawbacks of the PAL process lay in substrate inhibition of the enzyme, low enzyme specific activity, and enzyme instability. Screening regimes conducted to address these limitations resulted in the identification of an *R. graminis* strain that displayed greater enzyme-specific activity and stability than the *R. rubra* and *R. glutinis* strains used previously [58]. Subsequent mutagenesis and selection procedures conducted on this strain using the cinnamic acid analog phenylpropio-

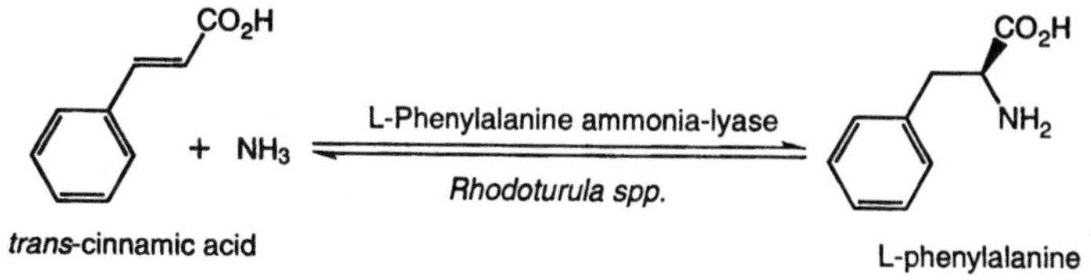

trans-cinnamic acid

L-phenylalanine

FIGURE 3 L-phenylalanine ammonia lyase transformation.

lic acid led to a further increase in PAL activity through a significant elevation in PAL gene expression [58]. Such strain and process development strategies enabled the Genex process to achieve yields of L-phenylalanine in excess of 50 g/L with close to 90% conversion of substrate to product [61]. Nevertheless, the biotransformation could not favorably compete with the economics of fermentative production and is not presently used in L-phenylalanine production.

4.3.2. Aspartate or Aromatic Aminotransferase Process

A number of reports from PEI, Nutrasweet, and Allelix, among others, have described the use of aminotransferases (transaminases) to produce L-phenylalanine (Figure 4). The reaction proceeds by a ''ping-pong'' mechanism whereby the amino group of a donor amino acid [66,67], typically aspartic acid, is exchanged with the keto group of phenylpyruvic acid to yield L-phenylalanine and the by-product, oxaloacetate. Aminotransferases are ubiquitous in nature, and microorganisms, typically possess multiple aminotransferases that are involved in the biosynthesis of a number of amino acids, often catalyzing the terminal step from the keto acid precursor [67–69]. The enzymes use pyridoxal 5'-phosphate as co-factor and typically possess high catalytic rates and stereoselectivity with broad substrate specificities [70].

This is exemplified by the multiple transaminases in *E. coli*, which can each catalyze the synthesis of L-tyrosine, L-aspartate, L-valine, L-isoleucine, and L-leucine in addition to L-phenylalanine [35]. Many aminotransferase-encoding genes have been cloned and used to overproduce the enzymes [35,71,72]. Often, significant sequence homology exists among the aminotransferases of individual strains, as well as between the corresponding enzymes of microbial and mammalian origin [35]. In addition, the tertiary structures of a number of aminotransferases have been determined, furthering the mechanistic understanding of the en-

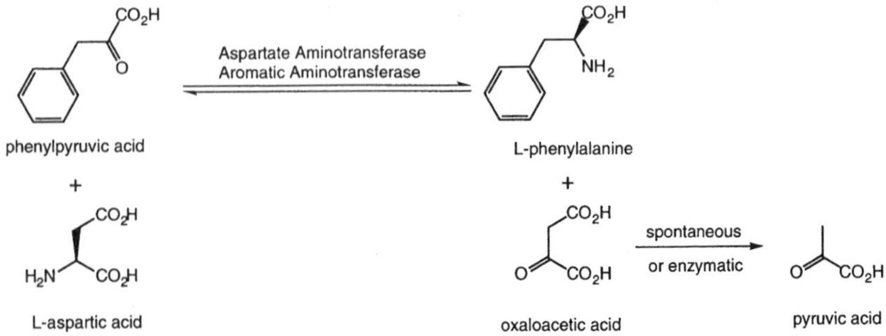

FIGURE 4 Transaminase route to L-phenylalanine.

zyme and prompting mutagenesis approaches for the alteration of enzyme properties [73]. Aminotransferases obtained from many microorganisms have been applied in L-phenylalanine biosynthesis [74–76]. In the case of *E. coli*, the highly homologous aspartate and aromatic aminotransferases, encoded, respectively, by the *aspC* and *tyrB* genes, have both been developed and applied in this bioconversion [77–79]. In general, the aspartate aminotransferase has been preferred because it has shown greater throughput and increased stability over the *tyrB* gene product [79,80]. A number of L-phenylalanine biosynthetic processes involving aminotransferases have been described using either isolated enzyme or whole cells in batched or immobilized systems [76,79,81,82]. In the mid-1980s, one such immobilized process was commercialized by PEI using the *E. coli* aspartate aminotransferase [79]. In this case, the PPA substrate was economically derived from chemically synthesized 5-benzilidene hydantoin. The highest yields in the transaminase synthesis are generally obtained from batch processes at the expense of the greater catalyst stability of the continuous systems. The cost of catalyst regeneration in batch systems is a drawback of the aminotransferase approach, as is the need for two primary metabolites as substrates. The principal drawback, however, lies in the reaction equilibrium, which, being close to 1, limits the yield of the reaction to 50% unless the keto acid by-product can be eliminated from further reaction. To overcome this limitation, commercial trans-amination approaches have almost invariably used aspartate as the amino donor, which, upon transamination, is converted to the relatively unstable oxaloacetate that spontaneously decarboxylates to yield pyruvate. This enables the reaction to proceed to a yield often exceeding 90%. The decarboxylation of oxaloacetate is more rapid at elevated temperatures and has prompted the development of transaminase bioconversions carried out in excess of 50°C [80]. Aminotransferases from thermophiles have been explored in this respect, as well as chemical and enzymatic steps to accelerate oxaloacetate decarboxylation and thereby optimize product yield and substrate conversion [83].

4.3.3.   Phenylalanine Dehydrogenase Process

A similar process with phenyl pyruvate as precursor has also been developed using the enzyme phenylalanine dehydrogenase. Unlike the aminotransferases, phenylalanine dehydrogenase can use ammonia as a substrate in the reductive amination of phenyl pyruvate, thereby realizing a cost advantage. In addition, the reaction equilibrium lies heavily in favor of phenylalanine synthesis. These advantages in the reaction mechanism are diminished by a dependence of the dehydrogenase on an NADH cofactor, which, in turn, requires enzymatic regeneration. This has been addressed through the use of a coupled enzyme system [84], where formate dehydrogenase is used along with phenylalanine dehydrogenase

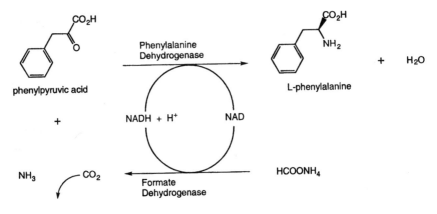

FIGURE 5   L-phenylalanine dehydrogenase process.

(Figure 5). The reaction proceeds to completion due to the evolution of carbon dioxide that is generated by the oxidation of formate. By the use of partially purified enzymes in membrane reactors, this reaction has achieved phenylalanine titers and substrate conversions comparable to the aminotransferase-based systems [84,85]. Despite the high efficiencies obtained in this bioconversion [86], commercial production of L-phenylalanine using this method has not been economically competitive with fermentative processes, mainly because of the costs of phenylpyruvic acid preparation. Nevertheless, in a similar process operated by Degussa that uses leucine dehydrogenase, this approach has been successfully applied to the synthesis of compounds of higher intrinsic value, such as the unnatural amino acid, L-*tert*-leucine [87].

4.4.   RESOLUTION-BASED L-PHENYLALANINE SYNTHESIS

4.4.1.   Amino Acid Amide Resolution

In contrast to the various fermentation and bioconversion approaches that have used enzyme stereoselectivity to synthesize L-phenylalanine as a single isomer, additional methods have been developed that rely on the same stereoselectivity to resolve racemic mixtures of chemically synthesized amino acids. One such general approach, which has been successfully commercialized for L-phenylalanine production, relies on cleavage of the L-isomer of a D,L-α-amino acid amide mixture by an enantiospecific amidase enzyme. The general procedure operated

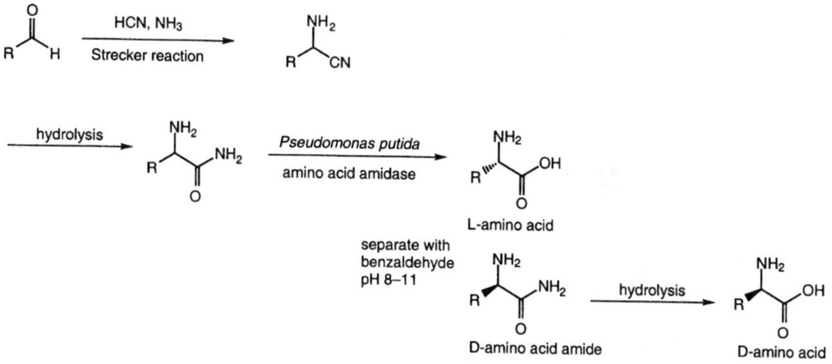

FIGURE 6 Amidase resolution method.

by DSM (Figure 6), uses an amidase-containing strain of *Pseudomonas putida* to specifically hydrolyze the L-isomer of the D,L-amino acid amide mixture. The racemic amino acid mixture is prepared inexpensively from an aldehyde precursor by the Strecker reaction [88,89], with alkaline hydrolysis of the resultant aminonitrile. Treatment with benzaldehyde after selective hydrolysis of the L-amino acid amide yields an insoluble Schiff base that forms between benzaldehyde and the residual D-amino acid amide. In this way, the residual D-amino acid amide may be readily separated [90] from the resultant L-amino acid and can then be chemically hydrolyzed and purified or racemized and recycled [91]. The advantages of this process include general applicability to the preparation of a range of amino acids [89,92], and the opportunity to recover the D-amino acid, if desired, from the reaction in addition to the L-amino acid. Conversely, in the preparation of only the L-amino acid by this approach, the need to recycle the unreacted substrate is considered a drawback of the process, necessitating additional complexity and cost. This aspect has been considered in a further development of this process by DSM, in which *P. putida* strains possessing D- or L-amino acid amidase as well as amino acid racemase activity have been identified. The in situ coupling of either amidase with the racemase activity avoids the recycling requirement of the process and facilitates direct conversion of the D,L-amino acid amide mix into the desired isomer [93].

4.4.2. Acylase Process

An earlier strategy involving stereoselective enzyme hydrolysis of chemically prepared racemic precursors has been operated successfully for many years by

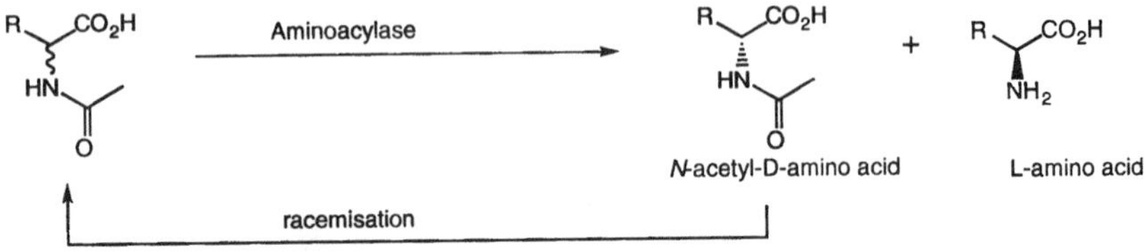

FIGURE 7 Acylase resolution method.

Tanabe Seiyaku. In this process (Figure 7), a DEAE-Sephadex immobilized L-specific aminoacylase selectively hydrolyses N-acetyl-L-amino acids from an N-acetyl-D,L-amino acid mixture [94,95]. After hydrolysis, the L-amino acid is isolated by selective crystallization and the remaining N-acetyl-D-amino acid is racemized, either chemically or enzymatically [96], and recycled. The preferred aminoacylase for this reaction was obtained from the mould *Aspergillus oryzae*, and is readily isolated and immobilized in the bioreactor [94]. The enzyme has broad substrate specificity and has been used by Tanabe in the production of L-methionine and L-valine in addition to L-phenylalanine. This process has also been further developed by Degussa/Rexim [97] to prepare bulk quantities of L-phenylalanine and a large number of L-amino acids. In addition, Tanabe has reported on the further development of the aminoacylase process through the coupling of acetamidocinnamate aminohydrolase (ACA) and aminotransferase activity. In this process, L-phenylalanine is produced in two steps from acetamidocinnamic acid through a phenylpyruvic acid intermediate using immobilized cells of two independent bacterial strains [81].

4.4.3. D,L-5-Monosubstituted Hydantoin-Based Strategies

The recycling aspect of resolution-based amino acid syntheses, such as the amidase and aminoacylase processes, is again considered and largely overcome in an alternative strategy using D,L-5-monosubstituted hydantoins. Inexpensive chemical routes, such as the Bucherer-Bergs synthesis or the condensation of aldehydes with hydantoin [98–101], are used to prepare the racemic hydantoin derivatives that are then resolved using either L- or D-enantiospecific hydantoinase and N-carbamoyl amino acid amidohydrolase enzymes to yield the desired amino acid isomer (Figure 8). In the case of phenylalanine, both L- and D-phenylalanine can be prepared from racemic D,L-5-benzylhydantoin. The advantage of the hydantoinase process is that in many cases, a dynamic resolution can be established whereby the unreacted isomer can be racemized in situ either chemically [102] or in the presence of a racemase enzyme [103] to enable the process to approach 100% yield. Enzymes with L- and D-specific hydantoinase activity

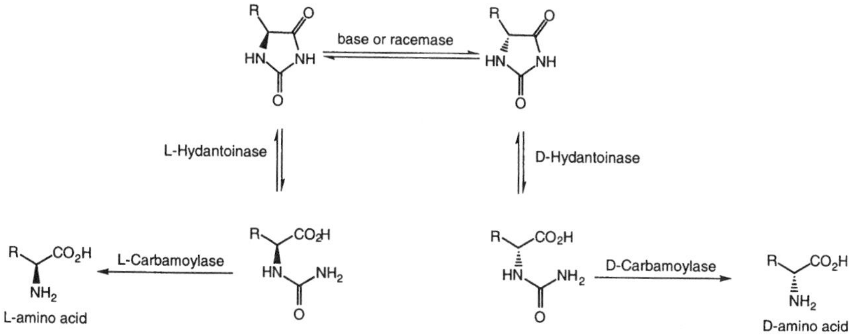

FIGURE 8   Hydantoin resolution method.

have been described in a number of microbial species, with the D-specific activity having been identified much more frequently, often attributable to dihydropyrimidinases involved in pyrimidine catabolism [104–108]. Substrate specificities vary significantly between enzyme isolates and contribute to the broad applicability of this approach [107–110]. Molecular cloning has been applied in at least one case to isolate a D-hydantoinase encoding gene from a thermophilic bacillus and to subsequently overproduce the enzyme [111]. The D- and L-specific N-carbamoyl amino acid amidohydrolases have also been shown to be widely distributed in microorganisms [107,108,110,112]. The bioconversion steps are typically carried out with whole cells in immobilized systems [104,112]; racemization of the unreacted hydantoin occurs under the normal alkaline conditions of the reaction [102]. In some cases, microbial racemases have been used to optimize the racemization step [103,113]. The hydantoinase process offers the advantage of relatively inexpensive precursors and high conversion rates through the in situ racemization of the substrate. In addition, through the broad substrate range of hydantoinase and carbamoylase enzymes, the process is broadly applicable in L- and D-amino acid biosynthesis. However, it is in D-amino acid production that large-scale commercial application of the hydantoinase process has been more significantly applied, most notably by Recordati and Kanegafuchi in the manufacture of D-phenyl glycine and D-p-hydroxyphenyl glycine as components of semisynthetic penicillins and cephalosporins [114–116], but also by Degussa and Ajinomoto in the production of additional D-amino acids including D-phenylalanine [116–118].

4.5. CONCLUSIONS

Numerous commercial processes have been developed for the large-scale commercial production of L-phenylalanine since the early 1980s, fueled by the enormous increase in L-phenylalanine demand for the dipeptide sweetener, aspartame. The most successful processes each use microbial enzymes at some stage, either in an enantioselective hydrolysis or biotransformation step using chemically derived precursors, or in whole-cell fermentations using microorganisms engineered through classical and recombinant strain improvement methodology. In general, the most economically competitive processes have resulted from large-scale fermentation approaches. This is mainly due to the incremental savings achieved in such processes through increased understanding of the physiology and molecular genetic regulation of L-phenylalanine biosynthesis in microbes and the low raw material costs attained through media and process development. Several highly efficient and elegant biotransformation approaches, such as the PAL and dehydrogenase processes, have been optimized through molecular genetic strategies but have been impacted by the drawbacks of greater precursor cost and, in some cases, limited biocatalyst stability. Nevertheless, a number of manufacturers, including DSM, Degussa/Rexim, and Tanabe, successfully produce L-phenylalanine through chemoenzymatic routes that are generally more versatile and are often part of more broadly applicable processes in natural and unnatural amino acid manufacture. This is apparent in the manufacture of D-phenylalanine, where, in contrast to the preponderance of fermentative routes to L-phenylalanine, the synthesis is currently carried out using the hydantoinase and amidase resolution strategies previously mentioned. Ultimately, through a greater understanding of microbial pathway engineering and the increasing availability of novel enzymatic activities, it is possible that the favorable economics of large-scale fermentation may also become more applicable to D-amino acid biosynthesis [119,120].

REFERENCES

1. Katsumata, R., Ozaki, A. *Eur. Pat. Appl.*, 1987, 0264, 914.
2. Shiio, I., Sugimoto, S., Kawamura, K. *Agric. Biol. Chem.* 1988, *52*, 2247.
3. Hagino, H., Nakayama, K. *Agric. Biol. Chem.* 1974, *38*, 2367.
4. Ikeda, M., Ozaki, A., Katsumata, R. *Appl. Microbiol. Biotechnol.* 1993, *39*, 318.
5. Tribe, D. E. *US Patent*, 1987, 4,681,852.
6. Hudson, G. S., Davidson, B. E. *J. Mol. Biol.* 1984, *180*, 1023.
7. Ozaki, A., Katsumata, R., Oka, T., Furuya, A. *Agric. Biol. Chem.* 1985, *49*, 2925.
8. Shiio, I., Sugimoto, S. *Agric. Biol. Chem.* 1981, *45*, 2197.
9. Nelms, J., Edwards, R. M., Warwick, J., Fotheringham, I. *Appl. Environ. Microbiol.* 1992, *58*, 2592.
10. Millar, G., Lewendon, A., Hunter, M., Coggins, J. R. *Biochem. J.* 1986, *237*, 437.
11. Tsuchida, T., Kubota, K., Morinaga, Y., Matsui, H., Enei, H., Yoshinaga, F. *Agric. Biol. Chem.* 1987, *51*, 2095.
12. Tokoro, Y., Oshima, K., Okii, M., Yamaguchi, K., Tanaka, K., Kinoshita, S. *Agric. Biol. Chem.* 1970, *34*, 1516.

13. Hwang, S. O., Gil, G. H., Cho, Y. J., Kang, K. R., Lee, J. H., Bae, J. C. *Appl. Microbiol. Biotech.* 1985, *22*, 108.
14. Nakayama, K., Sagamihara, Hagino, H. *US Patent*, 1973, 3,759,790.
15. Wallace, B. J., Pittard, J. *J. Bacteriol.* 1969, *97*, 1234.
16. Choi, Y. J., Tribe, D. E. *Biotechnol. Lett.* 1982, *4*, 223.
17. Haslam, E. *The Shikimate Pathway*; Wiley: New York, 1974.
18. Gibson, F., Pittard, J. *Curr Topics Cell Regul.* 1970, *2*, 29.
19. Brown, K. D. *Genetics* 1968, *60*, 31.
20. Brown, K. D. Somerville, R. L. *J. Bacteriol.* 1971, *108*, 386.
21. Camakaris, H., Pittard, J. *J. Bacteriol.* 1973, *115*, 1135.
22. Im, S. W. K., Davidson, H., Pittard, J. *J. Bacteriol.* 1971, *108*, 400.
23. Herrmann, K. M., Somerville, R. L. Eds. *Amino Acids: Biosynthesis and Genetic Regulation*; Addison-Wesley: Mass, 1983.
24. Shiio, I., Sugimoto, S., Miyajima, J. *J. Biochem.* 1974, *75*, 987.
25. Sugimoto, S., Shiio, I. *J. Biochem.* 1980, *87*, 881.
26. Hagino, H., Nakayama, K. *Agric. Biol. Chem.* 1975, *39*, 351.
27. Shiio, I., Sugimoto, S. *J. Biochem.* 1979, *86*, 17.
28. Sugimoto, S., Nakagawa, M., Tsuchida, T., Shiio, I. *Agric. Biol. Chem.* 1973, *37*, 2327.
29. Davies, W. D., Davidson, B. E. *Nucleic Acids Res.* 1982, *10*, 4045.
30. Ray, J. M., Yanofsky, C., Bauerle, R. *J. Bacteriol.* 1988, *170*, 5500.
31. Edwards, R. M., Taylor, P. P., Hunter, M. G., Fotheringham, I. G. *Eur. Pat.*, 1986, 229161.
32. Förberg, C., Eliaeson, T., Häggström, L. *J. Biotechnol.* 1988, *7*, 319.
33. Sugimoto, S., Yabuta, M., Kato, N., Seki, T., Yoshida, T., Taguchi, H. *J. Biotechnol.* 1987, *5*, 237.
34. Powell, J. T., Morrison, J. F. *Eur. J. Biochem.* 1978, *87*, 391.
35. Fotheringham, I. G., Dacey, S. A., Taylor, P. P., Smith, T. J., Hunter, M. G., Finlay, M. E., Primrose, S. B., Parker, D. M., Edwards, R. M. *Biochem. J.* 1986, *234*, 593.
36. Gething, M.-J., Davidson, B. E. *Eur. J. Biochem.* 1978, *86*, 165.
37. Dopheide, T. A. A., Crewther, P., Davidson, B. E. *J. Biol. Chem.* 1972, *247*, 4447.
38. Shiio, I., Sugimoto, S. *J. Biochem.* 1976, *79*, 173.
39. Sugimoto, S., Shiio, I. *J. Biochem.* 1974, *76*, 1103.
40. Sugimoto, S., Shiio, I. *Agric. Biol. Chem.* 1985, *49*, 39.
41. Shiio, I., Mori, M., Ozaki, H. *Agric. Biol. Chem.* 1982, *46*, 2967.
42. de Boer, L., Dijkhuizen, L. In *Advances in Biochemical Engineering/Biotechnology*; Fiechter, A., Ed., 1990; vol. 41; p. 1.
43. Ikeda, M., Katsumata, R. *Appl. Environ. Microbiol.* 1992, *58*, 781.
44. Backman, K. C., Ramaswamy, B. *US Patent*, 1988, *4*, 753,883.
45. Berry, A. *TIBTECH* 1996, *14*, 250.
46. Frost, J. W., Draths, K. M. *Ann. Rev. Microbiol.* 1995, *49*, 557.
47. Patnaik, R., Liao, J. C. *Appl. Environ. Microbiol.* 1994, *60*, 3903.
48. Patnaik, R., Spitzer, R. G., Liao, J. C. *Biotech. Bioeng.* 1995, *46*, 361.
49. Miller, J. E., Backman, K. C., O'Connor, M. J., Hatch, R. T. *J. Ind. Microbiol.* 1987, *2*, 143.
50. Postma, P. W. In *Escherichia coli and Salmonella typhimurium. Cellular and Mo-*

lecular Biology; Neidhardt, F. C., Ed.; American Society for Microbiology: Washington, D. C., 1987.

51. Flores, N., Xiao, J., Berry, A., Bolivar, F., Valle, F. *Nature Biotechnol.* 1996, *14*, 620.
52. Bledig, S. *PhD Thesis*, Warwick, 1994.
53. Draths, K. M., Pompliano, D. L., Conley, D. L., Frost, J. W., Berry, A., Disbrow, G. L., Staversky, R. J., Lievense, J. C. *J. Am. Chem. Soc.* 1992, *114*, 3956.
54. Whipp, M. J., Pittard, A. J. *J. Bacteriol.* 1977, *132*, 453.
55. Pi, J., Wookey, P. J., Pittard, A. J. *J. Bacteriol.* 1991, *173*, 3622.
56. Mulcahy, M. *PhD Thesis*, Dundee 1988.
57. Fotheringham, I. G., Ton, J., Higgins, C. *US Patent*, 1994, 5,354,672.
58. Orndorff, S. A., Constantino, N., Stewart, D., Durham, D. R. *Appl. Environ. Microbiol.* 1988, *54*, 996.
59. Vollmer, P. J., Schruben, J. J., Montgomery, J. P., Huei-Hsuing, Y. *US Patent*, 1986, *4*, 584,269.
60. Yamada, S., Nabe, K., Izuo, N., Nakamichi, K., Chibata, I. *Appl. Environ. Microbiol.* 1981, *42*, 773.
61. Hamilton, B. K., Hsiao, H. Y., Swann, W. E., Anderson, D. M., Delente, J. J. *Trends Biotechnol.* 1985, *3*, 64.
62. Gilbert, H. J., Tully, M. *J. Bacteriol.* 1982, *150*, 498.
63. Onishi, N., Yokozeki, K., Hirose, Y., Kubot, K. *Agric. Biol. Chem.* 1987, *51*, 291.
64. Evans, C. T., Hanna, K., Conrad, D., Peterson, W., Misawa, M. *Appl. Microbiol. Biotechnol.* 1987, *25*, 406.
65. Evans, C. T., Conrad, D., Hanna, K., Peterson, W., Choma, C., Misawa, M. *Appl. Microbiol. Biotechnol.* 1987, *25*, 399.
66. Kirsch, J. F., Eichele, G., Ford, G. C., Vincent, M. G., Jansonius, J. N. *J. Mol. Biol.* 1983, *174*, 497.
67. Jensen, R. A., Calhoun, D. H. *CRC Crit. Rev. Microbiol.* 1981, *8*, 229.
68. Tachiki, T., Tochikura, T. *Agric. Biol. Chem.* 1973, *37*, 1439.
69. Tachiki, T., Moriguchi, M., Tochikura, T. *Agric. Biol. Chem.* 1973, *39*, 43.
70. Christen, P., Metzler, D. E., Eds. *Transaminases*; Wiley: New York, 1984.
71. Kuramitsu, S., Ogawa, T., Ogawa, H., Kagamiyama, H. *J. Biochemistry* 1985, *97*, 993.
72. Sung, M.-H., Tanizawa, K., Tanaka, H., Kuramitsu, S., Kagamiyama, H., Hirotsu, K., Okamoto, A., Higuchi, T., Soda, K. *J. Biol. Chem.* 1991, *266*, 2567.
73. Kohler, E., Seville, M., Jager, J., Fotheringham, I., Hunter, M., Edwards, M., Jansonius, J. N., Kirschner, K. *Biochemistry* 1994, *33*, 90.
74. Bulot, E., Cooney, C. L. *Biotechnol. Lett.* 1985, *7*, 93.
75. Evans, C. T., Peterson, W., Choma, C., Misawa, M. *Appl. Microbiol. Biotechnol.* 1987, *26*, 305.
76. Nakamichi, K., Nabe, K., Nishida, Y., Tosa, T. *Appl. Microbiol. Biotechnol.* 1989, *30*, 243.
77. Then, J., Doherty, A., Neatherway, H., Marquardt, R., Deger, H. M., Voelskow, H., Wöhner, G., Präve, P. *Biotechnol. Lett.* 1987, *9*, 680.
78. Robinson, M., Marquardt, R., Then, J., Neatherway, H., Wöhner, G., Deger, H. M., Präve, P. *Biotechnol. Lett.* 1987, *9*, 673.

79. Calton, G. J., Wood, L. L., Updike, M. H., Lantz, L., Hamman, J. P. *Biotechnology* 1986, *4*, 317.
80. Lewis, D., Farrand, S. *Eur. Pat.*, 1991, 0152275.
81. Nakamichi, K., Nishida, Y., Nabe, K., Tosa, T. *Appl. Biochem. Biotechnol.* 1985, *11*, 367.
82. Ziehr, H., Kula, M. R., Schmidt, E., Wandrey, C., Klein, J. *Biotechnol. Bioeng.* 1987, *9*, 482.
83. Schutten, I., Harder, W., Dijkhuizen, L. *Appl. Microbiol. Biotechnol.* 1987, *27*, 292.
84. Hummel, W., Schmidt, E., Wandrey, C., Kula, M. R. *Appl. Microbiol. Biotechnol.* 1986, *25*, 175.
85. Hummel, W., Schutte, H., Schmidt, E., Wandrey, C., Kula, M. R. *Appl. Microbiol. Biotechnol.* 1987, *26*, 409.
86. Wandrey, C. In *Proceedings of the 4th European Congress on Biotechnology*; Neijssel, O. M., van der Meer, R. R., Luyben, K. C. A. M., Eds.; Elsevier: Amsterdam, 1987; vol. 4; p. 171.
87. Kragl, U., Vasic-Racki, D., Wandrey, C. *Chem. Ing. Tech.* 1992, *64*, 499.
88. Kamphuis, J., Boesten, W. H. J., Broxterman, Q. B., Hermes, H. F. M.; van Balken, J. A. M., Meijer, E. M., Schoemaker, H. E. *Adv. Biochem. Eng. Biotechnol.* 1990, *42*, 134.
89. Meijer, E. M., Boesten, W. H. J., Schoemaker, H. E., van Balken, J. A. M. In *Biocatalysis in Organic Synthesis*; Tramper, J., van der Plas, H. C., Linko, P., Eds.; Elsevier: Amsterdam, 1985; p. 135.
90. Boesten, W. H. J. *US Patent*, 1979, 4, 172, 846.
91. Boesten, W. H. J. *Eur. Pat. Appl.*, 1991, EP 442585.
92. Meijer, E. M., Kamphuis, J., Van Balken, J. A. M., Hermes, H. F. M., van den Tweel, W. J. J., Kloosterman, M., Boesten, W. H. J., Schoemaker, H. E. In *Trends in Drug Research*; Claassen, V., Ed.; Elsevier: Amsterdam, 1990; p. 363.
93. Clausen, K., Godtfredsen, E. S., Hermes, F. M., Ingvorsen, K., Meijer, M. E., Van Balkan, A. M. J. *Eur. Pat.*, 1989, 307023.
94. Tosa, T., Mori, T., Fuse, N., Chibata, I. *Enzymologia* 1966, *31*, 214.
95. Tosa, T., Mori, T., Fuse, N., Chibata, I. *Agric. Biol. Chem.* 1969, *33*, 1047.
96. Takahashi, T., Hatano, K. *Eur. Pat.*, 1989, 304,021.
97. Wandrey, C., Wichmann, R., Leuchtenberger, W., Kula, M.-R., Bückmann, A. *US Patent*, 1981, 4, 304, 858.
98. Ware, E. *Chem. Rev.* 1950, *46*, 403.
99. Euguchi, A., Hamaoka, T., Kawasaki, T., Kono, T. *Japan Patent*, 1979, 80 136279.
100. Kleeman, A., Samson, M. *Ger. Offen.*, 1980, DE 3043259 A1.
101. Drauz, K., Kleeman, A., Samson, M. *Chem. Ztg.* 1984, *12*, 391.
102. Bovarnick, M., Clark, H. T. *J. Am. Chem. Soc.* 1938, *60*, 2426.
103. Gross, C., Syldatk, C., Kackowiak, V., Wagner, F. *J. Biotechnol.* 1990, *14*, 363.
104. Olivieri, R., Fascetti, E., Angelini, L., Degen, L. *Biotechnol. Bioeng.* 1981, *23*, 2173.
105. Yokozeke, K., Nakamori, S., Euguchi, C., Yamada, K., Mitsugi, K. *Agric. Biol. Chem.* 1987, *51*, 355.
106. Möller, A., Syldatk, C., Schultze, M., Wagner, F. *Enzyme Microb. Technol.* 1988, *10*, 618.

107. Yokozeke, K., Hirose, Y., Kubota, K. *Agric. Biol. Chem.* 1987, *51*, 737.
108. Yamashiro, A., Kubota, K., Yokozeki, K. *Agric. Biol. Chem.* 1988, *52*, 2857.
109. Yamada, H., Takahashi, S., Kii, Y., Kumagai, H. *J. Ferment. Technol.* 1978, *56*, 484.
110. Yokozeke, K., Kubota, K. *Agric. Biol. Chem.* 1987, *51*, 721.
111. Mukohara, Y., Ishikawa, T., Watabe, K., Nakamura, H. *Biosci. Biotechnol. Biochem.* 1994, *58*, 1621.
112. Olivieri, R., Fascetti, E., Angelini, L., Degen, L. *Enzyme Microbiol. Technol.* 1979, *1*, 201.
113. Syldatk, C., Wagner, F. *Food Biotechnol.* 1990, *4*, 87.
114. Yamada, H., Shimizu, S. In *Biocatalysis in Organic Synthesis*; Tramper, J., van der Plas, H. C., Linko, P., Eds.; Elsevier: Amsterdam, 1985; p. 19.
115. Katayama, Y., Shimada, Y., Takahashi, S., Ueda, Y., Watanabe, K., Yamada, Y. *Eur. Pat.*, 0175312.
116. Yokozeke, K., Nakamori, S., Yamanaka, S., Eguchi, C., Mitsugi, K., Yoshinaga, F. *Agric. Biol. Chem.* 1987, *51*, 715.
117. Drauz, K., Makryleas, K. *Ger. Offen.*, 1990, DE 3918057.
118. Ishikawa, T., Kimura, H. *Jap. Pat. Appl.*, 1986, JP 86-200434.
119. Fotheringham, I., Bledig, S., Senkpeil, R., Taylor, P. Abstracts of the 1995 SIM Annual Meeting, 1995, San Jose, CA.
120. Dailey, J., Grinter, N., Nelson, R., Pantaleone, D., Senkpeil, R., Taylor, P., Ton, J., Yoshida, R., Fotheringham, I. Abstracts of the IBC Enzyme Technology Symposium, 1996, Lake Buena Vista, FL.

5
Carbohydrates in Synthesis

DAVID J. AGER
NSC Technologies, Mount Prospect, Illinois

5.1. INTRODUCTION

Carbohydrates are extremely important in nature, being used as building blocks for structures such as cell walls, as modifiers of solubility, and as means for storing energy. However, many carbohydrate derivatives are not abundant in the host organism, which despite the diverse array of functionality and stereochemistry only allows for a small number of these compounds to be available at a commercial scale [1]. This has to be counterbalanced against the breadth of sugar derivatives that are found in nature and, although only available at small scale (and often at a very high price), are cited as chirons [2]. In addition to sucrose, the carbohydrates produced at scale are D-glucose, D-sorbitol, D-lactose, D-fructose, D-mannitol, D-maltose, D-isomaltulose, D-gluconic acid, D-xylose, and L-sorbose [3,4]. Most of these carbohydrates are used as food additives rather than chemical raw materials.

The vast array of similar functionality often proves to be a major problem to the synthetic chemist who needs to differentiate between them. This can result in long, tedious, protection–deprotection sequences [1,5,6]. Although enzymes can be helpful, there is still much to be learned. Because carbohydrates are components of many pharmaceutical compounds, various methods have been developed for their synthesis [7,8], including methodology that is amenable to use at scale, e.g., the Sharpless epoxidation [9–13].

Of course, carbohydrates are used and manipulated at commercial scales. The isolation of sucrose and the enzymatic formation of high fructose corn syrup are obvious examples. However, although sucrose is available from a wide variety of sources at very large scale, it has found relatively little application in chemical transformations [14,15].

This chapter covers the uses of carbohydrates as "chiral pool" materials. Because hydroxy acids are closely related, they are also discussed here.

5.2. DISACCHARIDES

5.2.1. Sucrose

The disaccharide **1** is readily available in bulk quantaties from sugar cane or beet. Its major use is in the food industry as a sweetening agent. Despite numerous publications and a significant amount of research dedicated to sucrose, it has not found a place as a chiral chemical raw material. However, sucrose has been derivatized to provide useful food products that have become large-volume products (see following sections).

5.2.1.1. Sucrose Esters

Esters of this disaccharride, derived from fatty acids, have been developed as fat substitutes, of which the Proctor and Gamble product, Olestra, has been commercialized [16]. Although the market is still in its infancy, there is hope that this will be a large-volume commercial product. The esters are prepared by transesterification reactions [14].

5.2.1.2. Sucralose

Sucralose (TGS) (**2**), 4,1',6'-trichloro-4,1',6'-trideoxy-*galacto*-sucrose, is a trichloro disaccharide nonnutritive sweetener [17]. This compound was discovered through a systematic study in which sucrose derivatives were prepared. It was found that substitution of certain hydroxy groups by a halogen increased the sweetness potency dramatically [18,19]. Sucralose was chosen as the development candidate by Tate and Lyle [20–23].

5.2.1.2.1. CHEMICAL APPROACHES.

The synthesis of TGS (**4**) involves a series of selective protection and deprotection steps so that the 4-hydroxy group can be converted to chloro with inversion of configuration [21,24]. Differentiation between the primary hydroxy groups of the two sugar moieties is also required (Scheme 1) [25,26].

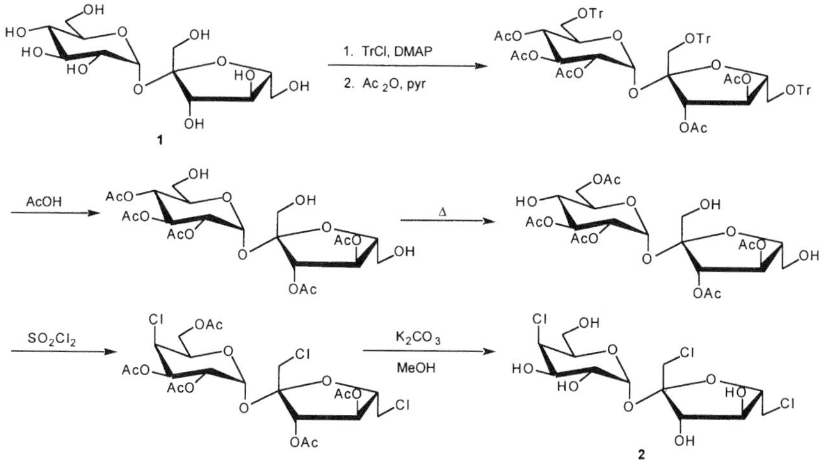

Scheme 1.

5.2.1.2.2. ENZYMATIC APPROACHES.

Another route to **2** utilizes both chemical and biocatalytic transformations and starts from glucose (**3**) (Scheme 2) [27,28].

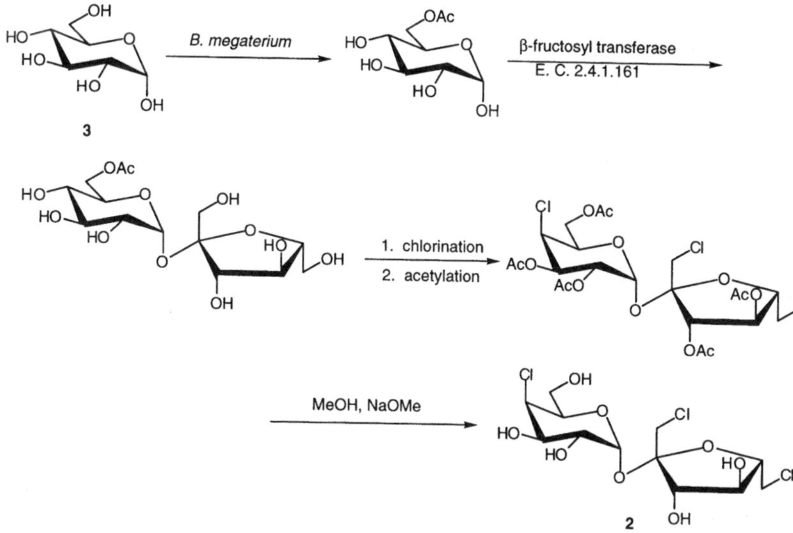

Scheme 2.

An enzymatic preparation to the useful intermediate **4** has been described from sucrose (Scheme 3) [29,30]. Chlorination of this intermediate **4** yields TGS (**2**).

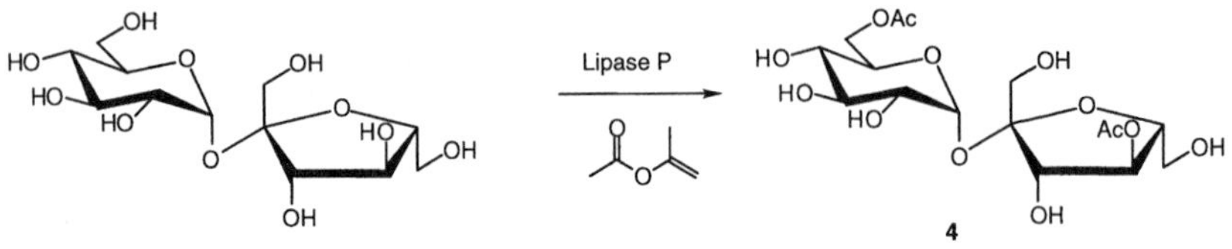

Scheme 3.

5.2.2.   Isomalt

Isomalt (**5**), also called Palatinit, is used as a low-calorie sweetener in some countries. It is a mixture of two compounds obtained by the hydrogenation of isomaltulose (**6**) (Scheme 4) [14].

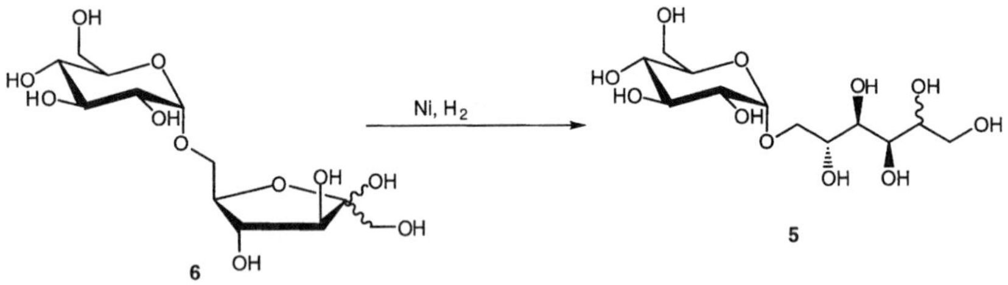

Scheme 4.

The disaccharide **6** is derived from sucrose by an enzymatic transformation catalyzed by *Protaminobacter rubrum* as well as other organisms [14,31].

5.3.   MONOSACCHARIDES AND RELATED COMPOUNDS

5.3.1.   D-Gluconic Acid

D-Gluconic acid (**7**) is produced at scale by the fermentative oxidation of D-glucose (**3**) with *Acetobacter, Pseudomonas,* or *Penicillium* sp. (Scheme 5) [32]. It is used as a processing aid.

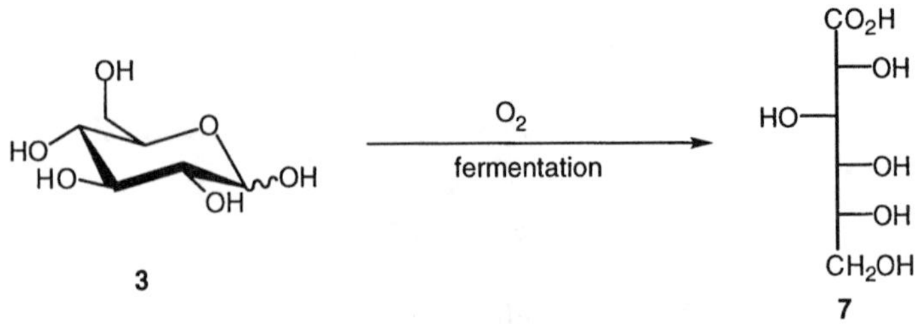

Scheme 5.

The same transformation can also be achieved by air oxidation in the presence of a number of metal catalysts as well as enzymes [4,32].

5.3.2. D-Sorbitol

Sorbitol (8) is used at scale in the manufacture of resins and surfactants [33]. It is produced by the reduction of D-glucose (3) (Scheme 6).

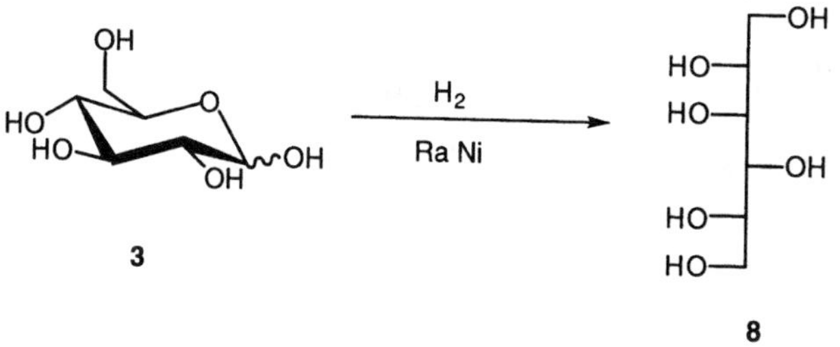

Scheme 6.

5.3.3. L-Ascorbic Acid

L-Ascorbic acid (9), or vitamin C, is produced at large scale by the Reichstein-Grussner process, which dates back to the 1920s (Scheme 7) [33]. It involves the enzymatic oxidation of D-sorbitol (8) that is, in turn, obtained from D-glucose (see 5.3.2).

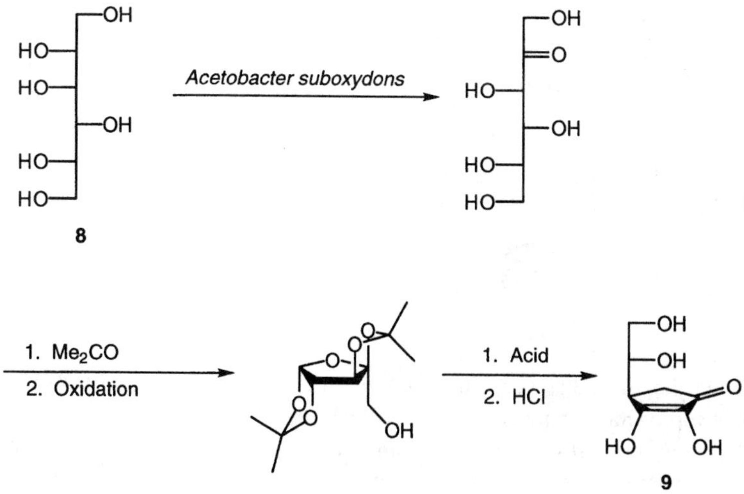

Scheme 7.

5.3.4. D-Erythrose

This valuable intermediate **10** is readily available by the oxidative cleavage of D-fructose (**11**) with hypochlorite (Scheme 8) [4].

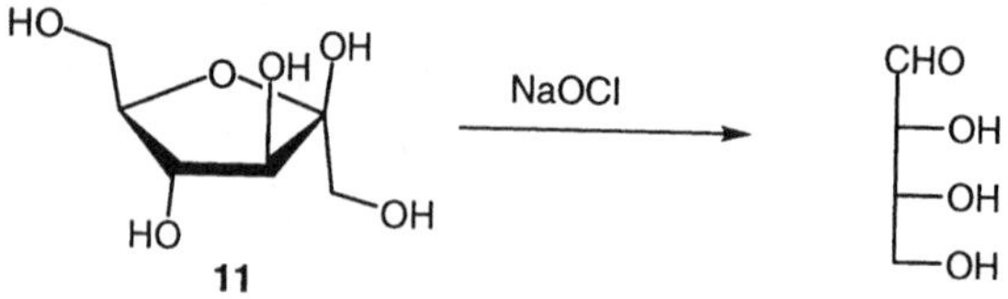

Scheme 8.

5.3.5. Nucleosides

These compounds are the building blocks of DNA and RNA and contain a carbohydrate unit together with a base. A number of derivatives have been prepared as pharmaceutical agents, perhaps the best known being azidothymidine (AZT) (**12**). Some of these compounds may be accessed from ribose [34]. In many cases, a nucleoside is used as the starting material for modification; a number of routes to **12** use this approach [35–37].

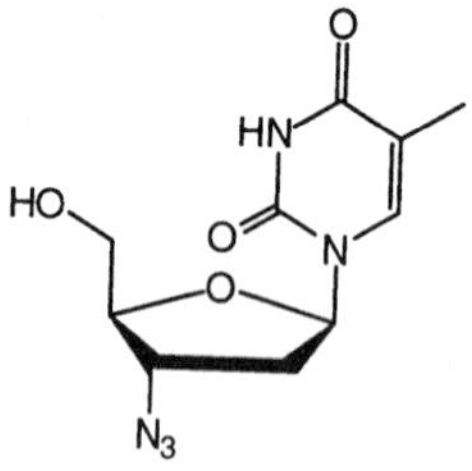

12

When the sugar has been modified, as in lamivudine (**13**) and GR95168 (**14**), most approaches do not use a carbohydrate precursor as the starting material, but instead rely on a resolution or asymmetric synthesis (see Chapter 6) [38]. Again, this illustrates the problems associated with the multitude of similar functional groups [39–41].

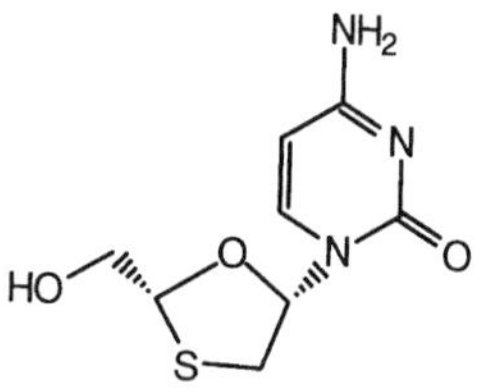

13

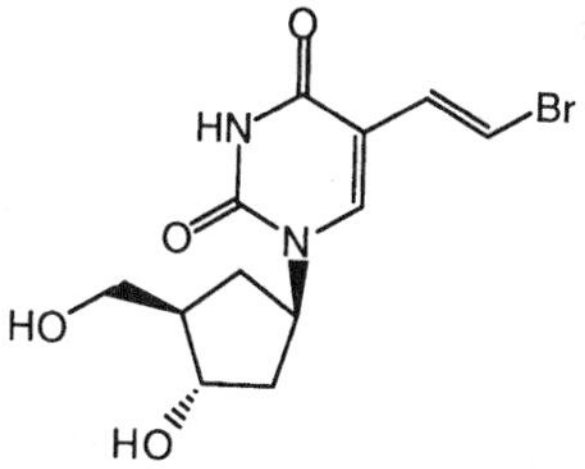

14

5.4. GLYCERALDEHYDE DERIVATIVES

5.4.1. S-Solketal

S-Solketal (**15**) can be obtained by the oxidative cleavage of D-mannitol (**16**) (Scheme 9). Of course, the use of heavy metals in the cleavage reaction does bring along environmental problems. The aldehyde **17**, although used for a wide variety of transformations, does suffer from epimerization problems; however, additions to the carbonyl group can be manipulated through the use of chelation control [1].

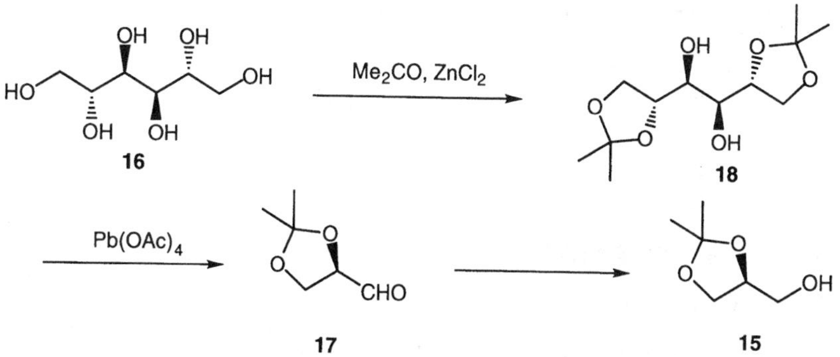

Scheme 9.

The alcohol **15** can also be prepared from protected glyceric acid [42]. This, in turn, can be obtained by cleavage of the protected diol **18** with bleach in the presence of a ruthenium catalyst (Scheme 10) [4,42].

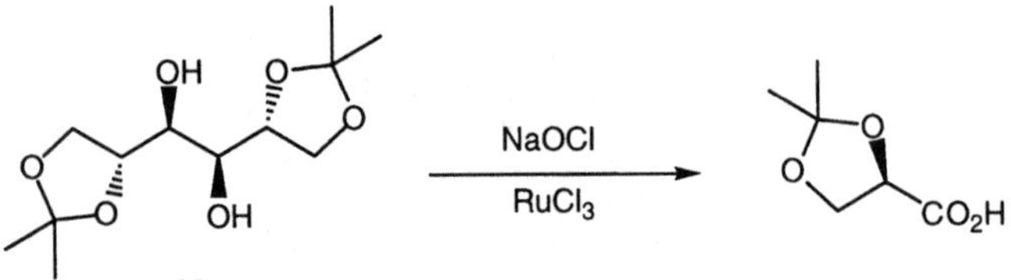

Scheme 10.

5.4.2. R-Solketal

This isomer of the aldehyde, **19**, can be accessed from L-ascorbic acid (**9**)(Scheme 11) [43].

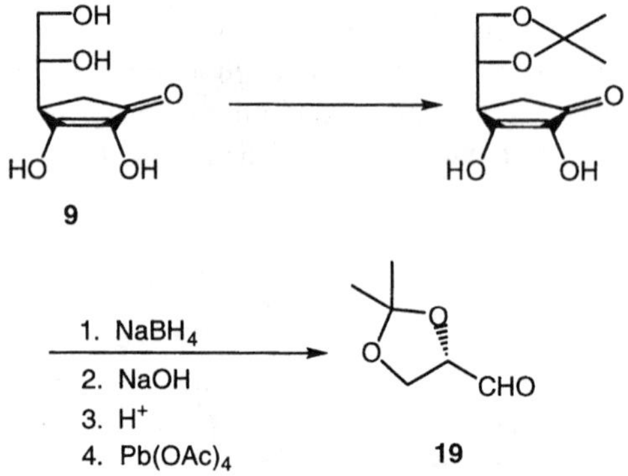

Scheme 11.

5.5. HYDROXY ACIDS

5.5.1. Tartaric Acid

Both isomers of this compound are available at scale, yet it has not found large-scale application as a chiral building block. Tartaric acid can be used as a resolution agent (see Chapter 8) and as a ligand for asymmetric catalysis. The R,R-isomer is available as a waste product of wine fermentation [4]. The S,S-isomer is a natural product but is produced by resolution and is much more expensive than its antipode [4].

5.5.2. Malic Acid

Again, this hydroxy acid is available by a wide variety of methods, including fermentation [33], but has not found a large-scale application as a building block.

5.5.3. Lactic Acid

Both isomers of this hydroxy acid are available from fermentation [33]. Although *R*-lactic acid can be used for the synthesis of herbicides, the development of alternative methods to *S*-2-chloropropinoic acid alleviate this need (see Chapter 3).

5.5.4. 3-Hydroxybutyric Acid

This hydroxy acid is available as the *R*-isomer from the biodegradable polymer, Biopol [33]. A number of derivatives are accessible from the polymer itself (Scheme 12) [44,45].

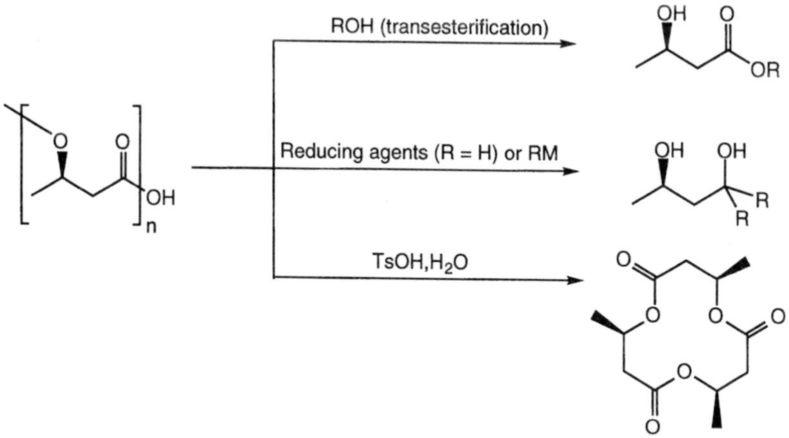

Scheme 12.

The *S*-isomer is available by yeast reduction of acetoacetate [1,46,47]. Both isomers are available by asymmetric hydrogenation of the β-keto ester (see Chapter 9) [48].

The monomeric unit, as illustrated by 3*R*-hydroxybutyric acid, is a precursor to a wide range of β-hydroxy acid derivatives (Scheme 13) [49].

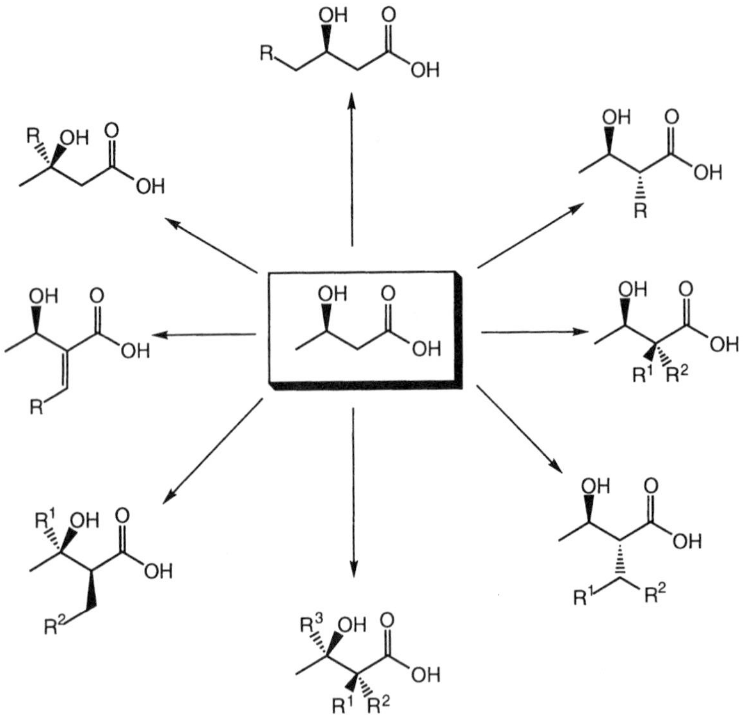

Scheme 13.

Although a wide range of β-hydroxy acids are available by reduction of the corresponding β-keto ester or acid (*vide supra*), and a number of schemes have been proposed to commercially important compounds such as captopril, this methodology and class of compounds must wait for a new opportunity to show its versatility [47].

5.5.5. Macrolide Antiobiotics

In some instances, macrolide antiobiotics may be considered as hydroxy acid derivatives. In addition, many of them have carbohydrates attached, often with unique structural features, as illustrated by erythromycin (**20**) [50]. Although some of these antibiotics are semisynthetic, all are derived by a fermentation process in which the antibiotic is formed as a secondary metabolite. This approach, of course, alleviates the need to perform complex carbohydrate chemistry.

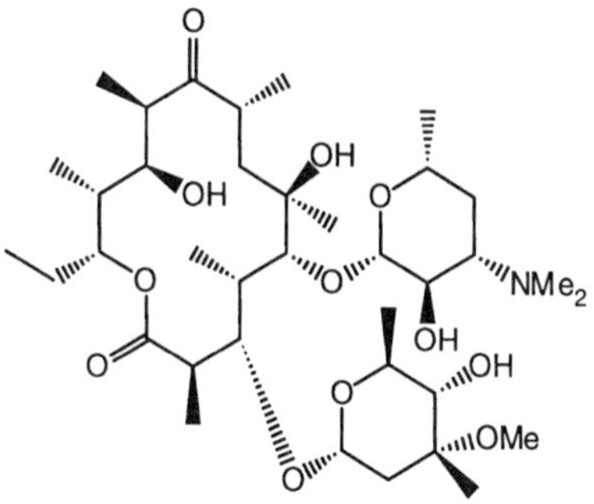

20

5.6. SUMMARY

Despite their widespread occurrence in nature and the fact that some members of the class are available at scale, carbohydrates and hydroxy acids have not found widespread application as synthons in large-scale fine chemical synthesis, with the exception of food ingredients. This is presumably due to the problems associated with the differentiation of very similar functional groups, although enzymatic methods are known, and the low atom efficiency when incorporated into a synthesis.

REFERENCES

1. Ager, D. J., East, M. B. *Asymmetric Synthetic Methodology;* CRC Press: Boca Raton, 1995.
2. Scott, J. S. in *Asymmetric Synthesis;* Morrison, J. D., Ed.; Academic Press: Orlando, 1984; *Vol. 4;* p. 1.
3. Lichtenthaler, F. W., Cuny, E., Martin, D., Rönninger, S., Weber, T. in *Carbohydrates as Organic Raw Materials;* Lichtenthaler, F. W., Ed.; VCH: Weinheim, 1991; p. 207.
4. Sheldon, R. A. *Chirotechnology: Industrial Synthesis of Optically Active Compounds;* Marcel Dekker: New York, 1993.
5. Hanessian, S., Banoub, J. *Synthetic Methods for Carbohydrates;* American Chemical Society: Washington, D. C., 1976.
6. Hanessian, S. *The Total Synthesis of Natural Products: The Chiron Approach;* Pergamon: Oxford, 1983.
7. Ager, D. J., East, M. B. *Tetrahedron* 1992, *48,* 2803.
8. Ager, D. J., East, M. B. *Tetrahedron* 1993, *49,* 5683.

9. Katsuki, T., Lee, A. W. M., Ma, P., Martin, V. S., Masamune, S., Sharpless, K. B., Tuddenham, D., Walker, F. J. *J. Org. Chem.* 1982, *7*, 1373.
10. Lee, A. W. M., Martin, V. S., Masamune, S., Sharpless, K. B., Walker, F. J. *J. Am. Chem. Soc.* 1982, *104*, 3515.
11. Ma, P., Martin, V. S., Masamune, S., Sharpless, K. B., Viti, S. M. *J. Org. Chem.* 1982, *47*, 1378.
12. Ko, S. Y., Lee, A. W. M., Masamune, S., Reed, L. A., Sharpless, K. B., Walker, F. J. *Science* 1983, *220*, 949.
13. Ko, S. Y., Lee, A. W. M., Masamune, S., Reed, L. A., Sharpless, K. B., Walker, F. J. *Tetrahedron* 1990, *46*, 245.
14. Schiweck, H., Munir, M., Rapp, K. M., Schneider, B., Vogel, M. in *Carbohydrates as Organic Raw Materials;* Lichtenthaler, F. W., Ed.; VCH: Weinheim, 1991; p. 57.
15. BeMiller, J. N. in *Carbohydrates as Organic Raw Materials;* Lichtenthaler, F. W., Ed.; VCH: Weinheim, 1991; p. 263.
16. van der Plank, P., Rozendaal, A. *Eur. Patent* 1988, EP 256, 585.
17. Ager, D. J., Pantaleone, D. P., Katritzky, A. R., Henderson, S. A., Prakash, I., Walters, D. E. *Angew. Chem. Int. Ed. Engl.* 1998, in press.
18. Hough, L., Phadnis, S. P. *Nature* 1976, *263*, 500.
19. Hough, L. *Chem. Soc. Rev.* 1985, *14*, 357.
20. Jenner, M. R., Waite, D. *US Patent* 1982, 4,343,934.
21. Jenner, M. R., Waite, D., Jackson, G., Williams, J. C. *US Patent* 1982, 4,362,869.
22. Hough, L., Phandis, S. P., Khan, R. A. *US Patent* 1984, 4,435,440.
23. Jenner, M. R. in *Sweeteners: Discovery, Molecular Design, and Chemoreception;* Walters, D. E., Orthoefer, F. T., DuBois, G. E., Eds.; American Chemical Society: Washington, D. C., 1991; p. 68.
24. Fairclough, P. H., Hough, L., Richardson, A. R. *Carbohydr. Res.* 1975, *40*, 285.
25. Jackson, G., Jenner, M. R., Waite, D., Williams, J. C. *UK Pat. Appl.* 1980, 2065648A.
26. Tully, W., Vernon, N. M., Walsh, P. A. *Eur. Pat. Appl.* 1987, EP 220907 A2.
27. Jones, J. D., Hacking, A. J., Cheetham, P. S. J. *Biotechnol. Bioeng.* 1992, *39*, 203.
28. Rathbone, E. B., Hacking, A. J., Cheetham, P. S. J. *US Patent* 1986, 4,617,269.
29. Dordick, J. S., Hacking, A. J., Khan, R. A. *US Patent* 1992, 5,128,248.
30. Dordick, J. S., Hacking, A. J., Khan, R. A. *US Patent* 1993, 5,270,460.
31. Crueger, W., Drath, L., Munir, M. *Eur. Patent* 1978, 1,099.
32. Röper, H. in *Carbohydrates as Organic Raw Materials;* Lichtenthaler, F. W., Ed.; VCH: Weinheim, 1991; p. 267.
33. Crosby, J. in *Chirality in Industry: The Commercial Manufacture and Applications of Optically Active Compounds;* Collins, A. N., Sheldrake, G. N., Crosby, J., Eds.; Wiley: Chichester, 1992; p. 1.
34. Dekker, C. A., Goodman, L. in *The Carbohydrates: Chemistry and Biochemistry;* Pigman, W., Horton, D., Eds.; Academic: New York, 1970; *Vol. IIA;* p. 1.
35. Chen, B.-C., Quinlan, S. L., Reid, G. J. *Tetrahedron Lett.* 1995, *36*, 7961.
36. Azhayev, A. V., Korpela, T. *World Patent* 1993, WO 9307162.
37. Czernecki, S., Valery, J. M. *Synthesis* 1991, 239.
38. Bray, B. L., Goodyear, M. D., Partridge, J. J., Tapolczay, D. J. in *Chirality in Industry II: Developments in the Commercial Manufacture and Applications of Optically*

Active Compounds; Collins, A. N., Sheldrake, G. N., Crosby, J., Eds.; Wiley: Chichester, 1997; p. 41.

39. Paulsen, H., Pflughaupt, K.-W. in *The Carbohydrates: Chemistry and Biochemistry;* Pigman, W., Horton, D., Eds.; Academic: New York, 1980; *Vol. IB;* p. 881.
40. Almond, M. R., Wilson, J. D., Rideout, J. L. *US Patent* 1990, 4,916,218.
41. Wilson, J. D. *US Patent* 1990, 4,921,950.
42. Emons, C. H. H., Kuster, B. F. M., Vekemans, J. A. J. M., Sheldon, R. A. *Tetrahedron: Asymmetry* 1991, 2, 359.
43. Jung, M. E., Shaw, T. J. *J. Am. Chem. Soc.* 1980, *102,* 6304.
44. Schnurrenberger, P., Seebach, D. *Helv. Chim. Acta* 1982, *65,* 1197.
45. Akita, S., Einaga, Y., Miyaki, Y., Fujita, H. *Macromolecules* 1976, *9,* 774.
46. Iimori, T., Shibasaki, M. *Tetrahedron Lett.* 1985, *26,* 1523.
47. Ohashi, T., Hasegawa, J. in *Chirality in Industry: The Commercial Manufacture and Applications of Optically Active Compounds;* Collins, A. N., Sheldrake, G. N., Crosby, J., Eds.; Wiley: Chichester, 1992; p. 269.
48. Ager, D. J., Laneman, S. A. *Tetrahedron: Asymmetry* 1997, *8,* 3327.
49. Seebach, D. *Angew. Chem. Int. Ed. Engl.* 1990, *29,* 1320.
50. Hanessian, S., Haskell, T. H. in *The Carbohydrates: Chemistry and Biochemistry;* Pigman, W., Horton, D., Eds.; Academic: New York, 1970; *Vol. IIA;* p. 139.

6
Terpenes: Expansion of the Chiral Pool

WEIGUO LIU
NSC Technologies, Mount Prospect, Illinois

6.1. INTRODUCTION

Nature has been, and continues to be, one of the greatest sources of chiral molecules that range from small amino acids to large proteins and nucleic acids to DNA. Synthetic organic chemists have long benefited from this resource by constantly being able to draw on members of these chiral natural products and converting them into more complex synthetic compounds, such as pharmaceuticals and other biologically active agents. During the past decade, an overwhelming amount of work has been conducted, with great success, on the use of amino acids and carbohydrates as chiral building blocks. In a similar manner, terpenes, especially monoterpenes, are also an important group of abundant natural products that contribute to nature's chiral pool. Many of these monoterpenes are widely distributed in nature and can be isolated from a variety of plants in high yields [1]. They usually possess one or two stereogenic centers within their structures, together with modest functionality that allows convenient structural manipulation and transformations. Many terpenes have constrained asymmetric carbocyclic ring systems that can be incorporated into the frameworks of complex polycyclic target molecules, or converted into chiral ligands for chiral auxiliaries, reagents, and resolving agents. Although amino acids and carbohydrates continue to be the major source of chiral building blocks in organic synthesis, terpenes are being used more and more frequently in asymmetric synthesis. In this chapter, we discuss some of the applications in the use of terpenes, especially monoterpenes, in the synthesis of chiral organic molecules. The discussion emphasizes the use of natural or unnatural monoterpenes as building blocks in the synthesis

83

of biologically active compounds and the stereoselective transformations and manipulations of these terpene molecules for asymmetric synthetic purposes.

6.2. ISOLATION

Some of the most abundant and readily available monoterpenes in nature's chiral pool include pinenes, menthones, menthols, carvones, camphor, and limonenes, which can be isolated in large quantities from various species and parts of plants found in different regions of the world [2]. These terpenes can also be interconverted through chemical and biological transformations, and can be derivatized into a variety of oxygenated forms by simple chemical manipulations [3]. Current industrial production of many optically active monoterpenes is still, performed through the extraction of plants, usually by steam distillation. Essential oils extracted from a variety of plants contain a mixture of a number of different monoterpenes that are separated and purified by repeated distillation [4]. Because the essential oil industry is highly localized by region and characterized by its diversity, the chemical and optical purities of commercial monoterpenes vary widely depending on region, the plant source, and production methods [5]. Unlike natural amino acids, which almost exclusively have the (S)-configuration, many monoterpenes exist in nature in the form of both enantiomers, which are often distributed in different plant species and in different quantities. Therefore, the enantiomeric excess (ee%) of a particular compound can vary according to the region of origin. Many terpenes are "scalemic," which means that both antipodes are made by a single source, although one usually predominates [6]. Chemists who use these terpenes for chiral synthetic purposes should always be cognizant of the quality and producer of the product, and, if possible, stay with just one for consistent, reliable product quality.

6.3. MONOTERPENES

6.3.1. α-Pinene

α-Pinene is perhaps the most abundant and chemically exploited monoterpene. It is isolated principally from a variety of pine trees, and both of its optical enantiomers occur in nature, although (+)-α-pinene (1) is the most abundant [7]. Bulk quantities of both optically active (+)-, and (−)-α-pinenes are available.* α-Pinenes obtained from natural sources have optical purities ranging from 80–90% ee, due to the coexistence of the other enantiomer. The best (+)-α-pinene is approximately 91.3% ee, whereas the (−)-α-pinene (2) is only 81.3%

---

* For example, this is available from Penta Manufacturing Co., Fairfield, NJ, as well as other fine chemical producers and dealers.

[8]. This kind of optical purity is usually not good enough for many chiral synthetic purposes. For the preparation of optically pure chiral borane reagent, diisopinocampheylborane (Ipc_2BH) (3), a simple and practical procedure has been developed to upgrade the optical purity of commercial pinenes to 99% ee [9]. For (+)-α-pinene, the commercial material (91.3% ee) was treated with $BH_3 \cdot THF$ or $BH_3 \cdot SMe_2$ to produce the hydroboration product Ipc_2BH (3) as a crystalline solid. Digestion of the solid suspended in THF with 15% excess (+)-α-pinene caused the (+)-isomer to become incorporated into the crystalline Ipc_2BH, while the (−)-isomer accumulates in solution. Filtration removes the (−)-isomer to give crystalline 3 that contains 99% (+)-α-pinene. Elimination of the alkyl groups from the trialkylboranes by treatment of the Ipc_2BH (3) with benzaldehyde and a catalytic amount of BF_3 at 100°C followed by distillation gives (+)-α-pinene (99.5% ee) (Scheme 1)[10]. The same process can be applied to upgrade (−)-α-pinene.

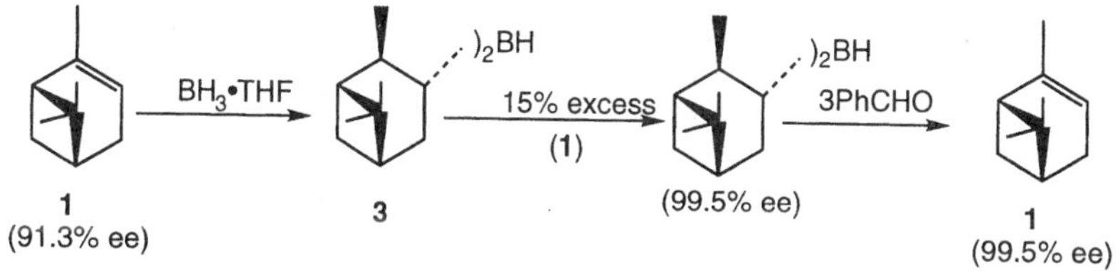

SCHEME 1.

Because of less abundant natural sources and its inferior optical purity, good quality (−)-α-pinenes are usually obtained by chemical isomerization of (−)-β-pinenes. The isomerization is brought about by various catalysts, including acids [11], bases [12], and metals [13]. For example, (−)-α-pinene (2) (92% ee) was obtained from commercial (−)-β-pinene (4) of the same optical purity in high yield through isomerization catalyzed by potassium 3-aminopropylamide (KAPA) (5) (Scheme 2) [14]. The (−)-α-pinene (2) obtained by this method can be further upgraded in optical purity by the previously mentioned hydroboration method.

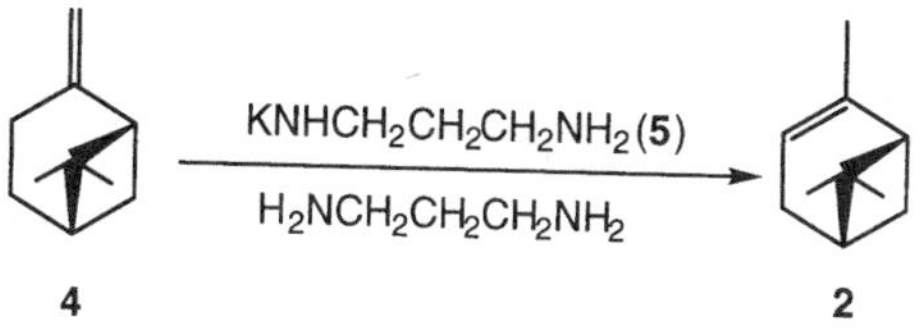

SCHEME 2.

6.3.2. β-Pinene

β-Pinene is an important raw material for various perfumes and polyterpene resins [7]. In contrast to α-pinene, only the (−)-isomer of β-pinene is isolated from nature in large quantities. The (+)-isomer of β-pinene (6) is produced mainly by chemical means from (+)-α-pinene (1) through the use of various isomerization methods [15–17]. One of these methods is to heat Ipc$_2$BH (3) from 1 with high optical purity to approximately 130°C, followed by treatment with 1-hexene and benzaldehyde to release the (+)-β-pinene (6) (> 99.5% ee) (Scheme 3) [10].

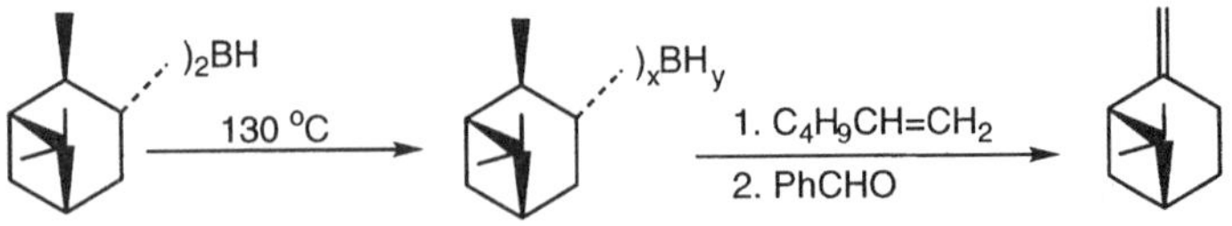

SCHEME 3.

6.3.2.1.   Other Monoterpenes Derived from Pinenes

Both α-, and β-pinenes are popular starting materials for the synthesis of other monoterpene chiral synthons such as carvone, terpineol, and camphor (see sections 6.3.6–6.3.8). Reactions leading to other monoterpenes are briefly summarized in Figure 1. Treatment of α-pinene with lead tetraacetate followed by rearrangement gives *trans*-verbenyl acetate (7), which is hydrolyzed to yield *trans*-verbenol (8) [18]. Subsequent oxidation of 8 gives verbenone (9), which can be reduced to give *cis*-verbenol (10). Verbenone (9) can also be converted into chrysanthenone (11) by ultraviolet irradiation [19]. The naturally occurring pinocamphones (12) and pinocampheols (13) can be prepared from α-pinene by conventional ring opening reactions of α-pinene epoxide (14) and the reduction [20,21]. One of the most effective methods for making *trans*-pinocarveol (15) is by sensitized photooxygenation of α-pinene followed by reduction of the resultant *trans*-pinocarveyl hydroperoxide (16) [22].

6.3.3.   Limonene

Limonene is another inexpensive and abundantly available chiral pool material. Both *d*-, and *l*-limonenes (17 and 18) are commercially available. *d*-Limonene (17) is the principal stereoisomer that can be isolated in large quantities from orange, caraway, dill, grape, and lemon oils, while *dl*-, and *l*-limonenes are obtained from a variety of terpentine oils [23]. Racemic limonenes can be synthesized by the dimerization of isoprene in a Diels-Alder condensation. Additionally, a 5-carbon to 4-carbon coupling under Diels-Alder conditions followed by Wittig addition of the methylene group has been used to produce racemic limonene (Scheme 4) [24].

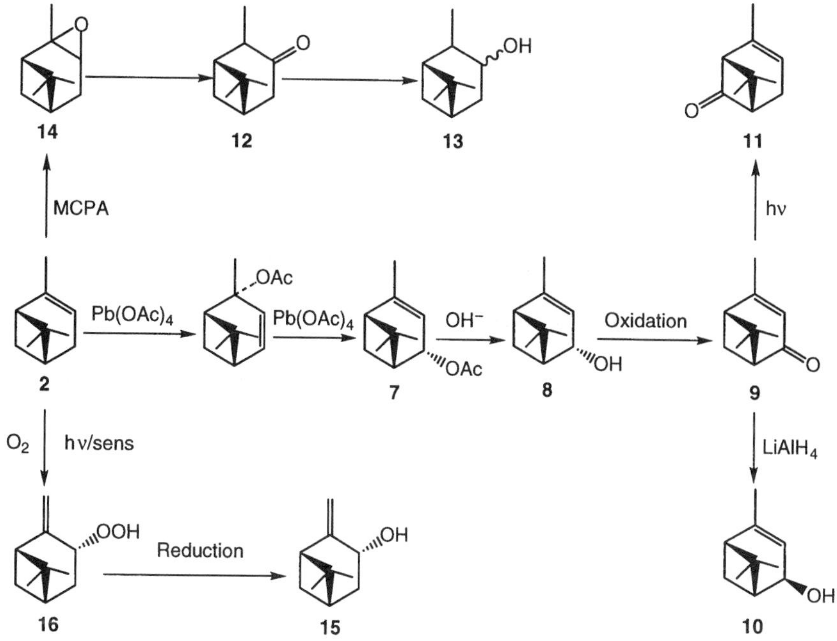

FIGURE I

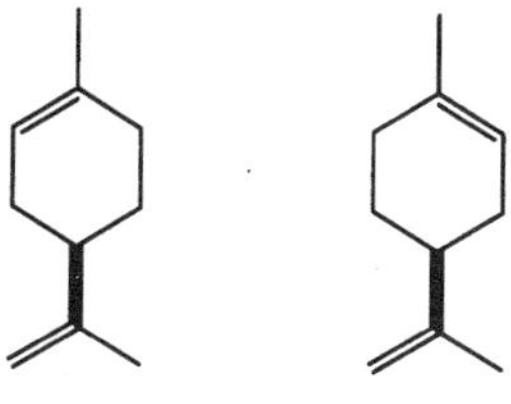

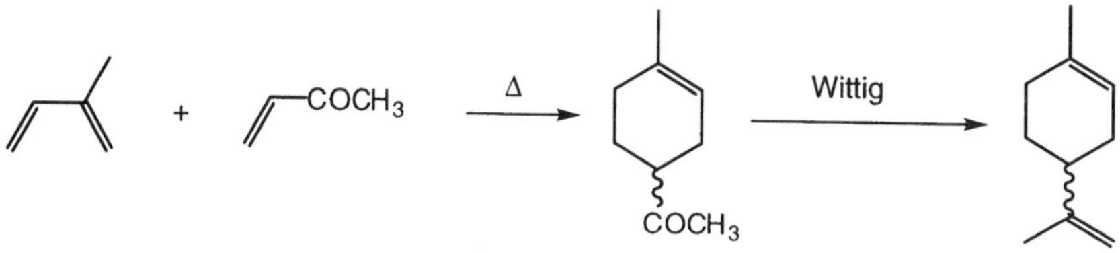

SCHEME 4.

The degree of chemical and optical purities of limonenes, again, depends on the exact source of the oil. Bulk *d*-limonene currently available in the United States has optical purities ranging from 96–98% ee.

6.3.3.1. Monoterpenes Derived from Limonene

Although limonene is used directly in the constitution of natural flavors and perfumes, its principle value lies in its use as a starting material for chemical synthesis and production of a variety of oxygenated *p*-menthane monoterpenes, a group of important organoleptic compounds and chiral building blocks in organic synthesis [1,25]. Many of these compounds are derived from a common intermediate, limonene epoxide, which is obtained through peroxyacid oxidation of limonene [26,27]. For example, acid catalyzed ring opening of (+)-limonene epoxide (**19**) gives a 1,2-diol, which, after acetylation, pyrolysis of the resultant diacetate, and hydrolysis, yields carveol, which in turn can be oxidized to give (−)-carvone (**20**) (Scheme 5) [28]. Other *p*-menthane types of monoterpenes that can be directly derived from limonene include α- and β-terpineol (**21**) and various menthadi-enols and menthen-diols (Figure 2) [29].

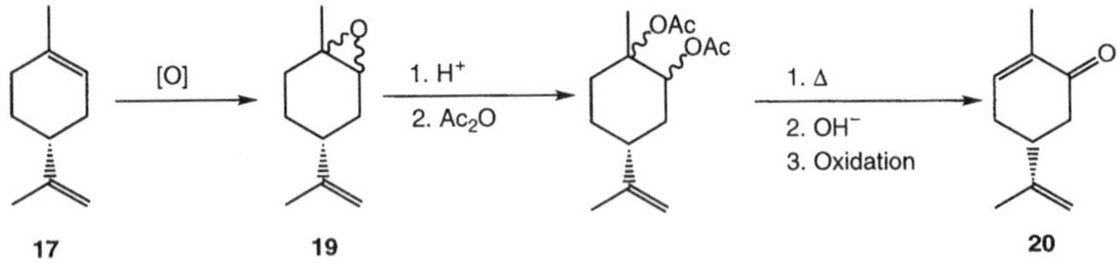

SCHEME 5.

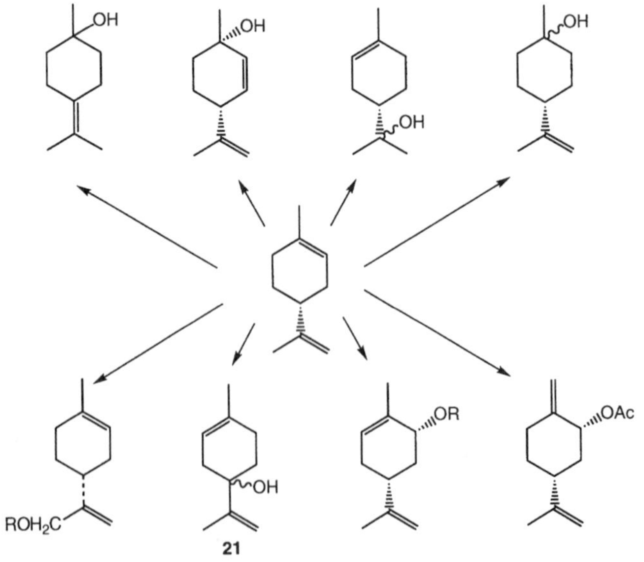

FIGURE 2

6.3.4. Menthones

p-Menthanes, which include menthol, menthone, terpineol, and carvone, are some of the best known monoterpene-based chiral synthons in organic synthesis. They are all relatively inexpensive and are commercially available in bulk quantities. Menthols and their corresponding ketones, menthones, were first isolated from peppermint oils of various species of *Mentha piperita L* [30]. Menthones can exist in two diastereomeric forms, each with two enantiomers. The ones with the methyl and isopropyl groups in a *trans*-orientation are termed menthones, whereas those with the two in a *cis*-orientation are isomenthones. *l*-Menthone (**22**), also denoted as (*1R,4S*)-(−)-menthone, is the most abundant stereoisomer and is obtained by the dry distillation of the wood of *Pinus palustris Mell* [31]. It can also be produced chemically by chromic acid oxidation of *l*-menthol (**23**) [32]. *l*-Menthone is commercially available in bulk quantities with optical purities of 90–98% ee [33].

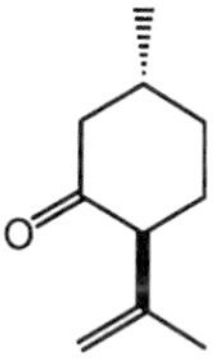

22 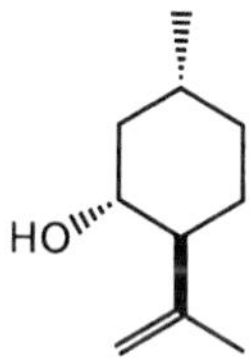   **23**

6.3.5. Menthols

There are four diastereomeric menthols according to the relative orientations of the methyl, isopropyl, and hydroxyl groups. They are known as menthol, isomenthol, neomenthol, and neoisomenthol, each of which exists in two enantiomeric forms. The most abundant and readily available is *l*-menthol, or (*1R,2S,5R*)-(−)-menthol (**23**), whereas other stereoisomers, although available from nature, are more economically produced through chemical derivatizations from menthone, limonene, and α-pinene. Commercial (−)-menthol from natural sources can be as pure as 99% in both chemical and optical purities. Synthetic menthol is currently made by an asymmetric isomerization (see Chapter 9) [34].

6.3.6. Carvones

Both enantiomeric forms of carvone are readily available. *l*-(−)-Carvone (**20**) is the major constituent of spearmint oil from *Mentha spicata*, whereas *d*-(+)-carvone is a major constituent of caraway oil from the fruit of *Carum carui* or dill from the fruit of *Anethum graveolens*. *dl*-Carvone is found in gingergrass oils.

Interestingly, the two enantiomers of carvone have different characteristic odors and flavors, suggesting that chirality plays an important role in organoleptic function [1]. Synthetic *l*-carvone (**20**) is produced on an industrial scale either from *d*-limonene (**17**) (Scheme 6) or from α-pinene. The conversion of **17** into **20** is accomplished in three steps through the intermediates *d*-limonene nitrosochloride (**24**) and *l*-carvoxime (**25**), with greater than 60% overall yield (Scheme 6) [35].

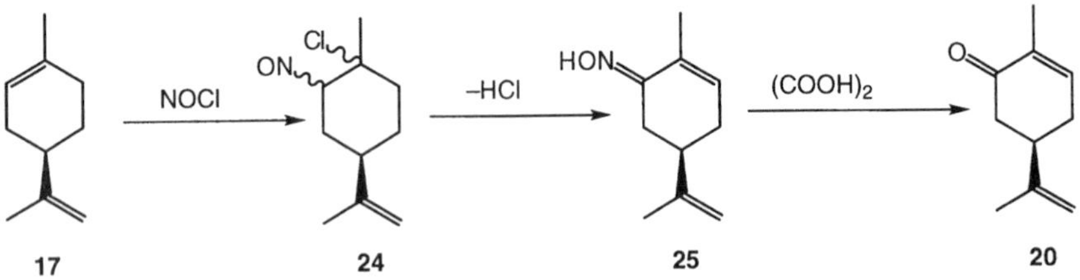

17   **24**   **25**   **20**

SCHEME 6.

An alternative route to *l*-(−)-carvone (**20**) was developed from the more abundant (−)-α-pinene (**2**), albeit with a somewhat lower overall yield (Scheme 7) [36]. The process involves hydroboration of **2** followed by oxidation of the resultant alcohol. Subsequent treatment of the *l*-isopinocamphone (**26**) with isopropenyl acetate and then anodic oxidation gives almost pure *l*-(−)-carvone (34% overall). Both processes are being applied on industrial-scale production, depending on the availability of the starting material. Optical purity of currently available commercial carvones ranges from 96–98% ee.

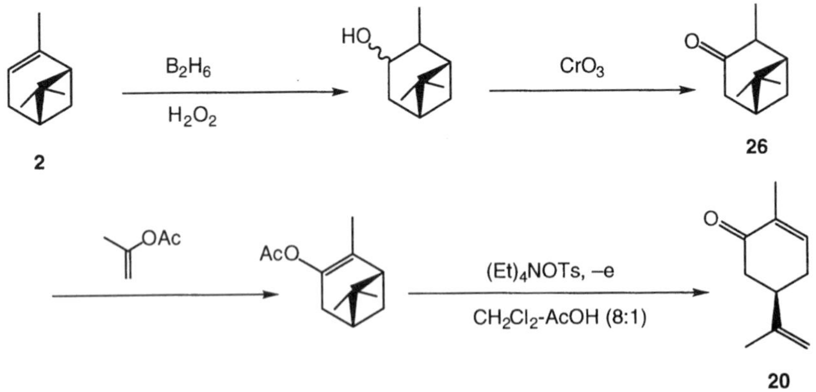

2   **26**   **20**

SCHEME 7.

6.3.7. Terpineol

Terpineol is mainly isolated from plants of the *Eucolaptus* species, and is a mixture of α-terpineol, β-terpineol, 4-terpineol, and 1,4- and 1,8-terpindiols. Both enantiomers of α-terpineol are commercially produced through hydration of α-pinenes [37].

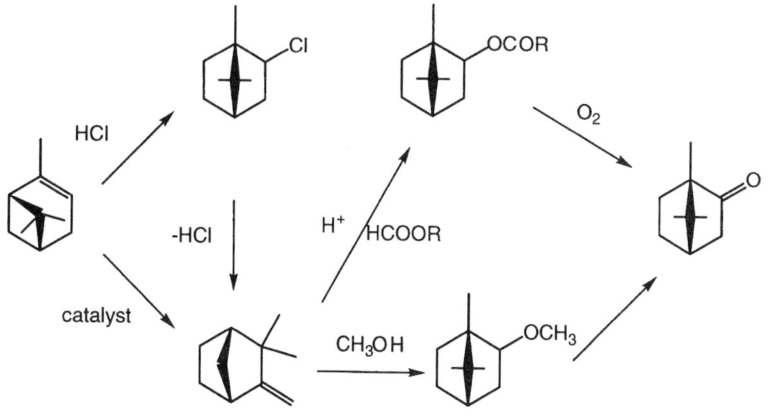

FIGURE 3

6.3.8. Camphor

Finally, camphor is perhaps the most well-known monoterpene. It certainly has seen the most widespread commercial applications than any other single terpene, with a current 50–60 million pounds per year market. The most abundant natural source of camphor is the wood of various species of *Cinnamomum camphora*, a tree common to the Far East [38]. Although both *d*-and *l*-camphor have been isolated from a number of natural sources, most commercial camphor is now obtained through the chemical transformation of pinene (Figure 3) [39]. Wagner-Meerwein rearrangement of α-pinene in HCl leads to bornyl chloride, which loses hydrogen chloride upon heating to give camphene (**27**). The latter is then converted to an isobornyl ester, which is oxidized to yield camphor. There have been many variations of this pinene-to-camphor conversion process, such as direct catalytic conversion of pinene to camphene, and methanolysis of camphene in the presence of strong cation exchange resin.

Racemic camphor can be produced by total synthesis from 1,1,2-trimethyl-cyclopentadiene and vinyl acetate through a Diels-Alder reaction (Scheme 8) [40]. Currently, approximately three quarters of the camphor sold in the United States is produced synthetically, and most is in the racemic form [41].

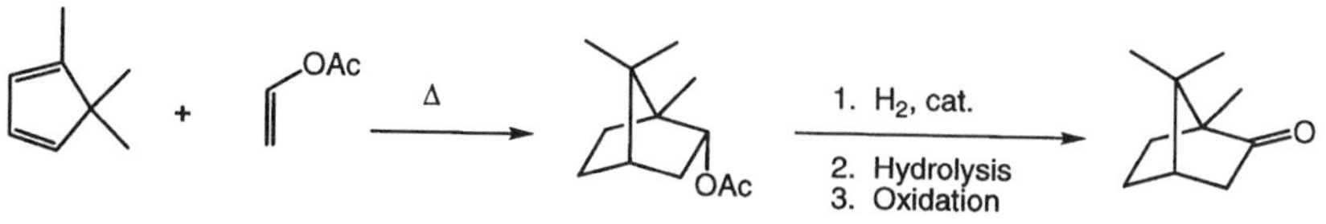

SCHEME 8.

6.4. REACTIONS OF MONOTERPENES

The most fascinating aspect of monoterpene chemistry is perhaps the variety of stereoselective organic reactions that take place around the chiral centers of these molecules. The architecture of many of the natural monoterpenes, especially cyclic and multicyclic monoterpenes, render them unique steric features and imposes strong influences of steric interactions and facial selectivities. These differentiated steric interactions around the molecule allow distinct stereospecific or stereoselective reactions to take place. Taking advantage of these steric features, synthetic organic chemists have designed various reactions to translate the chirality of these monoterpenes into various organic compounds generated in specific synthetic reactions. In these reactions, the chiral terpenes are used as chiral auxiliaries to direct the reaction to occur only in the preferred steric conformation, resulting in stereochemically pure or enriched products. As a result of these stereoselective transformations, a great number of chiral organic compounds or building blocks have been derived directly or indirectly from this class of natural products.

In addition, monoterpenes can provide some useful chiral reagents, such as the pinene-based organoborane reagents for chiral reductions that have been reviewed extensively [42]. The use of camphor-derived organic acids such as camphenesulfonic acid can be used for the resolution of racemic bases and is a common practice in industry (see Chapter 8).

6.4.1. Camphor as a Chiral Auxiliary

6.4.1.1. Alkylations

Camphor and camphor-derived analogs are used frequently as chiral auxiliaries in asymmetric synthesis. There have been numerous reports in the use of camphor imine as templates to direct enantioselective alkylation for the synthesis of α-amino acids, α-amino phosphonic acids, α-substituted benzylamines, and α-amino alcohols (e.g., Scheme 9) [43–47]. Enantiomeric excesses of the products range from poor to excellent, depending on the type of alkyl halides used.

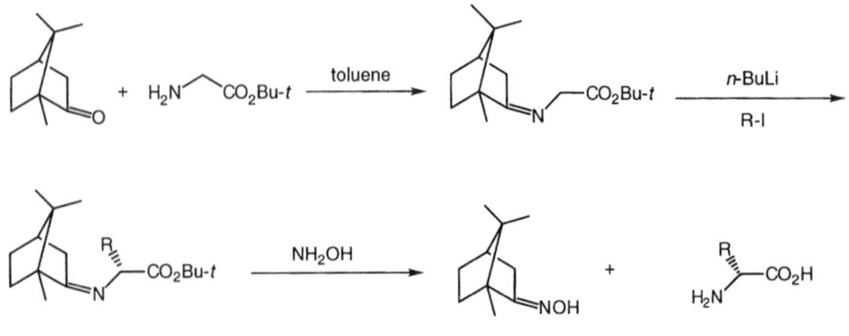

SCHEME 9.

Other camphor analogs with the C-8 methyl group replaced by larger and more hindered groups can significantly increase the stereoselectivity of the alkylation and result in much higher enantiomeric purity of the products [48].

6.4.1.2. Aldol Reaction

Camphor derivatives are also used as chiral auxiliaries in asymmetric aldol condensations (e.g., Scheme 10) [49–51].

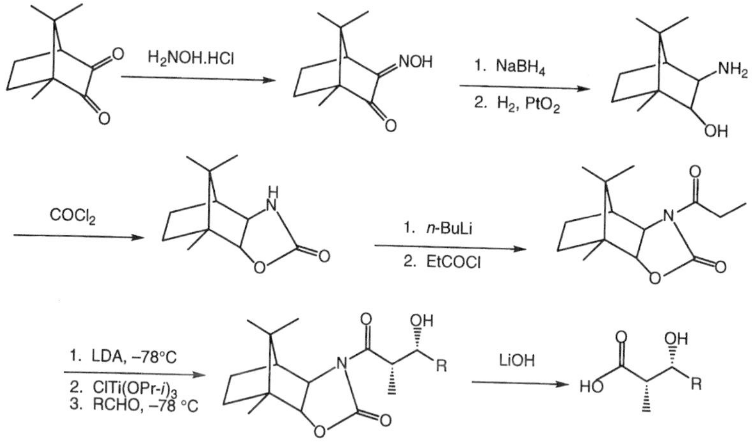

SCHEME 10.

As with most camphor-based chiral auxiliaries, the size and steric congestion at the C-8 position of the camphor moiety can determine its efficiency. Through the formation of a connection between C-8 and C-2, a novel camphor-based oxazinone auxiliary 28, which can be prepared from camphor in three steps [52], becomes highly effective in directing stereoselective aldol reactions (Scheme 11) [53].

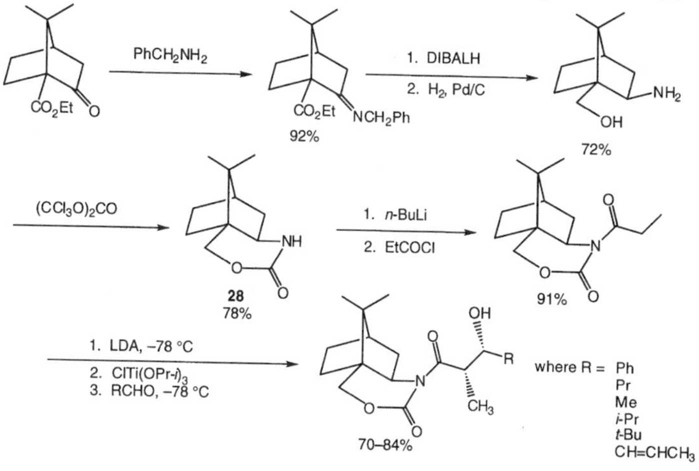

SCHEME 11.

6.4.1.3. Diels-Alder Reaction

Another application of camphor-based chiral auxiliaries is the use in stereoselective Diels-Alder cycloadditions (see Chapter 10), especially for the construction of quaternary carbon centers. An example is shown in Scheme 12 that employs a camphor-derived lactam **29** as the auxiliary [54].

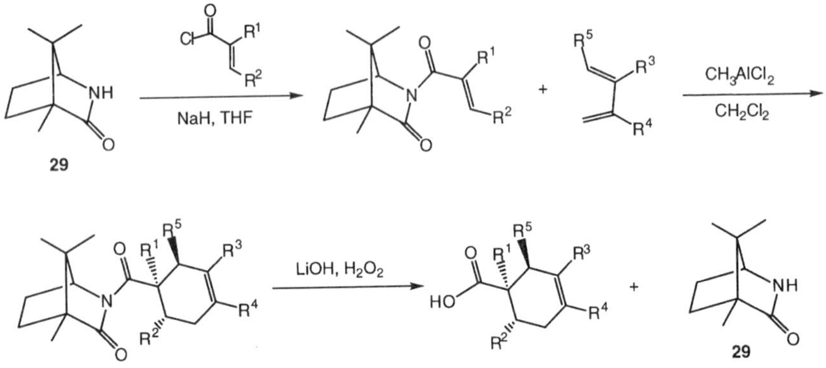

SCHEME 12.

6.4.2. Menthol as an Auxiliary

Optically active menthols are also commonly used as chiral auxiliaries in industrial production of fine chemicals. The most recent example of the use of this inexpensive and readily available monoterpene as chiral auxiliary for industrial-scale production of enantiomerically pure pharmaceuticals includes the synthesis of the anti–human immunodeficiency virus drug 3TC (**30**), or Lamivudine, marketed by Glaxo Wellcome. 3TC is an oxathiolane nucleoside with 2-hydroxymethyl-5-oxo-1,3-oxathiolane as the sugar portion and cytosine as the nucleobase. It contains two chiral centers and, thus, can exist in four stereoisomeric forms. The drug was originally developed as a racemic mixture of the β-anomer, but was later converted to the optically pure form due to favorable toxicology and pharmacokinetic profiles. There have been several independent routes leading to the synthesis of optically active 3TC that involve either classic resolution, enzymatic resolution, or asymmetric synthesis. The current production route uses (−)-menthol (**23**) as the chiral auxiliary (Scheme 13) [55]. The glycolate **31** is coupled with dithiodiene (**32**), the dimer of thioacetyl aldehyde, to produce the menthyl 4-oxo-1,3-oxathiolane-2-carboxylate (**33**). The presence of (−)-menthol as the chiral-directing group resulted in the stereoselective formation of the 2R-isomer. The lactol **33** is then treated with thionyl chloride to generate *in situ* the chloro derivative **34**, which is coupled with trimethylsilyl (TMS)-cytosine to yield, predominately, the β-nucleoside analog **35**. Again, the presence of the (−)-menthyl carboxylate group at the 5-position plays a major role in directing the stereoselectivity of this coupling reaction.

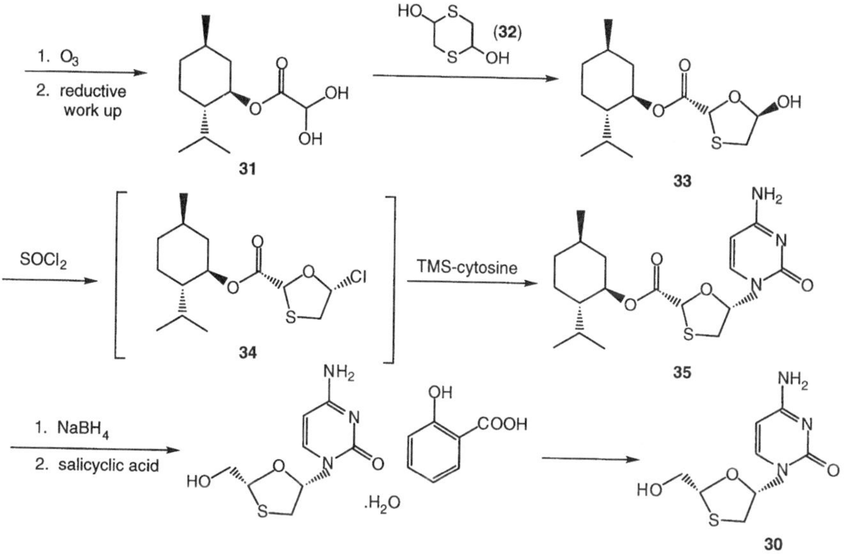

SCHEME 13.

6.4.3. Monoterpenes as Chirons

Despite their low cost and abundant availability, the applications of monoterpenes as chiral synthons or building blocks for synthesis of chiral fine chemicals on an industrial scale have lagged far behind amino acids and carbohydrates. Most of the work in this area is related to multistep total synthesis of complex natural products in laboratory scale. With the structures of new drug candidates in the research and development pipeline of pharmaceutical companies becoming bigger and more complicated, the application of more sophisticated chiral building blocks such as the terpenes will increase in the coming years. The major challenge lies in the design and development of practical chemical or biochemical reactions that effectively incorporate the simple monoterpenes into the target molecule of the desired products.

6.4.3.1. Robinson Annelation

One of the earliest industrial applications of monoterpenes is in the steroid synthesis. Robinson annelation is used frequently in these transformations as a key step.

The first example of this reaction was by Robinson himself to synthesize α-cyperone from dihydrocarvone [56]. This synthesis has been shown to follow a stereospecific course to give (+)-α-cyperone (36) from (+)-dihydrocarvone (37) (Scheme 14) [57–59]. Since then, numerous modifications and improvements have been made to apply this type of reaction for syntheses of a variety of natural and unnatural products [60].

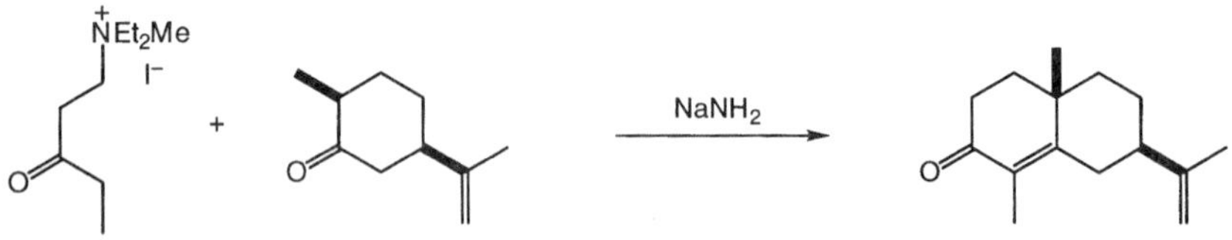

SCHEME 14.

(+)-α-Cyperone (36) can also be prepared stereoselectively in good yields from (−)-3-caranone (38) (Scheme 15) [61]. The stereoselectivity is derived from the bulky dimethylcyclopropane ring that directs the reaction away from the β-face, while the steric compression from the angular methyl group and avoidance of 1,3-diaxial interactions are probably the major force behind the selectivity of scission.

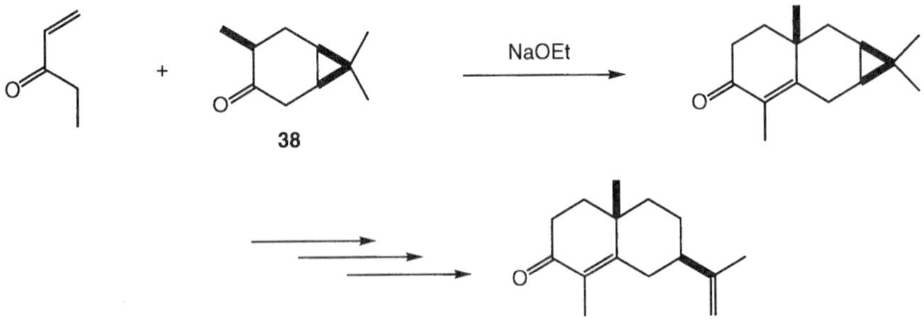

SCHEME 15.

It is much easier to prepare (+)-β-cyperone (39) by the Robinson method because it has only one asymmetric center and one enol form for alkylation (Scheme 16) [62].

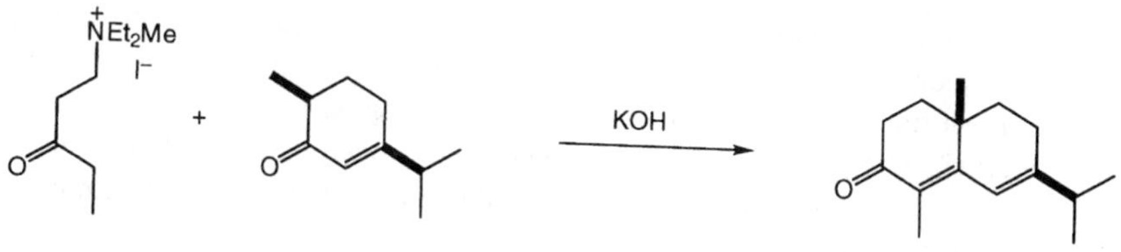

SCHEME 16.

The annelation process in the Robinson reaction can also be conducted in a stepwise mode; therefore, the alcohols from the aldol condensation step are often obtained (e.g., Scheme 17) [63].

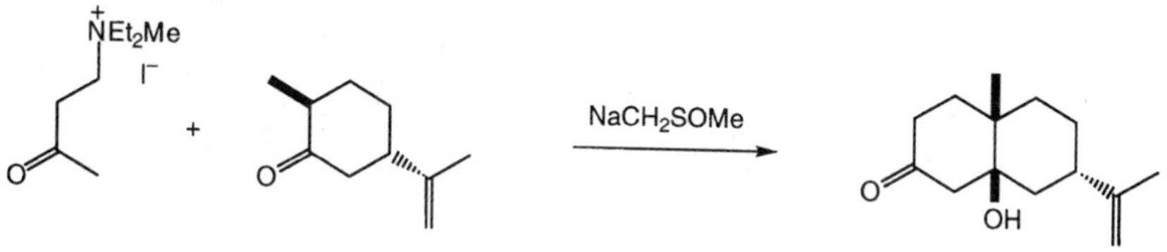

SCHEME 17.

The most abundant and widely used monoterpenes, pinenes and their analogs, are often upgraded to more sophisticated structures with a Robinson annelation as the key step. However, because of unusual stereochemical constraints imposed by the bicyclic ring, only certain conformational isomers of the pinane skeleton undergo the annelation. While the diketone (**40**) derived from *cis*-methylnopinone cyclized satisfactorily to the tricyclic ketone under Robinson conditions (Scheme 18), the epimeric diketone would not [64].

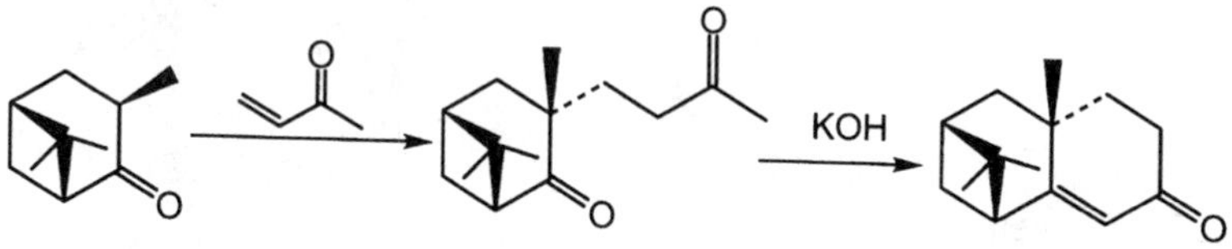

40

SCHEME 18.

6.4.3.2.   Steroid Synthesis

Because of their biological importance and widespread medical and industrial utilities, estrogens and related hormones are among the earliest targets in steroid total synthesis. Major efforts in the synthesis of these aromatic steroids were to set the correct chirality in the C,D rings since the A,B rings are mostly achiral.

The example uses monoterpenes as building blocks for synthesis of estrogen-type steroids (Scheme 19). Starting from the readily available d-(−)-camphor (41), direct functionalization at the C-9 methyl group was accomplished through three efficient steps to give (−)-π-bromocamphor (42) with complete retention of chirality. The bromo ketone was converted to the cyano ketone 43 by a number of transformations. The ketone 43 was then reacted with methyl vinyl ketone followed by POCl₃-pyridine treatment to give the tricyclenone 44, a critical C,D intermediate in aromatic steroid synthesis. At this stage, all the chiral centers in the final product are correctly set, and the tricyclo-ring system is together with the appropriate functionalities to allow completion of the steroid synthesis [65,66]. The construction of the A ring on the C,D intermediate can be effected by many alternative methods [67,68].

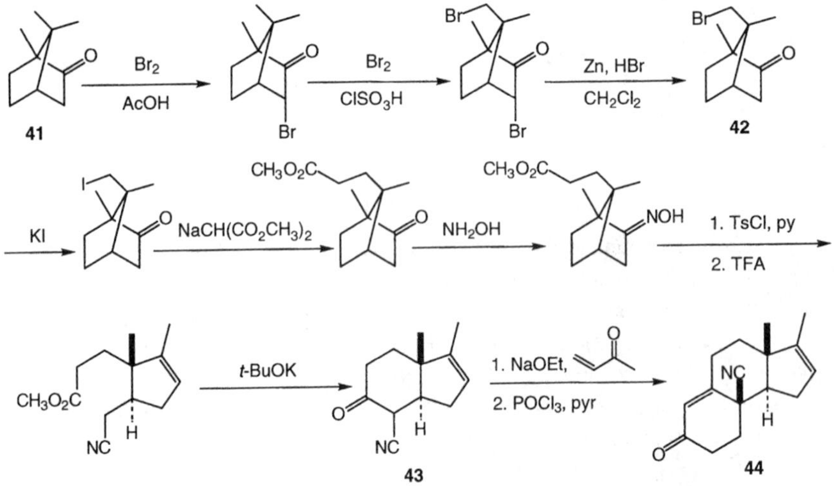

SCHEME 19.

6.4.3.3. Cinmethylin

Finally, an example of the use of monoterpenes as building blocks for production of chiral fine chemicals is demonstrated in the synthesis of cinmethylin (45), a herbicide developed by Shell Oil Company (Scheme 20) [69]. Enantiomerically pure S-terpinene-4-ol (46) was subjected to diastereoselective epoxidation by *tert*-butyl hydroperoxide in the presence of vanadium(III) acetylacetonate to generate the *cis*-epoxy-alcohol (47) [70]. The diastereoselectivity of this epoxidation is effected through chelation of vanadium with the 4-hydroxyl group to deliver the oxygen from the top face. The chiral epoxide 47 was opened by sulfuric acid followed by dehydration of the resultant 1,4-diol to give 2-(R)-*exo*-1,4-cineole (48), which reacts with o-methylbenzylbromide to give cinmethylin (45).

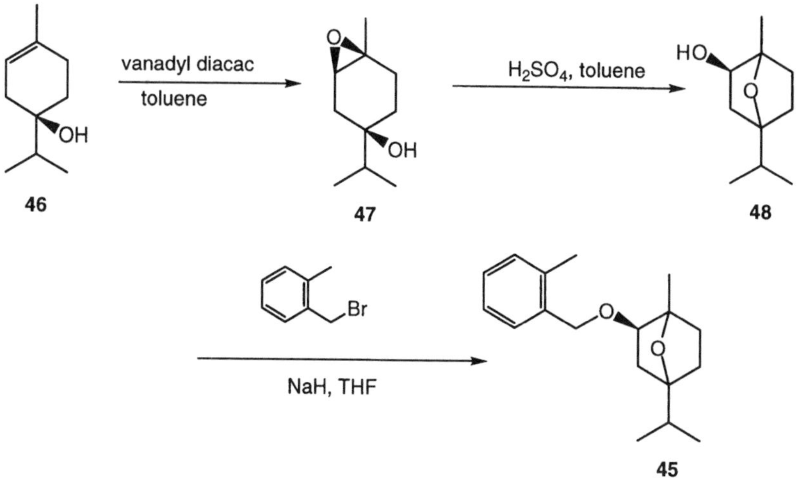

SCHEME 20.

The chiral epoxide intermediate **47** can also be used for the synthesis of dihydropinol (**48**) (Scheme 21) [71], and an analog of sobrerol [72], which is a mucolytic drug [73] marketed as a racemate, although differences in activity between the two enantiomers have been reported [74].

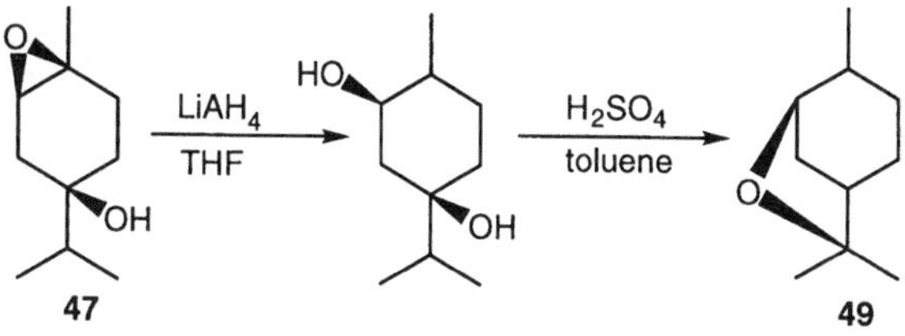

SCHEME 21.

6.5. SUMMARY

All of the optically active terpenes mentioned in this chapter are commercially available in bulk (> kg) quantities and are fairly inexpensive. Although many of them are isolated from natural sources, they can also be produced economically by synthetic methods. In fact, two thirds of these monoterpenes sold in the market today are manufactured by synthetic or semisynthetic routes. These optically active molecules usually possess simple carbocyclic rings with one or two stereogenic centers and have modest functionality for convenient structural manipulations. These unique features render them attractive as chiral pool materials for synthesis of optically active fine chemicals or pharmaceuticals. Industrial applications of these terpenes as chiral auxiliaries, chiral synthons, and chiral reagents have increased significantly in recent years. The expansion of the chiral pool into terpenes will continue with the increase in complexity and chirality of new drug candidates in the research and development pipeline of pharmaceutical companies.

REFERENCES

1. Erman, W. E. *Chemistry of Monoterpenes;* Marcel Dekker: New York, 1985.
2. Gildemeister, E., Hoffmann, F. *Die Atherischen Ole;* Akademie-Verlag: Berlin, 1960; *Vol. III.*
3. Newman, A. A. *Chemistry of Terpenes and Terpenoids;* Academic Press: New York, 1972.
4. Finnemore, H. *The Essential Oils;* Ernest Benn Ltd: London, 1927.
5. Guenther, E. *The Essential Oils;* van Nostrand: New York, 1948–1952; *Vol. 1–7.*
6. Pinder, A. R. *The Chemistry of Terpenes;* Wiley: New York, *1960.*
7. Banthorpe, D. V., Whittaker, D. *Chem. Rev.* 1966, *66,* 643.
8. Bir, G., Kaufmann, D. *Tetrahedron Lett.* 1987, *28,* 777.
9. Brown, H. C., Jadhav, P. K., Desai, M. C. *J. Org. Chem.* 1982, *47,* 4583.
10. Brown, H. C., Joshi, N. N. *J. Org. Chem.* 1988, *53,* 4059.
11. Settine, R. L. *J. Org. Chem.* 1970, *35,* 4266.
12. Bank, S., Rowe, C. A., Schriesheim, A., Naslund, C. A. *J. Am. Chem. Soc.* 1967, *89,* 6897.
13. Richter, F., Wolff, W. *Chem. Ber.* 1926, *59,* 1733.
14. Brown, C. A. *Synthesis* 1978, 754.
15. Lavalee, H. C., Singaram, B. *J. Org. Chem.* 1986, *51,* 1362.
16. Andrianone, M., Delmoond, B. *J. Chem. Soc. Chem. Commun.* 1985, 1203.
17. Zaidlewicz, M. *J. Organometal. Chem.* 1985, *239,* 139.
18. Whitham, G. H. *J. Chem. Soc.* 1961, 2232.
19. Hurst, J. J., Whitham, G. H. *J. Am. Chem. Soc.* 1960, *82,* 2864.
20. Nigam, I. C., Levi, L. *Can. J. Chem.* 1968, *46,* 1944.
21. Chloupek, F. J., Zweifel, G. *J. Org. Chem.* 1964, *29,* 2062.
22. Schenck, G. O., Eggert, H., Denk, W. *Annalen* 1953, *584,* 177.
23. Simonsen, J. L. *The Terpenes;* Cambridge University Press: Cambridge, 1974; *Vol. I;* p. 143.
24. Vig, O. P., Matta, K. L., Lal, A., Raj, I. *J. Ind. Chem. Soc.* 1964, *41,* 142.
25. Bates, R. B., Caldwell, E. S., Klein, H. P. *J. Org. Chem.* 1969, *34,* 2615.
26. Klein, E., Ohleff, G. *Tetrahedron* 1963, *19,* 1091.
27. Royals, E. E., Leffingwil, J. C. *J. Org. Chem.* 1966, *31,* 1937.
28. Linder, S. M., Greenspan, F. P. *J. Org. Chem.* 1957, *22,* 949.
29. Thomas, A. F. in *The Total Synthesis of Natural Products;* ApSimon, J. W., Ed., Wiley: New York, 1973; *Vol. 2;* p. 88.

30. Simonsen, J. L., Owen, L. N. in *The Terpenes;* Cambridge University Press: Cambridge, 1953; *Vol. I;* p. 230.
31. Budavari, S., Ed., *The Merck Index,* 12th ed. Merck & Co., Inc., Whitehouse Station, NJ, 1996, p. 5882.
32. Hussey, A. S. Baker, R. H. *J. Org. Chem.* 1960, *25,* 1434.
33. Brown, H. C., Garg, C. P. *J. Am. Chem. Soc.* 1961, *83,* 2952.
34. Akutagawa, S., Tani, K. in *Catalytic Asymmetric Synthesis;* Ojima, I., Ed.; VCH Publishers: New York 1993; p. 41.
35. Royals, E. E., Hornes, S. E. *J. Am. Chem. Soc.* 1951, *73,* 5856.
36. Shono, T., Nishiguchi, I., Yokoyama, T., Nitta, N. *Chem. Lett.* 1975, 433.
37. Merkel, D. *Die Ätherischen Ölen (Gildermeister Hoffmann);* Akademie-Verlag: Berlin, 1962; *Vol. IIIb;* p. 70.
38. Simonsen, J. L. *The Terpenes;* Cambridge University Press: Cambridge, 1957; *Vol. II;* p. 373.
39. Alder, K. *New Methods of Preparative Organic Chemistry;* Interscience: New York, 1948.
40. Vaughan, W. R., Perry, R. *J. Am. Chem. Soc.* 1953, *75,* 3168.
41. Budavar, S. E. *Merck Index,* 12th ed.; Merck: Whitehouse Station, NJ, 1996.
42. Srebnik, M., Ramachandran, P. V. *Aldrichimica Acta* 1987, *20,* 9.
43. McIntosh, J. M. Mishra, P. *Can. J. Chem.* 1986, *64,* 726.
44. Schöllköpf, U., Schütze, R. *Liebigs Ann. Chem.* 1987, 45.
45. Yaozhong, J., Guilan, L., Jinchu, L., Changyou, Z. *Synth. Commun.* 1987, *17,* 1545.
46. Yaozhong, J., Guilan, L., Jingen, D. *Synth. Commun.* 1988, *18,* 1291.
47. McIntosh, J. M., Cassidy, K. C., Matassa, L. *Tetrahedron* 1989, *45,* 5449.
48. Yaozhong, J., Peng, G., Guilan, L. *Synth. Commun.* 1990, *20,* 15.
49. Oppolzer, W., Blagg, J., Rodriguez, I., Walther, W. *J. Am. Chem. Soc.* 1990, *112,* 2767.
50. Boeckman, R. K., Johnson, A. T., Musselman, R. A. *Tetrahedron Lett.* 1994, *35,* 8521.
51. Kelly, T. R., Arvanitis, A. *Tetrahedron Lett.* 1984, *25,* 39.
52. Bartlett, P. D., Knox, L. H. *Org. Synth.* 1973, *Coll. Vol. 5,* 689.
53. Anh, K. H., Lee, S., Lim, A. *J. Org. Chem.* 1992, *57,* 5065.
54. Boeckman, R. K., Nelson, S. G., Gaul, M. D. *J. Am. Chem. Soc.* 1992, *114,* 2258.
55. Goodyear, M. D., Dwyer, P. O., Hill, M. L., Whitehead, A. J., Hornby, R., Hallett, P. *World Patent,* 1995, WO 95/29174
56. Adamson, P. S., McQuillin, F. C., Robinson, R., Simonsen, J. L. *J. Chem. Soc.* 1937, 1576.
57. Tanaka, A., Kamata, H., Yamashita, K. *Agric. Biol. Chem.* 1988, *52,* 2043.
58. McQuillin, F. J. *J. Chem. Soc.* 1955, 528.
59. Howe, R., McQuillin, F. J. *J. Chem. Soc.* 1955, 423.
60. Ho, T.-L. *Carbocycle Construction in Terpene Synthesis;* VCH: New York, 1988.
61. Fringuelli, F., Taticchi, A., Trazerso, G. *Gazz. Chim. Ital.* 1969, *99,* 231.
62. Roy, J. K. *Chem. Ind. (London)* 1954, 1393.
63. Humber, D. C., Pinder, A. R., Williams, R. A. *J. Org. Chem.* 1967, *32,* 2335.
64. Thomsa, A. F. *Pure Appl. Chem.* 1990, *62,* 1369.
65. Velluz, L., Nominé, G., Bucourt, R., Pierdet, A., Dufay, P. *Tetrahedron Lett.* 1961, 127.

66. Stevens, R. V., Caeta, C. A. *J. Am. Chem. Soc.* 1977, *99,* 6105.
67. Danishefsky, S., Cain, P. *J. Am. Chem. Soc.* 1975, *97,* 5282.
68. Danlewski, A. R. *J. Org. Chem.* 1975, *40,* 3135.
69. Payne, G. B. *US Patent,* 1984, 4,487,945.
70. Ohloff, G., Uhde, G. *Helv. Chim. Acta* 1965, *48,* 10.
71. Liu, W.-G., Rosazza, J. P. N. *Synth. Commun.* 1996, *26,* 2731.
72. Bovara, R., Carrea, G., Ferrara, L., Riva, S. *Tetrahedron: Asymmetry* 1991, *2,* 931.
73. Braga, P. C., Allegra, L., Bossi, R., Scuri, R., Castiglioni, C. L., Romandini, S. *Int. J. Clin. Pharm. Res.* 1987, *7,* 381.
74. Klein, E. A. *US Patent,* 1957, 2,815,378.

7
Substitution Reactions

DAVID J. AGER

NSC Technologies, Mount Prospect, Illinois

7.1. INTRODUCTION

This chapter covers reactions in which a stereogenic sp^3 center is inverted or undergoes a substitution reaction. To control the stereochemical outcome of a substitution reaction, an S_N2 mechanism usually has to be used. As a consequence, inversion of configuration is usually observed, although in some reactions two sequential substitutions can occur to give the overall appearance of retention. The use of an S_N2 reaction necessitates the establishment of the stereogenic center in the reactant. When an isolated center undergoes this type of reaction, it is usually to correct stereochemistry.

This chapter also includes reactions of epoxides and surrogates, as well as reactions at isolated centers. In the cases of vicinal functional groups, an asymmetric reaction can be used to introduce the vicinal functionalization and generate the stereogenic center, as with an epoxidation reaction [1].

Another type of substitution reaction is increasing in popularity—the use of an allylic substrate such as an allyl acetate, when the nucleophile is introduced with stereochemical control in the presence of a palladium catalyst and a chiral ligand. Reactions in which a chiral anion, whether derived from a chiral heteroatom group such as a sulfoxide or an auxiliary such as Evans's oxazolidinones, is not covered as the alkyl halide is usually relatively simple and the stereochemical selectivity is derived from the system itself.

7.2. S_N2 REACTIONS

Often, the purpose of a substitution reaction at a saturated center is to correct stereochemistry. An example is provided in the Sumitomo approach to pyrethroid alcohols, where the undesired isomer is inverted in the presence of the required one (Scheme 1) [2].

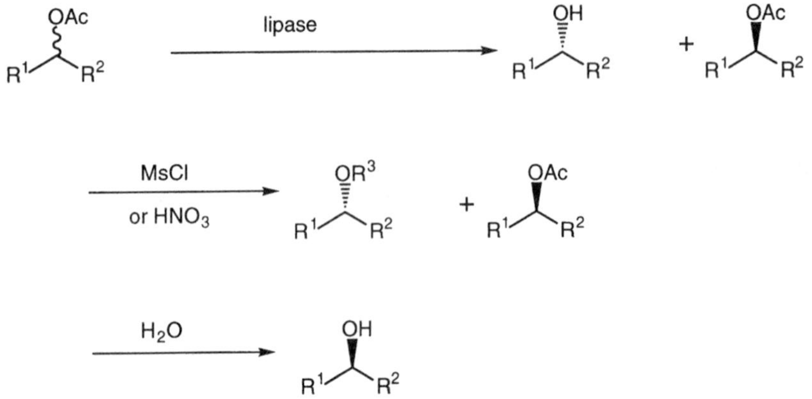

Scheme 1.

The phenoxypropionic acid herbicides (see Chapter 3) rely on an S_N2 reaction to convert the S-2-chloropropionic acid (**1**) to the R-2-aryloxy derivative **2** [3]. These compounds are also available from lactic acid by substitution reactions; R-lactic acid (**3**) requires two inversions, whereas S-lactate (**4**) only requires one (Schemes 2 and 3) [4].

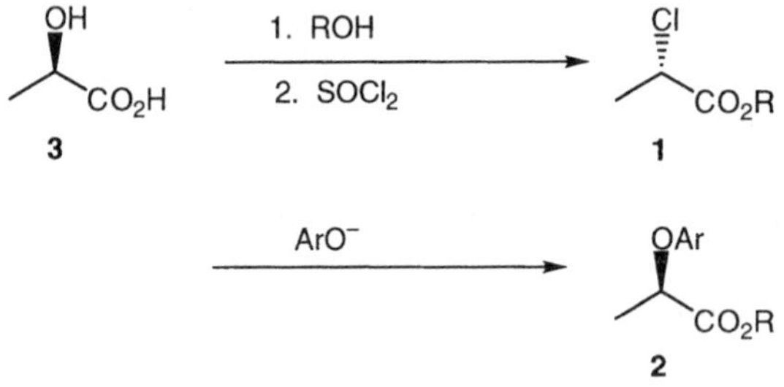

Scheme 2.

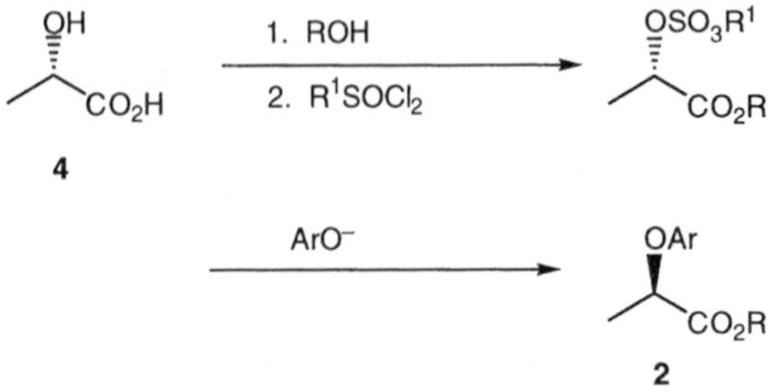

Scheme 3.

An additional example of the undesired isomer being inverted is provided in an approach to the pyrethroid, prallethrin (**5**) (Scheme 4) [5].

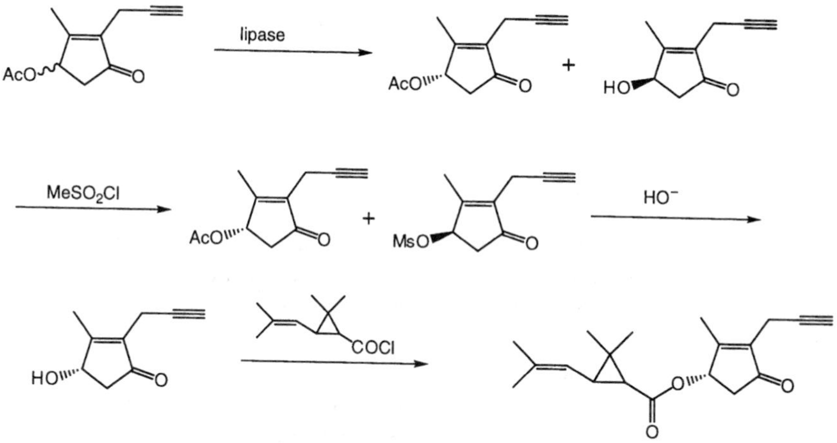

Scheme 4.

Reaction conditions must be carefully controlled to ensure that these types of substitution reactions proceed through an S_N2 mechanism and inversion occurs at the stereogenic center. In some examples, as with hydroxy groups, this is not a trivial undertaking [1]. An example for the inversion of the hydroxy group in menthol that uses dicyclohexylcarbodiimide to activate the hydroxy group (Scheme 5) [6].

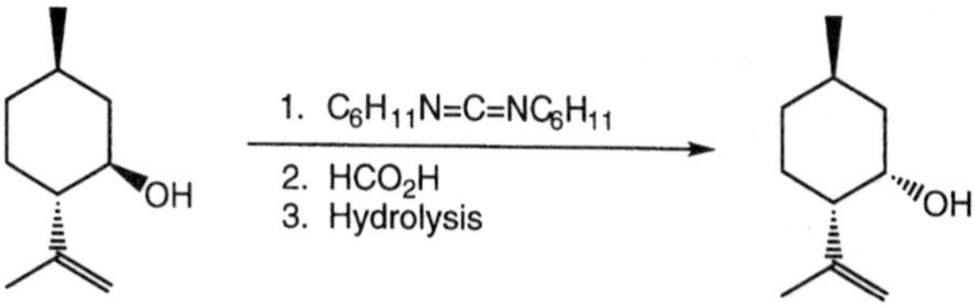

1. $C_6H_{11}N=C=NC_6H_{11}$
2. HCO_2H
3. Hydrolysis

Scheme 5.

7.3. EPOXIDE OPENINGS

Although considerable effort has been expended on asymmetric routes to the structural motif within a number of beta-blocker drugs, there has been little financial reward. Through the use of a chiral equivalent of epichlorohydrin (X = Cl), a substitution reaction at the sp^3 center followed by epoxide opening allows for entry into this class of drugs. The stereogenic center of the epoxide is retained throughout this sequence (Scheme 6) [3].

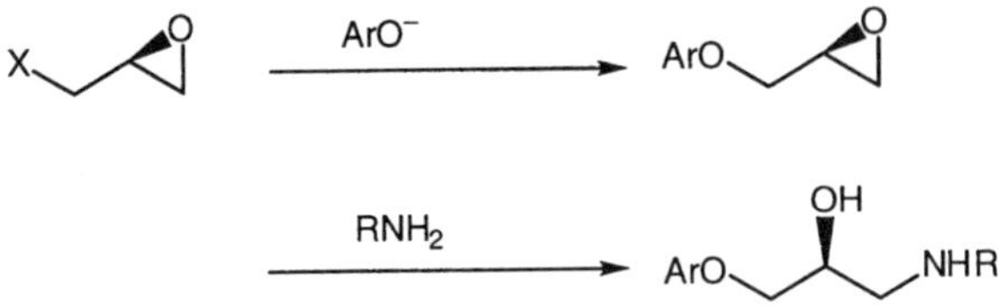

Scheme 6.

A considerable amount of work was required to optimize the leaving group and avoid racemization through a Payne rearrangement mechanism [7]. Of course, the Sharpless epoxidation of allyl alcohols is well known to access these 3-functionalized epoxides.

Manganese complexes provide sufficient induction for synthetic utility for the preparation of unsubstituted epoxides (Scheme 7) [8–13]. The manganese (III)salen complex 6 can also use bleach as the oxidant rather than an iodosylarene [14,15].

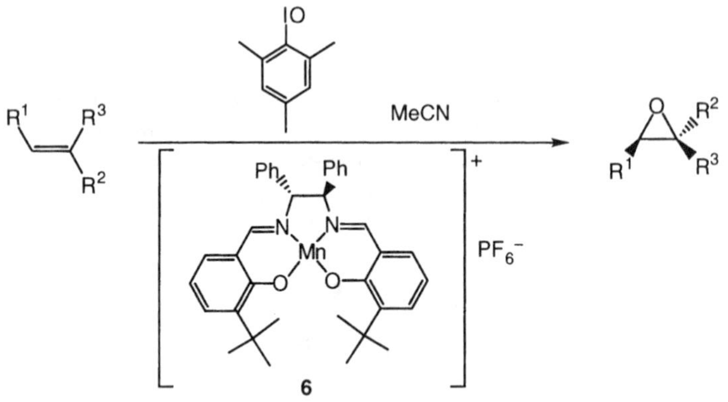

Scheme 7.

The regioselective control for the nucleophilic opening of an epoxide in an acyclic system is well known [1,16]. Under basic conditions, the nucleophile usually attacks the sterically less encumbered site, whereas under acidic conditions, the sterically more hindered site is favored [16–20]. The product invariably contains the functional groups in an *ancat*-disposition, when an S_N2 pathway is followed [1,16]. Epoxides react with a wide variety of nucleophiles [21].

7.3.1. HIV Protease Inhibitors

Human immunodeficiency virus (HIV) protease inhibitors have provided the pharmaceutical and fine chemical industry with a number of difficult problems to solve. Most have complex structures with a number of stereogenic centers that make synthetic sequences long. In addition, the dosages of these drugs are relatively high, and large volumes must be made. Because most may be considered peptidomimetics, amino acids are often used as a starting material. A number of candidates have relied on a homologation, formation of an epoxide and opening with a nitrogen nucleophile (Scheme 8) [22–30]. The recrystallization of the intermediate removes the unwanted diastereoisomer.

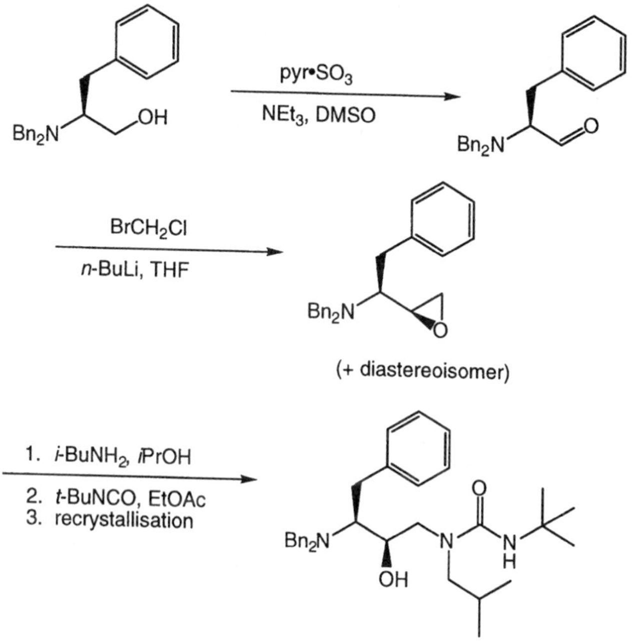

Scheme 8.

7.4.   CYCLIC SULFATE REACTIONS

Cyclic sulfates provide a useful alternative to epoxides now that it is viable to produce a chiral diol from an alkene. These cyclic compounds are prepared by reaction of the diol with thionyl chloride, followed by ruthenium catalyzed oxidation of the sulfur (Scheme 9) [31]. This oxidation has an advantage over previous procedures because it only uses a small amount of the transition metal catalyst [1,32,33].

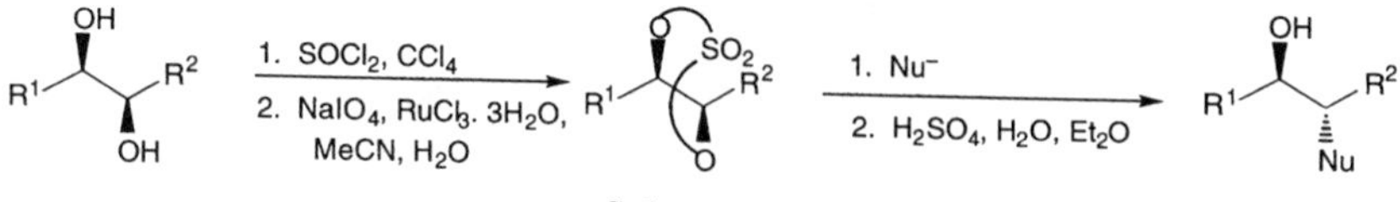

Scheme 9.

7.5. IODOLACTONIZATIONS

Intramolecular cyclization is a useful method for the preparation of lactones and cyclic ethers [34]. The most common examples are iodolactonization and iodoetherification, the former using a carboxylic acid derivative as the nucleophile and the latter relying on a hydroxy group. Thus, butyrolactones are available from γ,δ-unsaturated carboxylic acid derivatives [1,35,36], while unsaturated alcohols lead to cyclic ethers [37–40]. Lactones are also available from a wide variety of nucleophiles such as carbonates [41], orthoesters [42], or carbamates [43,44], which can all be used in place of a carboxylate anion [44,45].

The intermediate iodonium ion controls the relative stereochemistry of the cyclization. An asymmetric center present within the substrate, therefore, allows for enantioselectivity (Scheme 10) [46]. The reaction is very susceptible to the steric interactions within the transition state [43,46–48].

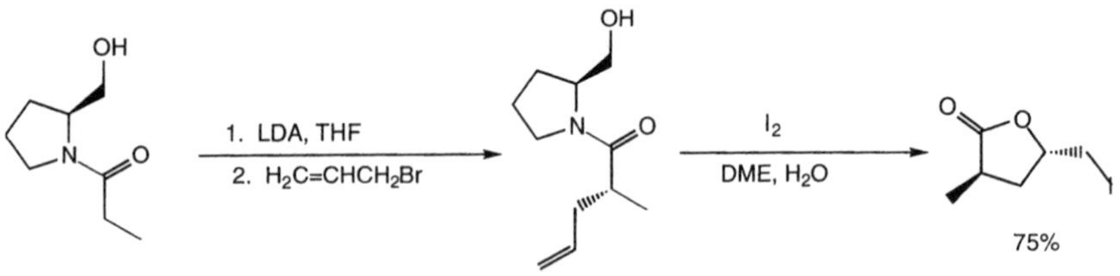

Scheme 10.

Variations on this cyclization methodology include the use of bromine rather than iodine; again, both lactones and ethers can be prepared [1,42,49–51]. An asymmetric bromolactonization procedure affords α-hydroxy acids with good asymmetric induction [52–57].

A further variation uses selenium or sulphur as the electrophilic species to induce cyclization. The diversity of organoselenium or organosulfur chemistry is then available for further modification of the product [1,58–70].

7.6. ALLYLIC SUBSTITUTIONS

Allylic systems can undergo nucleophilic substitution by either an S_N2 or S_N2' reactions [71–74]. The S_N2 reaction can compete effectively with an S_N2' reaction.

The use of a metal catalyst such as palladium also provides for some asymmetric induction when an allylic system is treated with a stabilized anion (Scheme 11) [75–80], or with other nucleophiles [75,80–84]. This approach also allows for the kinetic resolution of allyl acetates [81,85].

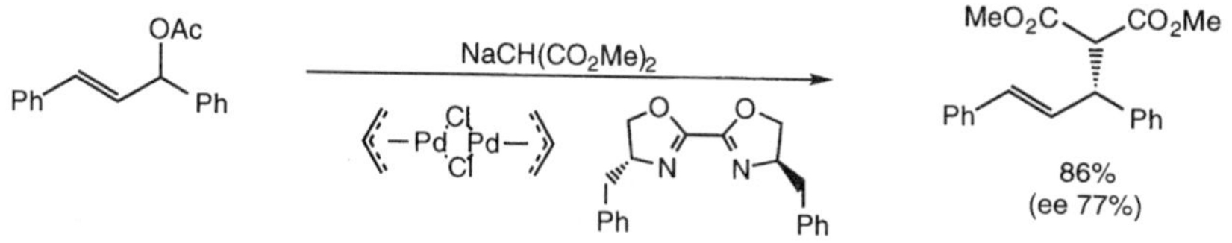

Scheme 11.

The mechanism of allylic alkylation reactions has been the subject of significant investigation [73,86]. The two major approaches were based on the use of a cyclic substrate and use of chiral substrates, and have given rise to the model shown in Scheme 12. The model also includes the following general observations:

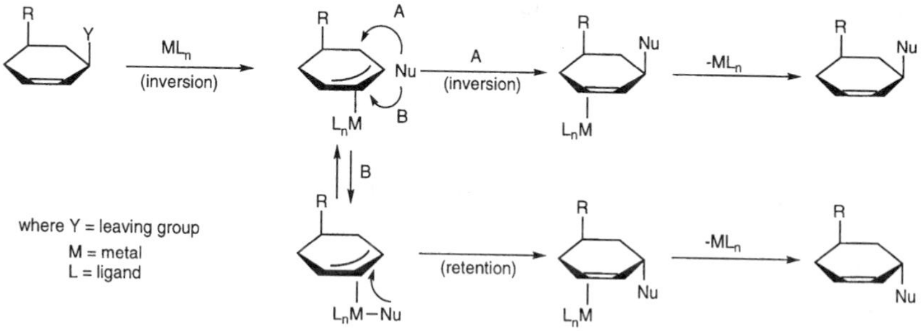

where Y = leaving group
M = metal
L = ligand

Scheme 12.

1. The allyl complex is generally formed with inversion of configuration.
2. If the leaving group is antiperiplanar to the metal, then formation of the π-allyl intermediate usually follows [87–89].
3. Attack by a "soft" nucleophile leads to inversion of configuration in the second step [90–99].
4. Attack by a "hard" nucleophile results in retention of configuration in the second step as the metal center is attacked first (pathway B) [100–103].

Palladium catalyzed allylation reactions proceed with net retention of configuration that is the result of two inversions [90–92,104–107]. Grignard reactions with nickel or copper catalysis proceed with net inversion [100,108–110].

The methodology suffers from competition between *syn* and *anti* mechanisms in acyclic cases. Indeed, the pathway may depend on the nature of nucleophile and substrate [71,111]. In addition, there is no clear-cut differentiation between "hard" and "soft" nucleophiles in the model previously described.

Use of a chiral catalyst for allylic substitution reactions allows for desymmetrization reactions and asymmetric reactions (Scheme 13) [112–115].

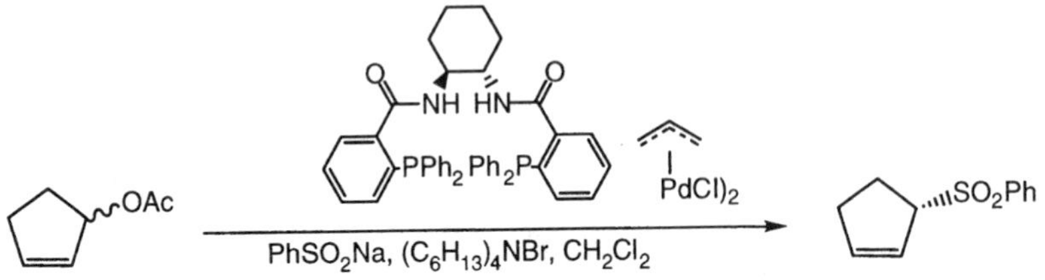

Scheme 13.

7.7. SUMMARY

Substitution reactions allow for the introduction or change of functional groups but rely on the prior formation of the stereogenic center. The approach can allow for the correction of stereochemistry. Reactions of epoxides and analogous systems such as cyclic sulphates allow for 1,2-functionality to be set up in a stereospecific manner. Reactions of this type have been key to the applications of asymmetric oxidations. The use of chiral ligands for allylic substitutions does allow for the introduction of a new stereogenic center. With efficient catalysts now identified, it is surely just a matter of time before this methodology is used at scale.

REFERENCES

1. Ager, D. J., East, M. B. *Asymmetric Synthetic Methodology;* CRC Press: Boca Raton, 1995.
2. Danda, H., Nagatomi, T., Maehara, A., Umemura, T. *Tetrahedron* 1991 *47,* 8701.
3. Sheldon, R. A. *Chirotechnology: Industrial Synthesis of Optically Active Compounds;* Marcel Dekker: New York, 1993.
4. Gras, G. *Ger. Patent,* 1979, 3,024,265.
5. Hirohara, H., Mitsuda, S., Ando, E., Komaki, R. in *Biocatalysts in Organic Synthesis;* Tramper, J., Van der Plas, H. C., Linko, P., Eds.; Elsevier: Amsterdam, 1985; p. 119.
6. Kaulen, J. *Ger. Patent* 1986, DE 3511210.
7. Klunder, J. M., Onami, T., Sharpless, K. B. *J. Org. Chem.* 1989, *54,* 1295.
8. Zhang, W., Loebach, J. L., Wilson, S. R., Jacobsen, E. N. *J. Am. Chem. Soc.* 1990, *112,* 2801.
9. Irie, R., Noda, K., Ito, Y., Matsumoto, N., Katsuki, T. *Tetrahedron Lett.* 1990, *31,* 7345.
10. Zhang, W., Loebach, J. L., Wilson, S. R., Jacobsen, E. J. *J. Am. Chem. Soc.* 1990, *112,* 2801.

11. Van Draanen, N. A., Arseniyadis, S., Crimmins, M. T., Heathcock, C. H. *J. Org. Chem.* 1991, *56*, 2499.
12. Okamoto, Y., Still, W. C. *Tetrahedron Lett.* 1988, *29*, 971.
13. Schwenkreis, T., Berkessel, A. *Tetrahedron Lett.* 1993, *34*, 4785.
14. Zhang, W., Jacobsen, E. N. *J. Org. Chem.* 1991, *56*, 2296.
15. Deng, L., Jacobsen, E. N. *J. Org. Chem.* 1992, *57*, 4320.
16. Parker, R. E., Isaacs, N. S. *Chem. Rev.* 1959, *59*, 737.
17. Boireau, G., Abenhaim, D., Bernardon, C., Henry-Basch, E., Sabourault, B. *Tetrahedron Lett.* 1975, 2521.
18. Schauder, J. R., Krief, A. *Tetrahedron Lett.* 1982, *23*, 4389.
19. Tanaka, T., Inoue, T., Kamei, K., Murakami, K., Iwata, C. *J. Chem. Soc. Chem. Commun.* 1990, 906.
20. Bartók, M., Láng, K. L. *Chem. Heterocycl. Comp.* 1984, *42 (pt.3)*, 1.
21. Williams, N. R. *Adv. Carbohydr. Chem. Biochem.* 1970, *25*, 109.
22. Liu, C., Ng, J. S., Behling, J. R., Yen, C. Y., Campbell, A. L., Fuzail, K. S., Yonan, E. E., Mehrota, D. V. *Org. Proc. Res. Develop.* 1997, *1*, 45.
23. Ng, J. S., Przybyla, C. A., Mueller, R. A., Vazquez, M. L., Getman, D. P. *US Patent* 1996, 5,583,238.
24. Ng, J. S., Przybyla, C. A., Liu, C., Yen, J. C., Muellner, F. W., Weyker, C. L. *Tetrahedron* 1995, *51*, 6397.
25. Bennett, F., Patel, N. M., Girijavallabhan, V. M., Ganguly, A. K. *Synlett* 1993, 703.
26. Vazquez, M. L., Bryant, M. L., Clare, M., DeCresenzo, G. A., Doherty, E. M., Freskos, J. N., Getman, D. P., Houseman, K. A., Julian, J. A., Kocan, G. P., Mueller, R. A., Shieh, H.-S., Stallings, W. C., Stegeman, R. A., Talley, J. J. *J. Med. Chem.* 1995, *38*, 581.
27. Ghosh, A. K., Hussain, K. A., Fidanze, S. *J. Org. Chem.* 1997, *62*, 6080.
28. Ahmad, S., Spergel, S. H., Barrish, J. C., DiMarco, J., Gougoutas, J. *Tetrahedron: Asymmetry* 1995, *6*, 2893.
29. Sham, H. L., Betebenner, D. A., Zhao, C., Wideburg, N. E., Saldivar, A., Kempf, D. J., Plattner, J. J., Norbeck, D. W. *J. Chem. Soc. Chem. Commun.* 1993, 1052.
30. Askin, D., Wallace, M. A., Vacca, J. P., Reamer, R. A., Volante, R. P., Shinkai, I. *J. Org. Chem.* 1992, *57*, 2771.
31. Gao, Y., Sharpless, K. B. *J. Am. Chem. Soc.* 1988, *110*, 7538.
32. Denmark, S. E. *J. Org. Chem.* 1981, *46*, 3144.
33. Lowe, G., Salamone, S. J. *J. Chem. Soc. Chem. Commun.* 1983, 1392.
34. Cardillo, G., Orena, M. *Tetrahedron* 1990, *46*, 3321.
35. Bartlett, P. A. *Tetrahedron* 1980, *36*, 3.
36. Bartlett, P. A., Richardson, D. P., Myerson, J. *Tetrahedron* 1984, *40*, 2317.
37. Williams, D. R., White, F. H. *Tetrahedron Lett.* 1986, *27*, 2195.
38. Baldwin, S. W., McIver, J. M. *J. Org. Chem.* 1987, *52*, 322.
39. Labelle, M., Morton, H. E., Guidon, Y., Springer, J. P. *J. Am. Chem. Soc.* 1988, *110*, 4533.
40. Kang, S. H., Lee, S. B. *Tetrahedron Lett.* 1993, *34*, 7579.
41. Bongini, A., Cardillo, G., Orena, M., Porzi, G., Sandri, S. *J. Org. Chem.* 1982, *47*, 4626.

42. Williams, D. R., Harigaya, Y., Moore, J. L., D'asa, A. *J. Am. Chem. Soc.* 1984, *106*, 2641.
43. Hirama, M., Uei, M. *Tetrahedron Lett.* 1982, *23*, 5307.
44. Kobayashi, S., Isobe, T., Ohno, M. *Tetrahedron Lett.* 1984, *25*, 5079.
45. Friesen, R. W. *Tetrahedron Lett.* 1990, *31*, 4249.
46. Najdi, S., Reichlin, D., Kurth, M. J. *J. Org. Chem.* 1990, *55*, 6241.
47. Kurth, M. J., Beard, R. L., Olmstead, M., Macmillan, J. G. *J. Am. Chem. Soc.* 1989, *111*, 3712.
48. Fuji, K., Node, M., Naniwa, Y., Kawabata, T. *Tetrahedron Lett.* 1990, *31*, 3175.
49. Fukuyama, T., Wang, C.-L. J., Kishi, Y. *J. Am. Chem. Soc.* 1979, *101*, 260.
50. Al-Dulayymi, J., Baird, M. S. *Tetrahedron Lett.* 1989, *30*, 253.
51. Jung, M. E., Lew, W. *J. Org. Chem.* 1991, *56*, 1347.
52. Jew, S.-s., Terashima, S., Koga, K. *Tetrahedron* 1979, *35*, 2337.
53. Terashima, S., Jew, S.-s. *Tetrahedron Lett.* 1977, 1005.
54. Terashima, S., Jew, S.-s. Koga, K. *Chemistry Lett.* 1977, 1109.
55. Terashima, S., Jew, S.-s., Koga, K. *Tetrahedron Lett.* 1977, 4507.
56. Terashima, S., Jew, S.-s., Koga, K. *Tetrahedron Lett.* 1978, 4937.
57. Jew, S.-s., Terashima, S., Koga, K. *Chem. Pharm. Bull.* 1979, *27*, 2351.
58. Nicolaou, K. C. *Tetrahedron* 1981, *37*, 4097.
59. Nicolaou, K. C., Barnette, W. E. *J. Chem. Soc. Chem. Commun.* 1977, 331.
60. Murata, S., Suzuki, T. *Tetrahedron Lett.* 1987, *28*, 4415.
61. Clive, D. L. J., Chittattu, G. *J. Chem. Soc. Chem. Commun.* 1977, 484.
62. Tiecco, M., Testaferri, L., Tingoli, M., Bartoli, D. *Tetrahedron* 1989, *45*, 6819.
63. Nicolaou, K. C., Sipio, W. J., Magolda, R. J., Claremon, D. A. *J. Chem. Soc. Chem. Commun.* 1979, 83.
64. Nicolaou, K. C., Seitz, S. P., Sipio, W. J., Blount, J. F. *J. Am. Chem. Soc.* 1979, *101*, 3884.
65. Bartlett, P. A., Holm, K. H., Morimoto, A. *J. Org. Chem.* 1985, *50*, 5179.
66. Nicolaou, K. C., Lysenko, Z. *J. Chem. Soc., Chem. Commun.* 1977, 293.
67. Nicolaou, K. C., Barnette, W. E., Magolda, R. L. *J. Am. Chem. Soc.* 1981, *103*, 3480.
68. Nicolaou, K. C., Barnette, W. E., Magolda, R. L. *J. Am. Chem. Soc.* 1981, *103*, 3486.
69. Nicolaou, K. C., Barnette, W. E., Magolda, R. L. *J. Am. Chem. Soc.* 1978, *100*, 2567.
70. Nicolaou, K. C., Claremon, D. A., Barnette, W. E., Seitz, S. P. *J. Am. Chem. Soc.* 1979, *101*, 3704.
71. Magid, R. M. *Tetrahedron* 1980, *36*, 1901.
72. Tsuji, J. *Pure Appl. Chem.* 1989, *61*, 1673.
73. Consiglio, G., Waymouth, R. M. *Chem. Rev.* 1989, *89*, 257.
74. Uguen, D. *Tetrahedron Lett.* 1984, *25*, 541.
75. Trost, B. M. *Acc. Chem. Res.* 1980, *13*, 385.
76. Muller, D., Umbricht, G., Weber, B., Pfaltz, A. *Helv. Chim. Acta* 1991, *74*, 232.
77. Leutenegger, U., Umbricht, G., Fahrni, C., von Matt, P., Pfaltz, A. *Tetrahedron* 1992, *48*, 2143.

78. von Matt, P., Pfaltz, A. *Angew. Chem. Int. Ed. Engl.* 1993, *32*, 566.
79. Reiser, O. *Angew. Chem. Int. Ed. Engl.* 1993, *32*, 547.
80. Trost, B. M., Van Vranken, D. L. *Chem. Rev.* 1996, *96*, 395.
81. Sawamura, M., Ito, Y. *Chem. Rev.* 1992, *92*, 857.
82. Trost, B. M. *Pure Appl. Chem.* 1979, *51*, 787.
83. Trost, B. M. *Tetrahedron* 1977, *33*, 2615.
84. Hayashi, T. *Pure Appl. Chem.* 1988, *60*, 7.
85. Hayashi, T., Yamamoto, A., Ito, Y. *J. Chem. Soc. Chem. Commun.* 1986, 1090.
86. Trost, B. M., Van Vranken, D. L., Bingel, C. *J. Am. Chem. Soc.* 1992, *114*, 9327.
87. Fiaud, J.-C., Aribi-Zouioneche, L. *J. Chem. Soc. Chem. Commun.* 1986, 390.
88. Fiaud, J.-C., Legros, J.-Y. *J. Org. Chem.* 1987, *52*, 1907.
89. Faller, J. W., Linebarrier, D. *Organometallics* 1988, *7*, 1670.
90. Trost, B. M., Verhoeven, T. R. *J. Org. Chem.* 1976, *41*, 3215.
91. Trost, B. M., Verhoeven, T. R. *J. Am. Chem. Soc.* 1980, *102*, 4730.
92. Trost, B. M., Weber, L. *J. Am. Chem. Soc.* 1975, *97*, 1611.
93. Keinan, E., Sahai, M., Roth, Z., Nudelman, A., Herzig, J. *J. Org. Chem.* 1985, *50*, 3558.
94. Trost, B. M., Keinan, E. *J. Am. Chem. Soc.* 1978, *100*, 7779.
95. Fiaud, J.-C. *J. Chem. Soc. Chem. Commun.* 1983, 1055.
96. Auburn, P. R., Whelan, J., Bosnich, B. *J. Chem. Soc. Chem. Commun.* 1986, 186.
97. Trost, B. M. Schmuff, N. R. *J. Am. Chem. Soc.* 1985, *107*, 396.
98. Trost, B. M., Verhoeven, T. R., Fortunak, J. M. *Tetrahedron Lett.* 1979, 2301.
99. Bäckvall, J.-E., Nordberg, R. E. *J. Am. Chem. Soc.* 1981, *103*, 4959.
100. Consiglio, G., Morandini, F., Piccolo, O. *J. Am. Chem. Soc.* 1981, *103*, 1845.
101. Sheffy, F. K., Stille, J. K. *J. Am. Chem. Soc.* 1983, *105*, 7173.
102. Matsushita, H., Negishi, E. *J. Chem. Soc. Chem. Commun.* 1982, 150.
103. Hayashi, T., Yamamoto, A., Hagihara, T. *J. Org. Chem.* 1986, *51*, 723.
104. Trost, B. M., Schmuff, N. R., Miller, M. J. *J. Am. Chem. Soc.* 1980, *102*, 5979.
105. Trost, B. M., Strege, P. E. *J. Am. Chem. Soc.* 1977, *99*, 1649.
106. Tanikaga, R., Jun, T. X., Kaji, A. *J. Chem. Soc. Perkin Trans. 1* 1990, 1185.
107. Mackenzie, P. B., Whelan, J., Bosnich, B. *J. Am. Chem. Soc.* 1985, *107*, 2046.
108. Felkin, H., Joly-Goudket, M., Davies, S. G. *Tetrahedron Lett.* 1981, *22*, 1157.
109. Gendreau, Y., Normant, J. F. *Tetrahedron Lett.* 1979, 1617.
110. Consiglio, G., Morandini, F., Piccolo, O. *J. Chem. Soc. Chem. Commun.* 1983, 112.
111. Hayashi, T., Yamamoto, A., Hayihara, T., Ito, Y. *Tetrahedron Lett.* 1986, *27*, 191.
112. Trost, B. M., Organ, M. G., O'Doherty, G. A. *J. Am. Chem. Soc.* 1995, *117*, 9662.
113. Trost, B. M., Lee, C. B., Weiss, J. M. *J. Am. Chem. Soc.* 1995, *117*, 7247.
114. Trost, B. M. *Pure Appl. Chem.* 1996, *68*, 779.
115. Trost, B. M. *Acc. Chem. Res.* 1996, *29*, 355.

8

Resolutions at Large Scale: Case Studies

WEIGUO LIU
NSC Technologies, Mount Prospect, Illinois

8.1. INTRODUCTION

Resolution and chirality are like twins born on the day Louis Pasteur separated crystals of salts of D- and L-tartaric acid under his microscope. Since then, the separation of each enantiomer from a racemic mixture has been the primary means to obtain optically actively organic compounds. Only recently, the fast and explosive new developments in asymmetric synthesis involving the use of organometallic catalysts, enzymes, and chiral auxiliaries have begun to challenge the resolution approach. Even so, owing to its simplicity, reliability, and practicality, resolution is so far still the most widely applied method for the production of optically pure fine chemicals and pharmaceuticals [1]. Through many years of evolution, the art of resolution has become a multidisciplinary science that includes diastereomeric, kinetic, dynamic, chromatographic, and enzymatic components. This chapter describes some examples of applications of the resolution approach for the production of chiral fine chemicals on an industrial scale.

The rapid development of chiral technology in recent years has provided process chemists an array of methodologies to choose from when considering the synthesis of optically active organic compounds. The choice may be a resolution, but, generally speaking, the nature of a 50% maximum theoretical yield makes the process less desirable when an alternative, asymmetric synthesis is available. In practice, however, and especially in industry, the major issue, and often the deciding factor behind the selection of the technology, is the practicality of the process for large-scale production. Also, at different stages of drug development there are different needs and priorities. In the early development stages, the quantity of material needed is relatively small and is often required on short notice. In these cases, simple resolution processes often win out owing to their swiftness and ease of scale up. For well-established bulk actives, lower costs become the primary goal of process development, and 100% theoretical yield reactions are almost a necessity. In these situations, resolution is less attractive, but can still be the method of choice if the unwanted enantiomer can be recycled in the process, or if the starting material is cheap and the resolution is performed very early in the synthesis, or both. The method of choice to synthesize optically active fine chemicals or pharmaceuticals is entirely dependant on the specific case of that project and should therefore be discussed on a case-by-case basis.

8.2.   CHEMICAL RESOLUTION

Classical resolution generally refers to the processes involving preferential crystallization of one enantiomer from a racemic mixture or the treatment of a racemate with a chiral reagent followed by crystallization of one of the resultant diastereoisomers. A very large number of commercially successful pharmaceuticals or their intermediates are primarily manufactured by these processes [2]. On the basis of their melting point diagram, approximately 5–10% of racemates are referred to as conglomerate or racemic mixtures that consist of a mechanical mixture of crystals of the two enantiomers in equal amounts—each individual crystal contains only one enantiomer (Figure 1). The remaining 90% of racemates are called racemic compounds and consist of crystals of an ordered array of R- and S-enantiomers, where each individual crystal contains equal amounts of both enantiomers. Racemic mixtures of conglomerates may be separated by direct crystallization, which usually involves seeding of supersaturated solution of the racemate with the desired enantiomer. The method of resolution by direct crystal-

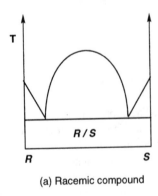

(a) Racemic compound

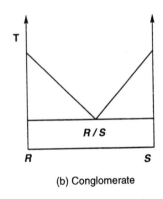

(b) Conglomerate

FIGURE 1

TABLE 1   Examples of Pharmaceuticals Resolved by
Classic Resolution

Pharmaceutical	Resolving agent
Ampicillin	D-Camphorsulfonic acid
Ethambutol	L-(+)-Tartaric acid
Chloramphenicol	D-Camphorsulfonic acid
Dextropropoxyphene	D-Camphorsulfonic acid
Dexbrompheniramine	D-Phenylsuccinic acid
Fosfomycin	r-(+)-Phenethylamine
Thiamphenicol	D-(−)-Tartaric acid
Naproxen	Cinchonidine
Diltiazem	r-(+)-Phenethylamine

Source: Data from Ref. 2.

lization has been used for industrial-scale manufacturing of several major pharmaceuticals such as α-methyl-L-dopa [3] and chloramphenicol [4]. The Merck process of direct crystallization for α-methyl-L-dopa production has been performed on the scale of several hundred tons per year [5].

An overwhelming majority of classic resolutions still involves the formation of diastereomeric salts of the racemate with a chiral acid or base (Table 1). These chiral resolving agents are relatively inexpensive and readily available in large quantities (Table 2). They also tend to form salts with good crystalline properties [6].

Chiral acids or bases are tested for salt formation with the target racemate in a variety of solvents. The selected diastereomeric salt must crystallize well,

TABLE 2   Commonly Used Resolving Agents

Chiral bases	Chiral acids
α-Methylbenzylamine	1-Camphor-10-sulfonic acid
α-Methyl-p-nitrobenzylamine	Malic acid
α-Methyl-p-bromobenzylamine	Mandelic acid
2-Aminobutane	α-Methozyphenylacetic acid
N-Methylglucamine	α-Methoxy-α-trifluoromethylphenylacetic acid
Cinchonine	2-Pyrrolidone-5-carboxylic acid
Cinchonidine	Tartaric acid
Ephedrine	2-Ketogulonic acid
Quinine	
Brucine	

Source: Data from Ref. 2.

and there must be an appreciable difference in solubility between the two diastereoisomers in an appropriate solvent. This type of selection process is still a matter of trial and error, although knowledge about the target racemate and the available chiral resolving agents, and experience with the art of crystallization, does provide significant help [7].

8.2.1. Naproxen

Naproxen was introduced to the market by Syntex in 1976 as a nonsteroidal antiinflammatory drug in an optically pure form. The original manufacturing process (Scheme 1) before product launch started from β-naphthol (1) which was brominated in methylene chloride to produce 1,6-dibromonaphthol (2). The labile bromine at the 1-position was removed with bisulfite to give 2-bromo-6-hydroxynaphthalene that was then methylated with methyl chloride in water-isopropanol to obtain 2-bromo-6-methoxynaphthalene (3) in 85–90% yield from β-naphthol. The bromo compound was treated with magnesium followed by zinc chloride. The resultant naphthylzinc was coupled with ethyl bromopropionate to give naproxen ethyl ester that was hydrolyzed to afford the racemic acid 4. The final optically active naproxen (5) was obtained by a classic resolution process. The racemic acid 4 was treated with cinchonidine to form diastereomeric salts. The S-naproxen-cinchonidine salt was crystallized and then released with acid to give S-naproxen (5) in 95% of the theoretical yield (48% chemical yield) [8,9].

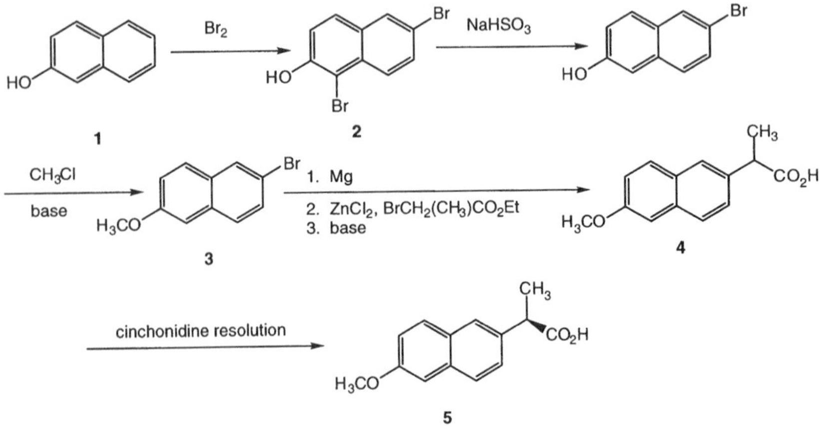

SCHEME 1.

Shortly after the product launch, the original process was modified (Scheme 2). Zinc chloride was removed from the coupling reaction, and ethyl bromopropionate was replaced with bromopropionic acid magnesium chloride salt. Most importantly, the resolution agent cinchonidine was replaced with N-alkylglucamine, which can be readily obtained from the reductive amination of D-glucose.

This new resolution process was equally efficient with greater than 95% theoretical yield, in addition to being much more cost effective due to the low cost of the resolving agent [10,11].

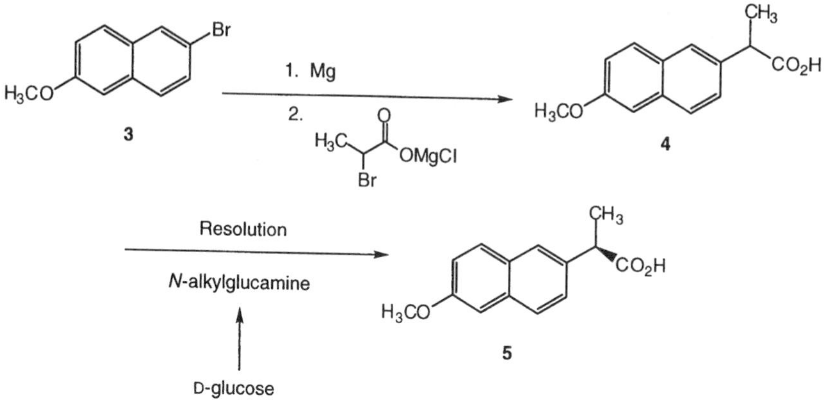

SCHEME 2.

Twenty years have passed since naproxen was introduced. During this period, it has become a billion dollar drug, lost its patent protection, and seen numerous alternative routes and improvements to its synthesis. Significant progress in the field of chiral technology created a long list of methods for generation of optically pure organic compounds. There are several asymmetric synthetic processes specifically designed for production of S-naproxen [12], including the well-known Zambon process (see Chapter 14) [13] and asymmetric hydrogenations (Chapter 9) [14–17].

Other approaches that have been suggested include catalytic asymmetric hydroformylation of 2-methoxy-6-vinylnaphthalene (6) using a rhodium catalyst on BINAPHOS ligand followed by oxidation of the resultant aldehyde 7 to yield S-naproxen (Scheme 3) [18]. However, the tendency of racemization of the aldehyde and the cogeneration of the linear aldehyde isomer make the process less attractive. Other modifications related to this process include catalytic asymmetric hydroesterification [19], hydrocarboxylation [20], and hydrocyanation [21].

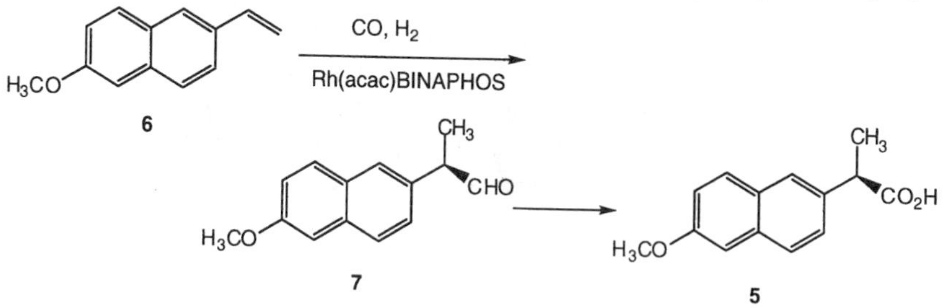

SCHEME 3.

One of the alternative routes of the chiral pool approach for production of S-naproxen was developed at Syntex in the 1980s (Scheme 4) [22,23]. The process starts with inexpensive ethyl L-lactate (8). Conversion of the 2-hydroxyl group to the mesylate followed by hydrolysis of the ethyl ester and conversion of the acid to acid chloride yields the chiral acylating agent 9 for the Grignard reagent derived from 2-bromo-6-methoxynapthalene to provide the optically pure ketone 10. Protection of the ketone with dimethyl-1,3-propanediol followed by rearrangement at 115–120°C gives the naproxen ester 11. Hydrolysis of this ester yields S-naproxen (5) with a 75% overall yield from 2-bromo-6-methoxynaphthalene.

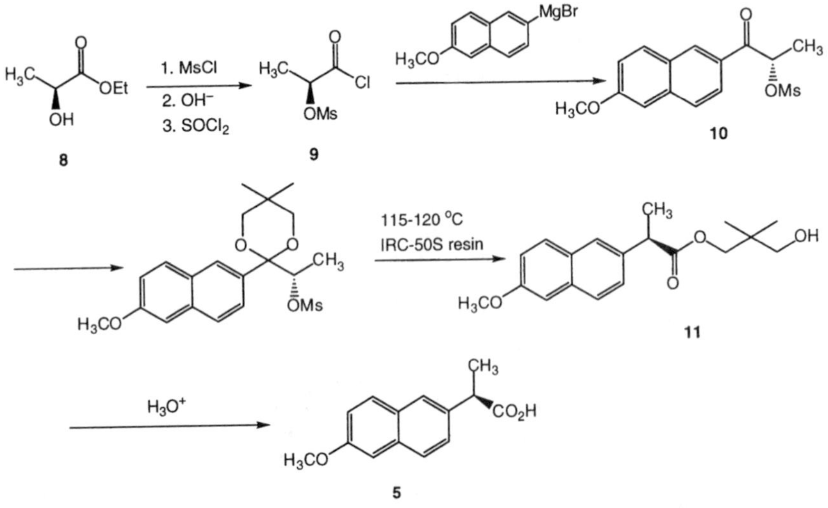

SCHEME 4.

Several biocatalytic processes for the production of (S)-(+)-naproxen (5) have also been developed (see Chapter 13). Direct isomerization of racemic naproxen (4) by a microorganism catalyst, *Exophialia wilhansil*, was reported to give the (S)-isomer 5 of (92%, 100% enantiomeric excess (ee%)) (Scheme 5) [24]. A one-step synthesis of (S)-(+)-naproxen (5) by microbial oxidation of 6-methoxy-2-isopropylnaphthalene (12) was developed by IBIS (Scheme 6) [25]. In both cases, typical bioprocess-related issues such as productivity, product isolation, and biocatalyst production have apparently prevented them from rapid commercialization.

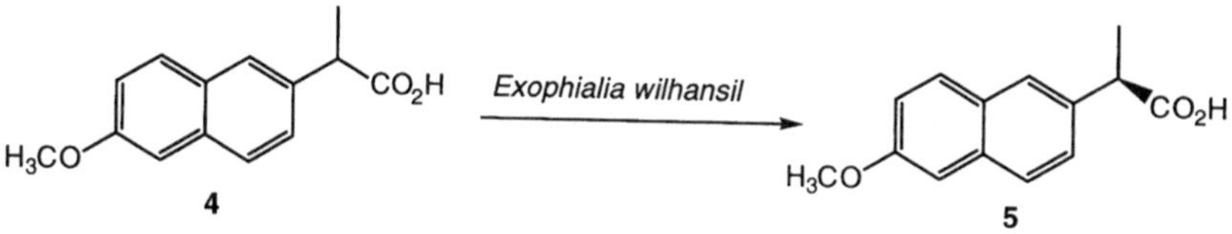

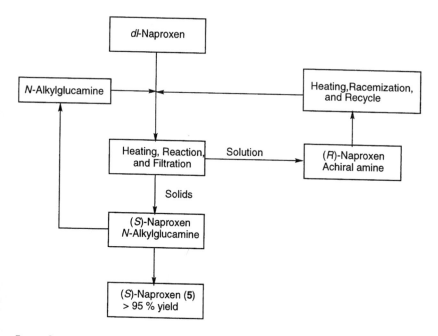

SCHEME 5.

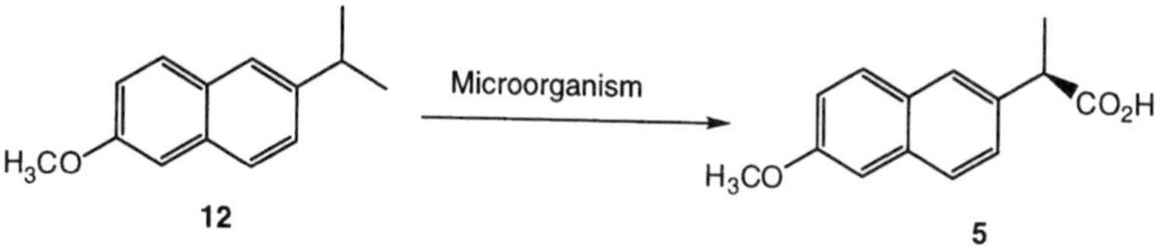

SCHEME 6.

During the same period, continuous process improvements on the resolution route to S-naproxen have achieved dramatic cost reductions. The breakthrough comes from in-process racemization and recycle of the R-naproxen by-product and the recovery of the resolving agent (Figure 2). In this novel and efficient resolution process, the racemic naproxen is reacted with half an equiva-

FIGURE 2

lent of the chiral N-alkylglucamine and another half an equivalent of an achiral amine [9]. Theoretically, there should be an equilibrium of four different salts in this mixture: the S-acid chiral amine salt, S-acid achiral amine salt, R-acid chiral amine salt, and R-acid achiral amine salt. However, only the S-acid chiral amine salt is insoluble in the system and crystallizes out. This process drives the equilibrium toward complete formation of S-naproxen N-alkylglucamine that is collected by filtration; acidification liberates S-naproxen (**5**). The mother liquor, which contains the unwanted acid and achiral amine, is heated. The amine base catalyzes the racemization of the R-acid; the resultant salt of the racemic naproxen and the achiral amine is then recycled to the resolution loop in a continuous operation. Although each individual cycle of the resolution yields the diastereomeric salt of 45–46%, the overall result from continuous operation produces S-naproxen of 99% ee in greater than 95% yield from the racemic mixture. The recovery of the resolving agent, the N-alkylglucamine, is greater than 98% per cycle. This effective combination of resolution, racemization, and recycle of both unwanted enantiomer and resolving agent dramatically increases the yield, lowers the cost, and cuts the waste stream, resulting in an ideal case for a Pope Peachy resolution [1]. Despite the availability of so many elegant asymmetric and chiral pool synthetic processes, the optimized resolution process is still the most cost-effective route for S-naproxen manufacture. This example demonstrates that classic resolution technology when combined with proper process engineering can be a superior method for manufacturing of chiral industrial chemicals [9].

8.2.2. Ibuprofen

Manufacture of optically pure (S)-$(+)$-ibuprofen (**13**), a nonsteroidal antiinflammatory drug similar to naproxen, is another example of the role of resolution in production of chiral fine chemicals, although from a somewhat different angle. Unlike naproxen, ibuprofen (**14**) was introduced to the market as a racemate almost 30 years ago [26,27]. At the time of the introduction, it was thought that both R- and S-isomers of ibuprofen had the same in vivo activity [28]. It has been demonstrated that the R-isomer is converted to the S-isomer in vivo [29] by a unique enzyme system called invertase [30]. Based on these data, ibuprofen has since been marketed as a racemate and has achieved sales of greater than a billion dollars per year. Only recently have the results from certain animal and human studies begun to reveal the possible superior performance of pure (S)-isomer (**13**) in comparison with the racemate **14** [31]. These findings have prompted several companies to invest in the research and development of **13** as a "chiral switch" to replace the racemate. As a result, a large number of synthetic processes have been developed for the production of (S)-$(+)$-ibuprofen (**13**). Many of these processes are the variations of those discussed in the synthesis of (S)-$(+)$-naproxen (**5**). For example, the electrocarboxylation and catalytic asymmetric hydrogenation process developed at Monsanto can be applied to the synthesis of (S)-(+)-ibuprofen [16,17,32].

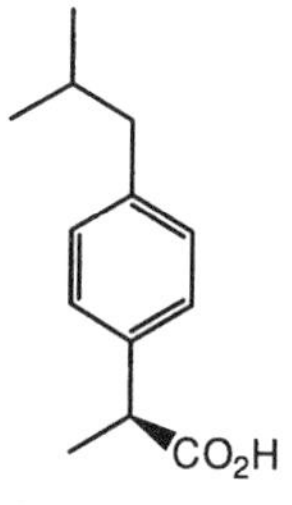

13 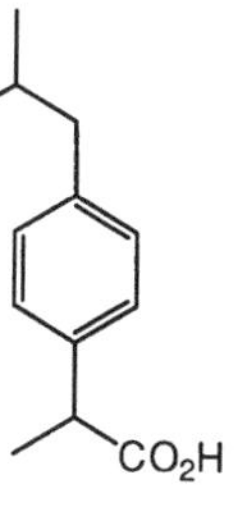   **14**

Other asymmetric synthetic processes used for the manufacturing of (S)-(+)-naproxen can also be applied to the production of (S)-(+)-ibuprofen; these include the Rh-phosphite catalyzed hydroformylation [33], hydrocyanation [21], and hydrocarboxylation reactions [20].

Since ibuprofen has been a successful drug on the market for almost 30 years with no patent protection since 1985, there is widespread competition for commercial production of this product throughout the world. As a result, several practical and economical industrial processes for the manufacture of racemic ibuprofen (**14**) have been developed and are in operation on commercial scales [34]. Most of these processes start with isobutylbenzene (**15**) and go through an isobutylstyrene [35–37] or an acetophenone intermediate [38]. The most efficient route is believed to be the Boots-Hoechst-Celanese process, which involves three steps from isobutylbenzene, all catalytic, and is 100% atom efficient (Scheme 7) [39,40].

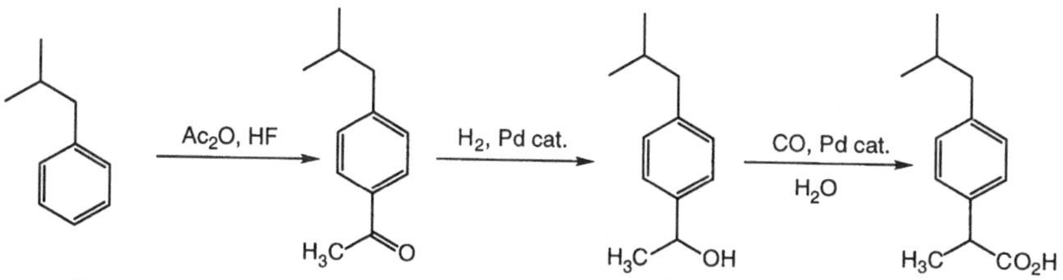

SCHEME 7.

Taking advantage of the ready availability of racemic ibuprofen, the resolution approach for production of (S)-(+)-ibuprofen becomes an attractive alternative. Merck's resolution process involves the formation of a diastereomeric salt of ibuprofen with (S)-lysine, an inexpensive and readily available natural amino acid [41]. The racemic ibuprofen is mixed with 1.0 equivalent of (S)-lysine in aqueous ethanol. The slurry is agitated to allow full dissolution. The supernatant, which is a supersaturated solution of ibuprofen–lysine salt, is separated from the solid and seeded with (S)-ibuprofen-(S)-lysine to induce crystallization. The precipitated solid is collected by filtration, and the mother liquor is recycled to the slurry of racemic ibuprofen and (S)-lysine. This process is continued until essentially all (S)-ibuprofen in the original slurry is recovered, resulting in the formation of greater than 99% ee (S)-ibuprofen-(S)-lysine. Subsequent acidification of the salt followed by crystallization gives enantiomerically pure (S)-ibuprofen. The undesired (R)-ibuprofen isomer can be recycled through racemization. A variety of methods have been reported for the racemization of ibuprofen, such as heating (R)-ibuprofen sodium salt in aqueous sodium hypochloride [42], the formation of anhydrides in the presence of thionyl chloride or acetic anhydride followed by heating and hydrolysis [43], or heating (R)-ibuprofen with a catalytic amount of palladium on carbon in a hydrogen atmosphere [44].

Another classic resolution process developed by Ethyl Corp. for (S)-ibuprofen production uses (S)-(−)-α-methylbenzylamine (MAB) as the chiral base for diastereomeric salt formation [45]. The difference in solubility between (S)-and (R)-ibuprofen MAB salts is so substantial that only half an equivalent of MAB is used for each mole of racemic ibuprofen and no seeding is needed. The process can also be performed in a wide range of solvents, and the unwanted (R)-ibuprofen can be recycled conveniently by heating the mother liquor in sodium hydroxide or hydrochloric acid. Other designer amines have been developed for resolution of ibuprofen with good stereoselectivities [46], but these chiral amines were prepared specifically for ibuprofen resolution and are thus unlikely to be economical for industrial production.

It is worth mentioning that ibuprofen is a racemic compound. Its eutectic composition determined from the binary melting point phase diagram (Figure 1) requires that the enantiomeric purity has to be 90% and higher to achieve optical purification by crystallization [47]. Even above this composition limit, crystallization is still rather inefficient [48]. However, the sodium salt of ibuprofen exhibits a much more favorable phase diagram in which the eutectic composition is approximately 60%. This means that ibuprofen of at least 60% optical purity can be upgraded through the sodium salt. A one-step crystallization of 76% (S)-ibuprofen after treatment with sodium hydroxide in acetone gave 100% ee (S)-ibuprofen [49].

Preparation of (S)-ibuprofen by enzyme-catalyzed enantioselective hydrolysis of racemic ibuprofen esters has been investigated by several companies such as Sepracor [50,51], Rhône-Poulenc [52,53], Gist-Brocades [54,55], and the Wisconsin Alumni Research Foundation [56]. These processes are usually performed under mild reaction conditions, yield highly optically pure product, and can be readily scaled up for industrial production. The disadvantage of these processes for (S)-ibuprofen production is the extra step needed to produce the corresponding ester of racemic ibuprofen, as well as the cost of producing the enzyme and microorganism catalysts.

Many of the available processes for the manufacturing of (S)-ibuprofen are highly efficient and cost competitive as far as chemistry is concerned. For example, the RuBINAP catalyzed asymmetric hydrogenation process is a superior route to (S)-ibuprofen than a resolution approach. However, in the broader economic environment, other non–process-related factors come into play. The risks associated with the unknowns of a new asymmetric catalysis technology and the capital investment for high-pressure hydrogenation vessels make potential producers reluctant to implement the new methodology. On the other hand, the existing, well-established processes and production facilities for racemic ibuprofen allow a step of classic resolution to be added with minimal front-end investment, thereby giving the racemic producers a competitive edge for production of the optically active form. Even the new players can easily purchase the bulk racemate and perform a simple resolution to obtain the optically pure form for packaging and sales. It is widely anticipated that classic resolution will be the ultimate method of choice for commercial manufacturing of (S)-(+)-ibuprofen if it becomes a successful product in the market place.

8.2.3. D-Proline

D-Proline (16) is an unnatural amino acid and an important chiral synthon for the synthesis of a variety of biologically active compounds. There are few chemical methods for asymmetric synthesis of this compound. Almost all processes for the production of D-proline at scale are based on resolution of dl-proline, and most of them involve the racemization of L-proline (17).

It is well known that catalytic amounts of aldehyde can induce racemization of α-amino acids through the reversible formation of Schiff bases [57]. Combination of this technology with a classic resolution leads to an elegant asymmetric transformation of L-proline to D-proline (Scheme 8) [58,59]. When L-proline is heated with one equivalent of D-tartaric acid and a catalytic amount of n-butyraldehyde in butyric acid, it first racemizes due to the reversible formation of the proline–butyraldehyde Schiff base. The newly generated D-proline forms an insoluble salt with D-tartaric acid and precipitates out of the solution, while the soluble L-proline is continuously being racemized. The net effect is the continuous transformation of the soluble L-proline to the insoluble D-proline-D-tartaric acid complex, resulting in near complete conversion. Treatment of the D-proline-D-tartaric acid complex with concentrated ammonia in methanol liberates the D-proline (16) (99% ee, with 80–90% overall yield from L-proline). This is a typical example of a dynamic resolution in which L-proline is completely converted to D-proline with simultaneous in situ racemization. As far as the process is concerned, this is an ideal case because no extra step is required for recycle and racemization of the undesired enantiomer, and a 100% chemical yield is achievable. The only drawback of this process is the use of stoichiometric amount of

D-tartaric acid, which is the unnatural form of tartaric acid and is relatively expensive. Fortunately, more than 90% of the D-tartaric acid is recovered at the end of the process as the diammonium salt that can be recycled after conversion to the free acid [60].

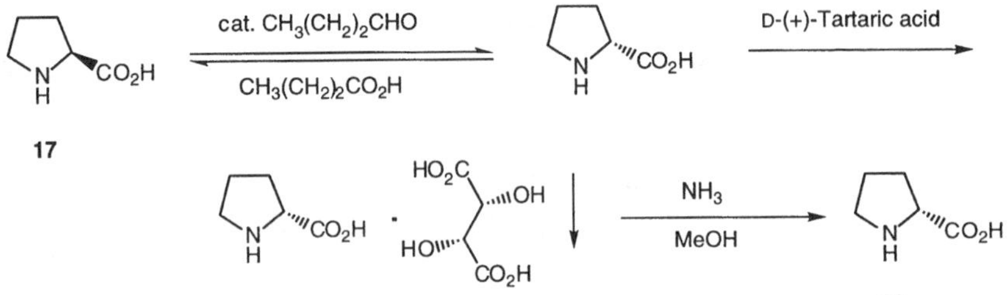

SCHEME 8.

8.2.4. Hydrolytic Kinetic Resolution of Epoxides

Chiral epoxides are extremely useful intermediates and building blocks for synthesis of a variety of optically active organic compounds, including pharmaceuticals and industrial fine chemicals. There have been several methods developed in recent years for asymmetric synthesis of chiral epoxides [61–63]. However, like other reactions, these chiral epoxidations are restricted to certain types of substrates. Optically active terminal epoxides are not effectively prepared by currently available methodologies [64], but are one of the most important and widely sought-after group of chiral compounds. In 1996, Jacobsen discovered a (salen)-Co(III) catalyzed enantioselective opening of racemic terminal epoxides by water [65]. This chemistry has since been applied successfully to the hydrolytic kinetic resolution (HKR) of terminal epoxides such as propylene oxide. The neat oil of racemic propylene oxide (1.0 mol) containing 0.2 mol % of (salen)Co(III)(OAc) complex (**18**) was treated with 0.55 equivalent of water at room temperature for 12 hours to afford a mixture of unreacted epoxide and propylene glycol (Scheme 9; R = Me). This mixture was then separated by fractional distillation to provide both compounds in high chemical and enantiomeric purity (> 98% ee) with nearly quantitative yield. The critical element in this resolution process is the chiral catalyst (**18**). At the end of the process after removing products, the catalyst was found reduced to the Co(II) complex (**19**), which can be recycled back to the active catalyst by treatment with acetic acid in air with no observable loss in activity or stereoselectivity (Scheme 10). This highly efficient hydrolytic kinetic resolution procedure has been applied to the preparation of a series of chiral terminal epoxides that have previously not been readily accessible in optically pure form (Table 3).

TABLE 3 Hydrolytic Kinetic Resolution of Terminal Epoxides with Water[a]

Entry	20, R	18 (mol %)	Water equivalent	Time (hr)	21 (%)	21 (ee %)	22 (%)	22 (ee %)	Relative rate (k_{rel})
1	CH_3	0.2	0.55	12	44	>98	50	98	>400
2	CH_2Cl	0.3	0.55	8	44	98	38	86	50
3	$(CH_2)_3CH_3$	0.42	0.55	5	46	98	48	98	290
4	$(CH_2)_5CH_3$	0.42	0.55	6	45	99	47	97	260
5	Ph	0.8	0.70	44	38	98	39*	98*	20
6	$CH=CH_2$	0.61	0.50	20	44	84	49	94	30
7	$CH=CH_2$	0.85	0.70	68	29	99	64	88	30

[a] See also Scheme 9.
* After recrystallization.

SCHEME 9.

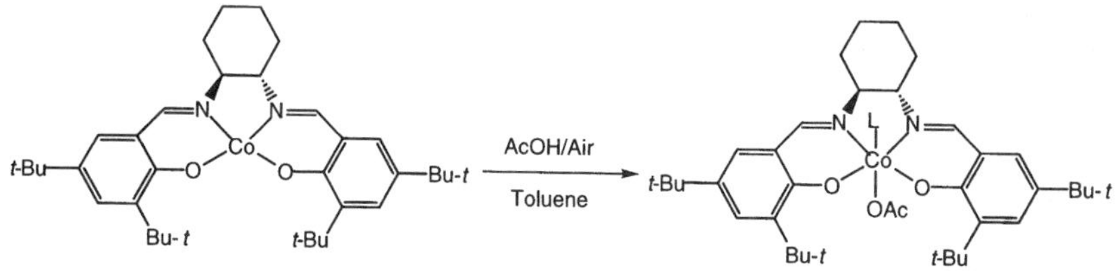

SCHEME 10.

Since most of the racemic epoxide substrates are readily available in bulk and are fairly inexpensive [66], the resolution can be an economically variable process for large-scale production of some of these optically active epoxides. The optically active diols produced from the hydrolytic kinetic resolution are also useful chiral building blocks in a variety of applications. In addition to the recyclable catalyst, water is the only reagent used in the reaction, and no solvent was added. There is virtually no waste stream from the process and the operation is simple. Chirex, Inc. (Dudley, UK) has licensed this technology and started producing some of the chiral epoxides and diols at scale [67]. The catalyst (salen)-Co(III)(OAc) has also been manufactured in multikilogram quantities. This simple, environmentally friendly, and highly efficient resolution process has a potential to offer a wide range of cost-effective chiral fine chemicals for existing and new products.

8.3.   ENZYMATIC RESOLUTIONS

The use of enzymes and microorganisms in organic synthesis, especially in the production of chiral organic compounds, has grown significantly in recent years and has been accepted as an effective and practical alternative for certain synthetic organic transformations. There have been plenty of excellent review articles, including Chapter 13 of this book, describing recent advances and various aspects of applications in this field. In general, biotransformations in organic synthesis has advanced from the stage of exploratory laboratory research to industrial-scale production and applications. Certain enzymes such as lipases and esterases are now considered as standard and routine reagents in organic synthesis. These enzymes are routinely used by scientists in both academic and industrial laboratories for resolution of racemic alcohols, esters, amides, amines, and acids through enantioselective hydrolysis, esterification, or acylation. Compared with classic diastereomeric resolutions, enzymatic resolutions are catalytic, highly selective, and environmentally benign. Their mild reaction conditions allow complex target compounds with multifunctional groups to be resolved effectively. However, in large-scale productions, process economics is often the overriding factor for the determination of the practicality of a technology. For an enzymatic resolution to be performed effectively on a large scale, issues such as solvent selection, volume through-put, product isolation, enzyme recycle, and the recycle of unwanted enantiomer have to be dealt with to achieve the acceptable economy. Very often, these evaluations consist of comparison with the results obtained using other alternative technologies such as classic resolution and asymmetric synthesis. Traditionally, research and development work on biocatalysis and biotransformations is conducted in a bioprocess laboratory since the isolation of new enzymes and handling of microorganisms and fermentation requires special sets of skills. With recent rapid advances in this field, a very large number of industrial enzymes have become commercially available, and are affordable in large quantities. This change has prompted many companies to set up biocatalysis groups within the chemical process development laboratories so that all available technologies relating to the preparation of chiral organic compounds are directly accessible for a specific development project and are coordinated based on the needs of the project.

8.3.1.   α,α-Disubstituted α-Amino Acids

α,α-Disubstituted α-amino acids have attracted increased attention in recent years. This group of nonproteinogenic amino acids induces dramatic conformational change when incorporated into peptides [68–71], and renders them more resistant to protease hydrolysis. They are found in natural peptide antibiotics [72–75], and some are potent inhibitors of amino acid decarboxylases [76–79]. Chiral α,α-disubstituted α-amino acids are also important building blocks for the synthesis of pharmaceuticals and other biological agents [80–84]. For the synthesis of an anticholesterol drug under development at the then Burroughs Wellcome Co. as 2164U90 (**22**), optically pure (**R**)-2-amino-2-ethylhexanoic acid (**23**) was needed as a critical building block (Scheme 11), and a campaign was launched for the development of a cost-effective process for the large-scale preparation of this chiral disubstituted amino acid and its analogs.

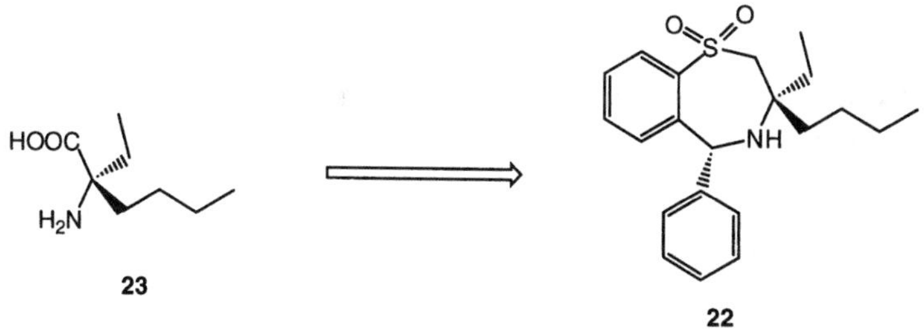

23

22

SCHEME 11.

A variety of methods exists for the synthesis of optically active amino acids, including asymmetric synthesis [85–93] and classic and enzymatic resolutions [94–97]. However, most of these methods are not applicable to the preparation of α,α-disubstituted amino acids due to poor stereoselectivity and lower activity at the α-carbon. Attempts to resolve the racemic 2-amino-2-ethylhexanoic acid and its ester through classic resolution failed. Several approaches for the asymmetric synthesis of the amino acid were evaluated, including alkylation of 2-aminobutyric acid using a camphor-based chiral auxiliary and chiral phase-transfer catalyst. A process based on Schöllkopf's asymmetric synthesis was developed (Scheme 12) [98]. Formation of piperazinone **24** through dimerization of methyl (S)-(+)-2-aminobutyrate (**25**) was followed by enolization and methylation to give (3S,6S)-2,5-dimethoxy-3,6-diethyl-3,6-dihydropyrazine (**26**) (Scheme 12). This dihydropyrazine intermediate is unstable in air and can be oxidized by oxygen to pyrazine **27**, which has been isolated as a major impurity.

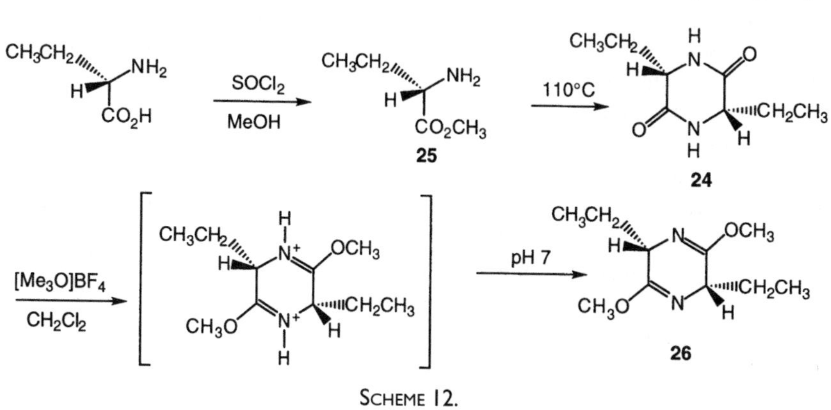

SCHEME 12.

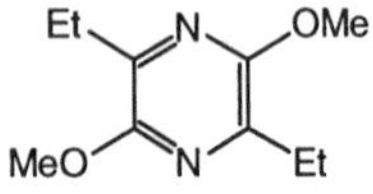

27

Operating carefully under nitrogen, compound **26** was used as a template for diastereoselective alkylation with a series of alkyl bromides to produce derivatives **28**, which were then hydrolyzed by strong acid to afford the dialkylated (*R*)-amino acids **29** (~90% ee), together with partially racemized 2-aminobutyric acid, which was separated from the product by ion-exchange chromatography (Scheme 13). This process provided gram quantities of the desired amino acids as analytical markers and test samples, but was not practical or economical for large-scale production.

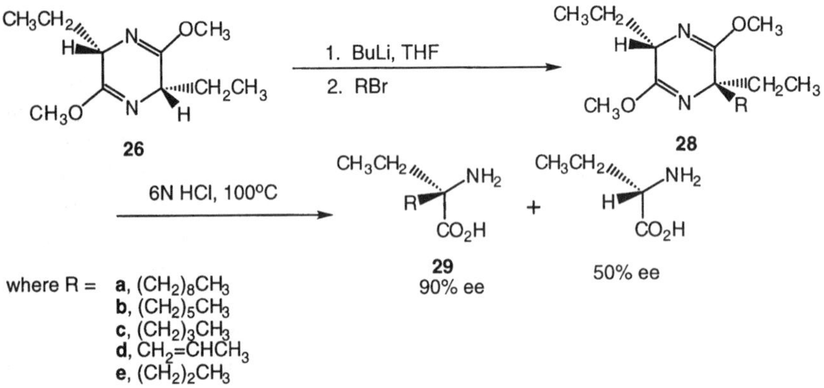

SCHEME 13.

Facing difficulties with chemical methods, development work was aimed at finding a biocatalytic process to resolve the amino acids. Typical commercial enzymes reported for resolution of amino acids were tested. Whole cell systems containing hydantoinase were found to produce only α-monosubstituted amino acids [99–105], the acylase catalyzed resolution of *N*-acyl amino acids has extremely low rates (often zero) of catalysis toward α-dialkylated amino acids [106,107], and the nitrilase system obtained from Novo Nordisk showed no activity toward the corresponding 2-amino-2-ethylhexanoic amide [108,109]. Finally, a large-scale screening of hydrolytic enzymes for enantioselective hydrolysis of racemic amino esters was conducted. Racemic α,α-disubstituted α-amino esters were synthesized by standard chemistry through alkylation of the Schiff's base of the corresponding natural amino esters (Scheme 14) [110], or through formation of hydantoins [111].

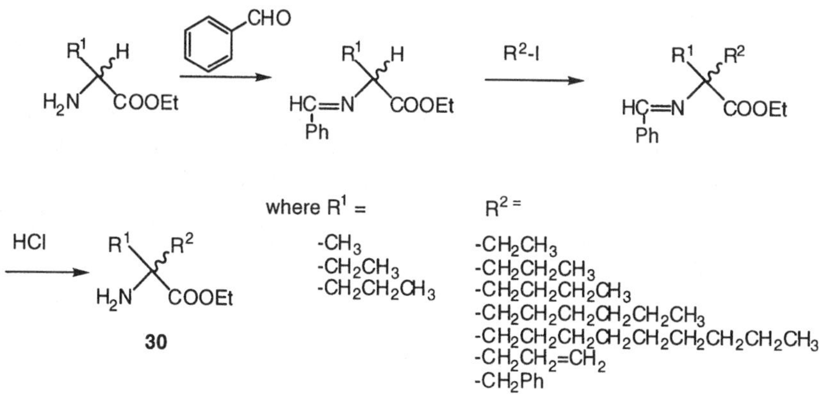

where R^1 =

-CH$_3$
-CH$_2$CH$_3$
-CH$_2$CH$_2$CH$_3$

R^2 =

-CH$_2$CH$_3$
-CH$_2$CH$_2$CH$_3$
-CH$_2$CH$_2$CH$_2$CH$_3$
-CH$_2$CH$_2$CH$_2$CH$_2$CH$_2$CH$_3$
-CH$_2$CH$_2$CH$_2$CH$_2$CH$_2$CH$_2$CH$_2$CH$_2$CH$_3$
-CH$_2$CH$_2$=CH$_2$
-CH$_2$Ph

SCHEME 14.

Initial enzyme screening was aimed at obtaining optically active 2-amino-2-ethylhexanoic acid (**29c**) or the corresponding amino alcohol. Enzymes reported for resolving α-H or α-Me analogs of amino acids failed to catalyze the corresponding reaction of this substrate [112], primarily due to the presence of the α-ethyl group that causes a critical increase in steric hindrance at the α-carbon. Of 50 different enzymes and microorganisms screened, pig liver esterase and *Humicola langinosa* lipase (Lipase CE, Amano) were the only ones found to catalyze the hydrolysis of the substrate.

Both enzymes catalyze the hydrolysis of the amino ester (**30**) enantioselectively (Scheme 15). At about 60% substrate conversion, the enantiomeric excess of recovered ester (**32**) from both reactions exceeds 98%. In addition, the acid product (**31**) (96–98% ee) was obtained by carrying the hydrolysis of the ester to 40%. The rates of hydrolysis become significantly slower when conversion approaches 50%, allowing a wide window for kinetic control of the resolution process. Both enzymes function well in a concentrated water/substrate (oil) two-phase system while maintaining high enantioselectivity, making this system very attractive for industrial processes.

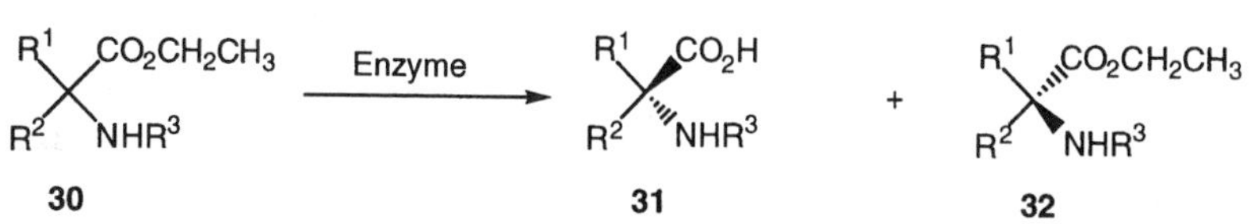

SCHEME 15.

Although pig liver esterase (PLE) catalyzes the hydrolysis of all amino esters tested in this work, it was only enantioselective toward esters (**30**, R^1 = ArCH$_2$, or *n*-Bu, R^2 = Et). This lack of correlation between enantioselectivity and substrate structure has been reported for many PLE-catalyzed reactions [113,114]. The lipase CE was obtained as a crude extract, containing approximately 10% of total protein. The active enzyme catalyzing the amino ester hydrolysis was isolated from the mixture and partially purified. This new enzyme, *Humicola* amino esterase, is present as a minor protein component, has a molecular weight of approximately 35,000 daltons, and does not show esterase activity toward *o*-nitrophenol butyrate or lipase activity to olive oil. However, it is highly effective in catalyzing the amino ester hydrolysis with very broad substrate specificity and high enantioselectivity. Various substrates, including aliphatic, aromatic, and cyclic amino esters, were resolved into optically active esters and acids (Table 4) with good E values [115]. Aliphatic amino esters with alkyl or alkenyl side chains as long as 10 carbon atoms were well accepted by the enzyme. Substrates with longer chain length's were limited by the solubility of the compounds rather than their binding with the enzyme. Resolutions were also extended

TABLE 4 Enantioselective Hydrolysis of α-Amino Esters Catalyzed by *Humicola* Amino Esterase

30, R^1	R^2	R^3	% Conversion	31 (%ee)	32 (%ee)	E value
$CH_3(CH_2)_8^-$	Et	H	50	91 (R)[d]	—	58
$CH_3(CH_2)_5^-$	Et	H	50	90 (R)[d]	99 (S)	58
$CH_3(CH_2)_5^-$	$CH_3(CH_2)_2^-$	H	37	92 (R)[c]	60 (S)	41
$ArCH_2^-$	Et	H	50	85 (R)[a]	—	35
$ArCH_2^-$	Me	H	74	32 (R)[a]	88 (S)	4.4
$ArCH_2^-$	H	H	68	—	72 (S)[b]	4.0
Ar−	H	H	62	—	78 (S)[b]	6.3
$CH_3(CH_2)_3^-$	Et	H	46	92 (R)[d]	—	58
			58	—	98 (S)[e]	
$CH_3(CH_2)_3^-$	Et	$-OCOCH_3$	NR			
$CH_3CH=CH-$	Et	H	42	94 (R)[d]	66 (S)	66
$CH_3(CH_2)_2^-$	Et	H	65	26 (R)[d]	53 (S)	2.6
Me	Et	H	72	36 (S)[a]	100 (R)	20
H	$CH_3(CH_2)_3^-$	H	40	72 (S)[b]	42 (R)	9.6
H	Et	H	55	53 (S)[b]	42 (R)	6.1
H	Et	$-OCOCH_3$	NR			

Absolute configurations were assigned by:
[a] comparison of optical rotation with literature data;
[b] comparison of high-performance liquid chromatography (HPLC) retention time with authentic samples;
[c] analogy to the results obtained from d;
[d] comparison of HPLC retention time of its diastereomeric derivative with those obtained through Schollkopf's asymmetric synthesis;
[e] x-ray crystallographic analysis.

to α,α-disubstituted amino esters in which the two alkyl groups differ in length by as little as a single carbon atom. The fact that the enzyme successfully catalyzes the resolution of straight chain aliphatic amino esters with two α-alkyl groups both larger than a methyl group is unique. These amino esters and their acids have been difficult to resolve by chemical and biochemical means because of the increased flexibility of the two large alkyl groups that become indistinguishable to most resolving agents. Unsubstituted amino esters underwent significant chemical hydrolysis under the experimental conditions, resulting in relatively lower E values.

When some amino esters are protected by N-acetylation, they become resistant to hydrolysis by the enzyme. Replacement of the α-amino group of the amino ester with a hydroxyl group also changes it from substrate to nonsubstrate. The enzyme showed high catalytic activity toward hydrolysis of phenylglycine ethyl ester, but was found incapable of catalyzing the hydrolysis of mandelic ethyl ester in which the amino group had been replaced with an α-hydroxyl group. The N-acetyl compounds and mandelic ester were not inhibitors, indicating that the free amino group is necessary for binding between the enzyme and substrates. It is most likely that the substrate is protonated at the amino group. The ammonium cation then binds to an anion on the enzyme's active site, forming a strong ionic bonding to facilitate the catalysis. This mechanism was further supported by the strong pH dependence of the enzymatic activity toward amino ester hydrolysis. Stability and relative activity of the amino esterase were measured over a range of pHs with ethyl 2-amino-2-ethylhexanoate (**30**, R^1 = Et, R^2 = n-Bu) as substrate, and the data were compared as in Figure 3. The enzyme has an optimum pH of 7.5 and is most stable at pH 8. At low pHs, both the stability and activity of the enzyme suffered a gradual loss; this is consistent with the effect of gradual

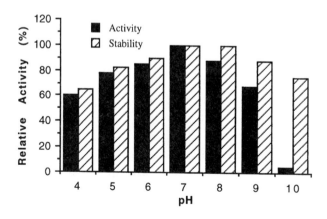

FIGURE 3

protein denaturation. At high pHs, however, the enzyme remained relatively stable, but its activity decreased sharply. The most significant decrease in activity occurs when the medium's pH is greater than 9.6 and coincides with the substrate's pKa. Clearly, when the substrate exists mostly in the form of a free base, its affinity with the enzyme dramatically diminishes at the active site where a charged ammonium cation is required for binding.

Initial scale up of the enzymatic resolution for production of kilogram quantities of (R)-2-amino-2-ethylhexanoic acid was performed in a batch process. The oil of ethyl 2-amino-2-ethylhexanoate was suspended in an equal volume of water containing the enzyme. The enantioselective hydrolysis of the ester proceeded at room temperature with titration of the produced acid by NaOH through a pH stat (Figure 4).

Significantly, the reaction mixture was worked up by removal of the unreacted ester by hexane extraction and concentration of the aqueous layer to obtain the desired (R)-amino acid. The process has a high through-put and was easy to handle on a large scale. However, because of the nature of a batch process, the enzyme catalyst could not be effectively recovered, adding significantly to the cost of the product. In the further scale up to 100-kg quantity productions, the resolution process was performed using Sepracor's membrane bioreactor module. The enzyme was immobilized by entrapment into the interlayer of the hollow fiber membrane. Water and the substrate amino ester as a neat oil or hexane solution were circulated on each side of the membrane. The ester was hydrolyzed enantioselectively by the enzyme at the membrane interface, and the chiral acid product diffused to the aqueous phase to be recovered by simple concentration or ion exchange chromatography (Figure 5). Optical purity of (R)-2-amino-2-ethylhexanoic acid obtained from this process ranged from 98–99% ee [116].

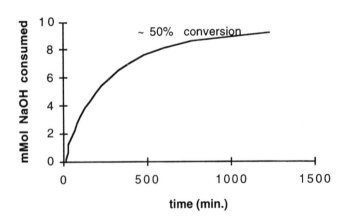

FIGURE 4

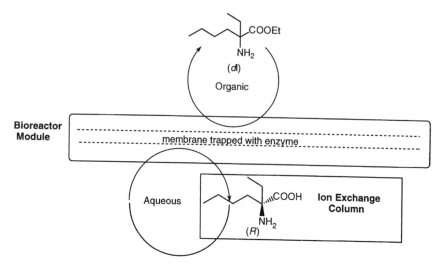

Figure 5

8.3.2. Preparation of D-Amino Acids Using Hydantoinase

D-Amino acids have played an increasingly important role as building blocks for pharmaceuticals and other biologically active agents such as unnatural therapeutic peptides [117], angiotensin-converting enzyme (ACE) inhibitors [118], and semi-synthetic β-lactam antibiotics [119]. D-Phenylglycine and D-p-hydroxyphenyl-glycine are used as raw materials in the production of some well-known antibiotics such as ampicillin, amoxicillin, cephalexin, and cefadroxil. D-Valine is a building block for the synthesis of the pyrethroid insecticide, fluvalinate [120]. Currently, most manufacturers of D-phenylglycine and D-p-hydroxyphenylglycine are still using the classic resolution process with camphorsulfonic acid and bromocamphorsulfonic acid, respectively, as resolving agents [85]. Both compounds are produced in thousands of tons per year. Although industrial productions of most of the D-amino acids are still conducted through classic resolution, enzymatic methods have begun to catch up [121]. At present, the most promising enzymatic processes for production of D-amino acids seem to be those catalyzed by D-specific hydantoinases.

D-Hydantoinases exist at different levels in many microorganism strains, and many of them have been cloned and overexpressed [103,122,123]. This group of enzymes catalyzes enantioselective ring cleavage of 5-substituted hydantoins to give the corresponding D-N-carbamylamino acids that can be further hydrolyzed chemically or enzymatically to the free D-amino acids. The chemical process for removal of the carbamoyl group is often performed under diazotation conditions using sodium nitrite and sulfuric or hydrochloric acid. Enzymatic cleavage of the carbamoyl residue has been reported [124]. Actually, the enzyme, *N*-carbamoyl D-amino acid amidohydrolase can be incorporated with hydantoinase to produce D-amino acids in one step from the corresponding racemic hydantoins [125]. Since many of the 5-substituted hydantoins undergo spontaneous racemization under the enzymatic reaction conditions (pH > 8), no extra steps are needed for racemization of the unwanted enantiomer, and complete conversion from racemic hydantóin to the D-amino acid can be achieved. The starting hydantoins are readily prepared according to the procedure of Bucherer and Bergs (Scheme 16) [126] by simple condensation of potassium cyanide and ammonium carbonate with the corresponding aliphatic or aromatic aldehydes, which are usually also the starting materials for synthesis of the corresponding racemic amino acids. Many of the D-hydantoinases have broad substrate specificities and can be used for production of a variety of D-amino acids [102,127].

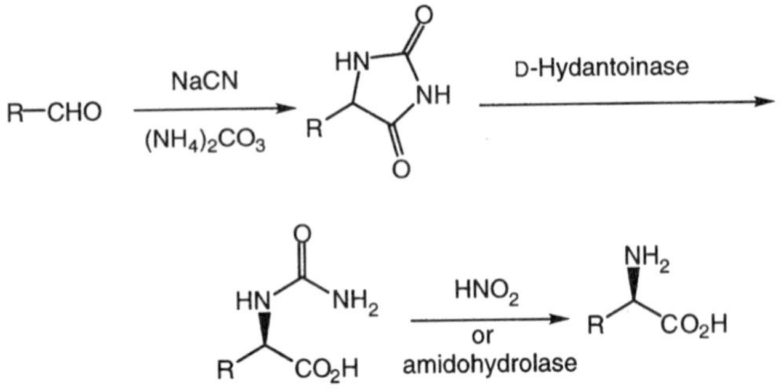

SCHEME 16.

D-Valine is produced from isobutyraldehyde through a Bucherer-Bergs reaction followed by a one-step enzymatic conversion of the hydantoin to the free D-amino acid [128]. The biocatalyst is *Agrobacterium radiobacter* that contains high levels of both D-hydantoinase and *N*-carbamoyl D-amino acid amidohydrolase [125]. A typical run of the biotransformation involves a 10% solution of the hydantoin with a weight ratio of hydantoin versus resting cells (dry weight) of 5. The incubation is usually performed at 40°C for 48 hours with continuous addition of 1 Normal NaOH to maintain a constant pH of 7.5. The final broth is worked up by centrifugation or ultrafiltration followed by simple concentration and precipitation with ethanol or ethanol/pyridine to give the optically pure D-valine, with an overall yield of approximately 60% from the starting aldehyde.

In the case of D-phenylglycine and D-*p*-hydroxphenylglycine manufacture, several producers are believed to have started the hydantoinase route (Scheme 17). Both enzymatic and classic resolution use, as the starting material, benzaldehyde for D-phenylglycine and phenol for D-p-hydroxyphenylglycine. Which one will eventually become the most cost-effective route is very much dependent on the process steps, cost of resolving agent or biocatalyst, and the final yield of product. The hydantoinase process has the advantage of not requiring a racemization step. If the step of chemical hydrolysis of the carbamoyl residue can be completely replaced by the *in situ* enzymatic hydrolysis, the enzymatic route will be much more simple and streamlined than the classic resolution. A few other issues that need to be worked out are increasing the activity and efficiency of the biocatalyst and recovering the product effectively from the aqueous solution. At least in the manufacture of D-p-hydroxyphenylglycine, where the classic resolution using bromocamphorsulfonic acid is not as well established as in the case of phenylglycine, the hydantoinase route has shown a competitive edge. Kanegafuchi uses a bacterial strain, *Bacillus brevis,* as the biocatalyst for enantioselective conversion of the racemic hydantoin to the N-carbamoyl-D-amino acid, which is then diazotized to give D-phenylglycine. Recordati applies a strain of *Agrobacterium radiobacter* that contains both hydantoinase and amidohydrolyase to convert the racemic hydantoin to D-phenylglycine in one step [5]. With continuous strain improvement and genetic engineering, the gap between enzymatic and classic resolution for the production of D-phenylglycine and D-p-hydroxyphenylglycine is getting closer.

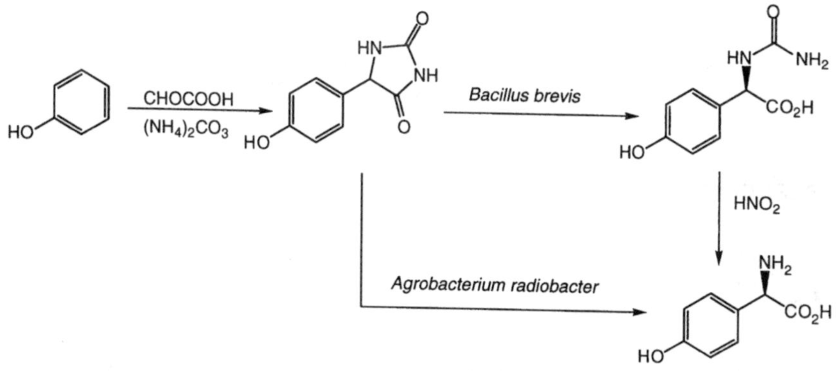

SCHEME 17.

In addition to being used in the whole-cell systems, D-hydantoinases are also commercially available as purified or partially purified enzymes in both free and immobilized forms. A chemist can easily test these enzymes for preparation of a specific target amino acid, and scale up the process to produce large-quantity materials. Boehringer Mannheim offers two types of immobilized D-hydantoinases, D-Hyd1 and D-Hyd2, in both research and bulk quantities. The enzymes are cloned from thermophilic microorganisms and expressed and overproduced in *Escherichia coli*. Both of them have broad substrate specificities, although D-Hyd1 is more active toward aromatic substituted hydantoins, whereas D-Hyd2 shows better activity to substrates with aliphatic substitution [129]. The combination of the two enzymes allows quick synthesis of a variety of D-amino acids in high chemical and optical yields, and relatively large quantities. Some D-amino acids reportedly prepared by these two enzymes include alanine, norleucine, homophenylalanine, phenylalanine, valine, isoleucine, methionine, serine, threonine, 2-thienylglycine, phenylglycine, and *p*-hydroxyphenylglycine. However, unlike the whole cells that could produce both D-hydantoinase and *N*-carbamoyl-amino acid amidohydrolase, these commercial hydantoinases only catalyze the cleavage of hydantoins to the *N*-carbamoyl-amino acid. A separate chemical hydrolysis step is necessary to obtain the free D-amino acids.

8.4. SUMMARY

Despite the revolutionary advances achieved recently in the field of catalytic asymmetric synthesis, resolution methods, both chemical and enzymatic, are still probably the most used methods for preparation of optically pure organic compounds. This is especially true on a large scale for the production of industrial fine chemicals. A very large number of chiral pharmaceuticals and pharmaceutical intermediates are manufactured by the process involving resolution. The reason behind the continued dominance of resolution in industrial production of optically pure fine chemicals is perhaps the reliability and scalability of these processes.

REFERENCES

1. Jacques, J., Collet A., Wilen, S. *S. Enantiomers, Racemates, and Resolutions;* Wiley: New York, 1981.
2. Bayley, C. R., Vaidya, N. A. in *Chirality in Industry;* Collins, A. N., Sheldrake, G. N., Crosby, J., Eds.; Wiley: New York, 1992; p. 71.
3. Reinhold, D. F., Firestone, R. A., Gaines, W. A., Chemerda, J. M., Sletzinger, M. *J. Org. Chem.* 1968, *33,* 1209.
4. Amiard, G. *Experientia* 1959, *15,* 1.
5. Sheldon, R. A. *Chirotechnology: Industrial Synthesis of Optically Acive Compounds;* Marcel Dekker: New York, 1993.
6. Wilen, S. H. *Tables of Resolving Agents;* University of Notre Dame: Notre Dame, IN, 1972.
7. Newman, P. *Optical Resolution Procedures;* Resolution Information Center, Manhattan College: New York, 1981.
8. Harrison, I. T., Lewis, B., Nelson, P., Rooks, W., Roszkowski, A., Tomolonis, A., Fried, J. H. *J. Med. Chem.* 1970, *13,* 203.

9. Harrington, P. J., Lodewijk, E. *Org. Process Res. Dev.* 1997, *1*, 72.
10. Holton, P. G. *US Patent* 1981, 4,246,193.
11. Felder, E., San Vitale, R., Pitre, D., Zutter, H. *US Patent* 1981, 4,246,164.
12. Sonawane, H. R., Bellur, N. S., Ahuja, J. R., Kulkarni, D. G. *Tetrahedron: Asymmetry* 1992, *3,* 163.
13. Giordano, C., Castaldi, G., Cavicchioli, S., Villa, M. *Tetrahedron* 1998, *45,* 4243.
14. Ohta, T., Takaya, H., Kitamura, M., Nagai, K., Noyori, R. *J. Org. Chem.* 1987, *52,* 3174.
15. Chan, A. S. C., Laneman, S. A. *US Patent* 1993, 5,202,473.
16. Chan, A. S. C., Laneman, S. A. *US Patent* 1992, 5,144,050.
17. Chan, A. S. C. *Chemtech.* 1993, *23,* 46.
18. Stille, J. K., Su, H., Brechot, P., Parinello, G., Hegedus, L. S. *Organometallics* 1991, *10,* 1183.
19. Hiyama, T., Wakasa, N., Kusumoto, T. *Synlett* 1991, 569.
20. Alper, H., Hamel, N. *J. Am. Chem. Soc.* 1990, *112,* 2803.
21. RajanBabu, T. V., Casalnuovo, A. L. *J. Am. Chem. Soc.* 1992, *114,* 6265.
22. Piccolo, O., Azzena, U., Melloni, G., Delogu, G., Valoti, E. *J. Org. Chem.* 1991, *56,* 183.
23. Piccolo, O., Speafico, F., Visentin, G. *J. Org. Chem.* 1985, *50,* 1946.
24. Reid, A. J., Phillips, G. T., Marx, A. F., Desmet, M. *J. Eur. Patent* 1989, 0 338 645-A.
25. Phillips, G. T., Robertson, B. W., Watts, P. D., Matcham, G. W. J., Bertola, M. A., Marx, A. F., Koger, H. S. *Eur. Pat. Appl.* 1986, 205 215.
26. Nicholson, J. S., Adams, S. S. *UK Patent* 1964, 971,700.
27. Nicholson, J. S. Adams, S. S. *US Patent* 1966, 3,228,831.
28. Adams, S. S., Bresloff, P., Mason, G. *Pharm. Pharmacol.* 1967, *28,* 256.
29. Adams, S. S., Cliffe, E. E., Lessel, B., Nicholson, J. S. *J. Pharm. Sci.* 1967, *56,* 1686.
30. Wechter, W. J., Loughhead, D. G., Reischer, R. J., Van Giessen, G. J., Kaiser, D. G. *Biochem. Biophys. Res. Commun.* 1974, *61,* 833.
31. Sunshine, A., Laska, E. M. *US Patent* 1989, 4,851,444.
32. Chan, A. S. C. *US Patent* 1991, 4,994,607.
33. Fowler, R., Connor, H., Bachl, R. A. *Chemtech* 1976, 722.
34. Rieu, J.-P., Boucherle, A., Cousse, H., Mouzin, G. *Tetrahedron* 1986, *42,* 4095.
35. Shimizu, I., Matsumura, Y., Tokumoto, Y., Uchida, K. *US Patent* 1987, 4,694,100.
36. Shimizu, I., Matsumaura, Y., Tokumoto, Y., Uchida, K. *US Patent* 1992, 5,097,061.
37. Tokumoto, Y., Shimizu, I., Inooue, S. *US Patent* 1992, 5,166,419.
38. Boots Pure Drug Co. *Fr. Patent* 1968, 1,545,270.
39. *Chem. Mkt. Rep.* 1993, 20.
40. *Chem. Mkt. Rep.* 1993, 41.
41. Tung, H.-H., Waterson, S., Reynolds, S. D. *US Patent* 1991, 4,994,604.
42. Trace, T. L. *US Patent* 1994, 5,278,338.
43. Larsen, R. D., Reider, P. *US Patent* 1990, 4,946,997.
44. Lin, R. W. *US Patent* 1994, 5,278,334.
45. Manimaran, T., Impastato, F. J. *US Patent* 1991, 5,015,764.
46. Nohira, H. *US Patent* 1994, 5,321,154.

47. Dwivedi, S. K., Sattari, S., Jamali, F., Mitchell, A. G. *Int. J. Pharm.* 1992, *87*, 95.
48. Manimaran, T., Stahly, P. G. *Tetrahedron: Asymmetry* 1993, *4*, 1949.
49. Manimaram, T., Stahly, P. G. *US Patent* 1993, 5,248,813.
50. Matson, S. L. *US Patent* 1989, 4,800,162.
51. Wald, S. A., Matson, S. L., Zepp, C. M., Dodds, D. R. *US Patent* 1991, 5,057,427.
52. Cobbs, C. S., Barton, M. J., Peng, L., Goswani, A., Malick, A. P., Hamman, J. P., Calton, G. J. *US Patent* 1992, 5,108,916.
53. Goswani, A. *US Patent* 1992, 5,175,100.
54. Bertola, M. A., Marx, A. F., Koger, H. S., Quax, W. J., Van der Laken, C. J., Phillips, G. T., Robertson, B. W., Watts, P. O. *US Patent* 1989, 4,886,750.
55. Bertola, M. A., De Smet, M. J., Marx, A. F., Phillips, G. T. *US Patent* 1992, 5,108,917.
56. Sih, C. J. *Eur. Pat. Appl.* 1986, 227 078.
57. Yamada, S., Hongo, C., Yoshioka, R., Chibata, I. *J. Org. Chem.* 1983, *48*, 843.
58. Shiraiwa, T., Shinjo, K., Kurokawa, H. *Chem. Lett.* 1989, 1413.
59. Shiraiwa, T., Shinjo, K., Kurokawa, H. *Bull. Chem. Soc., Jpn.* 1991, *64*, 3251.
60. Huang, T., Qian, X. *Shipin Yu Fajiao Gongye* 1987, *6*, 98.
61. Kolb, H. C., VanNieuwenhze, M. S., Sharpless, K. B. *Chem. Rev.* 1994, *94*, 2483.
62. Jacobsen, E. N. in *Comprehensive Organometallic Chemistry II;* Wilkinson, G., Stone, F. G. A., Abel, E. W., Hegedus, L. S., Eds.; Pergamon: New York, 1995; *Vol. 12;* Chapter 11.1.
63. Aggarwal, V. K., Ford, J. G., Thompson, A., Jones, R. V. H., Standen, M. *J. Am. Chem. Soc.* 1996, *118*, 7004.
64. Sinigalia, R., Michelin, R. A., Pinna, F., Strukul, G. *Organometallics* 1987, *6*, 728.
65. Tokunaga, M., Larrow, J. F., Kakiuchi, F., Jacobsen, E. N. *Science* 1997, *277*, 936.
66. Sato, K., Aoki, M., Ogawa, M., Hashimoto, T., Noyori, R. *J. Org. Chem.* 1996, *61*, 8310.
67. Pettman, R. B. *Chemspec Europe BACS Symposium* 1997, Manchester, p. 37.
68. Paul, P. K. C., Sukumar, M., Blasio, B. D., Pavone, V., Pedone, C., Balaram, P. *J. Am. Chem. Soc.* 1986, *108*, 6363.
69. Bardi, R., Piazzesi, A. M., Toniolo, C., Sukumar, M., Balaram, P. *Biopolymers* 1986, *25*, 1635.
70. Mutter, M. *Angew. Chem.* 1985, *97*, 639.
71. Barone, V., Lelj, F., Bavoso, A., Blasio, B. D., Grimaldi, P., Pavone, V., Pedone, C. *Biopolymers* 1985, *24*, 1759.
72. Khosla, M. C., Stackiwiak, K., Smeby, R. R., Bumpus, F. M., Piriou, F., Lintner, K., Fermandijan, S. *Proc. Natl. Acad. Sci. USA* 1981, *78*, 757.
73. Turk, J., Panse, G. T., Marshall, G. R. *J. Org. Chem.* 1975, *40*, 953.
74. Deeks, T., Crooks, T. P. A., Waigh, R. D. *J. Med. Chem.* 1983, *26*, 762.
75. Christensen, H. N., Handlogten, M. E., Vadgama, J. V., de la Cuesta, E., Ballesteros, P., Trigo, G. C., Avendano, C. *J. Med. Chem.* 1983, *26*, 1374.
76. Zembower, D. E., Gilbert, J. A., Ames, M. M. *J. Med. Chem.* 1993, *36*, 305.
77. Kollonitsch, J., Patchett, A. A., Marburg, S., Maycock, A. L., Perkins, L. M., Doldouras, G. A., Duggan, D. E., Aster, S. D. *Nature* 1978, *274*, 906.
78. Bhattacharjee, M. K., Snell, E. E. *J. Biol. Chem.* 1990, *265*, 6664.

79. Schirlin, D., Gerhart, F., Hornsperger, J. M., Hamon, M., Wagner, J., Jung, M. J. *J. Med. Chem.* 1988, *31*, 30.
80. Coppola, G. M., Schuster, H. F. *Asymmetric Synthesis: Construction of Chiral Molecules Using Amino Acids;* Wiley: New York, 1987.
81. Evans, D. A. in *Asymmetric Synthesis;* Morrison, J. D., Ed.; Academic Press: Orlando, 1984; *Vol. 3;* p. 1.
82. Mzengeza, S., Yang, C. M., Whitney, R. A. *J. Am. Chem. Soc.* 1987, *109*, 276.
83. Knudson, C. G., Palkowitz, A. D., Rappaport, H. *J. Org. Chem.* 1985, *50*, 325.
84. Lubell, W. D., Rappaport, H. *J. Am. Chem. Soc.* 1987, *109*, 236.
85. Williams, R. M., Im, M.-N. *J. Am. Chem. Soc.* 1991, *113*, 9276.
86. Schollkopf, U. *Pure Appl. Chem.* 1983, *55*, 1799.
87. Schollkopf, U., Busse, U., Lonsky, R., Hinrichs, R. *Liebigs Ann. Chem.* 1986, 2150.
88. Seebach, D., Boes, M., Naef, R., Schweizer, W. B. *J. Am. Chem. Soc.* 1983, *105*, 5390.
89. Seebach, D., Aebi, J. D., Naef, R., Weber, T. *Helv. Chim. Acta* 1985, *65*, 144.
90. O'Donnell, M. J., Bennett, W. D., Bruder, W. A., Jacobson, W. N., Knuth, K., LeClef, B., Polt, R. L., Bordwell, F. G., Mrozack, S. R., Cripe, T. A. *J. Am. Chem. Soc.* 1988, *110*, 8520.
91. Kolb, M., Barth, J. *Angew. Chem.* 1980, *92*, 753.
92. Goerg, G. I., Guan, X., Kant, J. *Tetrahedron Lett.* 1988, *29*, 403.
93. Obrecht, D., Spiegler, C., Schönholzer, P., Müller, K., Heimgartner, H., Stierli, F. *Helv. Chim. Acta* 1992, *75*, 1666.
94. Bosch, R., Brückner, H., Jung, G., Winter, W. *Tetrahedron* 1982, *38*, 3579.
95. Anantharamaiah, G. M., Roeske, R. W. *Tetrahedron Lett.* 1982, *23*, 3335.
96. Berger, A., Smolarsky, M., Kurn, N., Bosshard, H. R. *J. Org. Chem.* 1973, *38*, 457.
97. Kamphuis, J., Boesten, W. H. J., Broxterman, Q. B., Hermes, H. F. M., Meijer, E. M., Schoemaker, H. E. in *Adv. Biochem. Engineering Biotechnol.;* Flechter, A., Ed.; Springer-Verlag: Berlin, 1990; *Vol. 42.*
98. Schollkopf, U., Hartwig, W., Groth, U., Westphalen, K.-O. *Liebigs Ann. Chem.* 1981, 696.
99. Gross, C., Syldatk, C., Mackowiak, V., Wagner, F. J. *Biotechnology* 1990, *14*, 363.
100. Chevalier, P., Roy, D., Morin, A. *Appl. Microbiol. Biotechnol.* 1989, *30*, 482.
101. West, T. P. *Arch. Microbiol.* 1991, *156*, 513.
102. Shimizu, S., Shimada, H., Takahashi, S., Ohashi, T., Tani, Y., Yamada, H. *Agric. Biol. Chem.* 1980, *44*, 2233.
103. Syldatk, C., Wagner, F. *Food Biotechnol.* 1990, *4*, 87.
104. Runser, S., Chinski, N., Ohleyer, E. *Appl. Microbiol. Biotechnol.* 1990, *33*, 382.
105. Nishida, Y., Nakamicho, K., Nabe, K., Tosa, T. *Enzyme Micro. Technol.* 1987, *9*, 721.
106. Chenault, H. K., Dahmer, J., Whitesides, G. M. *J. Am. Chem. Soc.* 1989, *111*, 6354.
107. Keller, J. W., Hamilton, B. J. *Tetrahedron Lett.* 1986, *27*, 1249.
108. Kruizinga, W. H., Bolster, J., Kellogg, R. M., Kamphuis, J., Boesten, W. H. J., Meijer, E. M., Schoemaker, H. E. *J. Org. Chem.* 1988, *53*, 1826.
109. Schoemaker, H. E, Boesten, W. H. J., Kaptein, B., Hermes, H. F. M., Sonke, T.,

Broxterman, Q. B., van den Tweel, W. J. J., Kamphuis, J. *Pure Appl. Chem.* 1992, *64*, 1171.
110. Stein, G. A., Bronner, H. A., Pfister, K. *J. Am. Chem. Soc.* 1955, *77*, 700.
111. Ware, E. *Chem. Rev.* 1950, *46*, 403.
112. Yee, C., Blythe, T. A., McNabb, T. J., Walts, A. E. *J. Org. Chem.* 1992, *57*, 3525.
113. Toone, E. J., Jones, J. B. *Tetrahedron: Asymmetry* 1991, *2*, 201.
114. Lam, L. K. P., Brown, C. M., Jeso, B. D., Lym, L., Toone, E. J., Jones, J. B. *J. Am. Chem. Soc.* 1988, *110*, 4409.
115. Chen, C. S., Fujimoto, Y., Girdaukas, G., Sih, C. J. *J. Am. Chem. Soc.* 1982, *104*, 7294.
116. Liu, W., Ray, P., Beneza, S. A. *J. Chem. Soc. Perkin Trans.* 1995, *1*, 553.
117. Roth, H. J., Fenner, H. *Arzneistoffe-Pharmazeutische Chemie III;* Georg Thieme: Stuttgart, 1988.
118. Patchett, A. A. *Nature* 1980, *288*, 280.
119. Dürekheimer, W., Blumbach, J., Lattreil, R., Scheunemann, K. H. *Angew. Chem.* 1985, *97*, 183.
120. Kamphuis, J., Boesten, W. H. J., Schoemaker, H. E., Meijer, E. M. in *Conference Proceedings on Pharmaceutical Ingredients and Intermediates;* Reuben, B. G., Ed.; Expoconsult: Maarsen, 1991; p. 28.
121. Williams, R. M. *Synthesis of Optically Active α-Amino Acids;* Pergamon Press: Oxford, 1989.
122. Morin, A., Hummel, W., Kula, M.-R. *Appl. Microbiol. Biotechnol.* 1986, *25*, 91.
123. Morin, A., Touzel, J.-P., Lafond, A., Leblanc, D. *Appl. Microbiol. Biotechnol.* 1991, *35*, 536.
124. Olivieri, R., Fascetti, E., Angelini, L., Degen, L. *Enzyme Microb. Technol.* 1979, *1*, 201.
125. Olivieri, R., Fascetti, E., Angelini, L., Degen, L. *Biotechnol. Bioeng.* 1981, *23*, 2173.
126. Bucherer, H. T., Steiner, W. *J. Prakt. Chem.* 1934, *140*, 291.
127. Takahashi, S., Ohashi, T., Kii, Y., Kumagai, H., Yamada, H. *J. Ferment. Technol.* 1979, *57*, 328.
128. Battilott, M., Barberini, U. *J. Mol. Catal.* 1988, *43*, 343.
129. Keil, O., Schneider, M. P., Rasor, J. P. *Tetrahedron: Asymmetry* 1995, *6*, 1257.

9

Transition Metal Catalyzed Hydrogenations, Isomerizations, and Other Reactions

SCOTT A. LANEMAN
NSC Technologies, Mount Prospect, Illinois

9.1. INTRODUCTION

The desire to produce enantiomerically pure pharmaceuticals and other fine chemicals has advanced the field of asymmetric catalytic technologies. Since the independent discoveries of Knowles and Horner [1,2]; the number of innovative asymmetric catalysis for hydrogenation and other reactions has mushroomed. Initially, nature was the sole provider of enantiomeric and diastereoisomeric compounds; these form what is known as the ''chiral pool.'' This pool is comprised of relatively inexpensive, readily available, optically active natural products, such as carbohydrates, hydroxy acids, and amino acids, that can be used as starting materials for asymmetric synthesis [3,4]. Before 1968, early attempts to mimic nature's biocatalysis through noble metal asymmetric catalysis primarily focused on a heterogeneous catalyst that used chiral supports [5] such as quartz, natural fibers, and polypeptides. An alternative strategy was hydrogenation of substrates modified by a chiral auxiliary [6].

Knowles [1] and Horner [2] independently discovered homogeneous asymmetric catalysts based on rhodium complexes bearing a chiral monodentate tertiary phosphine. Continued efforts in this field have produced hundreds of asymmetric catalysts with a plethora of chiral ligands [7], dominated by chelating bisphosphines, that are highly active and enantioselective. These catalysts are beginning to rival biocatalysis in organic synthesis. The evolution of these catalysts has been chronicled in several reviews [8–13].

Asymmetric catalysis possesses many advantages over stoichiometric methodologies, of which the most important is chiral multiplication. A single chiral catalyst molecule can generate thousands of new stereogenic centers. Stoichiometric methods use resolution of racemates, or start from chiral pool materials. Resolution methods require the use of a resolving agent to form diastereoisomers that are then separated (see Chapter 8). This process can be quite wasteful since the undesired diastereoisomer must be either racemized or discarded. The same arguments can be applied to the recovery of the resolving agent. Utilization of the chiral pool in asymmetric synthesis can be limited by the availability of an inexpensive reagent that possesses the correct stereochemistry and structure similarity to the final target.

Enzymes cannot perform the range of reactions that organometallic catalysts can, and may be susceptible to degradation caused by heat, oxidation, and pH. Substrates not recognized by enzymes can be used in asymmetric catalysts, in which optimization of the enantioselectivities and overall chirality can be easily modified by change of the chiral ligands on the catalyst [11].

When coordinated to the metal, the chiral ligand plays an important role in the control of enantioselectivity during the course of the catalysis reaction. These ligands can contain chirality in the backbone (e.g., CHIRAPHOS, 1), at the phosphorus atom (e.g., DIPAMP, 2), or atropisomerically from C_2 symmetric axial configurations (e.g., BINAP, 3). Typically, most chiral ligands possess diphenylphosphine moieties (Figure 1).

The asymmetric outcome of the reaction results from a combination of steric repulsion imposed by the ligand and complex reaction kinetics. It is imperative that the ligands remain coordinated to the metal during the formation of the new stereogenic center to achieve high enantioselectivities. Chiral catalysis begins by enantiofacial differentiation of the substrate in the initial coordination to the metal center. A chiral environment about the metal is created by the positioning of the groups, usually phenyl, that result from the pucker in the phosphine-metallocycle, and act as a template for the coordinating substrate. Upon coordination of the prochiral substrate, the metal can form two diastereoisomeric complexes that are in rapid equilibrium. Enantioselectivity of the overall product is governed by the competing reaction rates at the metal center for each diastereoisomer [14]. The pucker and aryl array are illustrated by Rh(dipamp) (Figure 2), where the anisyl groups are showing their faces and the two phenyl groups are on edge [15].

Asymmetric catalysis has been most prevalent in the area of homogeneous hydrogenations. As previously stated, the advantages to produce considerable amounts of a single enantiomer or diastereoisomer from a small amount of chiral catalyst has a huge industrial impact. Natural and unnatural amino acids, particularly L-DOPA (12), have been produced by this method [15,16]. Catalysts based on rhodium and ruthenium have enjoyed the most success.

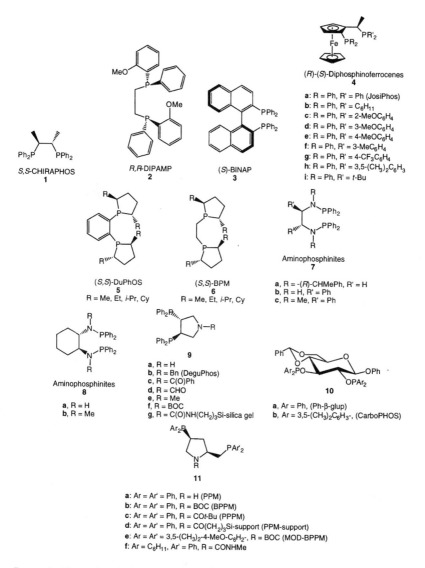

FIGURE I Examples of chiral ligands used as asymmetric hydrogenation catalysts.

Despite the explosion of asymmetric homogeneous catalysis technologies, very few have been able to make the jump from laboratory to manufacturing [10,17,18]. The focus of this chapter is to highlight and discuss the industrial feasibility of both old and new asymmetric hydrogenation technologies, as well

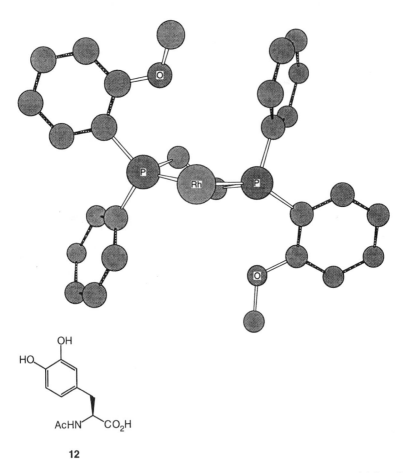

FIGURE 2   From the x-ray structure of [Rh(dipamp)COD]—hydrogens and 1,5-cycloocta-diene (COD) ligand are not shown [15].

as other transition metal catalyzed reactions, based on their respective chiral ligands for the synthesis of chiral intermediates. Several criteria must be filled for an asymmetric hydrogenation technology to be industrially feasible: the cost of catalyst preparation (cost of metal and chiral bisphosphine), catalyst efficiencies, catalyst selectivity expressed as enantiomeric excess (%ee), catalyst separation or recycle, reaction and operation conditions, and the need for specialized equipment. In each case, the feasibility of an asymmetric hydrogenation technology is also critical in the cost of the prochiral substrate to be reduced and isolation issues.

9.2. HOMOGENEOUS CATALYSIS: CHIRAL LIGANDS IMPLEMENTED IN INDUSTRIAL PROCESSES AND POTENTIAL PROCESSES

9.2.1. DIPAMP

As previously stated, Knowles et al. have developed an efficient asymmetric catalyst based on the chiral bisphosphine, R,R-DIPAMP (2), that has chirality at the phosphorus atoms and can form a 5-membered chelate ring with rhodium. [Rh(COD)(R,R-DIPAMP)]$^+$ BF$_4$$^-$ (13) has been used by Monsanto for the production of L-DOPA (12), a drug used for the treatment of Parkinson's disease, by an asymmetric reduction of the Z-enamide, 14a, in 96% ee (Scheme 1). The pure isomer of the protected amino acid intermediate, 15a, can be obtained upon crystallization from the reaction mixture as it is a conglomerate [15].

a, Ar = 3-AcO-4-MeOC$_6$H$_3$
b, Ar = Ph

SCHEME 1.

The chemistry of Rh(DIPAMP) and mechanism has been reviewed [9,14–16,19]. Marginally higher catalyst efficiencies are observed with higher alcohols compared with methanol, whereas the presence of water can result in the reduction of slurries. Filtration of the product can improve the %ee while the catalyst and D,L-product remain in the mother liquor. The catalyst stereoselectivities decreases as hydrogen pressure increases. [Rh(COD)(R,R-DIPAMP)]$^+$ BF$_4$$^-$ (13) affords the S-configuration of amino acids upon reduction of the enamide substrate. Reduction of enamides in the presence of base eliminates the pressure variances on the stereoselectivities, but the rate of reaction under these conditions is slow [15].

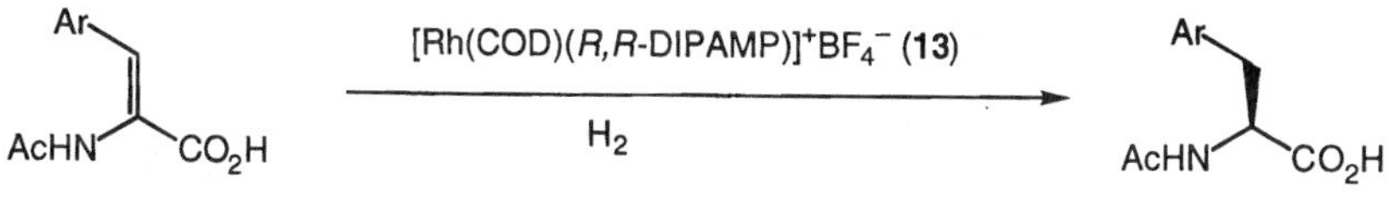

The advantages and disadvantages of Rh(DIPAMP) are summarized in Table 1. The catalyst precursor, 13, is air-stable, which simplifies handling operations on a manufacturing scale. Despite these advantages, ligand synthesis is very difficult. After the initial preparation of menthylmethylphenylphosphinate (16), the $(R)_P$-isomer is separated by two fractional crystallizations (Scheme 2). The yield of R,R-DIPAMP (2) is 18% based on 17 [16]. Another disadvantage is that the R,R-DIPAMP can racemize at 57°C, which can be problematic since the last step is performed at 70°C. Once the ligand is coordinated to rhodium, however, racemization is no longer a problem.

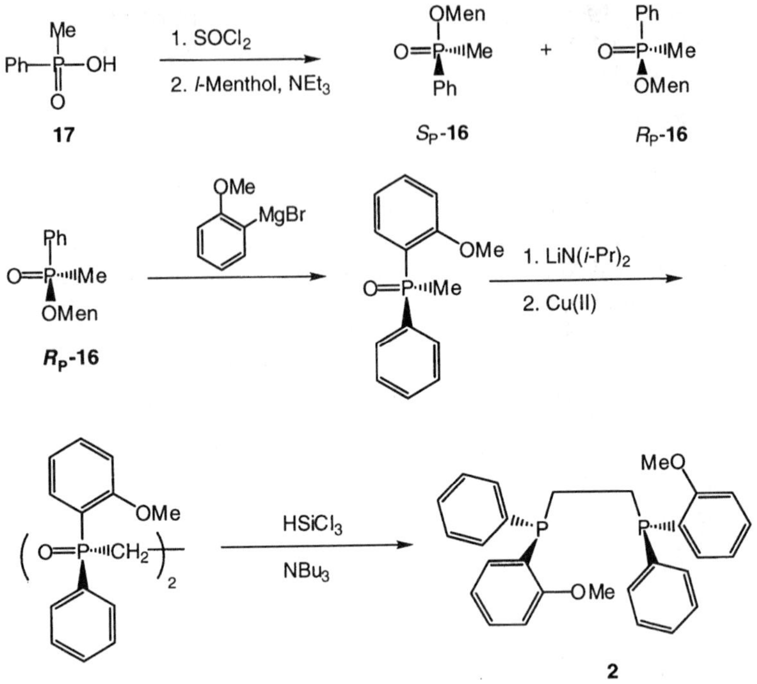

SCHEME 2.

Other routes to R,R-DIPAMP (**2**) have been reported [20–22]. At present, the most practical synthesis of DIPAMP involves the formation of a single diastereoisomer of **18** by the combination of PhP (NEt$_2$)$_2$ and (−)-ephedrine followed by formation of the borane adduct (Scheme 3) [20,23,24].

The cost of the ligand and catalyst preparation can be critical in the decision

TABLE I   Advantages and Disadvantages of Rh(DIPAMP)

Advantages	Disadvantages
High enantioselectivity for various aromatic enamides	Lower enantioselectivities of alkyl enamides
Rapid rates	Ligand synthesis difficult and in low yields
High catalyst turnovers	Free ligand racemizes at low temperature
Low hydrogen pressures	
Mild reaction temperatures	
Air stable catalyst precursor	

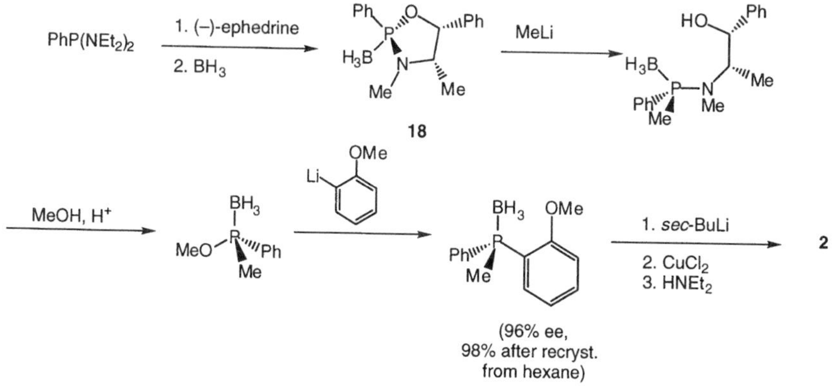

SCHEME 3.

to pursue technical transfer to manufacturing stage. Fortunately, the high reactivity of [Rh(COD)(R,R-DIPAMP)]$^+$ BF$_4^-$ (13) in most enamide reductions can offset the high-price, low-yield synthesis of DIPAMP.

[Rh(COD)(R,R-DIPAMP)]$^+$ BF$_4^-$ (13) has shown remarkable activity and enantioselectivity in the asymmetric hydrogenation of various enamides, enol acetates, and unsaturated olefins [16]. For the past 20 years, 13 has been the premier asymmetric homogeneous catalyst for amino acid synthesis, but a new generation of catalysts based on novel bisphosphines (see Chapter 18) has surpassed it in versatility.

9.2.2. BINAP

Asymmetric catalysis undertook a quantum leap with the discovery of ruthenium and rhodium catalysts based on the atropisomeric bisphosphine, BINAP (3). These catalysts have displayed remarkable versatility and enantioselectivity in the asymmetric reduction and isomerization of α-, β-, γ-keto esters, functionalized ketones, allylic alcohols and amines, α,β-unsaturated carboxylic acids, and enamides. Asymmetric transformation with these catalysts has been extensively studied and reviewed [8,13,25,26]. The key feature of BINAP is the rigidity of the ligand during coordination on a transition metal center, which is critical during enantiofacial selection of the substrate by the catalyst. Several industrial processes currently use these technologies, while a number of other opportunities show potential for scale up.

Each industrial or potential scale-up process that uses BINAP-ligated asymmetric homogeneous catalysts has a common major drawback despite their respective high reactivities and enantioselectivities—the cost to produce the chiral ligand and limited availability. Until 1995, Takasago Co. has been the only industrial producer of chiral BINAP because of the troublesome six-step procedure with an overall yield of 14% from 2-naphthol (**19**) (Scheme 4) [27], until significant improvements in the synthesis was developed by Merck (Scheme 5) [28,29]. The latter method has been improved further through modification of reaction parameters with the use of diphenylchlorophosphine rather than the pyrophoric diphenylphosphine [30].

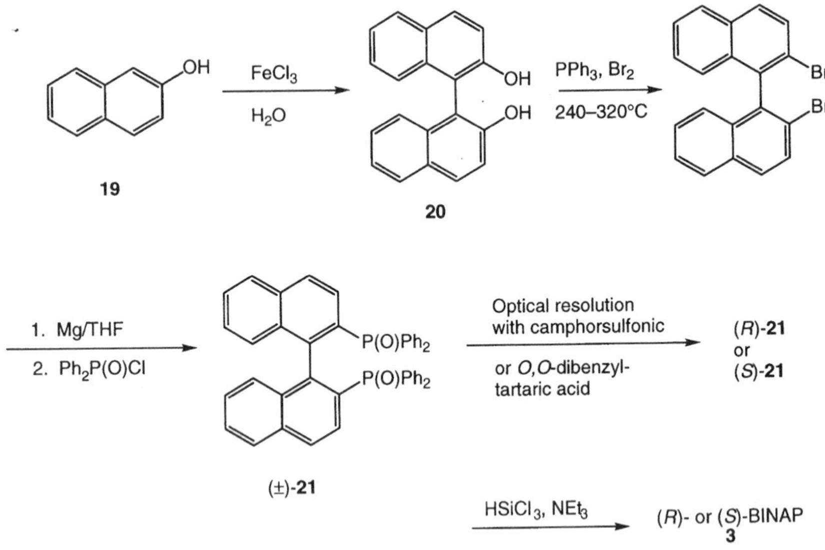

SCHEME 4.

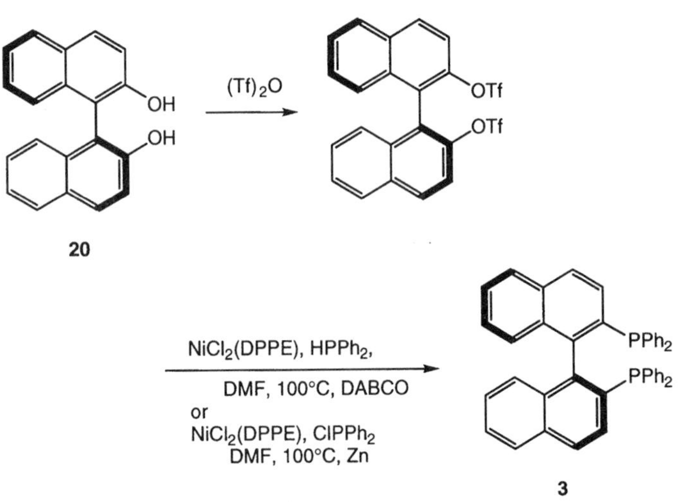

SCHEME 5.

9.2.2.1. Menthol

Menthol is used in many consumer products such as toothpaste, chewing gum, cigarettes, and pharmaceutical products, with an estimated worldwide consumption estimated at 4500 tons per year [31]. (−)-Menthol (22) is manufactured by Takasago Co. from myrcene (23), which is available from the cracking of inexpensive β-pinene (Scheme 6) [31,32]. The key step in the process is the asymmetric isomerization of N,N-diethylgeranylamine (24) catalyzed by either $[Rh(L_2)(S\text{-BINAP})]^+BF_4^-$ (where L is diene or solvent) or $[Rh(S\text{-BINAP})_2]^+BF_4^-$ to the diethyl enamine intermediate 25 in 96–99% ee [26,33]. Citronellal (26) is obtained in 100% ee after hydrolysis of the enamine intermediate; natural citronellal has an optical purity of 80% [25]. A stereospecific acid-catalyzed cyclization followed by reduction produces 22 [31].

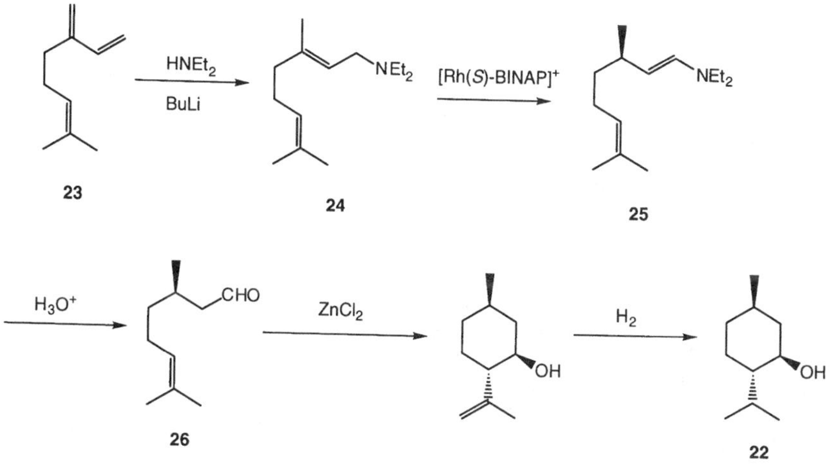

SCHEME 6.

The catalysts employed in the asymmetric isomerization of allylamines are very susceptible to water, oxygen, and carbon dioxide, as well as significant deactivation being observed by the presence of donor substances that include NEt_3, COD, 27 and 28. Unfortunately, commercial production of 24 is usually accompanied by formation of 0.5–0.7% of 28, and a thorough pretreatment of the substrate 24 is required for the reaction system to attain high turnover numbers (TON), especially when $[Rh(L_2)(S\text{-BINAP})]^+BF_4$ is used as the catalyst [31].

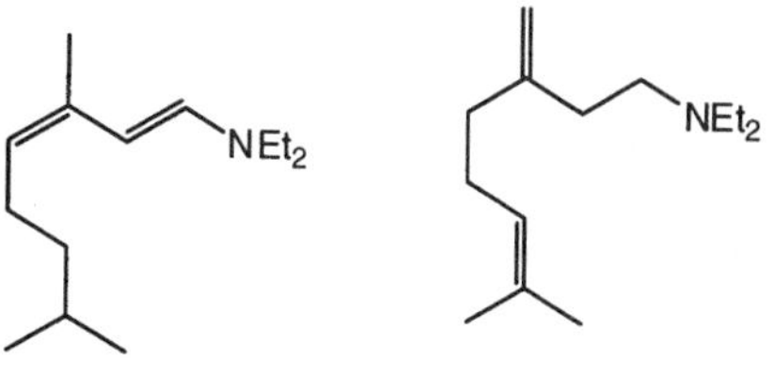

A significant improvement in the asymmetric isomerization came upon the discovery of [Rh(S-BINAP)₂]⁺BF₄⁻. This catalyst shows remarkable catalytic activity at high temperature (> 80°C) without deterioration of stereoselectivity (> 96% ee). The TON is improved to 400,000 through catalyst recycle (2% loss of catalyst per reuse). Additional improvements in the catalyst have been achieved by modification of the BINAP to p-TolBINAP (**29**) [31].

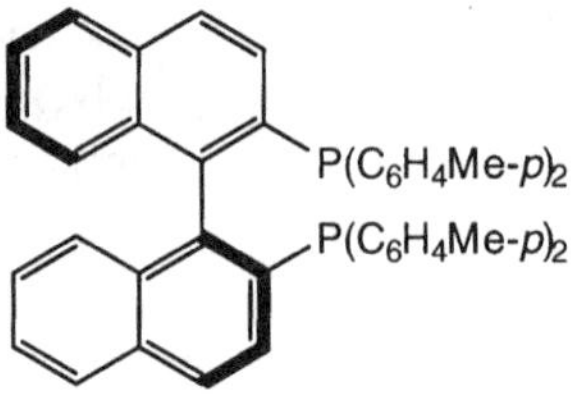

29

The stereochemical outcome of the enamine can be controlled by the configuration of the olefin geometry of the allylamine and the configuration of the chiral ligand (Scheme 7) [26,31].

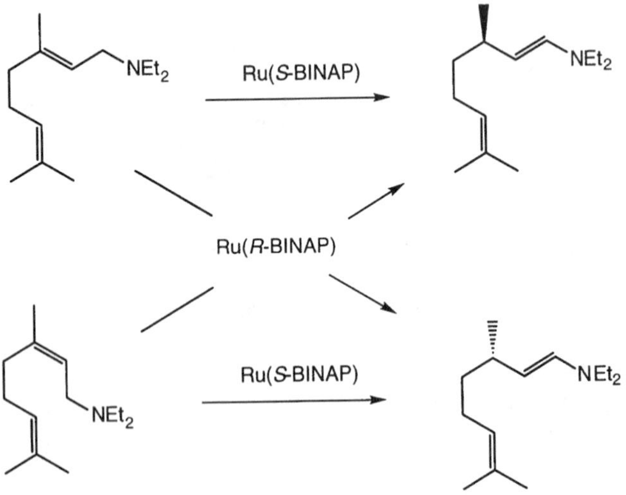

SCHEME 7.

Takasago has implemented their asymmetric isomerization technology to produce a variety of optically active terpenoids from allyl amines at various manufacturing scales, which is summarized in Table 2 [31].

TABLE 2 Optically Active Terpenoids Produced by Asymmetric Isomerization of Allylamines

Name	Structure	Usage	ee (%)	Production (tons/year)
(+)-Citronellal	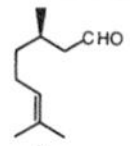	Intermediate	96–98	1500
(−)-Isopulegol	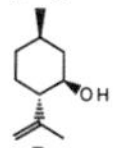	Intermediate	100	1100
(−)-Menthol	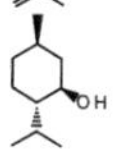	Pharmaceuticals Tobacco Household products	100	1000
(−)-Citronellol	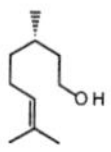	Fragrances	98	20
(+)-Citronellol	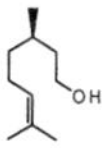	Fragrances	98	20
(−)-7-Hydroxy-citronellal	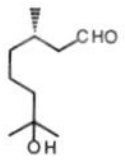	Fragrances	98	40
S-7-Methoxy-citronellal	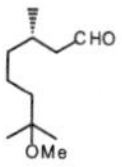	Insect growth regulator	98	10
S-3,7-Dimethyl-1-octanal	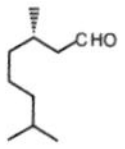	Insect growth regulator	98	7

Source: Data from Ref. 31.

9.2.2.2. Carbapenem Intermediates

Carbapenem antibiotics (**30**) can be manufactured from intermediates obtained by Ru(BINAP)-catalyzed reduction of α-substituted β-keto esters by a dynamic kinetic resolution (Scheme 8). 4-Acetoxy azetidinone (**31**) is prepared by a regio-selective RuCl₃-catalyzed acetoxylation reaction of **32** with peracetic acid [34]. This process has been successful in the industrial preparation of the azetidinone **31** in a scale of 120 tons per year [35].

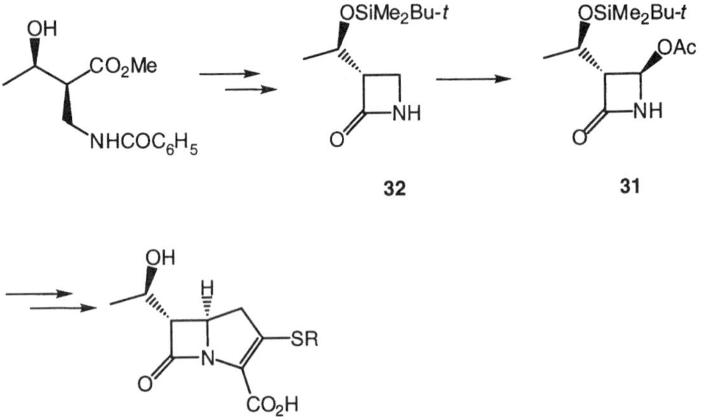

SCHEME 8.

Dynamic kinetic resolution can occur for α-substituted β-keto esters with epimerizable substituents provided that racemization of the antipodes **33** and **34** is rapid with respect to the Ru(BINAP)-catalyzed reduction, thereby potentially allowing the formation of a single diastereoisomer (Scheme 9). Deuterium labeling experiments have confirmed the rapid equilibrium of epimers at C-2 [36]. Generally, the configuration of C-3 is governed by the chirality of BINAP, whereas the configuration at C-2 is substrate specific.

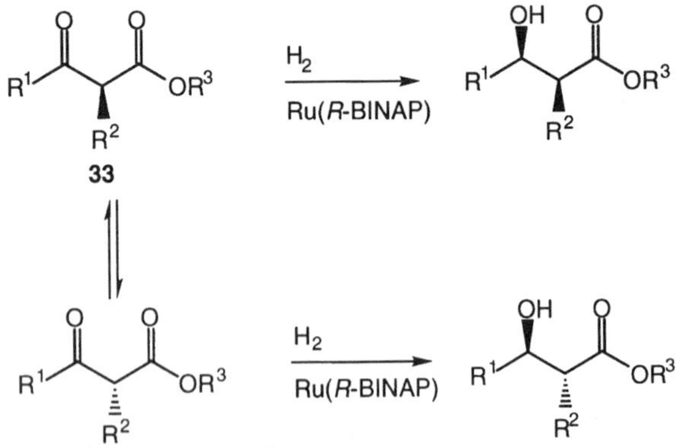

eg., R^1 = Me, R^2 = CH$_2$NHCOPh, R^3 = Et

SCHEME 9.

Catalyst optimization has only been reported for the [RuX(arene)(BI-NAP)]$^+$X$^-$ series by variation of the bisphosphines, halogen and coordinated arene ligands. The reductions of β-dicarbonyl compounds with Ru(BINAP-type) catalysts has been reviewed [37].

9.2.2.3. Other Intermediates

Asymmetric hydrogenation of racemic 2-substituted β-keto esters to produce 2-substituted β-hydroxy esters with two new chiral centers is a powerful method, and useful in the production of other pharmaceutical intermediates. The methodology can be used in the preparation of protected threonine derivatives **35**, where **35d** and **35e** are key intermediates for the anti-Parkinsonian agent, L-DOPS (**36**).

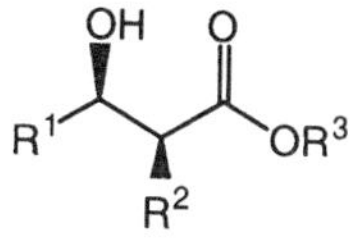

35

a, R^1 = Me, R^2 = NHCOMe, R^3 = Me
b, R^1 = Me, R^2 = NHCOMe, R^3 = t-Bu
c, R^1 = Me, R^2 = NHCOi-Pr, R^3 = Me
d, R^1 = 3,4-methylenedioxyphenyl, R^2 = NHCOMe, R^3 = Me
e, R^1 = 3,4-methylenedioxyphenyl, R^2 = NHCOPh, R^3 = Me
f, R^1 = Me, R^2 = Cl, R^3 = Me
g, R^1 = Ph, R^2 = Cl, R^3 = Et
h, R^1 = p-MeOPh, R^2 = Cl, R^3 = Me
i, R^1 = Me, R^2 = Me, R^3 = Me

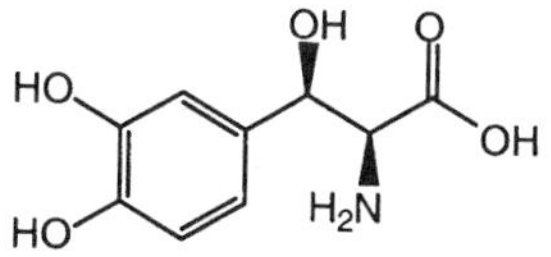

36

9.2.2.4. Nonsteroidal Antiinflammatory Drugs

(*S*)-Naproxen (**37**), a potent nonsteroidal antiinflammatory drug (NSAID), is the best-selling agent for arthritis and represents a billion-dollar-a-year market [27]. Extensive research is aimed at the formation of the active *S*-enantiomer since the *R*-isomer is a potent liver teratogen. An asymmetric reduction has been proposed for the manufacture of **37**, but has yet to be put into practice. Monsanto has reported a four-step manufacture process in development for naproxen (Scheme

10) [38]. Several parameters of the hydrogenation affect the enantioselectivities, which includes temperature, hydrogen pressure, solvent, base, and catalyst [39–41]. However, a significant increase of the substrate to catalyst ratio and decrease of the hydrogen pressure from 2000 psig as well as reduction in the cost of production of the alkene **38** is required for this technology to become economically feasible [27]. Catalyst cost could also be reduced by use of solid supports that have the potential to improve TON through catalyst recycle [42].

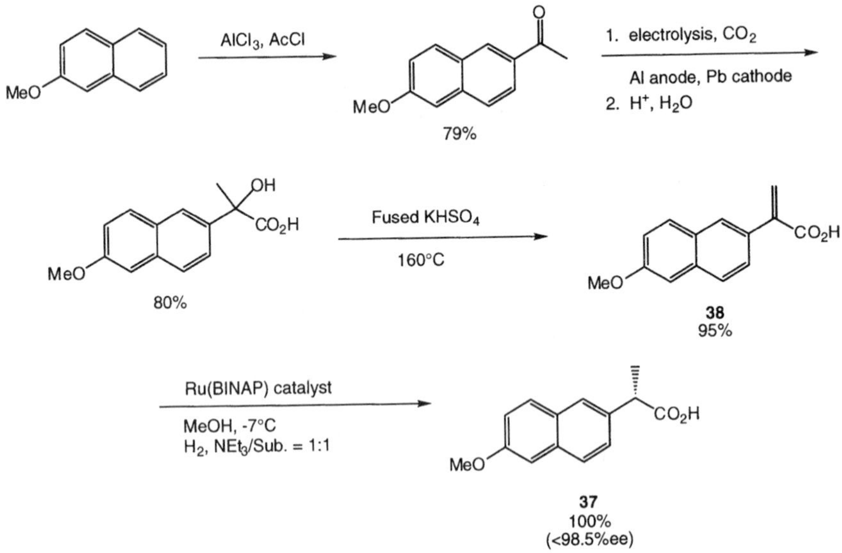

SCHEME 10.

The naproxen technology can be applied to ibuprofen, another popular analgesic and antiinflammatory drug sold over the counter with use volumes several times larger than naproxen. Ibuprofen is sold as a racemate, although extensive pharmacological studies have shown that only the *S*-isomer (**39**) has significant therapeutic effect [40]. For this reason, several pharmaceutical companies have been become interested in marketing *S*-ibuprofen as a premium analgesic [40]. The racemate of ibuprofen can be prepared by substitution of a heterogeneous catalyst (Pd/C) for the chiral catalyst.

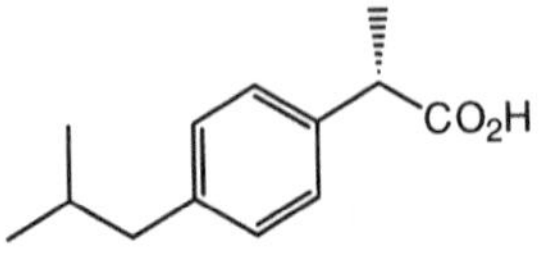

39

9.2.2.5. Levofloxacin

Another Ru(BINAP)-catalyzed asymmetric hydrogenation that has been performed at manufacturing scale involves the reduction of a functionalized ketone. The reduction of hydroxyacetone catalyzed by $[NH_2Et_2]^+[\{RuCl(p\text{-tol-}BINAP)\}_2(\mu\text{-Cl})_3]^-$ (**40**) proceeds in 94% ee (Scheme 11) [34]. The chiral diol (**41**) is incorporated into the synthesis of Levofloxacin (**42**), a quinolinecarboxylic acid that exhibits marked antibacterial activity. Current production of **41** is 40 tons per year by Takasago International Corp [34].

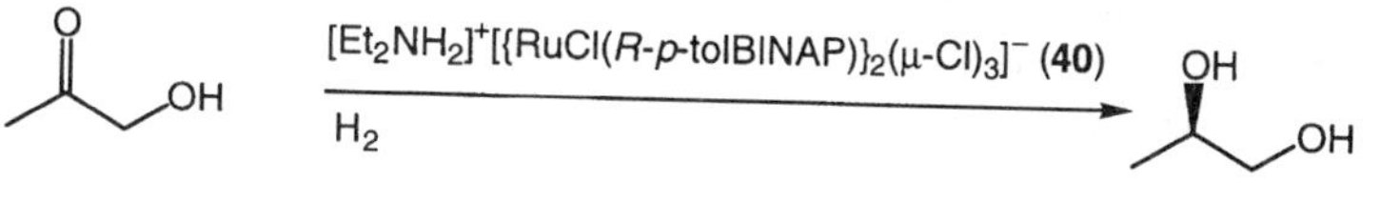

41
(94% ee)

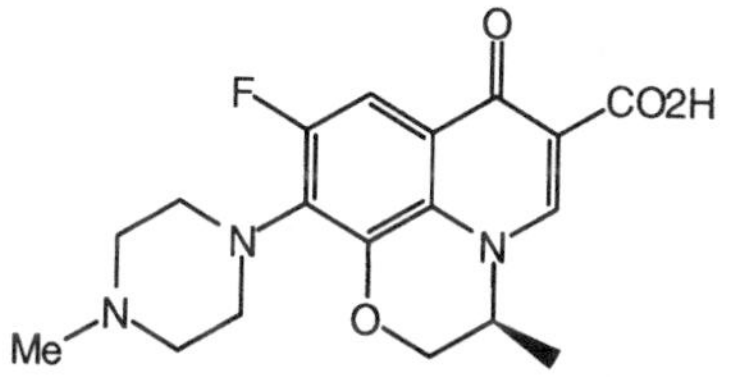

42

9.2.2.6. Potential Ru(BINAP) Technologies

The asymmetric hydrogenation of allylic alcohols has high potential for industrial scale up. $Ru(O_2CCF_3)_2(S\text{-BINAP})$ and $Ru(O_2CCF_3)_2(S\text{-}p\text{-tolBINAP})$ can catalyze the reduction of allylic alcohols in high enantioselectivities and reactivities [8]. Geraniol (**43**) and nerol (**44**) are hydrogenated in MeOH to give citronellol (**45**) in 96–99% ee with a substrate to catalyst ratio of 50,000 (Scheme 12) [43].

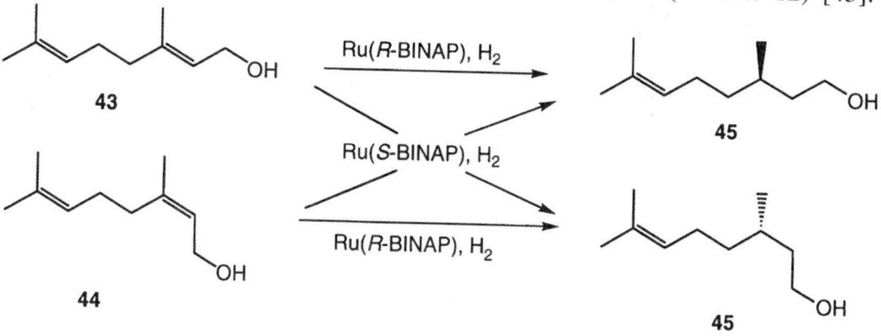

SCHEME 12.

In a similar manner to the asymmetric isomerization of allylic amines, the new stereogenic center can be controlled by either the substrate or the configuration of the ligand on ruthenium [43]. The highly efficient stereoselective transformations catalyzed by transition metals that contain BINAP have resulted in extensive efforts on the development of both new Ru-BINAP catalysts and chiral atropisomeric bisphosphines based on biaryl- and biheteroarylbackbones [37]. Coupled with the various classes of Ru(BINAP) catalysts and chiral bisphosphines, the number of efficient industrial asymmetric hydrogenations are sure to increase as optimization for fine precision and easy optimization in catalyst activity and enantioselectivity will be made easier [44–53].

9.2.3. Chiral Aminophosphines

A class of chiral bisphosphines that has not received much attention is aminophosphines (e.g., **7** and **8**) known as PNNP [7]. The chirality is derived from a 1,2-diamine. One advantage of aminophosphines is that they can be easily prepared by the reaction of a chiral amine and Ph_2PCl in the presence of a base [54]. However, the rhodium-catalyzed reductions must be performed with care as the P–N bonds are easily hydrolyzed or solvolyzed [15,55].

Rhodium complexes that contain these ligands have demonstrated moderate to high enantioselectivities (24–96%) in the reduction of enamides to protected amino acids (Scheme 1) [7,54]. Despite the moderate degree of asymmetric induction, an industrial process had been developed with this catalyst system by ANIC S.p.A. (EniChem) for the production of (S)-phenylalanine for the synthesis of aspartame [11,33]. The process uses cationic Rh-**7a** for the reduction of **14b** at 28 psig H_2 and 22°C for 3 hours substrate/catalyst (S/C) = 15,000) to give **15b** in 83.3% ee, which is enriched to 98.3% by recrystallization [56].

9.2.4. Ferrocenyldiphosphines

Most catalysts that have been developed for asymmetric catalysis contain chiral C_2-symmetric bisphosphines [7]. The development of chiral ferrocenylphosphines ventures away from this conventional wisdom. Chirality in this class of ligands can result from planar chirality due to 1,2-unsymmetrically ferrocene structure as well as from various chiral substituents. Two classes of ferrocenyldiphosphines exist: two phosphino groups substituted at the 1,1'-position about the ferrocene backbone (**46**); and both phosphino groups contained within a single cyclopentadienyl ring of ferrocene (**4**) [9].

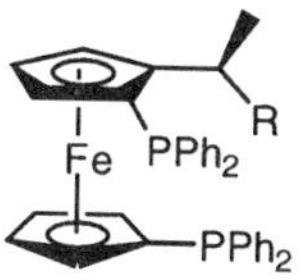

46

a, R = NMe$_2$
b, R = OH
c, R = OAc
d, R = H

Novartis (Ciba Geigy) has developed ferrocenyldiphosphines, **4**, for the asymmetric hydrogenation of sterically hindered N-aryl imines, specifically for the industrial manufacture of Metolachlor (**47**), a pesticide sold since March 1997 as the enantiomerically enriched form, and sold under the trade name DUAL MAGNUM [57–59]. Since its introduction in 1978, more than 20,000 tons per year of the racemic form has been sold [59]. Metolachlor exists as four diastereoisomers, a result of stereogenic carbon atom and restricted rotation about the phenyl-nitrogen bond. The highest biological activity is observed for diastereoisomers that contain the S-configuration at the stereogenic carbon, whereas chirality about the phenyl-nitrogen bond does not affect the biological activity [57]. The synthesis of **47** involves the asymmetric reduction by a catalyst formed in situ from [Ir(COD)Cl]$_2$ and **4h** to give **48** and subsequently **47** in 80% ee (Scheme 13) [58]. Fortunately, high ee's in agrochemicals are not as critical compared to pharmaceuticals, and the usage of the enantiomerically enriched herbicide leads to a 40% reduction in the environmental loading [59].

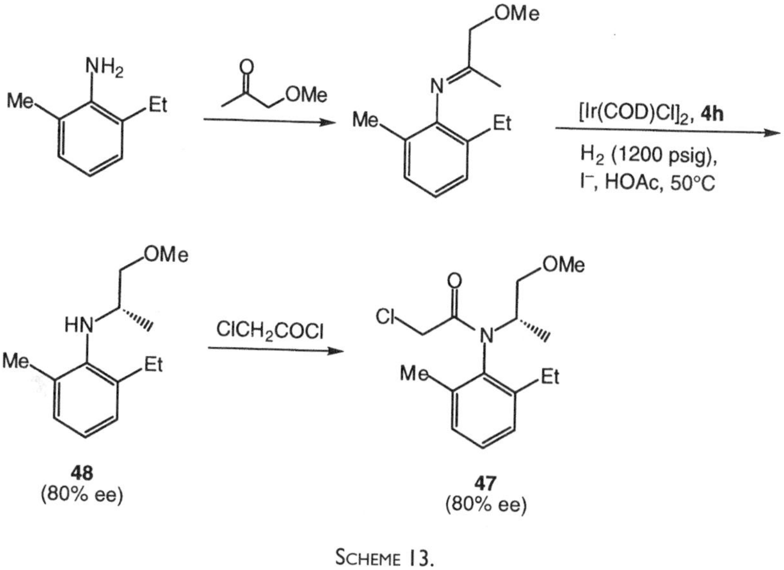

SCHEME 13.

The enantioselectivity and catalytic activity are highly influenced by the purity of imine, electronic effects of the ligand, and additives. In the early stage of process development, iridium catalysts with a variety of known chiral ligands (DIPAMP, DIOP, BPPM, BDPP, and BPPFOH) would deactivate if the purity of the imine was less than 99.1%, but deactivation is not observed with Ir-**4h**. The addition of both iodide anion and acetic acid additives are critical to the catalytic activity. High catalytic activities are not observed if either one of the additives is not present during the reduction [57]. The high catalyst cost can be offset by high catalytic activity (S/C = 500,000–1,000,000, and turnover frequencies (TOF) > 30,000/hr) and low catalyst usage. To date, these are the highest S/C ratios and catalytic rates ever reported for a catalytic process [57].

The syntheses of **4** are summarized in Scheme 14 [60–62].

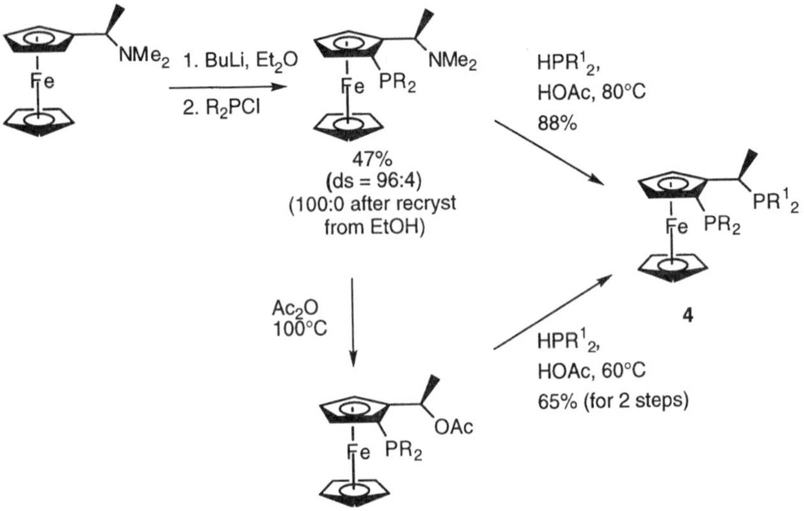

SCHEME 14.

Rhodium and palladium catalysts that contain **4** display high enantioselectivities for the asymmetric hydrogenation of enamides, itaconates, β-keto esters, asymmetric hydroboration, and asymmetric allylic alkylation [62–64], but this ligand system distinguishes itself from other chiral bisphophines in the asymmetric reduction of tetrahydropyrazines and tetra-substituted olefins. The reduction of tetrahydropyrazines produces the piperazine-2-carboxylate core, which is a common intermediate in the pharmaceuticals D-CPP-ene (**49**), an NMDA antagonist, Draflazine (**50**), a nucleoside transport blocker, and Indinavir (**51**), the well-known human immunodeficiency virus protease inhibitor. The chiral piperazine cores (**52**) can be prepared by the asymmetric reduction of tetrahydropyrazines (**53**) (Scheme 15) [65]. The preparation of the 2-substituted piperazine **53b** has been performed at a scale of nearly one ton [66].

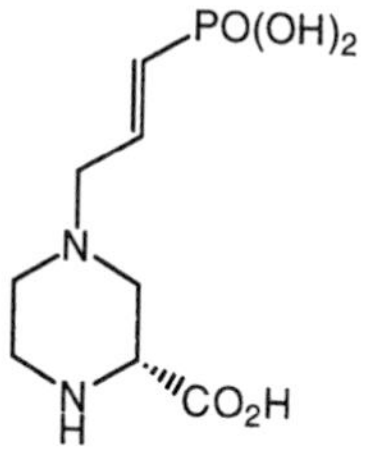

49

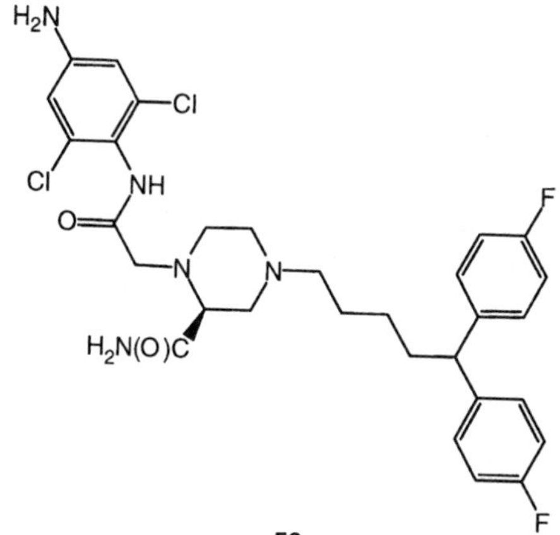

50

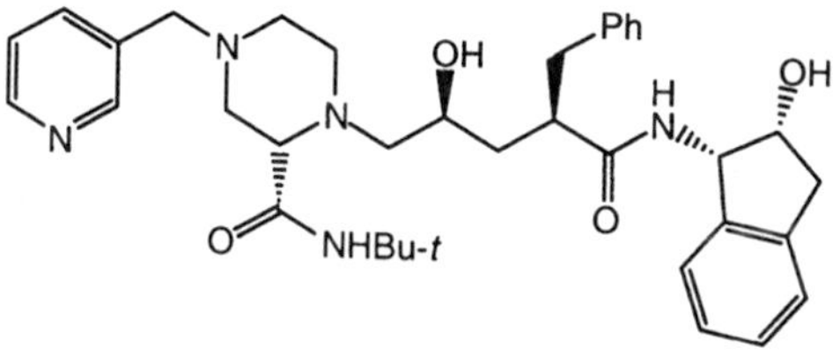

51

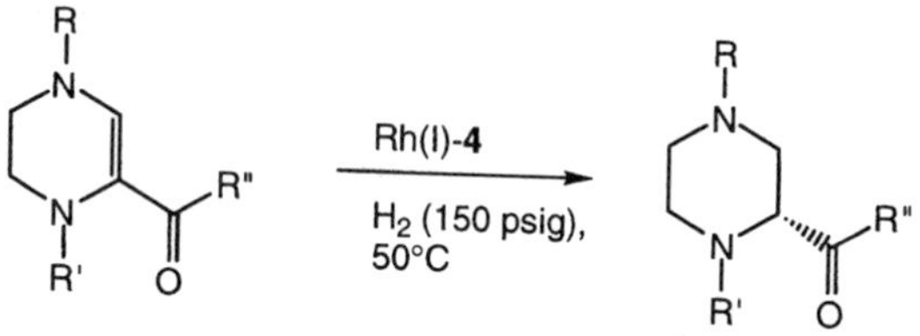

53 → **52**

Rh(I)-**4**

H$_2$ (150 psig), 50°C

a, R = CHO, R' = C(O)Et, R" = NHtBu
b, R = CHO, R' = C(O)CF$_3$, R" = NHtBu
c, R = H, R' = C(O)Me, R" = OMe
d, R = H, R' = C(O)Me, R" = NHtBu
e, R = C(O)Me, R' = C(O)Me, R" = OMe
f, R = C(O)Me, R' = C(O)Me, R" = NHtBu
g, R = BOC, R' = C(O)Me, R" = OMe
h, R = BOC, R' = C(O)Me, R" = NHtBu
i, R = BOC, R' = CBZ, R" = OMe
j, R = BOC, R' = CBZ, R" = NHtBu
k, R = BOC, R' = BOC, R" = OMe

SCHEME 15.

9.2.4.1.   Biotin

Biotin (**54**), a water-soluble vitamin with widespread application in the growing market for health and nutrition, acts as a cofactor for carboxylase enzymes and is essential fatty acid synthesis. The key step in the chemical synthesis of biotin is the asymmetric reduction of the tetra-substituted olefins **55** by in situ Rh(I)-**4i** catalyst (Scheme 16) [61,67]. The complete synthesis of *d*-(+)-biotin, is described elsewhere [65,68].

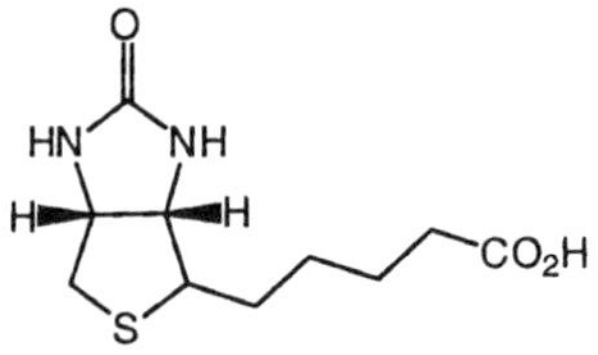

54

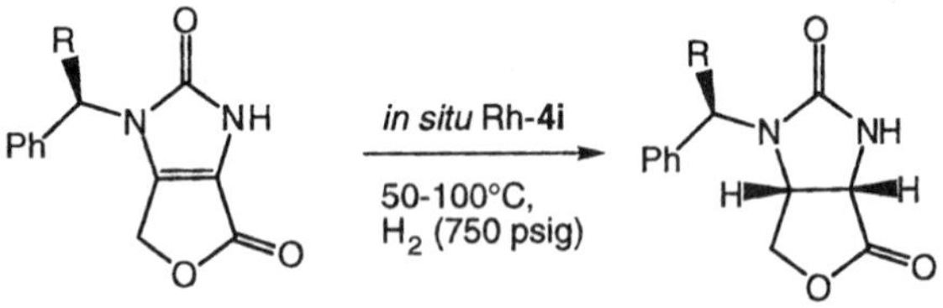

55

a, R = Me
b, R = H

ds **a** >99:1
    b >90:10

SCHEME 16.

9.2.5.   Carbohydrate Phosphinites

Carbohydrate backbones have been used as a chiral source in a series of phosphinite ligands. Several key features of these ligands are that the D-glucopyranoside backbone is very rigid and an inexpensive chiral source. The benzylidene protection of 4- and 6-hydroxy groups in **10a** is not critical for high enantioselectivity, but protection of the 1-hydroxy is required. Rhodium complexes that contain these ligands are enantioselective catalysts in the asymmetric hydrogenation of enamides [69].

Rhodium catalysts that contain **10a** have been reported for the industrial preparation of L-DOPA (Scheme 1). Similar to the Monsanto Rh(DIPAMP) process, the enamide **14a** has been reduced with [Rh(COD)(**10a**)]$^+$BF$_4^-$ in > 90% ee by VEB Isis-Chemie Zwickau [70].

Electronic amplification of the enantioselectivity is observed by modification of the aryl moieties of the phosphinite. Although [Rh(COD)(**10a**)]$^+$BF$_4^-$ is an efficient catalyst for the asymmetric reduction of α-acetamidocinnamate esters to protected phenylalanine amino acid esters, enantioselectivities decrease in the asymmetric hydrogenation of substituted phenylalanines. Substitution of the diphenylphosphinyl moieties for a more electron-donating 3,5-dimethyphenylphosphinyl produce ligands (**10b**) that can efficiently hydrogenate a wide variety of phenyl-substituted dehydroamino acid derivatives in high enantioselectives [71].

The ligands **10** produce L-amino acid derivatives by rhodium-catalyzed reduction of the corresponding enamide [71]. Amazingly, D-amino acid derivatives can be produced in the hydrogenation with ligands that contain the same D-glucopyranoside backbone, only the *O*-diarylphosphino moieties are on the 3-,4-hydroxy groups (**56** and **57**). The production of both D- and L-amino acids from the same inexpensive chiral source (D-glucose and D-glucopyranoside) is very advantageous.

OBz
Ar$_2$PO
Ar$_2$PO
OBz OMe
56
Ar = 3,5-(CH$_3$)$_2$C$_6$H$_3$-

OTBDMS
Ar$_2$PO
Ar$_2$PO
OMe
NHAc
57
Ar = 3,5-(CH$_3$)$_2$C$_6$H$_3$-

9.2.6. Pyrrolidine-Based Bisphosphines

9.2.6.1. 3,4-Bisphosphinopyrrolines (DeguPHOS)

A class of chiral bisphosphines based on 3,4-bis(diphenylphosphino)pyrrolidines (**9**) have been developed by Degussa and the University of Munich. Rhodium-bisphosphine catalysts of this class can reduce a variety enamides to chiral amino acid precursors with high enantioselectivities. These catalysts are extremely rapid and can operate with high substrate to catalyst ratios (10,000–50,000) under moderately high hydrogen pressure (150–750 psig). Contrary to other rhodium catalysts that contain chiral bisphosphines, the enantioselectivity of the pyrrolidine-bisphosphines catalysts are independent of the hydrogen pressure, although the rate of hydrogenation increases [72].

The asymmetric hydrogenation to *N*-acetyl-L-phenylalanine (S/C = 10,000) is catalyzed by [Rh(COD)(**9b**)]$^+$BF$_4^-$ at 40-kg scale in 99.5% ee and 100% conversion. The reduction is completed in 3–4 hours at 50°C under 150–225 psig hydrogen pressure. The TON can be further increased by the recycle of catalyst and ligand or by the attachment of the ligand to a solid support (**9g**) [72–74]. The asymmetric hydrogenation catalyst can be fine-tuned by the variation of the R group on the pyrrolidine backbone. The synthesis is summarized in Scheme 17 [72,73].

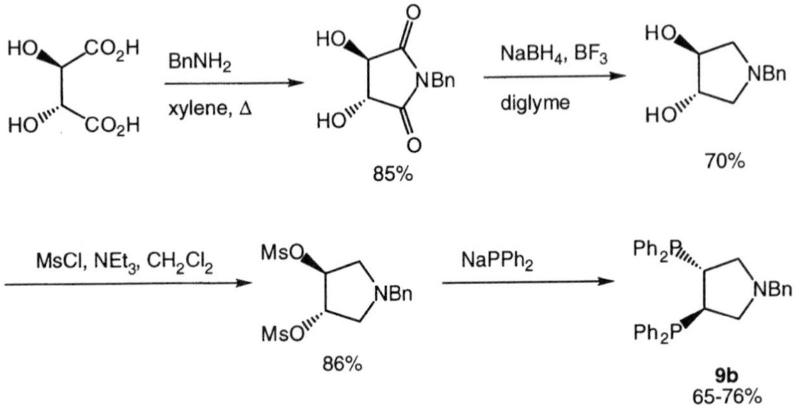

SCHEME 17.

9.2.6.2.   4-Phosphino-2-(Phosphinomethyl)Pyrrolidines (PPM and BPPM)

Another variation of the bisphosphinopyrrolidine ligand is chirality in the 2- and 4-positions (11) [75]. The synthesis of these ligands begins with readily available L-4-hydroxyproline (58) (Scheme 18) [75,76]. The synthesis of the antipode of BPPM requires a lengthy nine-step process from L-4-hydroxyproline that contains three stereocenter inversions [76].

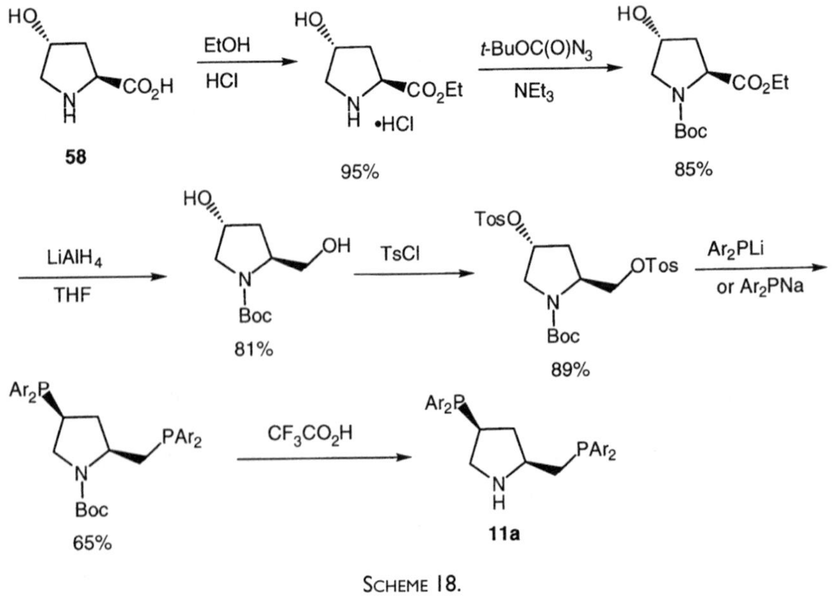

SCHEME 18.

Rhodium complexes that contain (2*S*,4*S*)-PPM ligands efficiently catalyze the asymmetric reduction of enamides to form protected amino acids with the *R*-configuration [75,77]. However, no large-scale process has been realized to date, although the potential exists as high substrate to catalyst ratios of 100,000 have been obtained in the hydrogenation of **59** (Scheme 19) [8,78].

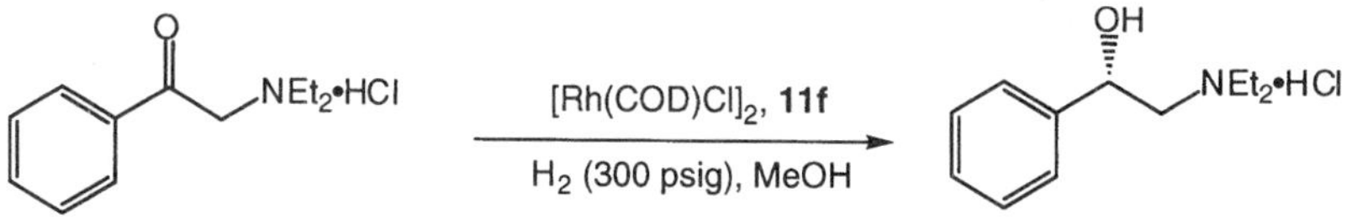

59   (95% ee)

SCHEME 19.

9.2.6.3. Bis(Phospholanes) (DuPHOS)

A chiral ligand system based on C_2-symmetric chiral bis(phospholanes) (**5** and **6**) has shown to be a powerful ligand in the asymmetric hydrogenation of various substrates that include enamides, enol acetates, C=N bond reduction, and β-keto esters. Detailed discussions of this class of ligands are included Chapter 18.

9.3. ASYMMETRIC HETEROGENEOUS CATALYSTS IMPLEMENTED IN INDUSTRY

The first forays into asymmetric organic synthesis used catalysis that were dominated by heterogeneous systems, which can boast a few catalytic systems that undergo asymmetric hydrogenation transformations with good to high enantioselectivities, but these have been overshadowed by the highly active and enantioselective homogeneous counterparts. Informative historical backgrounds and overviews on asymmetric heterogeneous catalytic systems have been published [5,6].

Strategies to induce chirality in a prochiral substrate included modification of existing heterogeneous catalysts by addition of naturally occurring chiral molecules, such as tartaric acid, natural amino acids, or alkaloids, and the implementation of chiral supports, that include quartz or natural fibers, for metallic catalysts. Both strategies have been successful on a limited basis.

Despite the convenience of handling and separation in heterogeneous catalysis, many other parameters have a strong influence on the stereochemical outcome: pressure, temperature, modifier, purity of substrates, high substrate specificity, catalyst preparation (which includes type, texture, and porosity of the support), dispersion, impregnation, reduction, and pretreatment of the metal [18].

Reproducibility of catalyst activities and enantioselectivities can often be a problem attributed to variations in catalyst preparations, and purity of the substrate [5,18].

The transformations that use asymmetric heterogeneous catalysis will be highlighted: β-keto esters and diketone reductions by Raney nickel catalyst modified with R,R-tartaric acid and NaBr, and α-keto acid reductions with cinchona modified Pt catalysts.

9.3.1. Ni/R,R-Tartaric Acid/NaBr

This catalyst system has been extensively studied. A variety of ketone substrates have been reduced with Ni/tartaric acid/NaBr catalysts with variable enantioselectivities, but the highest (> 85% ee) are obtained for the reductions of β-keto esters and β-diketones (Schemes 20 and 21) [5]. Asymmetric reduction of diketones results in the formation of meso and chiral diols. The highest meso/chiral diol ratio of 2/98 and enantioselectivities of 98% ee are obtained with modified Raney nickel catalysts treated by sonication [5].

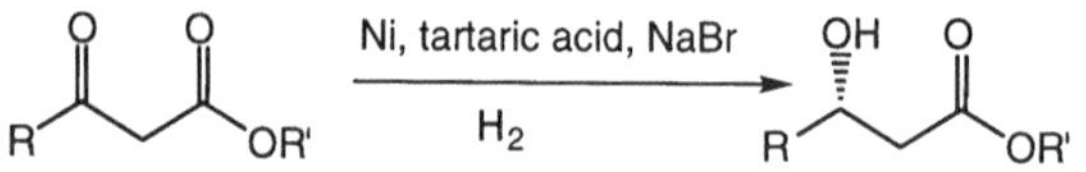

SCHEME 20.

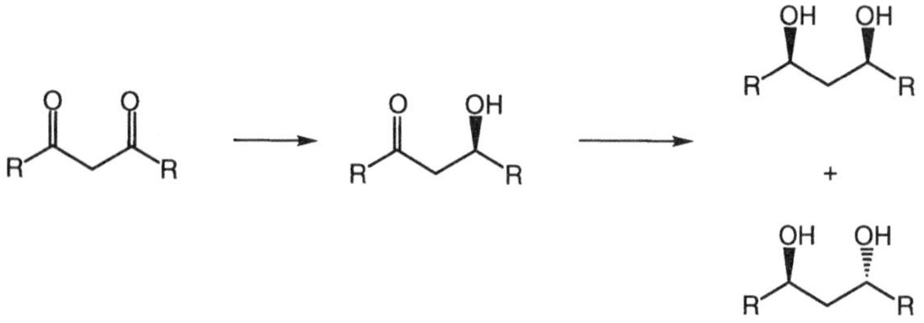

SCHEME 21.

Hoffmann-LaRoche has developed a process that utilizes R,R-tartaric acid/NaBr modified Raney nickel catalyst in the asymmetric hydrogenation of **60** to give **61** in 100% yield and 90–92% ee (6–100-kg scale) for the synthesis of the intermediate of tetrahydrolipstatin (**62**), a pancreatic lipase inhibitor (Scheme 22) [5].

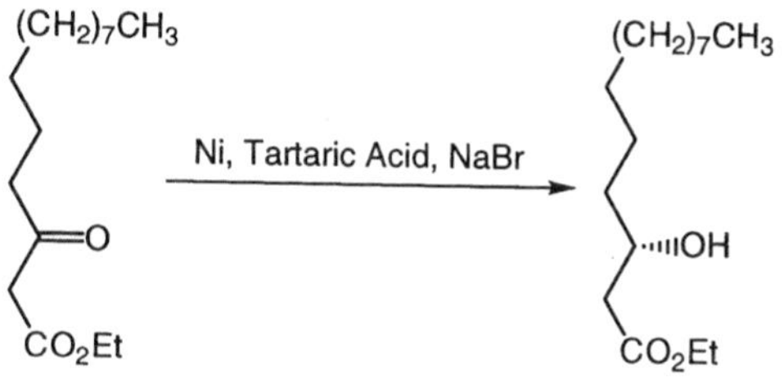

SCHEME 22.

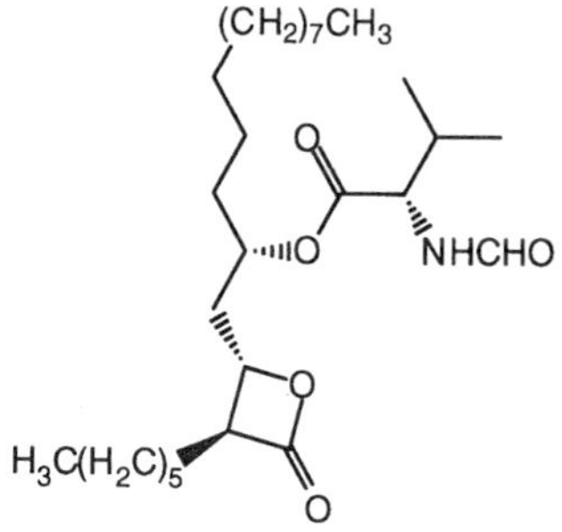

9.3.2. Cinchona-Modified Platinum

The stereoselective reduction of α-keto acid derivatives at a preparative scale are performed with cinchona-modified Pt catalyst. Enantioselectivities range from 57–95% for the reduction of α-keto acid derivatives [5] and are dependent on the preparation of the Pt catalyst [5,18,79,80]

Novartis (Ciba Geigy) has reported the synthesis of Benazepril (63) (Scheme 23), an angiotensin-converting enzyme inhibitor, via an intermediate prepared by cinchona-modified Pt asymmetric hydrogenation (10–200-kg scale, > 98%, 79–82 % ee) [5]. The low optical purities can be tolerated because enantio-enrichment is relatively easy in the latter stages of the synthesis [18].

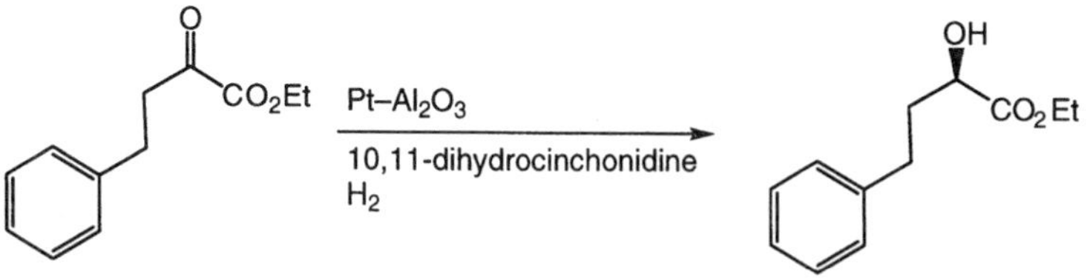

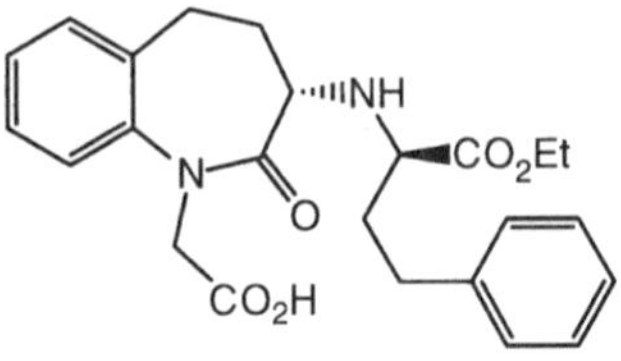

63

SCHEME 23.

9.4. ASYMMETRIC HYDROGEN TRANSFER

Asymmetric hydrogen transfer shows promise for use at industrial scale as ruthenium complexes that contain chiral vicinal diamino (**64**) or amino alcohol (**65**) ligands allow the reductions of substrates such as aryl ketones and imines to be achieved under mild conditions compared to the high pressures required when molecular hydrogen is used [13,81].

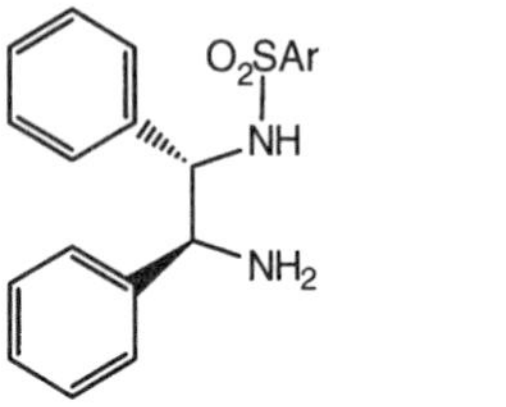

64

a, Ar = 4-CH$_3$C$_6$H$_4$ (TsDPEN)
b, Ar = 2,4,6-(CH$_3$)$_3$C$_6$H$_2$
c, Ar = 1-naphthyl

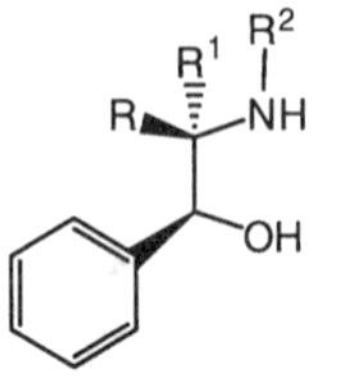

65

a, R = Ph, R^1 = H, R^2 = H
b, R = Ph, R^1 = H, R^2 = Me
c, R = Me, R^1 = H, R^2 = Me
d, R = Ph, R^1 = Me, R^2 = H
e, R = Me, R^1 = Me, R^2 = H

SCHEME 24.

9.4.1. Reductions of Aryl Ketones

The asymmetric reduction of aryl ketone can be achieved with ruthenium catalysts (Scheme 24), prepared separately or in situ by formation of [RuCl$_2$(arene)]$_2$ and ligand, in i-PrOH [81]. The high enantioselectivities and rate are very dependent upon the functionality of the substrate, η^6-arene and N-substitution of the diamino or amino alcohol ligands on ruthenium [81]. The hydrogen transfer reaction in i-PrOH is reversible, necessitating low concentrations, while extensive reaction times degrade the enantioselectivities of the alcohol [81]. This limitation can be overcome by the use of an azeotropic mixture of formic acid and NEt$_3$ (5:2), as the reaction is now irreversible [82].

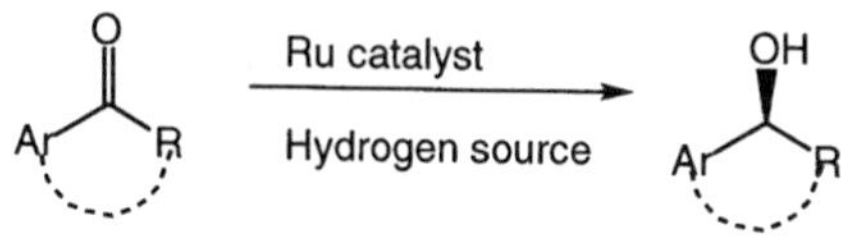

The asymmetric hydrogen transfer of aryl ketones can be accomplished with ruthenium catalysts that contain amino alcohols **65** in modest to high enantioselectivities [83]. With amino alcohol ligands, the optimal rate and stereoselectivities are produced from catalysts prepared in situ with [RuCl$_2$(η^6-C$_6$Me$_6$)]$_2$.

9.4.2. Reductions of Imines

Asymmetric reduction of imines by classic hydrogenation or hydrogen transfer has proven difficult, but selected imines have been reduced by chiral phosphine-Rh or -Ir to give secondary amines in moderate ee or by a chiral *ansa*-titanocene catalyst in 95–100% ee (S/C = 20) [84–90]. A series of preformed ruthenium catalysts [Ru(**64**)(arene)Cl] has successfully reduced a variety of cyclic imines **66** with high enantioselectivities in a 5:2 mixture of HCO$_2$H-NEt$_3$ (Scheme 25). These reductions can be performed in acetone even though the ruthenium catalyst is capable of catalyzing the hydrogen transfer of ketones [91].

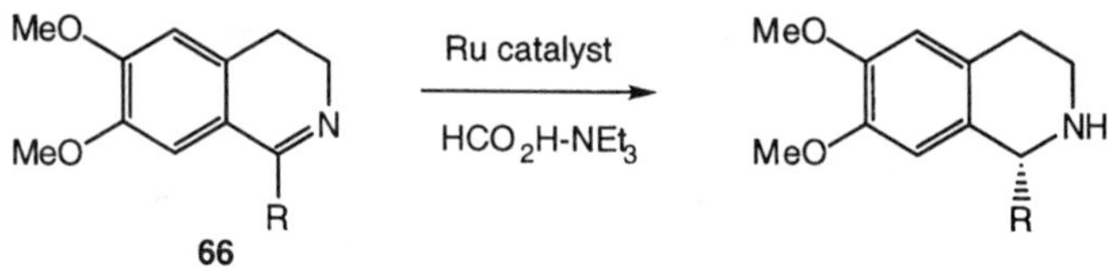

a, R = Me
b, R = 3,4-(CH$_3$O)$_2$C$_6$H$_3$CH$_2$
c, R = 3,4-(CH$_3$O)$_2$C$_6$H$_3$(CH$_2$)$_2$
d, R = Ph
e, R = 3,4-(CH$_3$O)$_2$C$_6$H$_3$

SCHEME 25.

9.5. HYDROFORMYLATION

Hydroformylation, also know as the oxo process, is the transition metal catalyzed conversion of olefins into aldehydes by the addition of synthesis gas (H$_2$/CO) and is the second largest homogeneous process in the world, from which over 12 billion pounds of aldehydes are produced each year [92]. Aldehydes are important intermediates in a variety of processes such as the production of alcohols, lubricants, detergents, and plasticizers. Several transition metals can catalyze the hydroformylation of olefins, but rhodium and cobalt have been used extensively. Despite high volumes of aldehyde products that are manufactured and a deep understanding of the catalytic mechanistic process, no successful industrial asymmetric hydroformylation process has been achieved. Rhodium and platinum catalysts that contain chiral bisphosphine and bisphosphinite ligands have provided the highest enantioselectivies in the study of asymmetric hydroformylation and are chronicled in several reviews [10,92–94].

9.6. HYDROSILYLATION

Optically active alcohols, amines, and alkanes can be prepared by the metal catalyzed asymmetric hydrosilylation of ketones, imines, and olefins [77,94,95]. Several catalytic systems have been successfully demonstrated, such as the asymmetric silylation of aryl ketones with rhodium and Pybox ligands; however, there are no industrial processes that use asymmetric hydrosilylation. The asymmetric hydrosilyation of olefins to alkylsilanes (and the corresponding alcohol) can be accomplished with palladium catalysts that contain chiral monophosphines with high enantioselectivities (up to 96% ee) and reasonably good turnovers (S/C = 1000) [96]. Unfortunately, high enantioselectivities are only limited to the asymmetric hydrosilylation of styrene derivatives [97]. Hydrosilylation of simple terminal olefins with palladium catalysts that contain the monophosphine, MeO-MOP (**67**), can be obtained with enantioselectivities in the range of 94–97% ee and regioselectivities of the branched to normal of the products of 66/43 to 94/6 (Scheme 26) [98,99].

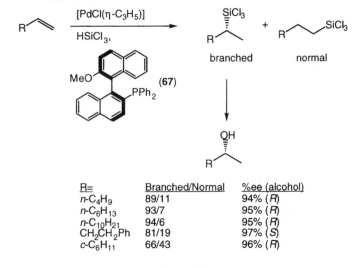

R=	Branched/Normal	%ee (alcohol)
n-C$_4$H$_9$	89/11	94% (R)
n-C$_6$H$_{13}$	93/7	95% (R)
n-C$_{10}$H$_{21}$	94/6	95% (R)
CH$_2$CH$_2$Ph	81/19	97% (S)
c-C$_6$H$_{11}$	66/43	96% (R)

SCHEME 26.

9.7. ASYMMETRIC CYCLOPROPANATIONS

Insecticides of the pyrethroid class, such as *trans*-chrysanthemic acid (**68**) have significant commercial value [100]. An asymmetric synthesis of **68** has been achieved through the use of a chiral copper carbenoid reaction (Scheme 27) [101,102]. With ethyl diazoacetate, equal amounts of the *cis*- and *trans*-cyclopropanes were formed. However, when the size of the alkyl group in the diazoester was increased, the geometric and enantioselectivity increased. The *l*-menthyl group was selected as this gave 93% of the *trans*-isomer with 94% ee.

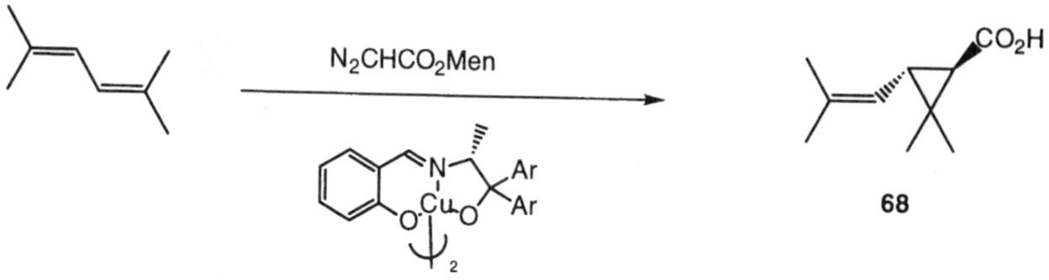

where Ar = 5-*t*-butyl-2-octyloxyphenyl

SCHEME 27.

9.8. CONCLUSIONS

The preparation of chiral material via asymmetric hydrogenation has come a long way. Asymmetric hydrogenation is a powerful and clean method to prepare chiral intermediates and products. However, few asymmetric hydrogenation processes have emerged at the manufacture scale. Since Monsanto's pioneering L-DOPA process, in which a key step is the asymmetric hydrogenation of an enamide to the protected amino acid, several other industrial asymmetric hydrogenation processes have been developed. L-DOPA and L-phenylalanine have been prepared at the manufacture scale from the asymmetric reduction of the corresponding enamide in moderate to high enantioselectivities by rhodium catalysts based on phosphine ligands derived from chiral diamines (PNNP), tartaric acid (DeguPHOS), and D-glucopyranoside (Ph-β-glup).

The discovery of the chiral atropisomeric ligand, BINAP, greatly expanded the number of asymmetric homogeneous hydrogenation catalyses. Rhodium and ruthenium complexes that contain BINAP and similar ligand systems have demonstrated an amazing versatility in the reduction of a wide variety of substrate classes in excellent stereoselectivities and reactivities. (−)-Menthol, a variety of terpenoids, carbapenem intermediates, 1,2-propanediol, and citranellol are manufactured by transition metal catalysis that contain BINAP. There are a number of potential industrial processes that could implement Rh- and Ru(BINAP) catalysts, such as naproxen. Several new ligands similar to BINAP have been synthesized, have demonstrated excellent catalytic reactivities and selectivities, and provide a means of catalytic "fine-tuning" for a specific process.

Phosphines ligands that have chirality from ferrocenes have been implemented in the iridium-catalyzed asymmetric hydrogenation of imine with moderate enantioselectivities for Novartis's manufacture of metolachlor. Electronic modifications of these ferrocenyl ligands have increased the enantioselectivity and catalyst reactivity for Lonza's asymmetric hydrogenation processes of biotin and 2-substituted piperazines, intermediates for several pharmaceutical drugs.

Several asymmetric hydrogenation technologies show high potential for use at an industrial scale. Electronic modifications of Ph-β-glup have produced rhodium catalysts that can reduce a wide variety of enamides to protected amino acids with high enantioselectivities. Alterations of the phosphinate positions of the same D-glucopyranoside backbone can produce rhodium catalysts that can produce the opposite antipode of protected amino acids. Noroyi et al. have developed ruthenium catalysts that contain chiral diamino or amino alcohol ligands that reduce aryl ketones with high stereoselectivities under hydrogen transfer conditions. A class of chiral bisphosphorane ligands (DuPHOS) has been developed by DuPont, whose transition metal complexes are very efficient in the asymmetric reductions of enamides, enol acetates, and β-keto esters.

Although asymmetric heterogeneous hydrogenation catalysts have been overshadowed by homogeneous catalysts, a few industrial processes have been developed, but the scope of these catalytic systems is narrow and highly substrate specific. The enantioselectivities that are obtained with heterogeneous systems are usually lower in comparison to homogeneous or biocatalysis, with some notable exceptions. Although heterogeneous systems are preferred to homogeneous systems in a commercial setting because of convenience with handling and separation, catalyst performance needs to be more reproducible. Raney nickel modified with R,R-tartaric acid and NaBr has been used by Hoffmann-LaRoche in the asymmetric hydrogenation of β-keto esters in moderately high ee's. Novartis (Ciba Geigy) has implemented a cinchona-modified Pt/Al_2O_3 catalyst in the asymmetric hydrogenation of α-keto esters for the industrial preparation of Benzaprin.

The stereoefficiencies of an asymmetric hydrogenation catalyst are not the sole criteria for an industrially feasible process. Many economic factors of a process that govern a successful industrial scale includes the cost of catalyst and ligands, substrate, operation, product through-put, equipment, catalyst activities, and stereoselectivities. Ideally, a low-cost process would have low catalyst loadings (high catalyst activity and TON) and high stereoselectivities in the reduction of an inexpensive substrate at convenient hydrogen pressures and high concentrations. Unfortunately, this is not always the case, as some processes are only stereoselective at high hydrogen pressures (2000 psig), which can be expensive to implement at scale. Other examples of high-cost scenarios include expensive catalysts that are only stereoselective at high catalyst loadings or low substrate concentrations. In addition, the cost and availability of the substrate needed for the asymmetric hydrogenation can be a key factor.

With the development of new catalysts and ligands, the number of asymmetric hydrogenation processes at industrial scale will continue to increase in the years to come.

REFERENCES

1. Knowles, W. S., Sabacky, M. J. *J. Chem. Soc. Chem. Commun.* **1968,** 1445.
2. Horner, L., Siegel, H., Buthe, H. *Angew. Chem. Int. Ed. Engl.* **1968,** *7,* 942
3. Crosby, J. *Tetrahedron* **1991,** *47,* 4789.
4. Ager, D. J., East, M. B. *Asymmetric Synthetic Methodology*; CRC Press: Boca Raton, 1995.
5. Blaser, H. *Tetrahedron: Asymmetry* **1991,** *2,* 843.
6. Harada, K. in *Asymmetric Synthesis*; Morrison, J. D., Ed.; Academic Press: Orlando, FL, 1985; *Vol. 5*; p. 345.
7. Kagan, H. B. in *Asymmetric Synthesis*; Morrison, J. D., Ed.; Academic Press: Orlando, FL, 1985; *Vol. 5*; p. 1.
8. Takaya, H., Ohta, T., Noyori, R. in *Catalytic Asymmetric Synthesis*; Ojima, I., Ed.; VCH Publishers. New York, NY, 1993; p. 1.
9. Koenig, K. E. in *Asymmetric Synthesis*; Morrison, J. D., Ed.; Academic Press: Orlando, FL, 1985; *Vol. 5*; p. 71.
10. Ojima, I., Clos, N., Bastos, C. *Tetrahedron* **1989,** *45,* 6901.
11. Nugent, W. A., RajanBabu, T. V., Burk, M. J. *Science* **1993,** *259,* 479.
12. Noyori, R. *Science* **1990,** *248,* 1194.
13. Noyori, R. *Tetrahedron* **1994,** *50,* 4259.
14. Halpern, J. in *Asymmetric Synthesis*; Morrison, J. D., Ed.; Academic Press: Orlando, 1985; *Vol. 5*; p. 41.
15. Knowles, W. S. *Acc. Chem. Res.* **1983,** *16,* 106.
16. Vineyard, B. D., Knowles, W. S., Sabacky, M. J., Bachman, G. L., Weinkauff, D. J. *J. Am. Chem. Soc.* **1977,** *99,* 5946.
17. Fryzuk, M. D., Bosnich, B. *J. Am. Chem. Soc.* **1977,** *99,* 6262.
18. Blaser, H.-U., Jalett, H.-P., Spindler, F. *J. Mol. Catalysis A: Chemical* **1996,** *107,* 85.
19. Koenig, K. E., Sabacky, M. J., Bachman, G. L., Christoffel, W. C., Barnstorff, H. D., Friedman, R. B., Knowles, W. S., Stults, B. R., Vineyard, B. D., Weinkauff, D. J. *Ann. N.Y. Acad. Sci.* **1980,** *333,* 16.
20. Juge, S., Stephan, M., Laffitte, J. A., Genet, J. P. *Tetrahedron Lett.* **1990,** *31,* 6357.

21. Schmidt, U., Riedl, B., Griesser, H., Fitz, C. *Synthesis* **1991**, 655.
22. Juge, S., Genet, J. P. *Tetrahedron Lett.* **1989**, *30*, 2783.
23. Juge, S., Stephan, M., Merdes, R., Genet, J. P., Halut-Desportes, S. *J. Chem. Soc. Chem. Commun.* **1993**, 531.
24. Juge, S., Genet, J. P. *US Patent*, **1991**, 5,043,465.
25. Noyori, N. *Chem. Soc. Rev.* **1989**, *18*, 187.
26. Noyori, R., Takaya, H. *Acc. Chem. Res.* **1990**, *23*, 345.
27. Takaya, H., Akutagawa, S., Noyori, R. *Org. Synth.* **1989**, *67*, 20.
28. Cai, D., Payack, J. F., Bender, D. R., Hughes, D. L., Verhoeven, T. R., Reider, P. J. *J. Org. Chem.* **1994**, *59*, 7180.
29. Cai, D., Payack, J. F., Verhoeven, T. R. *US Patent*, **1995**, 5399771.
30. Ager, D. J., East, M. B., Eisenstadt, A., Laneman, S. A. *J. Chem. Soc. Chem. Commun.* **1997**, 2359.
31. Akutagawa, S., Tani, K. in *Catalytic Asymmetric Synthesis*; Ojima, I., Ed.; VCH Publishers: New York, NY, 1993; p. 41.
32. Otsuka, S., Tani, K. in *Asymmetric Synthesis*; Morrison, J. D., Ed.; Academic Press: Orlando, FL, 1985; *Vol. 5*; p. 171.
33. Kagan, H. B. *Bull. Soc. Chim. Fr.* **1988**, *5*, 846.
34. Kumobayashi, H., Miura, T. *Chiral USA '96*, Boston, MA, 1996.
35. Akutagawa. *Applied Catalysis A: General* **1995**, *128*, 171.
36. Noyori, R., Ikeda, T., Ohkuma, T., Widhalm, M., Kitamura, M., Takaya, H., Akutagawa, S., Sayo, N., Saito, T., Taketomi, T., Kumobayashi, H. *J. Am. Chem. Soc.* **1989**, *111*, 9134.
37. Ager, D. J., Laneman, S. A. *Tetrahedron: Asymmetry* **1997**, *8*, 3327.
38. Zhang, X., Mashima, K., Koyano, K. Sayo, N., Kumobayashi, H., Akutagawa, S., Takaya, H. *J. Chem. Soc. Perkin Trans.* **1994**, *1*, 2309.
39. Chan, A. S. C., Laneman, S. A. *Unpublished results*.
40. Chan, A. S. C. *CHEMTECH* **1993**, *March*, 46.
41. Chan, A. S. C., Laneman, S. A. *US Patent*, **1993**, 5,202,473.
42. Wan, K. T., Davis, M. E. *Nature* **1994**, *370*, 449.
43. Takaya, H., Ohta, T., Sayo, N., Kumobayashi, H., Akutagawa, S., Inoue, S., Kasahara, I., Noyori, R. *J. Am. Chem. Soc.* **1987**, *109*, 1596.
44. Noyori, R., Ohkuma, T., Kitamura, M. *J. Am. Chem. Soc.* **1987**, *109*, 5856.
45. Ikariya, T., Ishii, Y., Kawano, H., Arnai, T., Saburi, M., Yoshikawa, S., Akutagawa, S. *J. Chem Soc. Chem. Commun.* **1985**, 922.
46. Mashima, K., Kusano, K., Ohta, T., Noroyi, R., Takaya, H. *J. Chem. Soc. Chem. Commun.* **1989**, 1208.
47. Kitamura, M., Tokunaga, M., Ohkuma, T., Noroyi, R. *Tetrahedron Lett.* **1991**, *32*, 4163.
48. Heiser, B., Broger, E. A., Crameri, Y. *Tetrahedron: Asymmetry* **1991**, *2*, 51.
49. Hoke, J. B., Hollis, L. S., Stern, E. W. *J. Organometal. Chem.* **1993**, *455*, 193.
50. Genet, J. P., Pinel, C., Ratovelomanana-Vidal, V., Mallart, S., Pfister, X., Bischoff, L., De Andrade, M. C. C., Darses, S., Galopin, C., Laffitte, J. A. *Tetrahedron: Asymmetry* **1994**, *5*, 675.
51. Chan, A. S. C., Laneman, S. A. *US Patent*, **1992**, 5,144,050.

52. Shao, L., Takeuchi, K., Ikemoto, M., Kawai, T., Ogasawara, M., Takeuchi, H., Kawano, H., Saburi, M. *J. Organometal. Chem.* **1992**, *435*, 133.
53. Noyori, N., Kitamura, M., Sayo, N., Kumobayashi, H., Giles, M. F. US Patent, **1993**, 5,198,562.
54. Fiorini, M., Giongo, G. M. *J. Mol. Catal.* **1979**, *5*, 303.
55. Pracejus, G., Pracejus, H. *Tetrahedron Lett.* **1977**, *39*, 3497.
56. Fiorini, M., Riocci, M., Giongo, M. *Eur. Pat. Appl.* **1983**, 0077099 A2.
57. Spindler, F., Pugin, B., Jalett, H.-P., Buser, H.-P., Pittelkow, U., Blaser, H.-U. in *Chem. Ind.*; Malz, J., Ed., Marcel Dekker: New York, **1996**; *Vol. 68*; p. 153.
58. Stinson, S. C. *Chem. Eng. News* **1997**, *75*, 34.
59. Blaser, H.-U., Spindler, F. *Chimia* **1997**, *51*, 297.
60. Hayashi, T., Mise, T., Fukushima, M., Kagotani, M., Nagashima, N., Hamada, Y., Matsumoto, A., Kawakami, S., Monishi, M., Yamomoto, K., Kumada, M. *Bull. Chem. Soc. Jpn.* **1980**, *53*, 1138.
61. Brieden, W. *World Patent*, **1996**, WO 96/16971 A1.
62. Togni, A., Breutel, C., Schnyder, A., Spindler, F., Landert, H., Tijani, A. *J. Am. Chem. Soc.* **1994**, *116*, 4062.
63. Barbaro, P., Pregosin, P. S., Salzmann, R., Albinati, A., Kunz, R. *Organometal.* **1995**, *14*, 5160.
64. Breutel, C., Pregosin, P. S., Salzmann, R., Togni, A. *J. Am. Chem. Soc.* **1994**, *116*, 4067.
65. Brieden, W. *Chiral USA '97*, Boston, MA, **1997**, 45.
66. Brieden, W. Personal communication, **1997**.
67. Bader, R. R., Baumeister, P., Blaser, H.-U. *Chimia* **1996**, *30*, 9.
68. Imwinkelried, R. *Chimia* **1997**, *51*, 300.
69. Selke, R. *J. Organometal. Chem.* **1989**, *370*, 249.
70. Vocke, W., Hanel, R., Flother, F.-U. *Chem. Techn.* **1987**, *39*, 123.
71. RajanBabu, T. V., Ayers, T. A., Casalnuovo, A. L. *J. Am. Chem. Soc.* **1994**, *116*, 4101.
72. Andrade, J. G., Prescher, G., Schaefer, A., Nagel, U. in *Chem. Ind.*; Kosik, J., Ed.; Marcel Dekker: New York, **1990**; p. 33.
73. Nagel, U., Kinzel, E., Andrade, J., Prescher, G. *Chem. Ber.* **1986**, *119*, 3326.
74. Pugin, B., Muller, M., Spindler, S. US Patent, **1993**, 5,306,853.
75. Achiwa, K. *J. Am. Chem. Soc.* **1976**, *98*, 8265.
76. Baker, G. L., Fritschel, S. J., Stille, J. R., Stille, J. K. *J. Org. Chem.* **1981**, *46*, 2954.
77. Ojima, I., Kogure, T., Yoda, N. *J. Org. Chem.* **1980**, *45*, 4728.
78. Takeda, H., Tachinami, T., Aburatani, M., Takahashi, H., Morimoto, T., Achiwa, K. *Tetrahedron Lett.* **1989**, *30*, 363.
79. Garland, M., Blaser, H.-U. *J. Am. Chem. Soc.* **1990**, *112*, 7048.
80. Augustine, R. L., Tanielyan, S. K., Doyle, L. K. *Tetrahedron: Asymmetry* **1993**, *4*, 1803.
81. Hashiguchi, S., Fujii, A., Takehara, J., Ikariya, T., Noyori, R. *J. Am. Chem. Soc.* **1995**, *117*, 7562.

82. Fujii, A., Hashiguchi, S., Uematsu, N., Ikariya, T., Noyori, R. *J. Am. Chem. Soc.* **1996**, *118*, 2521.

83. Takehara, J., Hashiguchi, S., Fujii, A., Inoue, S., Ikariya, T., Noroyi, R. *J. Chem. Soc. Chem. Commun.* **1996**, 233.

84. Lee, N. E., Buchwald, S. L. *J. Am. Chem. Soc.* **1994**, *116*, 5985.

85. Bakos, J., Toth, I., B., H., Marko, L. *J. Organometal. Chem.* **1985**, *279*, 23.

86. Becalski, A. G., Cullen, W. R., Fryzuk, M. D., James, B. R., Kang, G.-J., Rettig, S. J. *Inorg. Chem.* **1991**, *30*, 5002.

87. Bakos, J., Orosz, A., Heil, B., Laghmari, M., Lhoste, P., Sinou, D. *J. Chem. Soc. Chem. Commun.* **1991**, 1684.

88. Chan, Y., Ng, C., Osborn, J. A. *J. Am. Chem. Soc.* **1990**, *112*, 9600.

89. Morimoto, T., Achiwa, K. *Tetrahedron: Asymmetry* **1995**, *6*, 2661.

90. Hoveyda, A. H., Morken, J. P. *Angew. Chem. Int. Ed. Engl.* **1996**, *35*, 1262.

91. Uematsu, N., Fujii, A., Hashiguchi, S., Ikariya, T., Noyori, R. *J. Am. Chem. Soc.* **1996**, *118*, 4916.

92. Stanley, G. G. Carbonylation Processes by Homogeneous Catalysis. in *Encyclopedia of Inorganic Chemistry*; King, R. B., Ed.; J. Wiley: New York, 1994; *Vol. 2*; p. 575.

93. Consiglio, G. in *Catalytic Asymmetric Synthesis*; Ojima, I., Ed.; VCH Publishers: New York, 1993; p. 273.

94. Ojima, I., Hirai, K. in *Asymmetric Synthesis*; Morrison, J. D., Ed.; Academic Press: Orlando, 1985; *Vol. 5*; p. 103.

95. Brunner, H., Nishiyama, H., Itoh, K. in *Catalytic Asymmetric Synthesis*; Ojima, I., Ed.; VCH Publishers: New York, 1993; p. 303.

96. Kitayama, K., Uozumi, Y., Hayashi, T. *J. Chem. Soc. Chem. Commun.* **1995**, 1533.

97. Kitayama, K., Tsuji, H., Uozumi, Y., Hayashi, T. *Tetrahedron Lett.* **1996**, *37*, 4169.

98. Uozumi, Y., Hayashi, T. *J. Am. Chem. Soc.* **1991**, *113*, 9887.

99. Uozumi, Y., Kitayama, K., Hayashi, T., Yanagi, K., Fukuyo, E. *Bull. Chem. Soc. Jpn.* **1995**, *68*, 713.

100. Matsui, M., Yamamoto, I. in *Naturally Occurring Insecticides*; Jacobson, M., Crosby, D. G., Eds.; Marcel Dekker: New York, 1971; p. 3.

101. Nozaki, H., Moriuti, S., Takaya, H., Noyori, R. *Tetrahedron Lett.* **1966**, 5239.

102. Aratani, T. *Pure Appl. Chem.* **1985**, *57*, 1839.

10

Pericyclic Reactions

MICHAEL B. EAST
NSC Technologies, Mount Prospect, Illinois

10.1. INTRODUCTION

The literature abounds with examples of pericyclic (and pericyclic-type*) reactions that have been used to synthesize a broad range of target molecules, including natural products [1–3]. Although few of these reactions have been scaled up outside of the laboratory, their potential to provide a complex target molecule with a minimal number of transformations is unparalleled in the organic chemist's repertoire. For example, in the Diels-Alder reaction, up to six stereogenic centers can be controlled in a single reaction [4]. Recent progress, especially with regard to asymmetric catalysts, should encourage the use of these reactions at scale.

This chapter has been organized into sections based on the major reaction types. Hence, the hetero-Diels-Alder reaction has been included as a subsection of the Diels-Alder reaction. Although there are numerous reports of asymmetric pericyclic reactions, our discussion concentrates on reactions that are most likely to allow for scale up.

10.2. THE DIELS-ALDER REACTION

The Diels-Alder reaction, now more than 60 years old, has regained new prominence with the ability to use asymmetric technology to control relative and abso-

---

* Some transformations do not proceed in a concerted manner, but follow a stepwise mechanism. These reactions have been included where the rules governing pericyclic reactions can also predict the stereochemical outcome of these reactions.

lute stereochemistry while creating two new carbon–carbon bonds [1–3,5,6]. Because of the diversity of dienes and dienophiles, the application of asymmetric methodology to the Diels-Alder reaction has lagged behind other areas such as aldol reactions [3]. However, considerable advances have been made in recent years through the use of chiral dienophiles and catalysts [3,6–21]

10.2.1. Regiochemistry and Stereochemistry

Cycloadditions that involve two unsymmetrical reactants can lead to regioisomers. The regioselectivity of these adducts can be predicted with a high degree of success through the use of frontier molecular orbital theory [22–25]. The *ortho* product* is usually the preferred isomer from 1-substituted dienes, while 2-substituted dienes provide the *para* isomer as the major adduct. However, when a Lewis acid is used as a catalyst in the reaction, the ratio of these isomers can alter dramatically, and occasionally can be reversed [22].

In addition, Diels-Alder adducts are formed through two types of approach that lead to *endo* or *exo* isomers. The *endo* isomer is usually favored over the *exo* isomer, although the *exo* isomer is generally the thermodynamically preferred product. This is known as the Alder, or *endo* rule and can be attributed to the additional stability gained by secondary molecular orbital overlap during the cycloaddition [22,26,27]. Again, the use of a Lewis acid catalyst can alter the *endo/exo* ratio and has even been shown to give the thermodynamic *exo* adduct as the major product [28].

Complete stereoselectivity occurs in the Diels-Alder reaction through *syn* addition of the dienophile to the diene. Hence, the reaction of dimethyl fumarate and dimethyl maleate with cyclopentadiene yield the *trans* and *cis* adduct, respectively (Scheme 1) [29].

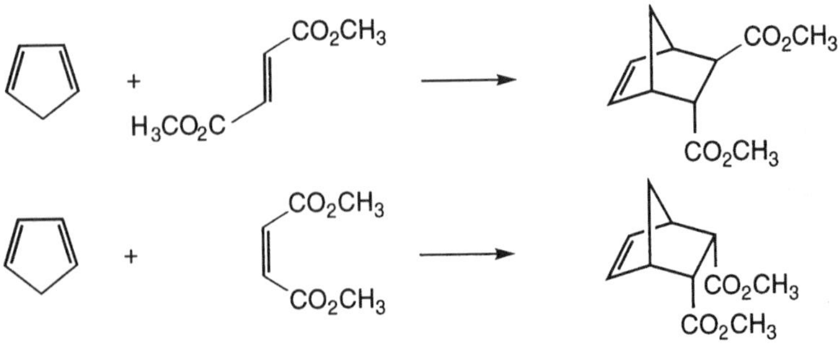

SCHEME 1.

---

* This nomenclature follows the analogy of disubstituted aromatic systems.

10.2.2. Catalysts

The financial advantage of using a catalyst on an industrial scale is obvious. In addition to increased regiochemistry and stereochemistry, the addition of a Lewis acid catalyst often allows for a dramatic increase in the rate of Diels-Alder reactions; this has been attributed to changes in energy of molecular orbitals when complexed so that there is a lower activation energy [22,30]. Although highly diastereoselective thermal Diels-Alder cycloadditions are known [31], the addition of a Lewis acid appears to be essential to maximize selectivity for the vast majority of Diels-Alder reactions [22,32–34]. The use of a catalyst allows the reaction to be run at lower temperatures and, thus, allows for greater differentiation of the diastereomeric transition states.

A large number of Lewis acids have been advocated for the catalysis of the Diels-Alder reaction [3]. The vast majority are based on zinc, aluminum, titanium, and boron derivatives. It should be noted that the use of some metals, such as zinc, cause significant disposal problems in waste water streams and can add significantly to costs. Polymerization can be minimized by use of Lewis acids where strongly electron-withdrawing substituents, such as chloride, have been replaced by more moderate ligands, such as alkoxy, or weaker Lewis acids such as lanthanide complexes [34–42]. Although these changes allow for the preservation of fragile functional groups and avoid substrate decomposition, they also compromise the Lewis acidity, and the catalyst may be less effective. In addition to Lewis acids, proteins, antibodies, enzymes, and radicals have been found to catalyze the Diels-Alder cycloaddition [43].

10.2.2.1. Chiral Lewis Acids

The discovery that chiral Lewis acids can catalyze the asymmetric Diels-Alder reaction is a major milestone for the scale up and use of this reaction on an industrial scale. The use of such a catalyst obviates the need to use a chiral auxiliary on the diene or dienophile. The vast majority of chiral auxiliaries that have been used in the Diels-Alder reaction are either expensive or not commercially available. In addition, the chemical steps needed to attach and remove the chiral auxiliary increase the cost and complexity of the synthesis. Chiral catalysts may also be recovered, further decreasing cost [44]. Research in this area is very active, and catalysts based on a number of metals (Table 10.1) have shown encouraging asymmetric induction [21]. Our understanding of the role these catalysts play in the asymmetric induction of Diels-Alder reactions is increasing and more general reagents should appear [27,45–51].

The reaction of methacrolein with cyclopentadiene catalyzed by a chiral menthoxyaluminum complex gives adducts with ee's of up to 72%, but with other dienophiles, little, if any, induction was noted [85,86]. A chiral cyclic amido aluminum complex (1) catalyzes the cycloaddition of cyclopentadiene with the

TABLE 10.1 Chiral Catalysts Used in the Diels-Alder Reaction

Ligand system	Metal	Reference	Ligand system	Metal	Reference
	Ti	52		Al, B, Ti	11, 12, 53–59
	Al	44, 50, 60, 61		Al, B, Ti	13, 49, 62–66
	Al, Sn, Ti	56		Ti	14
	Cu, Fe, Mg	8, 67, 68		Cu	69
	Cu	16, 70		Cr	15
	Al	71		Al, Sn, Ti	56, 72–76
	Ti	77–79		Al, B, Sn, Ti	53, 56, 80–82
	B	49, 83, 84		Al	56
	Al	19	R*OMCl₂	Al	85, 86
	Co	87		B	88

trans-crotyl derivative (**2**) in good yield and enantioselectivity (Scheme 2) [44]. This chiral catalyst can also be easily recovered.

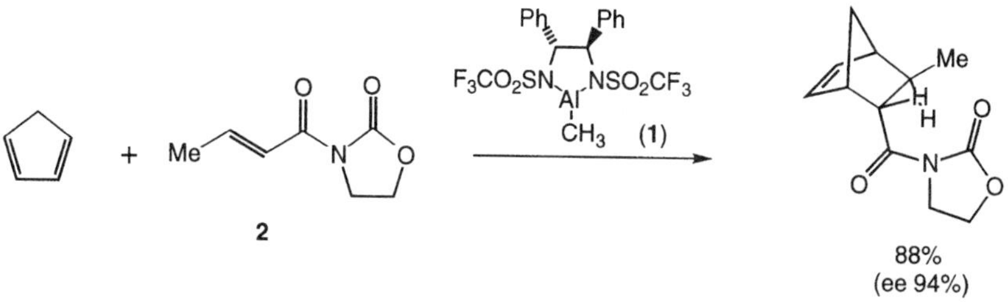

88%
(ee 94%)

SCHEME 2.

High ee's have been observed in analogous reactions with titanium catalysts derived from tartaric acid [53,72,81] and bisoxazolines [17,67,87,89,90].

To date, the vast majority of successful chiral catalysts are based on tartaric acid, BINAP, or oxazolidinone derivatives (Table 10.1). As derivatives of both these compounds are commercially available, scale up should not present a problem. If the observed asymmetric induction is found to be low with catalysts based on tartaric acid or oxazolidinones, the sterically hindered titanium BINAP-type complexes should allow for increased selectivity. In addition, nontoxic metal counterions such as iron and aluminum do not appear to compromise the asymmetric induction.

10.2.2.2. Chiral Bases

In addition to Lewis acid catalysts, chiral bases catalyze the Diels-Alder reaction [91,92]. Although the use of bases as catalysts does show promise, especially for acid-sensitive functionality, the current level of asymmetric induction may not be acceptable for scale up. In addition, only a limited number of dienes and dienophiles have been shown to undergo the chiral base catalyzed Diels-Alder reaction.

10.2.2.3. Biological Catalysts

Although the use of enzymes as catalysts in the Diels-Alder reaction has not been particularly successful [43,93,94], abzymes can be used to control regioselectivity and stereoselectivity in the Diels-Alder reaction [43,95]. Biological catalysts have not been used on significant scale in the Diels-Alder reaction. Furthermore, the use of abzymes would require a large investment initially, and any subtle change in the substituents on either the diene or dienophile may dramatically reduce asymmetric induction and yield.

10.2.3. Chiral Dienophiles

Before the advent of chiral catalysts, the most widely used approach to chiral adducts involved the use of chiral dienophiles and, of these, acrylates, derived from a chiral alcohol or amine and acryloyl chloride, are the most common (Table 10.2.) [3,4]. The use of a chiral auxiliary or group on the dienophile can provide for face selectivity. High *endo–exo* selectivities have been achieved with bicyclic adducts, together with high asymmetric induction. Chiral dienophiles can be classified as either type I or type II reagents (Figure 1) [4]. Type I reagents, such as chiral acrylates, incorporate a chiral group in a simple and straightforward manner. Type II reagents, where the chiral group is one atom closer to the double bond, are more difficult to synthesize and recycle. In addition, significant stereoselection is only observed in the presence of a Lewis acid (see 10.2.2.); this is especially true of type I reagents [22,32–34]. However, for type II dienophiles such as the sulfoxide **3**, thermal Diels-Alder conditions can provide high asymmetric induction and excellent overall yields (e.g., Scheme 3) [123].

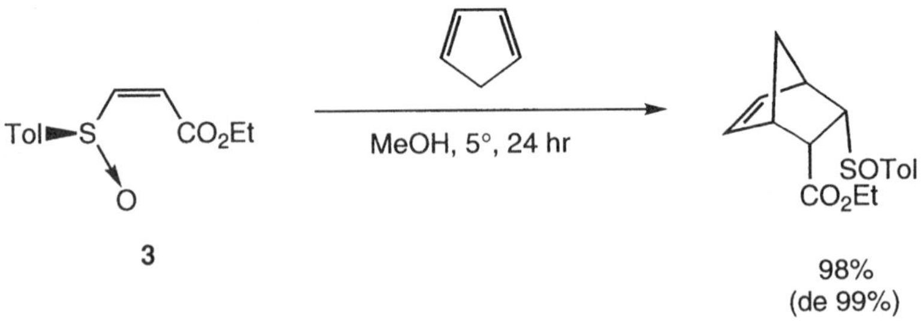

Tol—S   CO$_2$Et   MeOH, 5°, 24 hr   SOTol
      ‖   CO$_2$Et
      O

3

98%
(de 99%)

SCHEME 3.

Although a large number of chiral dienophiles have been developed (Table 10.2), their ability to provide high asymmetric induction appears to be limited to specific dienes. However, there are some dienophiles that tolerate a wider variety of dienes, including menthol derivatives [108,109] camphor derivatives [6,36,37,96,98–104,171,172], and oxazolidinones [111,156,173,174]. It should be noted that even these auxiliaries would require an efficient recycle protocol for economic scale up.

The ethyl aluminum dichloride catalyzed synthesis of (R)-(+)-cyclohex-3-enecarboxylic acid, using galvinoxyl to inhibit polymerization, has been successfully scaled up to the kilogram level [97]. An improved synthesis of the chiral auxiliary, *N*-acryloylbornane-10,2-sultam, was also described together with a recycle protocol.

10.2.4. Chiral Dienes

Chiral dienes have proved to be less popular in asymmetric Diels-Alder reactions than their chiral dienophile counterparts. This is primarily due to the problem of

TABLE 10.2 Chiral Dienophiles Used in the Diels-Alder Reaction

Dienophile	Catalyst	Reference	Dienophile	Catalyst	Reference
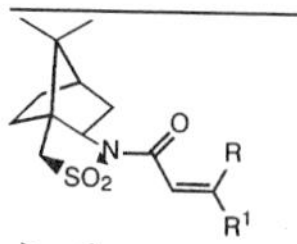	TiCl$_4$	6, 96, 97	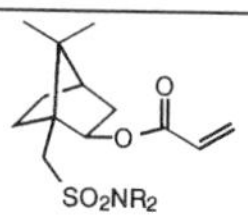	TiCl$_2$(OiPr)$_2$ EtAlCl$_2$ Me$_2$AlCl	98
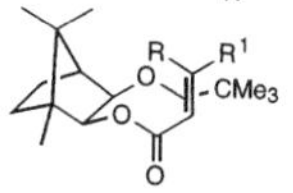	TiCl$_2$(OiPr)$_2$	36, 37	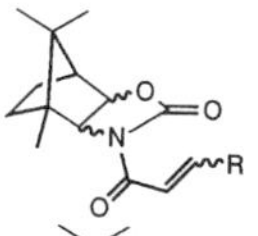	Et$_2$AlCl BF$_3$·OEt$_2$ TiCl$_4$ TiCl$_2$(OiPr)$_2$	99–101
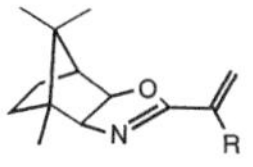		102–104	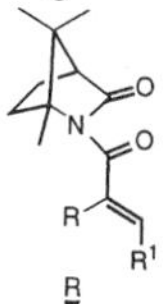	MeAlCl$_2$	105
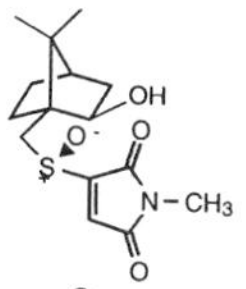	ZnCl$_2$	106	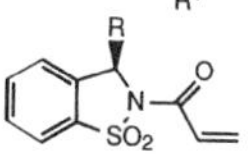	Me$_2$AlCl EtAlCl$_2$	35, 107
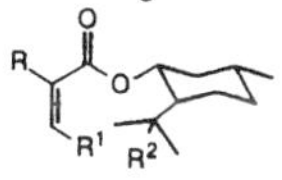	TiCl$_4$ Et$_2$AlCl Clay	37, 98, 108– 117	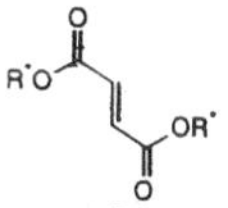	Et$_2$AlCl iBu$_2$AlCl AlCl$_3$	118–120
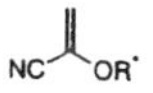	ZnX$_2$	121, 122	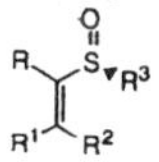	ZnX$_2$ MgBr$_2$·OEt$_2$ SiO$_2$ LiClO$_4$ Et$_2$AlCl BF$_3$·OEt$_2$ Eu(fod)$_3$ TiCl$_4$	7, 31, 123– 137
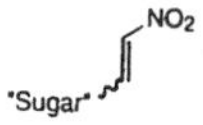	—	138	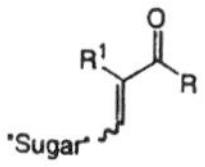	— Et$_3$Al EtAlCl$_2$ Et$_2$AlCl Cp$_2$TiCl$_2$	122, 139– 143
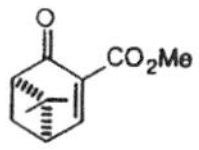	ZnCl$_2$ BF$_3$·OEt$_2$ SnCl$_4$ Et$_2$AlCl	144	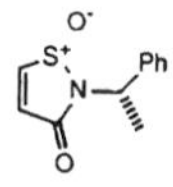	—	145
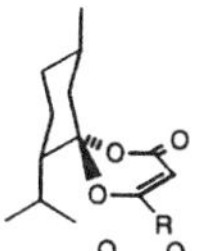	—	146	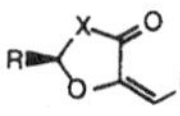 where X = O, NAc	—	147–150
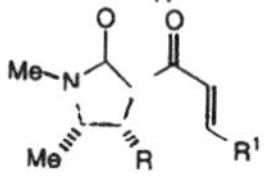	Et$_2$AlCl	151	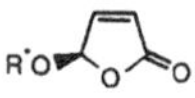	—	152, 153

TABLE 10.2   Continued

Dienophile	Catalyst	Reference	Dienophile	Catalyst	Reference
(cyclopentenone, R)	BF₃ · OEt₂	10	*(MenO₂C ⟶ CO₂Men, AlCl₃)*	AlCl₃	154, 155
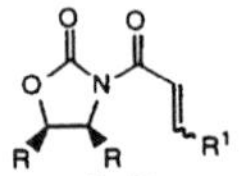 *(oxazolidinone acryloyl, R, R, R¹)*	Et₂AlCl Me₂AlCl	111, 156– 158	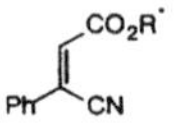 *(Ph, CN, CO₂R*)*	—	159
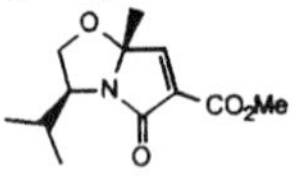 *(bicyclic, CO₂Me)*	ZnCl₂	160	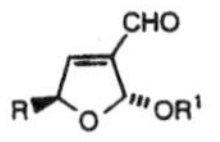 *(R, CHO, OR¹ furanone)*	Eu(fod)₃	161
(pyrrolidinone acryloyl, R¹, R)	— Eu(fod)₃ t-BuMe₂SiOTf	162, 163	*(R, succinimide N–CH₃, TiCl₄)*	TiCl₄	164–166
(OEt acetal)	TiCl₄	164–166	*(R¹, tBu, OH)*	ZnCl₂ Ti(OiPr)₄ BF₃ · OEt₂	167
(OEt / NR₂)*	BF₄⁻·OEt₃⁺	168	*(Ph₃CO, cyclopentanedione)*	ZnCl₂ TiCl₂(OiPr)₂ BF₃ · OEt₂ Et₂AlCl EtAlCl₂ iBu₂AlCl TiCl₄	169
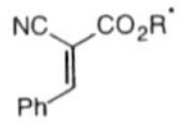 *(NC, CO₂R*, Ph)*	EtAlCl TiCl₄	170			

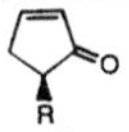

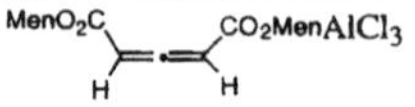

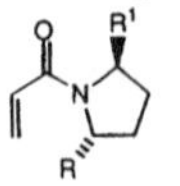

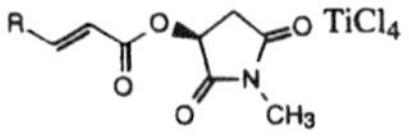

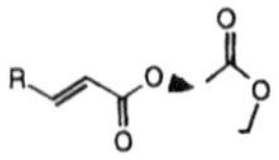

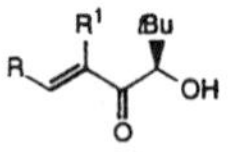

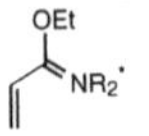

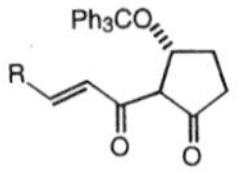

[a] The major products are endo.

designing a molecule that incorporates a chiral moiety such as the formation of a chiral isoprenyl ether or vinyl ketene acetal [175–178]. In addition, diastereoselectivities often are not high [51,179–187], as illustrated by the cycloaddition of the chiral butadiene (4) with acrolein (Scheme 4). Improved stereoselection is observed through the use of double asymmetric induction, although this is a somewhat wasteful protocol [32,51,167,188].

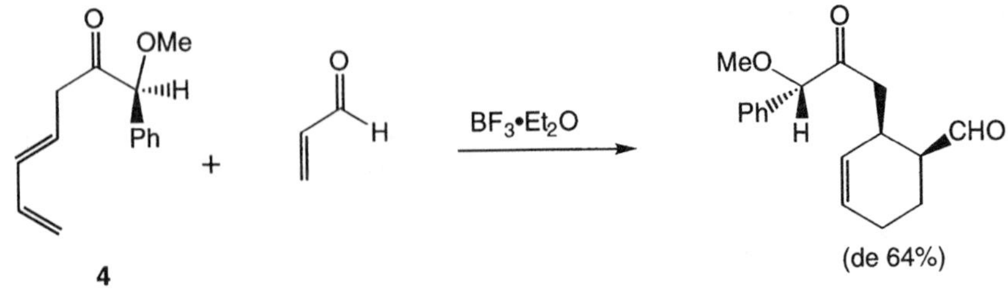

SCHEME 4.

Type I   Type II

FIGURE I   Type I and II dienophiles for the Diels-Alder reaction.

Sugar substitued dienes are readily accessible, and although asymmetric induction may be low, the desired product is often easily isolated by crystallization [180,182,183,189,190]. Chiral dienophiles are usually more difficult to prepare, and it is often difficult to recycle the chiral auxiliary. Hence, Diels-Alder reactions involving chiral dienophiles should be avoided unless the chiral auxiliary becomes incorporated into the target molecule.

10.2.5. Cooperative Blocking Groups

The use of a chiral fumarate ester allows for asymmetric induction irrespective to the approach of the dienophile to the diene. In particular, dimethyl fumarate (**5**) has been advocated for large scale because of its ready availability, low cost, excellent yields, and high asymmetric induction [32,119,120,191–196]. Although other more exotic chiral auxiliaries may be used [32], the use of **5** coupled with a homogeneous Lewis acid catalyst at low temperatures allows for remarkably high diastereoselectivity with a number of dienes (Scheme 5) [120,155, 197].

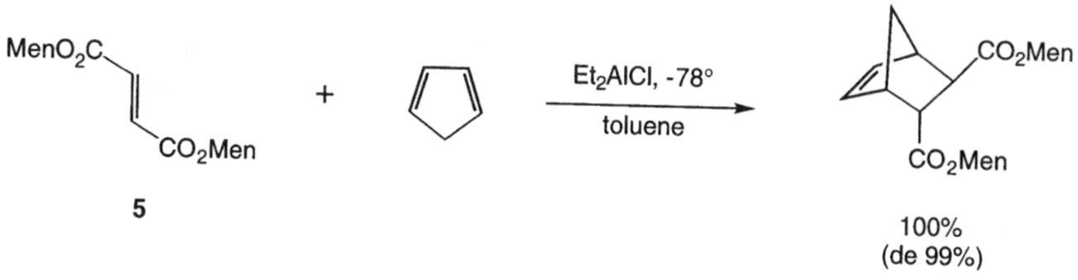

SCHEME 5.

10.2.6. Solvent Effects

The choice of solvent has had little, if any, effect on the majority of Diels-Alder reactions [198,199]. Although the addition of a Lewis acid might be expected to show more solvent dependence, generally there appears to be little effect on asymmetric induction [109,120]. However, a dramatic effect of solvent polarity has been observed for chiral metallocene triflate complexes [200]. The use of polar solvents, such as nitromethane and nitropropane, leads to a significant improvement in the catalytic properties of a copper Lewis acid complex in the hetero Diels-Alder reaction of glyoxylate esters with dienes [201].

10.2.7. Inverse Electron Demand Diels-Alder Reactions

The asymmetric cycloaddition of electron-rich dienophiles with electron-poor dienes, although discovered relatively recently, has provided some encouraging results (Scheme 6) [202–211]. It should be noted that a multistep synthesis is often required to synthesize the diene [203,208], and a recycle protocol for the chiral auxiliary may not be possible [202,206,208].

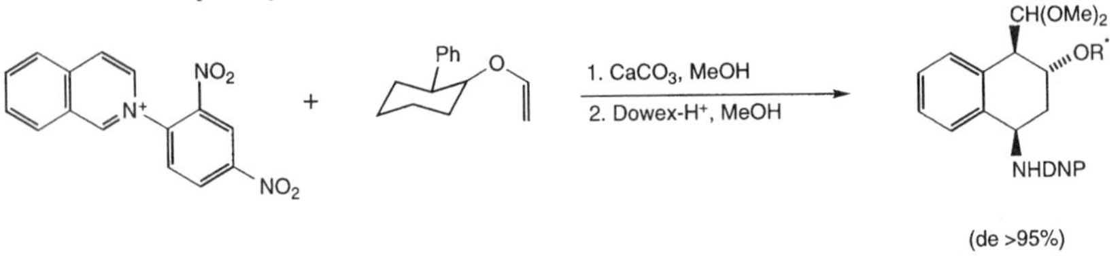

SCHEME 6.

10.2.8. Intramolecular Diels-Alder Reactions

The intramolecular Diels-Alder reaction has provided a large number of valuable intermediates for the synthesis of polycyclic compounds and has been used in the synthesis of a number of natural products [32,212–216]. A distinct advantage of this reaction is the ability to construct the reactant to provide for either *endo* or *exo* attack, which allows for excellent stereoselection [213,217–222].

The simplicity and power of this reaction is illustrated by the spontaneous cyclization of the derivative **6**, formed from the condensation of (*R*)-citronellal (**7**) and 5-*n*-pentyl-1,3-cyclohexanedione, which occurs with complete stereochemical control (Scheme 7) [223].

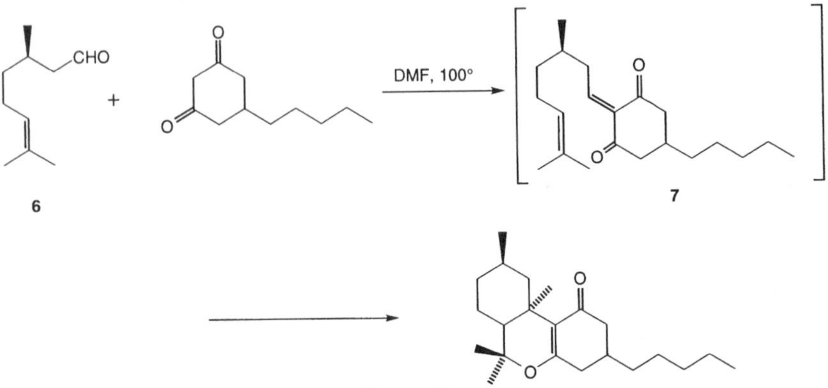

SCHEME 7.

Lewis acid catalysts have been used in this approach [213,224–226]. There are also a number of examples of hetero intramolecular Diels-Alder reactions [213,226–229].

10.2.9. Hetero Diels-Alder Reactions

Although the reactions of carbonyl compounds with dienes follow a stepwise addition, the overall stereochemical outcome mimics that of the Diels-Alder reaction [1,2,5,230–232]. The products of the hetero Diels-Alder reaction have been used as substrates for a wide variety of transformations and natural products syntheses [230,233–246]. Many hetero Diels-Alder reactions use chiral auxiliaries or catalysts that have been success for the Diels-Alder reaction.

As with the Diels-Alder reaction, the addition of a Lewis acid appears to be essential to maximize selectivity [247–251]. Again, it should be noted that the use some metals, such as zinc, cause significant disposal problems in waste water streams and can add significantly to costs. The type of the Lewis acid used can determine the relative stereochemical outcome of the reaction (Scheme 8) [247,252].

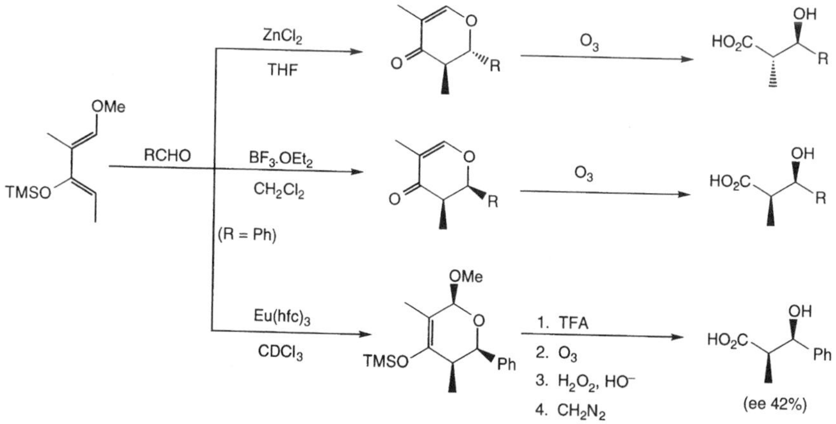

SCHEME 8.

Chiral catalysts do provide high degrees of asymmetric induction [58,253–260]. Double asymmetric induction was observed in the synthesis of homochiral pyranose derivatives using a chiral auxiliary–chiral catalyst combination [261, 262].

10.3. CLAISEN-TYPE REARRANGEMENTS

The Claisen rearrangement has been instrumental in the synthesis of a number of natural products [263–273]. A number of useful derivatives have been prepared using the Claisen-type rearrangement, including enol ethers [274], amides [275–277], esters and orthoesters [278–280], acids [281,282], oxazolines [283], ketene acetals [284,285], and thioesters [286]. Many of these variants use a cyclic primer to control relative and absolute stereochemistry. The Claisen and oxy Cope provide the best candidates for scale up because of the nonreversible nature of these reactions.

10.3.1. Cope Rearrangement

The [3,3]-sigmatropic rearrangement of a 1,5-hexanediene is known as the Cope rearrangement and usually proceeds through a chair transition state. Generally, a large substituent at C-3 (or C-4) prefers to adopt an equatorial-like confirmation [287,288]. As the reaction is concerted, chirality at C-3 (or C-4) is transferred to the new chiral center at C-1 (or C-6). The reaction can be catalyzed by transition metals [289]. The use of a palladium catalyst allows for the reaction to be conducted at room temperature instead of extremely high temperatures (Scheme 9) [290,291].

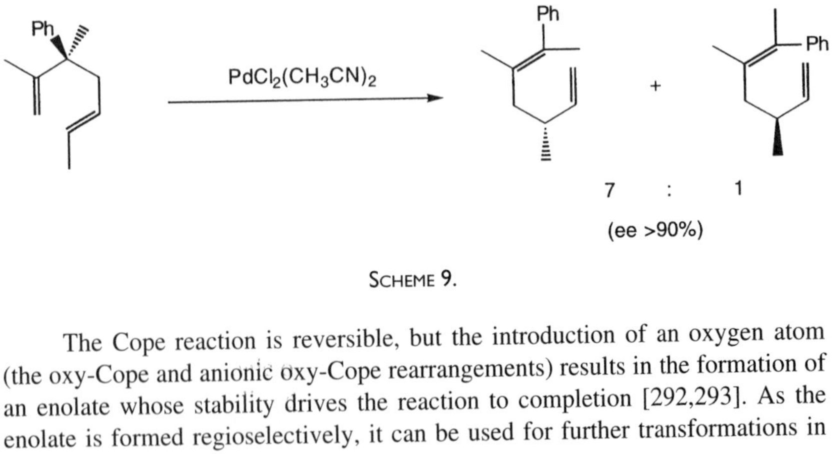

SCHEME 9.

The Cope reaction is reversible, but the introduction of an oxygen atom (the oxy-Cope and anionic oxy-Cope rearrangements) results in the formation of an enolate whose stability drives the reaction to completion [292,293]. As the enolate is formed regioselectively, it can be used for further transformations in situ [294–296].

10.3.2. Claisen Rearrangement

The [3,3]-sigmatropic rearrangement of allyl vinyl ethers is the Claisen rearrangement and is mechanistically analogous to the Cope rearrangement. As the product of the reaction is a carbonyl compound, the rearrangement usually goes to completion. The use of a large substituent, such as a bulky silyl group, that can occupy the equatorial position allows for good stereoselectivity (Scheme 10) [297].

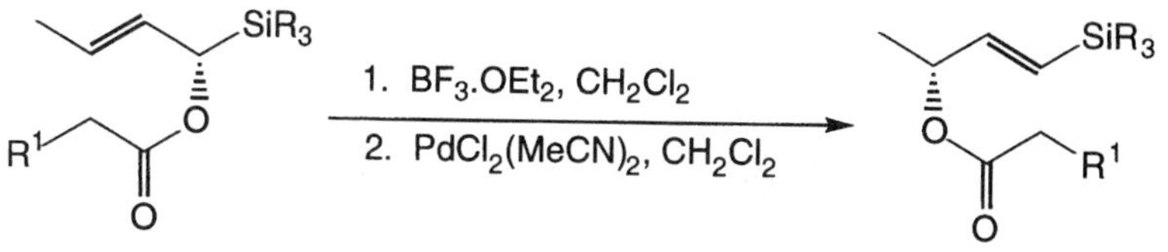

8

SCHEME 10.

The presence of π-electron donating substituents at the 2-position of the vinyl portion of the ether allows for significant acceleration of the Claisen rearrangement [298–302]. Aliphatic Claisen rearrangements can proceed in the presence of organoaluminum compounds [270,303,304], although other Lewis acids have failed to show reactivity [270,305–308]. Useful levels of (Z)-stereoselection and asymmetric induction have been obtained by use of bulky chiral organoaluminum Lewis acids [309–311].

The Claisen reaction can be catalyzed by palladium [312], thus, the [3,3]-sigmatropic rearrangement of a bulky allylsilane derivative **8** proceeds with high selectivity (Scheme 10) [313–315].

10.3.3. Ester Enolate Claisen Rearrangement

The most synthetically useful Claisen rearrangement, the ester enolate reaction of allyl esters, requires only relatively mild reaction conditions and is most amenable to scale up [281,316]. The geometry of the initially formed enol ether substrate is controlled by the choice of solvent system (Scheme 11) [282,317–319]. Thus, the methodology provides a useful alternative to an aldol approach.

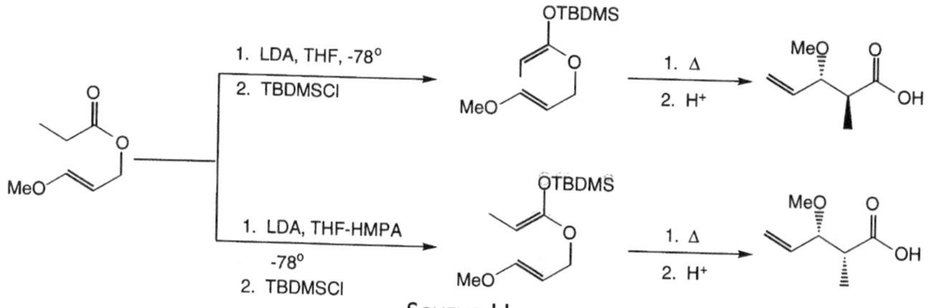

SCHEME 11.

The Claisen ester enolate reaction has proven extremely useful in the synthesis of a large number of natural products [3]. In addition, the rearrangement has been extended to allow the preparation of a number of useful intermediates such as, α-alkoxy esters [79,313,315,320–327], α-phenylthio esters [323,328, 329], α- and β-amino acids [324,330–334], α-fluoro esters [335], cycloalkenes [336,337], tetronic acids [338], and dihydropyrans [339–341].

The use of a heteroatom α to the ester carbonyl group allows for the formation of a chelate with the metal counterion; hence, the geometry of the ester enolate can be assured [320–322,342,343]. This approach was used in the rearrangement of the glycine allylic esters **9** to γ,δ-unsaturated amino acids in good yields and excellent diastereoselectivity (Scheme 12) [342]. The enantioselectivity could be reversed by using quinidine instead of quinine.

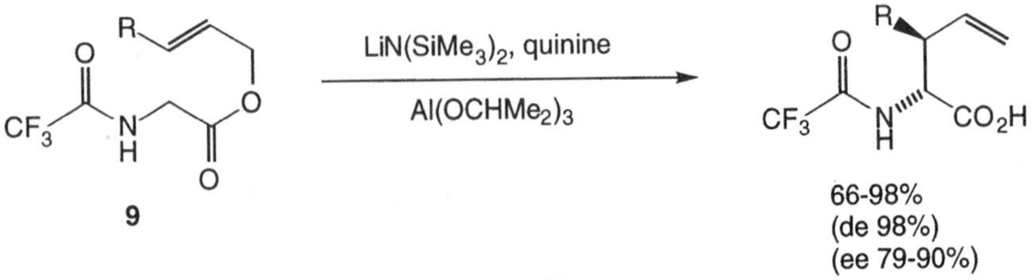

66-98%
(de 98%)
(ee 79-90%)

SCHEME 12.

10.4. THE ENE REACTION

The ene reaction [3,6,344–349], the addition of a carbon–carbon or carbon–oxygen double bond with concomitant transfer of an allylic hydrogen, can allow for chirality transfer [350–353]. The reaction has similarities to the Diels-Alder reaction in that a σ-bond is formed at the expense of a π-bond. In addition, the use of a Lewis acid as a catalyst allows for control of the relative stereochemistry (Scheme 13) [354–356]. Large-scale reactions will be complicated by the need to use either high temperatures or Lewis acids. In addition, thermal and Friedel-Crafts-type degradation products may be problematic with the use of these conditions [345,357].

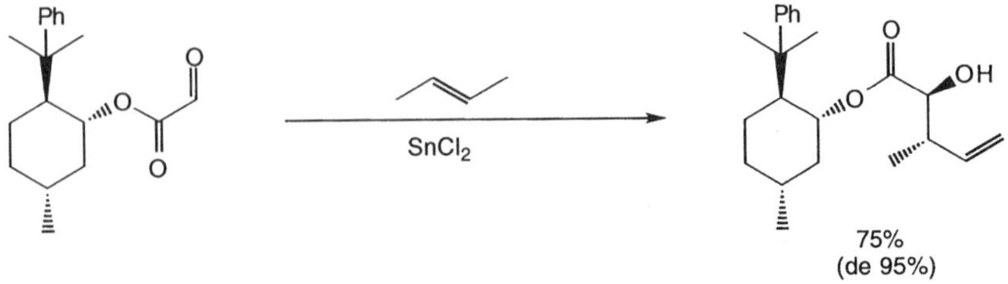

75%
(de 95%)

SCHEME 13.

Lewis acid–catalyzed reactions can be conducted at much lower temperatures than the uncatalyzed reaction, which can require temperatures in excess of 200°C [6,260,344,347,354,357–373].

The methodology allows for a selective preparation of cyclic compounds

[345,374–378], as well as acyclic ones (Scheme 14) [347,367,379–383]. Stereochemical control for acyclic reactions can be increased with the use of a cyclic primer [353,356,359,384–393].

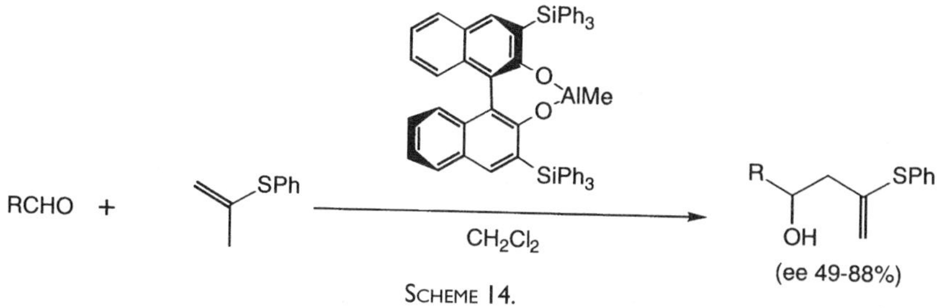

RCHO +

SCHEME 14.

(ee 49-88%)

Intramolecular ene reactions can be highly diastereoselective [385,394–400]. An example is provided by the synthesis of the corynantheine derivative **10** (Scheme 15) [394].

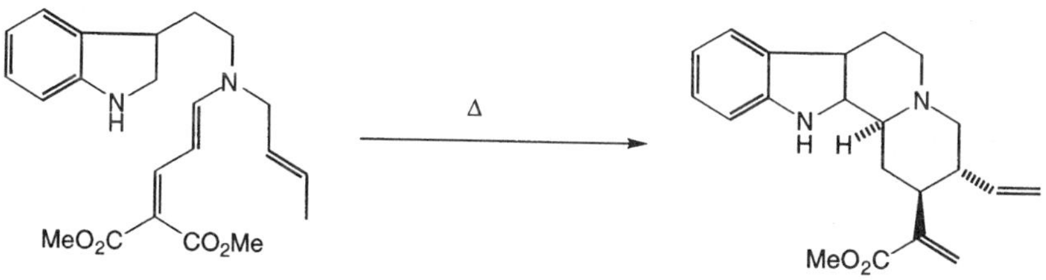

SCHEME 15.

10.5. DIPOLAR CYCLOADDITIONS

Dipolar cycloadditions, closely related to the Diels-Alder reaction [1,2], result in the synthesis of a five-membered adduct, including cyclopentane derivatives [401–410].

The cycloaddition of a nitrile oxide with a chiral allylic ether affords an isoxazoline with selectivity for the *pref*-isomer. This selectivity increases with the size of the alkyl substituent and is insensitive to the size of the allyl oxygen substituent. On the other hand, allyl alcohols tend to form the *parf*-isomer preferentially, although the selectivity is often low [411–418]. The product of dipolar cycloadditions based on nitrile oxides, the isoxazoline moiety, can be converted into a large variety of functional groups under relatively mild conditions [3]. Among other products, the addition can be used to prepare β-hydroxy ketones (Scheme 16) [419]. The isoxazoline moiety can be used to control the relative stereochemistry through chelation control [420,421].

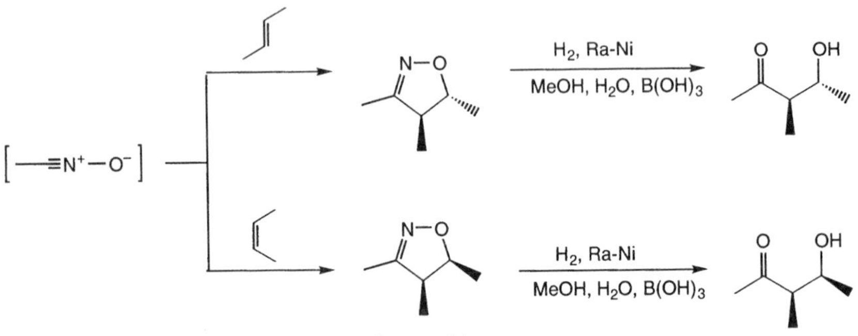

SCHEME 16.

10.6. [2,3]-SIGMATROPIC REARRANGEMENTS

There are two classes of [2,3]-sigmatropic rearrangements, anionic and neutral (Figure 2) [422,423]. The Wittig rearrangement provides the most useful example of anionic rearrangements, whereas the Evans rearrangement is the most important example of neutral rearrangements [422,424]. Numerous other examples such as allylic sulfenates, amine oxides, and selenoxides are known, but the electronic and stereochemical arguments used for the two major reactions allow similarities and extrapolations to be drawn with confidence [422]. The [2,3]-rearrangement often needs to be conducted at low temperature to minimize competition from a [3,3]-rearrangement or nonconcerted process adding to the problem of scale up [425–428].

10.6.1. Wittig Rearrangement

The Wittig rearrangement is primarily used in the transformation of an allylic ether to an α-allyl alcohol (Scheme 17) [429,430]. The transition state geometry plays an important role to determine the reaction outcome, which, in turn, is dependent on the stereochemistry of the double bond (Figure 3) [422,426,429, 431–439].

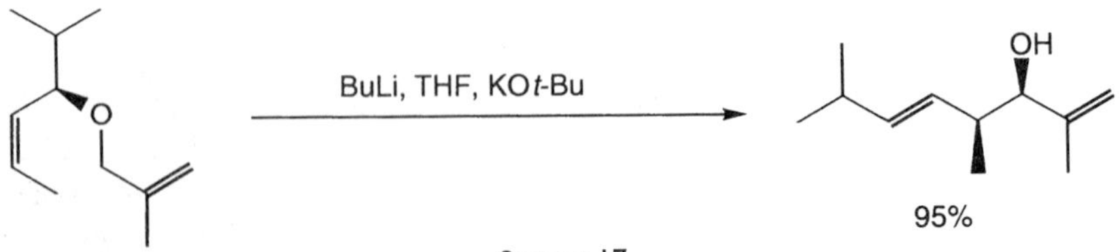

SCHEME 17.

The Wittig rearrangement has been used in the synthesis of a wide range of compounds that rely on this versatile protocol [440], ranging from large antibiotics [441,442] to dihydrofurans [443,444] to steroid derivatives [445–447].

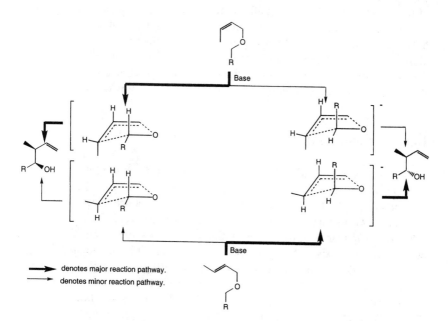

FIGURE 2   X, Y, Z = heteroatoms.

However, the use of strongly basic conditions are the major hindrance to the use of this reaction at scale.

10.6.2. Evans Rearrangement

The other major [2,3]-rearrangement, the Evans rearrangement [448], can be extended to synthesize functionalized allyl alcohols [449,450]. The Evans re-

→ denotes major reaction pathway.
→ denotes minor reaction pathway.

FIGURE 3   Reaction pathways for the Wittig rearrangement.

arrangement involves the rearrangement of an allylic sulfoxide to an allyl alcohol [448]. The reverse reaction is also possible, the conversion of an allyl alcohol to the analogous sulfoxide. The "push–pull" nature of the two rearrangements (Scheme 18), coupled with the transfer of chirality, provides a method to invert an allyl alcohol together with the isomerization of the alkene (Scheme 19) [422,451].

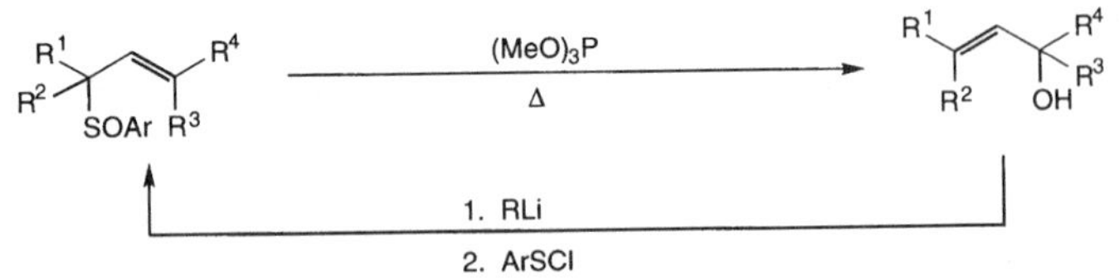

SCHEME 18.

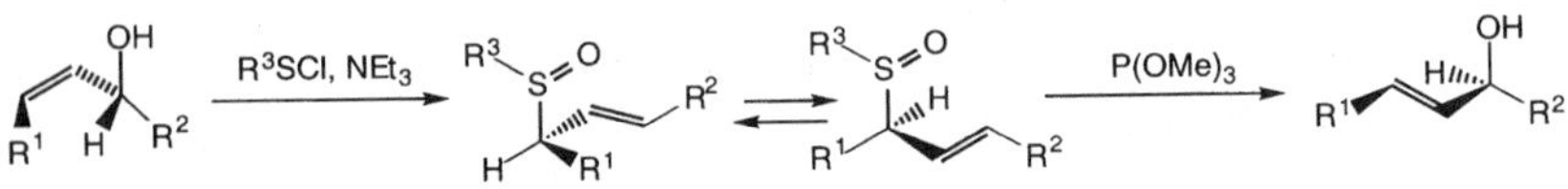

SCHEME 19.

The Evans rearrangement can be driven to completion by the addition of a thiophile, such as trimethylphosphite (Scheme 18) [424,452–455]. This strategy allows the chemistry of the allyl phenyl sulfoxide, or other sulfur precursor, to be exploited before the allyl alcohol is unmasked [448,456–459]. The addition of phenylsulfenyl chloride to an alkene, followed by the elimination of hydrogen chloride and subsequent rearrangement, provides a useful synthesis of allyl alcohols [458,460]. The [2,3]-Evans sigmatropic rearrangement is concerted and allows for stereochemical transfer [461,462]. The reverse reaction, formation of the allyl sulfoxide, results from the treatment of an allyl alcohol using a base followed by arylsulfenyl chloride to produce the allyl sulfoxide [463,464].

10.7. OTHER PERICYCLIC REACTIONS

Although photochemical cycloadditions can prove difficult to scale up, they do offer access to cycloadducts not directly accessible by other methods [465,466]. The initial studies into asymmetric synthesis by photochemical means were in the solid phase or organized assemblies, and few examples were known in solution [467,468]. Circular polarized light, a chiral agent, has the potential to induce asymmetric synthesis, although useful ee's have not yet been obtained [466].

The condensation of an imine with a Reformatsky-type reagent and tandem reactions can result in asymmetric induction [3,195,469–472]. The reaction of a ketene with an electron-rich alkene results in a [2 + 2] cycloaddition, although other systems can also be used [473–475]. The stereochemistry of the adduct is *cis*, and functionalized ketenes can also be used. The ketene can be generated in situ (Scheme 20) [476].

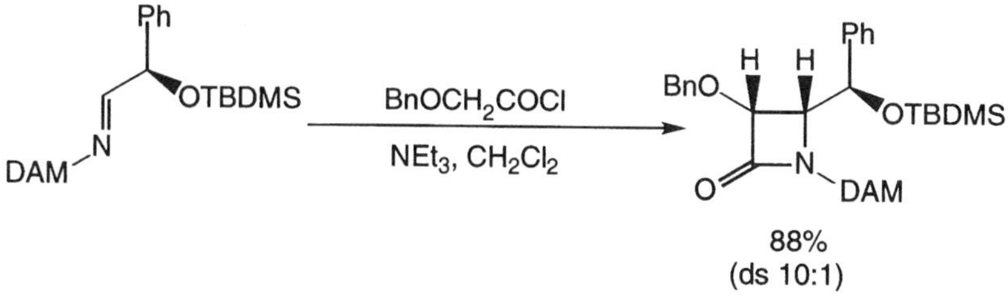

SCHEME 20.

The cycloaddition of ketene with chloral in the presence of catalytic quantities of quinidine or quinine leads to the oxetanones in high optical and chemical yield (Scheme 21). This reaction is practiced on an industrial scale with the chiral building blocks malic and citramalic acids being formed by hydrolysis [477].

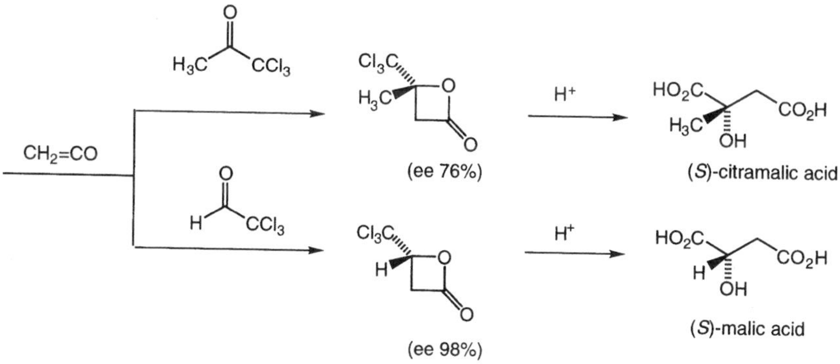

SCHEME 21.

10.8. SUMMARY

Many types of pericyclic and cycloaddition reactions have been documented. Although there are no general guidelines for the asymmetric preparation, reagents such as chiral catalysts are providing more general routes. Many of the reactions discussed rely on the use of low temperature. Although it is expensive to conduct low-temperature reactions on an industrial scale, reactions that need temperatures as low as $-105\,°C$ can be conducted. It should be noted that at such temperatures, only stainless steel vessels that require neutral or basic conditions can be used at these extreme temperatures. In addition, reactions that involve the use of a metal can cause contamination problems in waste water or the product.

Other sections of this chapter should be accessed as many chiral auxiliaries and catalysts appear to cross over from one type of pericyclic reaction to another. It should also be noted that although a large number of pericyclic reactions are known, large differences in asymmetric induction can occur with only subtle changes in reagents. This is especially true of the Diels-Alder reaction.

The Claisen rearrangement, Cope rearrangement, and associated variants are powerful tools that can be used to create a number of new chiral centers in an expeditious manner, but the use of heavy metals such as mercury should be avoided. Of these reactions, the Ireland Claisen ester enolate reaction provides the most versatile synthetic pathway with minimal scale up problems.

Our understanding of the chiral catalyst has greatly increased in recent years, and we are beginning to see catalysts that are less substrate dependent. The advantages of chiral catalysts on an industrial scale are obvious. With the enormous potential of pericyclic reactions, particularly as more than one carbon–carbon bond and a number of stereogenic centers can all be created at the same time, it can only be a matter of time before they become a key weapon in the process scale up strategy to chiral compounds.

REFERENCES

1. Desimoni, G., Tacconi, G., Barco, A., Polinni, G. P. In *Natural Product Synthesis Through Pericyclic Reactions*; American Chemical Society: Washington, D. C., 1983; pp. 119.
2. Kocovsky, P., Turecek, F., Hajicek, J. In *Synthesis of Natural Products: Problems of Stereoselection*; CRC Press: Boca Raton, 1986; Vol. 1.
3. Ager, D. J., East, M. B. *Asymmetric Synthetic Methodology*; CRC: Boca Raton, 1995.
4. Masamune, S., Choy, W., Petersen, J. S., Sita, L. R. *Angew. Chem. Int. Ed. Engl.* 1985, *24*, 1.
5. Danishefsky, S. *Acc. Chem. Res.* 1981, *14*, 400.
6. Oppolzer, W. *Angew. Chem. Int. Ed. Engl.* 1984, *23*, 876.
7. Alonso, I., Carretero, J. C., Ruano, J. L. G. *J. Org. Chem.* 1993, 3231.
8. Evans, D. A., Lectka, T., Miller, S. J. *J. Am. Chem. Soc.* 1993, *115*, 6460.
9. Carretero, C. J. C., Ruano, G. J. L., Lorente, A., Yuste, F. *Tetrahedron: Asymmetry* 1993, *4*, 177.
10. Hoffmann, H., Bolte, M., Berger, B., Hoppe, D. *Tetrahedron Lett.* 1993, *34*, 6537.
11. Bao, J., Wulff, W. D., Rheongold, A. L. *J. Am. Chem. Soc.* 1993, *115*, 3814.
12. Kobayashi, S., Hachiya, I., Ishitani, H., Araki, M. *Tetrahedron Lett.* 1993, *34*, 2581.
13. Corey, E. J., Loh, T.-P. *Tetrahedron Lett.* 1993, *34*, 3979.
14. Corey, E. J., Roper, T. D., Ishihara, K., Sarakinos, G. *Tetrahedron Lett.* 1993, *34*, 8399.
15. Uemura, M., Hayashi, Y., Hayashi, Y. *Tetrahedron Asymmetry* 1993, *4*, 2291.
16. Evans, D. A., Lectka, T., Miller, S. J. *Tetrahedron Lett.* 1993, *34*, 7027.

17. Nakanishi, S., Kumeta, K., Sawai, Y., Takata, T. *J. Organomet. Chem.* 1996, *515*, 99.
18. Arai, Y., Masuda, T., Tsutomu, Y., Shiro, M. *J. Chem. Soc. Perkin Trans.* 1996, *1*, 759.
19. Bao, J., Wulff, W. D., Rheingold, A. L. *J. Am. Chem. Soc.* 1993, *115*, 3814.
20. Posner, G. H., Carry, J. C., Anjeh, T. E. W., French, A. N. *J. Org. Chem.* 1992, *57*, 7012.
21. Kagan, H. B., Riant, O. *Chem. Rev.* 1992, *92*, 1007.
22. Fleming, I. *Frontier Orbitals and Organic Chemical Reactions*; Wiley: New York, 1976.
23. Houk, K. N. *Acc. Chem. Res.* 1975, *8*, 361.
24. Fringuelli, F., Taticchi, A. *Dienes in the Diels-Alder Reaction*; Wiley: New York, 1990.
25. Yamauchi, M., Honda, Y., Matsuki, N., Watanabe, T., Date, K., Hiramatsu, H. *J. Org. Chem.* 1996, *61*, 2719.
26. Houk, K. N., Strozier, R. W. *J. Am. Chem. Soc.* 1973, *95*, 4094.
27. Chao, T.-M., Baker, J., Hehre, W. J., Kahn, S. D. *Pure Appl. Chem.* 1991, *63*, 283.
28. Brion, F. *Tetrahedron Lett.* 1982, *23*, 5299.
29. Sauer, J., Wuest, H., Mielert, A. *Chem. Ber.* 1964, *97*, 3183.
30. Nelson, D. J. *J. Org. Chem.* 1986, *51*, 3185.
31. Koizumi, I., Hakamada, L., Yoshii, E. *Tetrahedron Lett.* 1984, *25*, 87.
32. Paquette, L. A. In *Asymmetric Synthesis*; Morrison, J. D., Ed.; Academic Press: Orlando, 1984; Vol. 3; pp. 455.
33. Poll, T., Metter, J. O., Helmchen, G. *Angew. Chem. Int. Ed. Engl.* 1985, *24*, 112.
34. Yamamoto, Y., Suzuki, I. *J. Org. Chem.* 1993, *58*, 4783.
35. Oppolzer, W., Wills, M., Kelly, M. J., Signer, M., Blagg, J. *Tetrahedron Lett.* 1990, *31*, 5015.
36. Oppolzer, W., Chopius, C. *Tetrahedron Lett.* 1983, *24*, 4665.
37. Oppolzer, W., Chapius, C., Dao, G. M., Reichlin, D., Godel, T. *Tetrahedron Lett.* 1982, *23*, 4781.
38. Bednarski, M., Danishefsky, S. *J. Am. Chem. Soc.* 1983, *105*, 3716.
39. Bonnesen, P. V., Puckett, C. L., Honeychuck, R. V., Hersh, W. H. *J. Am. Chem. Soc.* 1989, *111*, 6070.
40. Grieco, P. A., Nunes, J. J., Gaul, M. D. *J. Am. Chem. Soc.* 1990, *112*, 4595.
41. Smith, D. A., Houk, K. N. *Tetrahedron Lett.* 1991, *32*, 1549.
42. Danishefsky, S., Bednarski, M. *Tetrahedron Lett.* 1984, *25*, 721.
43. Pindur, U., Lutz, G., Otto, C. *Chem. Rev.* 1993, *93*, 741.
44. Corey, E. J., Imwinkelried, R., Pikul, S., Xiang, Y. B. *J. Am. Chem. Soc.* 1989, *111*, 5493.
45. Gothelf, K. V., Jorgensen, K. A. *J. Org. Chem.* 1995, *60*, 6847.
46. Gothelf, K. V., Hazell, R. G., Jorgensen, K. A. *J. Am. Chem. Soc.* 1995, *117*, 4435.
47. de Pascual-Teresa, B., Gonzalez, J., Asensio, A., Houk, K. N. *J. Am. Chem. Soc.* 1995, *117*, 4347.
48. Haase, C., Sarko, C. R., Dimare, M. *J. Org. Chem.* 1995, *60*, 1777.
49. Corey, E. J., Loh, T.-P., Roper, T. D., Azimioara, M. D., Noe, M. C. *J. Am. Chem. Soc.* 1992, *114*, 8290.

50. Corey, E. J., Sarshar, S. *J. Am. Chem. Soc.* 1992, *114*, 7938.
51. Trost, B. M., O'Krongly, D., Belletire, J. L. *J. Am. Chem. Soc.* 1980, *102*, 7595.
52. Maruoka, K., Murase, N., Yamamoto, H. *J. Org. Chem.* 1993, *58*, 2938.
53. Yamamoto, H., Maruoka, K., Furuta, K., Naruse, Y. *Pure Appl. Chem.* 1989, *61*, 419.
54. Terada, M., Mikami, K., Nakai, T. *Tetrahedron Lett.* 1991, *32*, 935.
55. Kaufmann, D., Boese, R. *Angew. Chem. Int. Edn.* 1990, *29*, 545.
56. Ketter, A., Glahsl, G., Herrmann, R. *J. Chem. Research (S)* 1990, 278.
57. Hattori, K., Yamamoto, H. *J. Org. Chem.* 1992, *57*, 3264.
58. Motoyama, Y., Terada, M., Mikami, K. *Synlett* 1995, 967.
59. Kobayashi, S., Araki, M., Hachiya, I. *J. Org. Chem.* 1994, *59*, 3758.
60. Corey, E. J., Sarshar, S., Lee, D.-H. *J. Am. Chem. Soc.* 1994, *116*, 12089.
61. Corey, E. J. *Tetrahedron Lett.* 1991, *32*, 7517.
62. Takasu, M., Yamamoto, H. *Synlett* 1990, 194.
63. Seerden, J.-P. G., Scheeren, H. W. *Tetrahedron Lett.* 1993, *34*, 2669.
64. Sartor, D., Saffrich, J., Helmchen, G. *Synlett* 1990, 197.
65. Sartor, D., Saffrich, J., Helmchen, G., Richards, C. J., Lambert, H. *Tetrahedron: Asymmetry* 1991, *2*, 639.
66. Corey, E. J., Loh, T. P. *J. Am. Chem. Soc.* 1991, *113*, 8966.
67. Corey, E. J., Imai, N., Zhang, H.-Y. *J. Am. Chem. Soc.* 1991, *113*, 728.
68. Corey, E. J., Ishihara, K. *Tetrahedron Lett.* 1992, *33*, 6807.
69. Davies, I. W., Senanayake, C. H., Larsen, R. D., Verhoeven, T. R., Reider, P. J. *Tetrahedron Lett.* 1996, *37*, 1725.
70. Knol, J., Meetama, A., Feringa, B. L. *Tetrahedron: Asymmetry* 1995, *6*, 1069.
71. Rebiere, F., Riant, O., Kagan, H. B. *Tetrahedron: Asymmetry* 1990, *1*, 199.
72. Narasaka, K., Iwasawa, N., Inoue, M., Yamada, T., Nakashima, M., Sugimori, J. *J. Am. Chem. Soc.* 1989, *111*, 5340.
73. Corey, E. J. *Tetrahedron Lett.* 1991, *32*, 6289.
74. Narasaka, K., Tanaka, H., Kanai, F. *Bull. Chem. Soc. Jpn* 1991, *64*, 387.
75. Engler, T. A., Letavic, M. A., Takusagawa, F. *Tetrahedron Lett.* 1992, *33*, 6731.
76. Cativiela, C., Lopez, P., Mayoral, J. A. *Tetrahedron: Asymmetry* 1991, *2*, 1295.
77. Devine, P. N., Oh, T. *J. Org. Chem.* 1992, *57*, 396.
78. Devine, P. N., Oh, T. *Tetrahedron Lett.* 1991, *32*, 883.
79. Oh, T., Wrobel, Z., Devine, P. N. *Synlett* 1992, 81.
80. Furuta, K., Shimizu, S., Miwa, Y., Yamamoto, H. *J. Org. Chem.* 1989, *54*, 1481.
81. Furuta, K., Miwa, Y., Iwanaga, K., Yamamoto, H. *J. Am. Chem. Soc.* 1988, *110*, 6254.
82. Furuta, K., Kanematsu, A., Yamamoto, H., Takaoka, S. *Tetrahedron Lett.* 1989, *30*, 7231.
83. Ishihara, K., Gao, Q., Yamamoto, H. *J. Org. Chem.* 1993, *58*, 6917.
84. Gao, Q., Maruyama, T., Mouri, M., Yamamoto, H. *J. Org. Chem.* 1992, *57*, 1951.
85. Takemura, H., Komeshima, N., Takashito, I., Hashimoto, S.-I., Ikota, N., Tomioka, K., Koga, K. *Tetrahedron Lett.* 1987, *28*, 5687.
86. Hashimoto, S.-I., Komeshima, N., Koga, K. *J. Chem. Soc. Chem. Commun.* 1979, 437.
87. Lautens, M., Lautens, J. C., Smith, A. C. *J. Am. Chem. Soc.* 1990, *112*, 5627.

88. Hawkins, J. M., Loren, S. *J. Am. Chem. Soc.* 1991, *113*, 7794.
89. Kuendig, E. P., Bourdin, B., Bernardinelli, G. *Angew. Chem.* 1994, *106*, 1931.
90. Wada, E., Pei, W., Kanemasa, S. *Chem. Lett.* 1994, 2345.
91. Riant, O., Kagan, H. B. *Tetrahedron Lett.* 1989, *30*, 7403.
92. Okamura, H., Nakamura, Y., Iwagawa, T., Nakatani, M. *Chem. Lett.* 1996, 193.
93. Van der Eycken, J., Vandewalle, M., Heinemann, G., Laumen, K., Schneider, M. P., Kredel, J., Sauer, J. *J. Chem. Soc. Chem. Commun.* 1989, 306.
94. Janssen, A. J. M., Klunder, A. J. H., Zwanenburg, B. *Tetrahedron Lett.* 1990, *31*, 7219.
95. Braisted, A. C., Schultz, P. G. *J. Am. Chem. Soc.* 1990, *112*, 7430.
96. Oppolzer, W., Rodriguez, I., Blagg, J., Bernardinelli, G. *Helv. Chim. Acta* 1989, *72*, 123.
97. Thom, C., Kocienski, P., Jarowicki, K. *Synthesis* 1993, 475.
98. Oppolzer, W., Chapuis, C., Bernardinelli, G. *Tetrahedron Lett.* 1984, *25*, 5885.
99. Tanaka, K., Uno, H., Osuga, H., Suzuki, H. *Tetrahedron: Asymmetry* 1993, *4*, 629.
100. Banks, M. R., Blake, A. J., Alexander, J., Cadogan, J. I. G., Doyle, A. A., Gosney, I., Hodgson, P. K. G., Thorburn, P. *Tetrahedron* 1996, *52*, 4079.
101. Palomo, C., Berree, F., Linden, A., Villalgordo, J. M., Facultad, Q. *J. Chem. Soc. Chem. Commun.* 1994, 1861.
102. Langlois, Y., Pouilhes, A. *Tetrahedron: Asymmetry* 1991, *2*, 1223.
103. Kouklovsky, C., Pouilhes, A., Langlois, Y. *J. Am. Chem. Soc.* 1990, *112*, 6672.
104. Pouilhes, A., Uriarte, E., Kouklovsky, C., Langlois, N., Langlois, Y., Chiaroni, A., Riche, C. *Tetrahedron Lett.* 1989, *30*, 1395.
105. Boeckman, R. K., Nelson, S. G., Gaul, M. D. *J. Am. Chem. Soc.* 1992, *114*, 2258.
106. Arai, Y., Matsui, M. Koizumi, T., Shiro, M. *J. Org. Chem.* 1991, *56*, 1983.
107. Oppolzer, W., Seletsky, B. M., Bernardinelli, G. *Tetrahedron Lett.* 1994, *35*, 3509.
108. Oppolzer, W., Kurth, M., Reichlin, D., Chapuis, C., Mohnhaupt, M., Moffatt, F. *Helv. Chim. Acta* 1981, *64*, 2802.
109. Oppolzer, W., Kurth, M., Reichlin, D., Moffatt, F. *Tetrahedron Lett.* 1981, *22*, 2545.
110. Oppolzer, W., Chapuis, C., Kelly, M. J. *Helv. Chim. Acta* 1983, *66*, 2358.
111. Evans, D. A., Chapman, K. T., Bisaha, J. *J. Am. Chem. Soc.* 1984, *106*, 4261.
112. Corey, E. J., Ensley, H. E. *J. Am. Chem. Soc.* 1975, *97*, 6908.
113. Cativiela, C., Figueras, F., Fraile, J. M., Garcia, J. I., Mayoral, J. A. *Tetrahedron: Asymmetry* 1991, *2*, 953.
114. Cativiela, C. L. P., Mayoral, J. A. *Tetrahedron: Asymmetry* 1990, *1*, 61.
115. Gras, J. L., Poncet, A., Nouguier, R. *Tetrahedron Lett.* 1992, *33*, 3323.
116. Takayama, H., Iyobe, A., Koizumi, T. *J. Chem. Soc. Chem. Commun.* 1986, 771.
117. Ohkata, K., Kubo, T., Miyamoto, K., Ono, M., Yamamoto, J., Akiba, K.-Y. *Heterocycl.* 1994, *38*, 1483.
118. Waldmann, H., Drager, M. *Tetrahedron Lett.* 1989, *30*, 4227.
119. Walborsky, H. M., Barash, L., Davis, T. C. *Tetrahedron* 1963, *19*, 2333.
120. Furuta, K., Iwanaga, K., Yamamoto, H. *Tetrahedron Lett.* 1986, *27*, 4507.
121. Reymond, J.-L., Vogel, P. *J. Chem. Soc. Chem. Commun.* 1990, 1070.
122. Chen, Y., Vogel, P. *Tetrahedron Lett.* 1992, *33*, 4917.
123. Ronan, B., Kagan, H. B. *Tetrahedron: Asymmetry* 1991, *2*, 75.

124. Lopez, R., Carretero, J. C. *Tetrahedron: Asymmetry* 1991, *2*, 93.
125. Fuji, K., Tanaka, K., Abe, H., Itoh, A., Node, M., Taga, T., Miwa, Y., Shiro, M. *Tetrahedron: Asymmetry* 1991, *2*, 179.
126. Alonso, I., Carretero, J. C., Garcia Ruano, J. L. *Tetrahedron Lett.* 1991, *32*, 947.
127. Fuji, K., Tanaka, K., Abe, H., Matsumoto, K., Taga, T., Miwa, Y. *Tetrahedron: Asymmetry* 1992, *3*, 609.
128. Arai, Y., Kuwayama, S.-I., Takeuchi, Y., Koizumi, T. *Tetrahedron Lett.* 1985, *26*, 6205.
129. Arai, Y., Matsui, M., Koizumi, T, J. *Chem. Soc. Perkin Trans. I* 1990, 1233.
130. Takahashi, T., Kotsubo, H., Iyobe, A., Namiki, T., Koizumi, T. *J. Chem. Soc. Perkin Trans.* 1990, *I*, 3065.
131. Takahashi, T., Iyobe, A., Arai, Y., Koizumi, T. *Synthesis* 1989, 189.
132. Takayama, H., Hayashi, K., Koizumi, T. *Tetrahedron Lett.* 1986, *27*, 5509.
133. Takayama, H., Iyobe, A., Koizumi, T. *Chem. Pharm. Bull.* 1987, *35*, 433.
134. Alonso, I., Carretero, J. C., Ruano, J. L. G. *Tetrahedron Lett.* 1989, *30*, 3853.
135. Alonso, I., Cid, M. B., Carretero, J. C., Ruano, J. L., Hoyos, M. A. *Tetrahedron: Asymmetry* 1991, *2*, 1193.
136. Carreno, M. C., Ruano, J. L. G., Urbano, A. *Tetrahedron Lett.* 1989, *30*, 4003.
137. Arai, Y., Koizumi, T. *Sulfur Rep.* 1993, *15*, 41.
138. Serrano, J. A., Caceres, L. E., Roman, E. *J. Chem. Soc. Perkin Trans.* 1992, *I*, 941.
139. Kim, K. S., Cho, I. H., Joo, Y. H., Yoo, I. J., Song, J. H., Ko, J. H. *Tetrahedron Lett.* 1992, *33*, 4029.
140. Jurczak, J., Tkacz, M. *Synthesis* 1979, 42.
141. Horton, D., Machinami, T., Takagi, Y. *Carbohydrate Res.* 1983, *121*, 135.
142. Horton, D., Machinami, T. *J. Chem. Soc. Chem. Commun.* 1981, 88.
143. Franck, R. W., John, T. V., Olejniczak, K., Blount, J. F. *J. Am. Chem. Soc.* 1982, *104*, 1106.
144. Liu, H.-J., Chew, S. Y., Browne, E. N. C. *Tetrahedron Lett.* 1991, *32*, 2005.
145. Waldner, A. *Tetrahedron Lett.* 1989, *30*, 3061.
146. Sato, M., Orii, C., Sakaki, J.-I., Kaneko, C. *J. Chem. Soc. Chem. Commun.* 1989, 1435.
147. Roush, W. R., Brown, B. B. *Tetrahedron Lett.* 1989, *30*, 7309.
148. Roush, W. R., Essenfield, A. P., Warmus, J. S., Brown, B. B. *Tetrahedron Lett.* 1989, *30*, 7305.
149. Mattay, J., Mertes, J., Maas, G. *Chem. Ber.* 1989, *122*, 327.
150. Kneer, G., Mattay, J., Raabe, G., Kruger, C., Lauterwein, J. *Synthesis* 1990, 599.
151. Jensen, K. N., Roos, G. H. P. *Tetrahedron: Asymmetry* 1992, *3*, 1553.
152. De Jong, J. C., van Bolhuis, F., Feringa, B. L. *Tetrahedron: Asymmetry* 1991, *2*, 1247.
153. De Jong, J. C., Jansen, J. F. G. A., Feringa, B. L. *Tetrahedron Lett.* 1990, *31*, 3047.
154. Aso, M., Ikeda, I., Kawabe, T., Shiro, M., Kanematsu, K. *Tetrahedron Lett.* 1992, *33*, 5789.
155. Ikeda, I., Honda, K., Osawa, E., Shiro, M., Aso, M., Kanematsu, K. *J. Org. Chem.* 1996, *61*, 2031.
156. Evans, D. A., Chapman, K. T., Bisaha, J. *J. Am. Chem. Soc.* 1988, *110*, 1238.

157. Kimura, K., Murata, K., Otsuka, K., Ishizuka, T., Haratake, M., Kunieda, T. *Tetrahedron Lett.* 1992, *33*, 4461.
158. Chapuis, C., Bauer, T., Jezewski, A., Jurczak, J. *Pol. J. Chem.* 1994, *68*, 2323.
159. Cativiela, C., Mayoral, J. A., Avenoza, A., Peregrina, J. M., Lahoz, F. J., Gimeno, S. *J. Org. Chem.* 1992, *57*, 4664.
160. Meyers, A. I., Busacca, C. A. *Tetrahedron Lett.* 1989, *30*, 6977.
161. Rehnberg, N., Sundin, A., Magnusson, G. *J. Org. Chem.* 1990, *55*, 5477.
162. Lamy-Schelkens, H., Ghosez, L. *Tetrahedron Lett.* 1989, *30*, 5891.
163. Waldmann, H., Dräger, M. *Liebigs Ann. Chem.* 1990, 681.
164. Linz, G., Weetmen J., Hady, A. F. A., Helmchan, G. *Tetrahedron Lett.* 1989, *30*, 5599.
165. Helmchen, G. H., A. F. A., Hartmann, H., Karge, R., Krotz, A., Sartor, K., Urmann, M. *Pure Appl. Chem.* 1989, *61*, 409.
166. Poll, T., Hady, A. F. A., Karge, R., Linz, G., Weetman, J., Helmchen, G. *Tetrahedron Lett.* 1989, *30*, 5595.
167. Masamune, S., Reed, L. A. I., Davis, J. T., Choy, W. *J. Org. Chem.* 1983, *48*, 4441.
168. Jung, M. E., Vaccaro, W. D., Buszek, K. R. *Tetrahedron Lett.* 1989, *30*, 1893.
169. Tomioka, K. H., N., Suenaga, T., Koga, K. *J. Chem. Soc. Perkin Trans.* 1990, *I*, 426.
170. Avenoza, A., Cativiela, C., Mayoral, J. A., Peregrina, J. M., Sinou, D. *Tetrahedron: Asymmetry* 1990, *1*, 765.
171. Yang, T.-K., Chen, C.-J., Lee, D.-S., Jong, T.-T., Jiang, Y.-Z., Mi, A.-Q. *Tetrahedron: Asymmetry* 1996, *7*, 57.
172. Oppolzer, W., Chapuis, C., Dupuis, D., Guo, M. *Helv. Chim. Acta* 1985, *68*, 2100.
173. Evans, D. A., Chapman, K. T., Bisaha, J. *Tetrahedron Lett.* 1984, *25*, 4071.
174. Evans, D. A., Chapman, K. T., Hung, D. T., Hawaguchi, A. T. *Angew. Chem. Int. Ed. Engl.* 1987, *26*, 1184.
175. Thiem, R., Rotscheidt, K., Breitmaier, E. *Synthesis* 1989, 836.
176. Konopelski, J. P., Boehler, M. A. *J. Am. Chem. Soc.* 1989, *111*, 4515.
177. Rieger, R., Btreitmaier, E. *Synthesis* 1990, 697.
178. Lyssikatos, J. P., Bednarski, M. D. *Synlett* 1990, 230.
179. McDougal, P. G., Jump, J. M., Rojas, C., Rico, J. G. *Tetrahedron Lett.* 1989, *30*, 3897.
180. Beagley, B., Larsen, D. S., Pritchard, R. G., Stoodley, R. J. *J. Chem. Soc. Perkin Trans.* 1990, *I*, 3113.
181. Giuliano, R. M., Jordan, A. D. J., Gauthier, A. D., Hoogstein, K. *J. Org. Chem.* 1993, *58*, 4979.
182. Larsen, D. S., Stoodley, R. J. *J. Chem. Soc. Perkin Trans.* 1989, *I*, 1841.
183. Gupta, R. C., Larsen, D. S., Stoodley, R. J., Slawin, A. M. Z., Williams, D. J. *J. Chem. Soc. Perkin Trans.* 1989, *I*, 739.
184. Bird, C. W., Lewis, A. *Tetrahedron Lett.* 1989, *30*, 6227.
185. Bloch, R., Chaptal-Gradoz, N. *Tetrahedron Lett.* 1992, *33*, 6147.
186. Kozikowski, A. P., Nieduzak, T. R. *Tetrahedron Lett.* 1986, *27*, 819.
187. Defoin, A., Pires, J., Tissot, I., Tschamber, T., Bur, D., Zehnder, M., Streith, J. *Tetrahedron: Asymmetry* 1991, *2*, 1209.

188. Masamune, S., Reed, L. A. I., Choy, W. *J. Org. Chem.* 1983, *48*, 1137.
189. Gupta, R. C., Raynor, C. M., Stoodley, R. J., Slawin, A. M. Z., Williams, D. J. *J. Chem. Soc. Perkin Trans.* 1988, *1*, 1773.
190. Larsen, D. S., Stoodley, R. J. *J. Chem. Soc., Perkin Trans.* 1990, *1*, 1339.
191. Corey, E. J. *Pure Appl. Chem.* 1990, *62*, 1209.
192. Tolbert, L. M., Ali, M. B. *J. Am. Chem. Soc.* 1981, *103*, 2104.
193. Walborsky, H. M., Barash, L., Davis, T. C. *J. Org. Chem.* 1961, *26*, 4778.
194. Misumi, A., Iwanaga, K., Furuta, K., Yamamoto, H. *J. Am. Chem. Soc.* 1985, *107*, 3343.
195. Tolbert, L. M., Ali, M. B. *J. Am. Chem. Soc.* 1982, *104*, 1742.
196. Tolbert, L. M., Ali, M. B. *J. Am. Chem. Soc.* 1984, *106*, 3806.
197. Charlson, J. L., Chee, G., McColeman, H. *Can. J. Chem.* 1995, *73*, 1454.
198. Reichardt, C. *Solvents and Solvent Effects in Organic Chemistry*; VCH: Weinheim, 1988.
199. Sauer, J., Sustmann, R. *Angew. Chem. Int. Eng. Ed.* 1980, *19*, 779.
200. Jaquith, J. B., Guan, J., Wang, S., Collins, S. *Organometallics* 1995, *14*, 1079.
201. Johannsen, M., Joergensen, K. A. *Tetrahedron* 1996, *52*, 7321.
202. Choudhury, A., Franck, R. W., Gupta, R. B. *Tetrahedron Lett.* 1989, *30*, 4921.
203. Posner, G. H., Wettlaufer, D. G. *Tetrahedron Lett.* 1986, *27*, 667.
204. Arnold, T., Reissig, H.-U. *Synlett* 1990, 514.
205. Boger, D. L., Corbett, W. L., Curran, T. T., Kasper, A. M. *J. Am. Chem. Soc.* 1991, *113*, 1713.
206. Backvall, J.-E., Rise, F. *Tetrahedron Lett.* 1989, *30*, 5347.
207. Mattay, J., Kneer, G., Mertes, J. *Synlett* 1990, 145.
208. Posner, G. H., Wettlaufer, D. G. *J. Am. Chem. Soc.* 1986, *108*, 7373.
209. Greene, A. E., Charbonnier, F., Luche, M. J., Moyano, A. *J. Am. Chem. Soc.* 1987, *109*, 4752.
210. Boger, D. L., Mullican, M. D. *Tetrahedron Lett.* 1982, *23*, 4551.
211. Marko, I. E., Evans, G. R., Seres, P., Chelle, I., Janousek, Z. *Pure. Appl. Chem.* 1996, *68*, 113.
212. Ihara, M., Katsumata, A., Egashira, M., Suzuki, S., Tokunaga, Y., Fukumoto, K. *J. Org. Chem.* 1995, *60*, 5560.
213. Fallis, A. G. *Can. J. Chem.* 1984, *62*, 183.
214. Takano, S. *Pure Appl. Chem.* 1987, *59*, 353.
215. Brieger, G., Bennett, J. N. *Chem. Rev.* 1980, *80*, 63.
216. Jung, M. E., Zimmerman, C. N., Lowen, G. T., Khan, S. I. *Tetrahedron Lett.* 1993, *34*, 4453.
217. Coe, J. W., Roush, W. R. *J. Org. Chem.* 1989, *54*, 915.
218. Roush, W. R., Kageyama, M., Riva, R., Brown, B. B., Warmus, J. S., Moriarty, K. J. *J. Org. Chem.* 1991, *56*, 1192.
219. Pyne, S. P., Spellmeyer, D. C., Chen, S., Fuchs, P. L. *J. Am. Chem. Soc.* 1982, *104*, 5728.
220. Pyne, S. G., Hensel, M. J., Fuchs, P. L. *J. Am. Chem. Soc.* 1982, *104*, 5719.
221. Gibbs, R. A., Bartels, K., Lee, R. W. K., Okamura, W. H. *J. Am. Chem. Soc.* 1989, *111*, 3717.
222. Davidson, A. H., Moloney, B. A. *J. Chem. Soc. Chem. Commun.* 1989, 445.

223. Tietze, L.-F., von Kiedrowski, G., Harms, K., Clegg, W., Sheldrick, G. *Angew. Chem. Int. Ed. Engl.* 1980, *19*, 134.
224. Bailey, M. S., Brisdon, B. J., Brown, D. W., Stark, K. M. *Tetrahedron Lett.* 1983, *24*, 3037.
225. Roush, W. R., Gillis, H. R., Ko, A. I. *J. Am. Chem. Soc.* 1982, *104*, 2269.
226. Levin, J. I. *Tetrahedron Lett.* 1989, *30*, 2355.
227. Sisko, J., Weinreb, S. M. *Tetrahedron Lett.* 1989, *30*, 3037.
228. Uyehara, T., Suzuki, I., Yamamoto, Y. *Tetrahedron Lett.* 1990, *31*, 3753.
229. Annunziata, R., Cinquini, M., Cozzi, F., Riamondi, L. *Tetrahedron Lett.* 1989, *30*, 5013.
230. Danishefsky, S. J. *Aldrichimica Acta* 1986, *19*, 59.
231. Kametani, T., Chu, S.-D., Honda, T. *Heterocycles* 1987, *25*, 241.
232. Petrzilka, M., Grayson, J. I. *Synthesis* 1981, 753.
233. Danishefsky, S., DeNinno, M. P., Phillips, G. B., Zelle, R. E., Lartey, P. A. *Tetrahedron* 1986, *42*, 2809.
234. Fraser-Reid, B., Rahman, M. A., Kelly, D. R., Srivastava, R. M. *J. Org. Chem.* 1984, *49*, 1835.
235. Wender, P. A. *J. Am. Chem. Soc.* 1987, *109*, 4390.
236. Danishefsky, S. J., Pearson, W. H., Harvey, D. F., Maring, C. J., Springer, J. P. *J. Am. Chem. Soc.* 1985, *107*, 1256.
237. Danishefsky, S. J., Myles, D. C., Harvey, D. F. *J. Am. Chem. Soc.* 1987, *109*, 862.
238. Danishefsky, S. J., Armistead, D. M., Wincott, F. E., Selnick, H. G., Hungate, R. *J. Am. Chem. Soc.* 1987, *109*, 8117.
239. Egbertson, M., Danishefsky, S. J. *J. Org. Chem.* 1989, *54*, 11.
240. Danishefsky, S. J., Selnick, H. G., Armistead, D. M., Wincott, F. E. *J. Am. Chem. Soc.* 1987, *109*, 8119.
241. Danishefsky, S. J., Armistead, D. M., Wincott, F. E., Selnick, H. G., Hungate, R. *J. Am. Chem. Soc.* 1989, *111*, 2967.
242. Tietze, L. F., Schnieder, C. *J. Org. Chem.* 1991, *56*, 2476.
243. Danishefsky, S. J., DeNinno, M. P. In *Trends in Synthetic Carbohydrate Chemistry*; Horton, D., Hawkins, L. D., Eds.; American Chemical Society: Washington, D. C., 1989; pp. 160.
244. Danishefsky, S. J., DeNinno, M. P., Audia, J. E., Schulte, G. In *Trends in Synthetic Carbohydrate Chemistry*; Horton, D., Hawkins, L. D., Eds.; American Chemical Society: Washington, D. C., 1989; pp. 176.
245. Danishefsky, S. J., DeNinno, M. P. *Angew. Chem. Int. Ed. Engl.* 1987, *26*, 15.
246. Golebiowski, A., Jurczak, J. *J. Chem. Soc. Chem. Commun.* 1989, 263.
247. Danishefsky, S., Larson, E. R., Askin, D. *J. Am. Chem. Soc.* 1982, *104*, 6457.
248. Belanger, J., Landry, N. L., Pare, J. R. J., Jankowski, K. *J. Org. Chem.* 1982, *47*, 3649.
249. Danishefsky, S., Bednarski, M., Izawa, T., Maring, C. *J. Org. Chem.* 1984, *49*, 2290.
250. Danishefsky, S., Kerwin, J. F. J., Kobayashi, S. *J. Am. Chem. Soc.* 1982, *104*, 358.
251. Larson, E. R., Danishefsky, S. *J. Am. Chem. Soc.* 1982, *104*, 6458.
252. Danishefsky, S. J., Pearson, W. H., Harvey, D. F. *J. Am. Chem. Soc.* 1984, *106*, 2456.

253. Maruoka, K., Yamamoto, H. *J. Am. Chem. Soc.* 1989, *111*, 789.
254. Maruoka, K., Itoh, T., Shirasaka, T., Yamamoto, H. *J. Am. Chem. Soc.* 1988, *110*, 310.
255. Wada, E., Yasuoka, H., Kanemasa, S. *Chem. Lett.* 1994, 1637.
256. Faller, J. W., Smart, C. J. *Tetrahedron Lett.* 1989, *30*, 1189.
257. Katagiri, N., Makino, M., Tamura, T., Takahiro, K., Kaneko, C. *Chem. Pharm. Bull* 1996, *44*, 850.
258. Mikami, K. Kotera, O., Motoyama, Y., Sakaguchi, H. *Synlett* 1995, 975.
259. Inanaga, J., Sugimoto, Y., Hanamoto, T. *New J. Chem.* 1995, *19*, 707.
260. Johannsen, M., Joergensen, K. A. *J. Org. Chem.* 1995, *60*, 5757.
261. Bednarski, M., Danishefsky, S. *J. Am. Chem. Soc.* 1983, *105*, 6968.
262. Page, P. C. B., Prodger, J. C. *Synlett* 1991, 84.
263. Ziegler, F. E. *Acc. Chem. Res.* 1977, *10*, 227.
264. Ziegler, F. E. *Chem. Rev.* 1988, *88*, 1423.
265. Rhoads, S. J., Rawlins, N. R. *Org. Reactions* 1975, *22*, 1.
266. Bennett, G. B. *Synthesis* 1977, *589*.
267. Blechert, S. *Synthesis* 1989, 71.
268. Tarbell, D. S. *Org. Reactions* 1944, *2*, 1.
269. Tarbell, D. S. *Chem. Rev.* 1940, *27*, 495.
270. Takai, K., Mori, I., Oshima, K., Nozaki, H. *Tetrahedron Lett.* 1981, *22*, 3985.
271. Hansen, H.-J., Schmid, H. *Tetrahedron* 1974, *30*, 1959.
272. Ferrier, R. J., Vethavijasar, N. *J. Chem. Soc. Perkin Trans.* 1973, *I*, 1791.
273. Mandai, T., Ueda, M., Hasegawa, S.-I., Kawada, M., Tsuji, J. *Tetrahedron Lett.* 1990, *31*, 4041.
274. Reetz, M. T., Gansäuer, A. *Tetrahedron* 1993, *49*, 6025.
275. Felix, D., Gschwend-Steen, D., Wick, A. E., Eschenmoser, A. *Helv. Chim. Acta* 1969, *52*, 1030.
276. Ito, S., Tsunoda, T. *Pure Appl. Chem.* 1990, *62*, 1405.
277. Nubbemeyer, U. *Synthesis* 1993, 1120.
278. Jones, G. B., Huber, R. S., Chau, S. *Tetrahedron* 1993, *49*, 369.
279. Mulzer, J., Scharp, M. *Synthesis* 1993, 615.
280. Bao, R., Valverde, S., Herradón, B. *Synlett* 1992, 217.
281. Ireland, R. E., Muller, R. H. *J. Am. Chem. Soc.* 1972, *94*, 5897.
282. Ireland, R. E., Willard, A. K. *Tetrahedron Lett.* 1975, 3975.
283. Kurth, M., Soares, C. J. *Tetrahedron Lett.* 1987, *28*, 1031.
284. Funk, R. L., Abelman, M. M., Munger, J. D. *Tetrahedron* 1986, *42*, 2831.
285. Nubbemeyer, U., Öhrlein, R., Gonda, J., Ernst, B., Bellus, D. *Angew. Chem. Int. Ed. Engl.* 1991, *30*, 1465.
286. Beslin, P., Perrio, S. *Tetrahedron* 1993, *49*, 3131.
287. Dewar, M. J. S., Jie, C. *Acc. Chem. Res.* 1992, *25*, 537.
288. Hill, R. K., Gilman, N. W. *J. Chem. Soc. Chem. Commun.* 1967, 617.
289. Lutz, R. P. *Chem. Rev.* 1984, *84*, 205.
290. Overman, L. E., Jacobsen, E. J. *J. Am. Chem. Soc.* 1982, *104*, 7225.
291. Overman, L. E., Knoll, F. M. *J. Am. Chem. Soc.* 1980, *102*, 865.
292. Paquette, L. A. *Angew. Chem., Int. Ed. Engl.* 1990, *29*, 609.
293. Paquette, L. A., Maynard, G. D. *Angew. Chem. Int. Ed. Engl.* 1991, *30*, 1368.

294. Paquette, L. A., Oplinger, J. A. *Tetrahedron* 1989, *45*, 107.
295. Wender, P. A., Sieburth, S. M., Petraitis, J. J., Singh, S. K. *Tetrahedron* 1981, *37*, 3967.
296. Uma, R., Rajogopalan, K., Swaminathan, S. *Tetrahedron* 1986, *42*, 2757.
297. Mikami, K., Maeda, T., Kishi, N., Nakai, T. *Tetrahedron Lett.* 1984, *25*, 5151.
298. Koreeda, M., Luengo, J. I. *J. Am. Chem. Soc.* 1985, *107*, 5572.
299. Denmark, S. E., Harmata, M. A. *J. Am. Chem. Soc.* 1982, *104*, 4972.
300. Denmark, S. E., Harmata, M. A., White, K. S. *J. Am. Chem. Soc.* 1989, *111*, 8878.
301. Denmark, S. E., Rajendra, G., Marlin, J. E. *Tetrahedron Lett.* 1989, *30*, 2469.
302. Denmark, S. E., Marlin, J. E. *J. Org. Chem.* 1991, *56*, 1003.
303. Nonoshita, K., Banno, H., Maruoka, K., Yamamoto, H. *J. Am. Chem. Soc.* 1990, *112*, 316.
304. Maruoka, K., Banno, H., Nonoshita, K., Yamamoto, H. *Tetrahedron Lett.* 1989, *30*, 1265.
305. van der Baan, J. L., Bickelhaupt, F. *Tetrahedron Lett.* 1986, *27*, 6267.
306. Hill, R. K., Khatri, H. N. *Tetrahedron Lett.* 1978, 4337.
307. Maruoka, K., Banno, H., Yamamoto, H. *J. Am. Chem. Soc.* 1990, *112*, 7791.
308. Yamamoto, H., Maruoka, K. *Pure Appl. Chem.* 1990, *62*, 2063.
309. Masse, C. E., Panek, J. S. *Chemtracts. Org. Chem.* 1995, *8*, 242.
310. Maruoka, K., Saito, S., Yamamoto, H. *J. Am. Chem. Soc.* 1995, *117*, 1165.
311. Boeckman, R. K. J., Need, M. J., Gaul, M. D. *Tetrahedron Lett.* 1995, *36*, 803.
312. Sugiura, M., Yanagisawa, M., Nakai, T. *Synlett* 1995, 447.
313. Panek, J. S., Sparks, M. A. *J. Org. Chem.* 1990, *55*, 5564.
314. Panek, J. S., Clark, T. D. *J. Org. Chem.* 1992, *57*, 4323.
315. Sparks, M. A., Panek, J. S. *J. Org. Chem.* 1991, *56*, 3431.
316. Pereira, S., Srebnik, M. *Aldrichimica Acta* 1993, *26*, 17.
317. Ireland, R. E., Wilcox, C. S. *Tetrahedron Lett.* 1977, 2839.
318. Ireland, R. E., Wipf, P., Armstrong, J. D. *J. Org. Chem.* 1991, *56*, 650.
319. Narula, A. S. *Tetrahedron Lett.* 1981, *22*, 2017.
320. Sato, T., Tajima, K., Fujia, T. *Tetrahedron Lett.* 1983, *24*, 729.
321. Fujisawa, T., Kohama, H., Tajima, K., Sato, T. *Tetrahedron Lett.* 1984, *25*, 5155.
322. Kallmerten, J., Gould, T. J. *Tetrahedron Lett.* 1983, *24*, 5177.
323. Ager, D. J., Cookson, R. C. *Tetrahedron Lett.* 1982, *23*, 3419.
324. Tsunoda, T., Tatsuki, S., Shiraishi, Y., Akasaka, M., Itô, S. *Tetrahedron Lett.* 1993, *34*, 3297.
325. Tadano, K.-i., Minami, M., Ogawa, S. *J. Org. Chem.* 1990, *55*, 2108.
326. Burke, S. D., Pacofsky, G. J. *Tetrahedron Lett.* 1986, *27*, 445.
327. Barlett, P. A., Tanzella, D. J., Barstow, J. F. *J. Org. Chem.* 1982, *47*, 3941.
328. Lythgoe, B., Milner, J. R., Tideswell, J. *Tetrahedron Lett.* 1975, 2593.
329. Richardson, S. K., Sabol, M. R., Watt, D. S. *Synth. Commun.* 1989, *19*, 359.
330. Barlett, P. A., Holm, K. H., Morimoto, A. *J. Org. Chem.* 1985, *50*, 5179.
331. Barlett, P. A., Barstow, J. F. *Tetrahedron Lett.* 1982, *23*, 623.
332. Barlett, P. A., Barstow, J. F. *J. Org. Chem.* 1982, *47*, 3933.
333. Baumann, H., Duthaler, R. O. *Helv. Chim. Acta* 1988, *71*, 1025.
334. Dell, C. P., Khan, K. M., Knight, D. W. *J. Chem. Soc. Perkin Trans.* 1994, *I*, 341.
335. Welch, J. T. Plummer, J. S., Chou, T.-S. *J. Org. Chem.* 1991, *56*, 353.

336. Danishefsky, S., Tsuzuki, K. *J. Am. Chem. Soc.* 1980, *102*, 6891.
337. Danishefsky, S., Funk, R. L., Kerwin, J. F. *J. Am. Chem. Soc.* 1980, *102*, 6889.
338. Brandage, S., Flodman, L., Norberg, A. *J. Org. Chem.* 1984, *49*, 927.
339. Ireland, R. E., Wuts, P. G. M., Ernst, B. *J. Am. Chem. Soc.* 1981, *103*, 3205.
340. Burke, S. D., Cobb, J. E. *Tetrahedron Lett.* 1986, *27*, 4237.
341. Ireland, R. E., Thaisrivongs, S., Vanier, N., Wilcox, C. S. *J. Org. Chem.* 1980, *45*, 48.
342. Kazmaier, U., Krebs, A. *Angew. Chem. Int. Ed. Engl.* 1995, *34*, 2012.
343. Kazmaier, U., Maier, S. *J. Chem. Soc. Chem. Commun.* 1995, 1991.
344. Snider, B. B. *Acc. Chem. Res.* 1980, *13*, 426.
345. Oppolzer, W. *Pure Appl. Chem.* 1981, *53*, 1181.
346. Oppolzer, W. *Angew. Chem. Int. Ed. Engl.* 1989, *28*, 38.
347. Mikami, K., Shimizu, M. *Chem. Rev.* 1992, *92*, 1021.
348. Thomas, B. E., Houk, K. N. *J. Am. Chem. Soc.* 1993, *115*, 790.
349. Hoffmann, H. M. R. *Angew. Chem. Int. Ed. Engl.* 1969, *8*, 556.
350. Lehmkuhl, H. *Bull. Soc. Chim. France* 1981, 87.
351. Oppolzer, W., Jacobsen, E. *J. Tetrahedron Lett.* 1986, *27*, 1141.
352. Oppolzer, W., Robbiani, C. Battig, K. *Tetrahedron* 1984, *40*, 1391.
353. Oppolzer, W., Begley, T., Ashcroft, A. *Tetrahedron Lett.* 1984, *25*, 825.
354. Mikami, K., Shimizu, M., Nakai, T. *J. Org. Chem.* 1991, *56*, 2952.
355. Whitesell, J. K., Deyo, D., Bhattacharya, A. *J. Chem. Soc. Chem. Commun.* 1983, 802.
356. Whitesell, J. K., Bhattacharya, A., Aguilar, D. A., Henke, K. *J. Chem. Soc. Chem. Commun.* 1982, 989.
357. Terada, M., Sayo, N., Mikami, K. *Synlett* 1995, 411.
358. Shimizu, M., Mikami, K. *J. Org. Chem.* 1992, *57*, 6105.
359. Mikami, K., Loh, T.-P., Nakai, T. *Tetrahedron Lett.* 1988, *29*, 6305.
360. Houston, T. A., Tanaka, Y., Koreeda, M. *J. Org. Chem.* 1993, *58*, 4287.
361. Mikami, K., Terada, M., Nakai, T. *J. Am. Chem. Soc.* 1989, *111*, 1940.
362. Mikami, K., Terada, M., Nakai, T. *J. Am. Chem. Soc.* 1990, *112*, 3949.
363. Mikami, K., Yajima, T., Takasaki, T., Matsukawa, S., Terada, M., Uchimaru, T., Maruta, M. *Tetrahedron* 1996, *52*, 85.
364. Terada, M., Mikami, K. *J. Chem. Soc. Chem. Commun.* 1995, 2391.
365. Kitamoto, D., Imma, H., Nakai, T. *Tetrahedron Lett.* 1995, *36*, 1861.
366. Mikami, K., Motoyama, Y., Terada, M. *Inorg. Chim. Acta* 1994, *222*, 71.
367. Mikami, K., Matsukawa, S. *Tetrahedron Lett.* 1994, *35*, 3133.
368. Terada, M., Mikami, K. *J. Chem. Soc. Chem. Commun.* 1994, 833.
369. Mikami, K., Terada, M., Narisawa, S., Nakai, T. *Org. Synth.* 1993, *71*, 14.
370. Terada, M., Matsukawa, S., Mikami, K. *J. Chem. Soc. Chem. Commun.* 1993, 327.
371. Van der Meer, F. T., Feringa, B. L. *Tetrahedron Lett.* 1992, *33*, 6695.
372. Mikami, K. *Pure Appl. Chem.* 1996, *68*, 639.
373. Faller, J. W., Liu, X. *Tetrahedron Lett.* 1996, *37*, 3449.
374. Sarko, T. K., Gosh, S. H., Subba Rao, P. S. V., Satapathi, T. K. *Tetrahedron Lett.* 1990, *31*, 3461.
375. Oppolzer, W., Keller, T. H., Kuo, D. L., Pachinger, W. *Tetrahedron Lett.* 1990, *31*, 1265.

376. Oppolzer, W. *Pure Appl. Chem.* 1990, *62*, 1941.
377. Takacs, J. M., Anderson, L. G., Greswell, M. W., Takacs, B. E. *Tetrahedron Lett.* 1987, *28*, 5627.
378. Mikami, K., Takahashi, K., Nakai, T. *Tetrahedron Lett.* 1989, *30*, 357.
379. Mikami, K., Terada, M., Narisawa, S., Nakai, T. *Synlett* 1992, 255.
380. Tanino, K., Nakamura, T., Kuwajima, I. *Tetrahedron Lett.* 1990, *31*, 2165.
381. Snider, B. B., Cartaya-Mari, C. P. *J. Org. Chem.* 1984, *49*, 1688.
382. Maruoka, K., Hoshino, Y., Shirasaka, T., Yamamoto, H. *Tetrahedron Lett.* 1988, *29*, 3967.
383. Mikami, K., Narisawa, S., Shimizu, M., Terada, M. *J. Am. Chem. Soc.* 1992, *114*, 6566.
384. Kim, B. H., Lee, J. Y., Kim, K., Whang, D. *Tetrahedron: Asymmetry* 1991, *2*, 27.
385. Oppolzer, W., Pitteloud, R. *J. Am. Chem. Soc.* 1982, *104*, 6478.
386. Mikami, K., Kaneko, M., Loh, T.-P., Terada, M., Nakai, T. *Tetrahedron Lett.* 1990, *31*, 3909.
387. Benner, J. P., Gill, G. B., Parrott, S. J., Wallace, B., Belay, M. J. *J. Chem. Soc. Perkin Trans.* 1984, *1*, 314.
388. Whitesell, J. K., Nabona, K., Deyo, D. *J. Org. Chem.* 1989, *54*, 2258.
389. Whitesell, J. J., Carpenter, J. F., Yaser, H. K., Machajewski, T. *J. Am. Chem. Soc.* 1990, *112*, 7653.
390. Whitesell, J. K., Bhattacharya, A., Buchanan, C. M., Chen, H. H., Deyo, D., James, D., Liu, C.-L., Minton, M. A. *Tetrahedron* 1986, *40*, 2993.
391. Snider, B. B., Rodini, D. J. *Tetrahedron Lett.* 1980, *21*, 1815.
392. Snider, B. B., Rodini, D. J., Cann, R. S. E., Sealfon, S. *J. Am. Chem. Soc.* 1979, *101*, 5283.
393. Snider, B. B., van Straten, J. W. *J. Org. Chem.* 1979, *44*, 3567.
394. Tietze, L. F., Wichmann, J. *Angew. Chem. Int. Ed. Engl.* 1992, *31*, 1079.
395. Duncia, J. V., Lansbury, P. T., Miller, T., Snider, B. B. *J. Am. Chem. Soc.* 1982, *104*, 1930.
396. Snider, B. B., Duncia, J. V. *J. Org. Chem.* 1981, *46*, 3223.
397. Snider, B. B., Duncia, J. V. *J. Am. Chem. Soc.* 1980, *102*, 5926.
398. Hiroi, K., Umemura, M. *Tetrahedron Lett.* 1992, *33*, 3343.
399. Narasaka, K., Yujiro, S., Yamada, J. *Isr. J. Chem.* 1991, *31*, 261.
400. Murase, N., Ooi, T., Yamamoto, H. *Synlett* 1991, 857.
401. Trost, B. M., King, S., Nanninga, T. N. *Chemistry Lett.* 1987, 15.
402. Trost, B. M., Mignani, S. M. *Tetrahedron Lett.* 1985, *26*, 6313.
403. Trost, B. M., Lynch, J., Renaut, P., Steinman, D. H. *J. Am. Chem. Soc.* 1986, *108*, 284.
404. Trost, B. M., Bonk, P. J. *J. Am. Chem. Soc.* 1985, *107*, 1778.
405. Trost, B. M., Miller, M. L. *J. Am. Chem. Soc.* 1988, *110*, 3687.
406. Trost, B. M., Seoane, P., Mignani, S., Acemoglu, M. *J. Org. Chem.* 1989, *54*, 7487.
407. Trost, B. M., Acemoglu, M. *Tetrahedron Lett.* 1989, *30*, 1495.
408. Molander, G. A., Andrews, S. W. *Tetrahedron Lett.* 1989, *30*, 2351.
409. Trost, B. M. *Angew. Chem. Int. Ed. Engl.* 1986, *25*, 1.
410. Horiguchi, Y., Suchiro, I., Sasaki, A., Kuwajima, I. *Tetrahedron Lett.* 1993, *34*, 6077.

411. Houk, K. N., R., M. S., Wu, Y.-D., Rondan, N. G., Jager, V., Schohe, R., Fronczek, F. R. *J. Am. Chem. Soc.* 1984, *106*, 3880.
412. Kozikowski, A. P., Ghosh, A. K. *J. Org. Chem.* 1984, *49*, 2762.
413. Vasella, A. *Helv. Chim. Acta* 1977, *60*, 426.
414. Kozikowski, A. P., Ghosh, A. K. *J. Am. Chem. Soc.* 1982, *104*, 5788.
415. Huber, R., Vasella, A. *Tetrahedron* 1990, *46*, 33.
416. Kanemasa, S., Tsuruoka, T., Wada, E. *Tetrahedron Lett.* 1993, *34*, 87.
417. Hoffmann, R. W. *Chem. Rev.* 1989, *89*, 1841.
418. Curran, D. P., Kim, B. H. *Synthesis* 1990, 312.
419. Curran, D. P. *J. Am. Chem. Soc.* 1983, *105*, 5826.
420. Annunziata, R., Cinquini, M., Cozzi, F., Restelli, A. *J. Chem. Soc. Chem. Commun.* 1984, 1253.
421. Kamimura, A., Marumo, S. *Tetrahedron Lett.* 1990, *31*, 5053.
422. Hoffmann, R. W. *Angew. Chem. Int. Ed. Engl.* 1979, *18*, 563.
423. Tomooka, K., Watanabe, M., Nakai, T. *Tetrahedron Lett.* 1990, *31*, 7353.
424. Evans, D. A., Andrews, G. C., Sims, C. L. *J. Am. Chem. Soc.* 1971, *93*, 4956.
425. Baldwin, J. E., DeBernardis, J., Patrick, J. E. *Tetrahedron Lett.* 1970, 353.
426. Baldwin, J. E., Patrick, J. E. *J. Am. Chem. Soc.* 1971, *93*, 3556.
427. Rautenstrauch, V. *J. Chem. Soc. Chem. Commun.* 1970, 4.
428. Schollkopf, U., Fellenberger, V., Rizk, M. *Liebigs Ann. Chem.* 1970, *734*, 106.
429. Felkin, H., Frajerman, C. *Tetrahedron Lett.* 1977, 3485.
430. Tsai, D. J.-S., Midland, M. M. *J. Am. Chem. Soc.* 1985, *107*, 3915.
431. Marshall, J. A., Jenson, T. M. *J. Org. Chem.* 1984, *49*, 1707.
432. Still, W. C., Mitra, A. *J. Am. Chem. Soc.* 1978, *100*, 1927.
433. Mikami, K., Kimura, Y., Kishi, N., Nakai, T. *J. Org. Chem.* 1983, *48*, 279.
434. Mikami, K., Fujimoto, K., Kasuga, T., Nakai, T. *Tetrahedron Lett.* 1984, *25*, 6011.
435. Mikami, K., Takahashi, O., Tabei, T., Nakai, T. *Tetrahedron Lett.* 1986, *27*, 4511.
436. Mikami, K., Maeda, T., Nakai, T. *Tetrahedron Lett.* 1986, *27*, 4189.
437. Rautenstrauch, V., Buchi, G., Wuest, H. *J. Am. Chem. Soc.* 1974, *96*, 2576.
438. Hoffmann, R., Brückner, R. *Angew. Chem. Int. Ed. Engl.* 1992, *31*, 647.
439. Antoniotti, P., Tonachini, G. *J. Org. Chem.* 1993, *58*, 3622.
440. Nakai, T., Mikami, K. *Org. React.* 1994, *46*, 105.
441. Balestra, M., Wittman, M. D., Kallmerten, J. *Tetrahedron Lett.* 1988, *29*, 6905.
442. Bruckner, R. *Tetrahedron Lett.* 1988, *29*, 5747.
443. Marshall, J. A., Robinson, E. D., Zapata, A. *J. Org. Chem.* 1989, *54*, 5854.
444. Marshall, J. A., Wang, X.-j. *J. Org. Chem.* 1990, *55*, 2995.
445. Mikami, K., Kawamoto, K., Nakai, T. *Tetrahedron Lett.* 1985, *26*, 5799.
446. Midland, M. M., Kwon, Y. C. *Tetrahedron Lett.* 1985, *26*, 5021.
447. Fujimoto, Y., Ohhana, M., Terasawa, T., Ikekawa, N. *Tetrahedron Lett.* 1985, *26*, 3239.
448. Evans, D. A., Andrews, G. C. *Acc. Chem. Res.* 1974, *7*, 147.
449. Burgess, K., Henderson, I. *Tetrahedron Lett.* 1989, *30*, 4325.
450. Sato, T., Otera, J., Nozaki, H. *J. Org. Chem.* 1989, *54*, 2779.
451. Goldmann, S., Hoffmann, R. W., Maak, N., Geueke, K. *J. Chem. Ber.* 1980, *113*, 831.
452. Grieco, P. A. *J. Chem. Soc. Chem. Commun.* 1972, 702.

453. Evans, D. A., Andrews, G. C. *J. Am. Chem. Soc.* 1972, *94*, 3672.
454. Babler, J. H., Haack, R. A. *J. Org. Chem.* 1982, *47*, 4801.
455. Hoffmann, R. W., Goldmann, S., Maak, N., Gerlach, R., Frickel, F., Steinbatch, G. *Chem. Ber.* 1980, *113*, 819.
456. Ogiso, A., Kitazawa, E., Kurabayashi, M., Sato, A., Takahashi, S., Noguchi, H., Kuwano, H., Kobayashi, S., Mishima, H. *Chem. Pharm. Bull.* 1978, *26*, 3117.
457. Evans, D. A., Andrews, G. C., Fujimoto, T. T., Wells, D. *Tetrahedron Lett.* 1973, 1389.
458. Masaki, Y., Hashimoto, K., Sakuma, K., Kaji, K. *J. Chem. Soc. Chem. Commun.* 1979, 855.
459. Evans, D. A., Andrews, G. C., Fujimoto, T. T., Wells, D. *Tetrahedron Lett.* 1973, 1385.
460. Masaki, Y., Hashimoto, K., Kaji, K. *Tetrahedron Lett.* 1978, 4539.
461. Kojima, K., Koyama, K., Ameniya, S. *Tetrahedron* 1985, *41*, 4449.
462. Miller, J. G., Kurz, W., Untch, K. G., Stork, G. *J. Am. Chem. Soc.* 1974, *96*, 6774.
463. Bickart, P., Carson, F. W., Jacobus, J., Miller, E. G., Mislow, K. *J. Am. Chem. Soc.* 1968, *90*, 4869.
464. Baudin, J.-B., Julai, S. A. *Tetrahedron Lett.* 1989, *30*, 1963.
465. Leblanc, Y., Fitzsimmons, B. J. *Tetrahedron Lett.* 1989, *30*, 2889.
466. Inoue, Y. *Chem. Rev.* 1992, *92*, 741.
467. Kagan, H. B., Fiaud, J. C. *Top. Stereochem.* 1978, *10*, 175.
468. Kaupp, G., Haak, M. *Angew. Chem. Int. Ed. Engl.* 1993, *32*, 694.
469. Sayo, N., Kimara, Y., Nakai, T. *Tetrahedron Lett.* 1982, *23*, 3931.
470. Koch, H., Runsink, J., Scharf, H.-D. *Tetrahedron Lett.* 1983, *24*, 3217.
471. Zamojski, A., Jarosz, S. *Tetrahedron* 1982, *38*, 1447.
472. Zamojski, A., Jarosz, S. *Tetrahedron* 1982, *38*, 1453.
473. Engler, T. A., Ali, M. H., Velde, D. V. *Tetrahedron Lett.* 1989, *30*, 1761.
474. Marouka, K., Concepcion, A. B., Yamamoto, H. *Synlett* 1992, 31.
475. Müller, F., Mattay, J. *Chem. Rev.* 1993, *93*, 99.
476. Kobayashi, Y., Takemoto, Y., Ito, Y., Terashima, S. *Tetrahedron Lett.* 1990, *31*, 3031.
477. Wynberg, H., Staring, E. G. J. *J. Am. Chem. Soc.* 1982, *104*, 166.

11

Asymmetric Reduction of Prochiral Ketones Catalyzed by Oxazaborolidines

MICHEL BULLIARD
PPG-SIPSY, Avrille, France

11.1. INTRODUCTION

A number of the new drugs under development are chiral. The two main reasons for this choice by pharmaceutical companies are the regulatory constraints to assess the properties of single enantiomers, even if the racemic mixture is developed, and the scientific evidence that enantiomers often have very different and sometimes opposite pharmacological effects. Among the numerous techniques available today to industrial chemists, asymmetric synthesis has been used successfully to obtain chiral compounds. From an industrial point of view, asymmetric catalysis is becoming the preferred approach because of its low environmental impact and high potential productivity.

In particular, asymmetric reduction of ketones to alcohols has become one of the more useful reactions. To achieve the selective preparation of one enantiomer of the alcohol, chemists first modified the classical reagents with optically active ligands; this led to modified hydrides. The second method consisted of reaction of the ketone with a classical reducing agent in the presence of a chiral catalyst. The aim of this chapter is to highlight one of the best practical methods that could be used on an industrial scale: the oxazaborolidine catalyzed reduction [1].

11.2. STOICHIOMETRIC REACTIONS

The first solution to the problem of reducing a prochiral ketone was to use a conventional reducing agent modified with stoichiometric amounts of chiral ligand. Among the reagents that illustrate this approach, those prepared from lithium aluminum hydride (LAH) should be noted first. Weigel at Eli Lilly [2] demonstrated the scope and limitation of the Yamagushi-Mosher reagent prepared from LAH and Chirald [(2S, 3R)-(+)4-dimethylamino-1,2-diphenyl-3-methyl-2-butanol] or its enantiomer, ent-Chirald, in the course of the synthesis of an analog of the antidepressant, fluoxetine (Prozac).

Another method has been described by Noyori. Optically active binaphtol reacts with LAH to afford BINAL-H (**1**) [3]. This reagent has proven to be an extremely efficient hydride donor achieving very high enantiomeric excesses at a low temperature.

The Yamaguchi-Mosher reagent and BINAL-H (**1**) have mostly been used over the last few years. Although these reagents have proved to be efficient, their manipulation is not easy and requires very low temperatures, sometimes below −78°C, to secure good stereoselectivity.

Another excellent way to produce enantiomerically pure secondary alcohols is to use chiral organoboranes. After the work pioneered by Brown, many reagents have been developed, such as β-3-pinanyl-9-borabicyclo-[3.3.1]-nonane (Alpine-Borane) (**2**) and chlorodiisopinocampheylborane (Ipc$_2$BCl) (**3**) [4]. The latter is now commercially available at scale, and is a useful reagent to reach high chemical and optical yields. A good example of its efficacy has been reported in the preparation of the antipsychotic, BMS 181100, from Bristol-Myers.

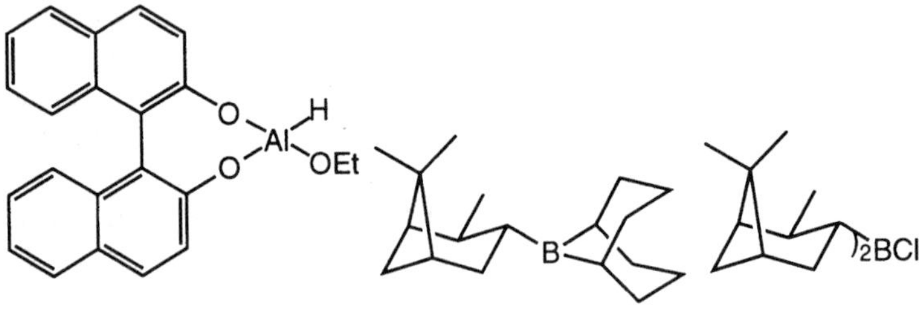

| **1** | **2** | **3** |

Many reagents such as glucoride, other modified borohydrides, or Masamune's borolanes have also been tried successfully [5].

11.3. CATALYTIC APPROACH

All of the previously mentioned reagents, despite their effectiveness, have an important drawback. They are used in stoichiometric amounts and are therefore often costly. Their use involves separation and purification steps that can be troublesome. Very often, efficient recycling of the ligand cannot be achieved. Both researchers and industrial chemists have designed catalytic reagents to overcome this limitation.

The most well-known technology is probably homogeneous catalytic hydrogenation. Numerous articles on asymmetric hydrogenation have appeared over the last few years, and a chapter of this book refers to this technology. It constitutes one of the best techniques available today, although it is sometimes more elegant than useful because of technical or commercial drawbacks. These processes can be particularly difficult to develop on an industrial scale and require a long chemical development time that hampers their use for the rapid preparation of quantities for clinical trials. Although enzymes and yeasts have been shown to be useful on a laboratory scale, they suffer disadvantages that limits their usefulness on a large scale: the high dilution required and troublesome work-ups. Moreover, enzymes can lack broad application, and the outcome of an enzymatic resolution is often difficult to predict.

Apart from these techniques, the most interesting method recently described is the Itsuno-Corey reduction. This catalytic reaction is general, highly predictable, and can rapidly be scaled up.

11.3.1. Catalyst Structure and the Effect on the Enantiomeric Excess

Building on the excellent work of Itsuno [6a,b], who first described the use of oxazaborolidine as a homochiral ligand, and of Kraatz [7], Corey was the first to report the enantioselective reduction of ketones to chiral secondary alcohols in the presence of an oxazaborolidine in substoichiometric amounts [8a,b]. This general method was named the CBS method (Scheme 1).

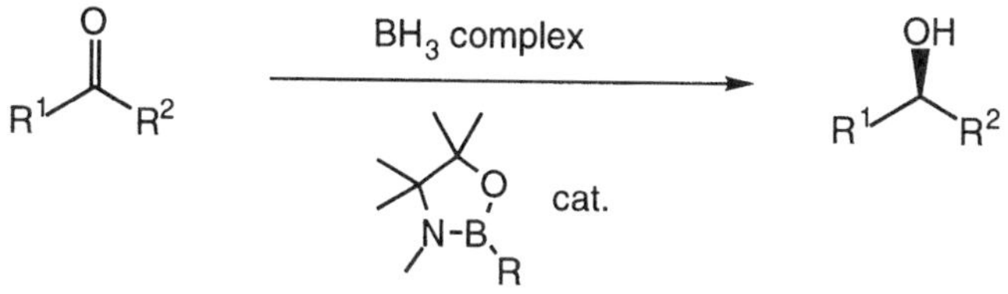

SCHEME 1.

Since that time, numerous articles dealing with this versatile reagent have appeared in the literature. Many new catalysts have been designed, usually by replacement of the diphenylprolinol moiety with other derivatives of various natural or unnatural amino acids. A large number of new oxazaborolidines have been reported. The enantiomeric excess varies substantially from one amino alcohol to another. Generally, the carbon adjacent to the nitrogen bears the chirality simply because amino alcohols are easily prepared from chiral amino acids belonging

TABLE I Effect of Catalyst Structure in the Reduction of Acetophenone

	ee (%)	Reference		ee (%)	Reference
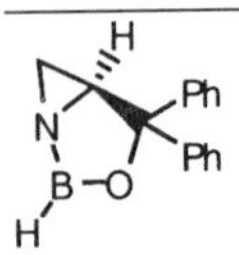	94	11	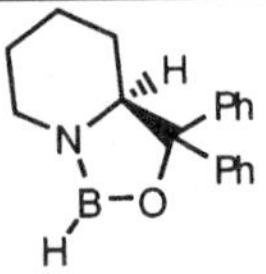	87	15
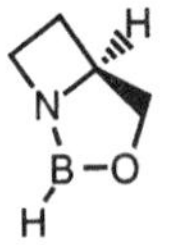	0	12	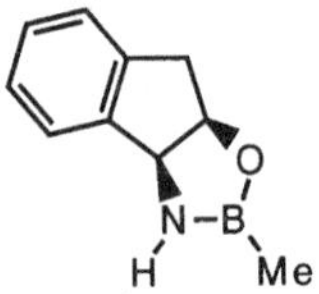	86 87 (1 eq.)	16 17
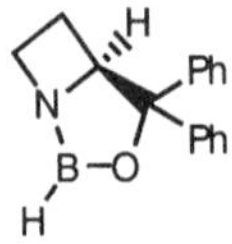	95–98	13, 14	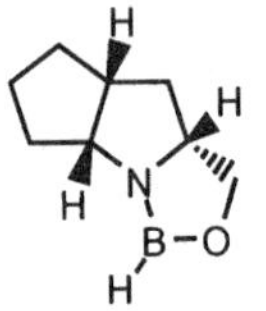	33	
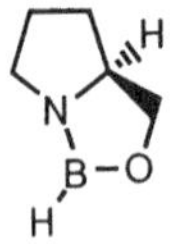	48	12	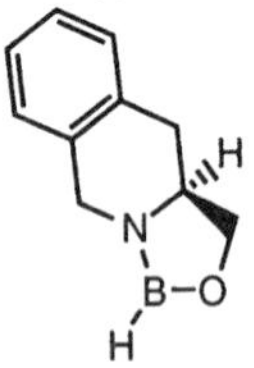	71	
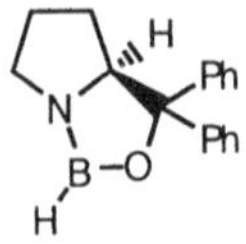	97	8			

to the chiral pool. There are only few examples of catalysts in which the carbon adjacent to the oxygen is the stereogenic center.

In the case of acetophenone reduction, it appears that amino alcohols that are sterically hindered at the carbon adjacent to the alcohol lead to much better results. A dramatic effect has been found in the case of prolinol and diphenylprolinol: the enantiomeric excess increases from 50% to 97%. Cyclic amino alcohols in which the nitrogen is in a 4- or 5-membered ring have exceptional catalytic properties and lead to very good enantiomeric excesses. Diphenylprolinol (4) is a very good choice because of its availability and performance.

Diphenyl oxazaborolidine, as reported by Quallich at Pfizer, is possibly a good alternative because it has the advantage of both enantiomers of the *erythro* aminodiphenylethanol being cheap and commercially available [9]. However, its performance is somewhat lower than that of diphenylprolinol. On the industrial

TABLE 2 Effect of Catalyst Structure in the Reduction of Acetophenone

ee (%)	Reference		ee (%)	Reference
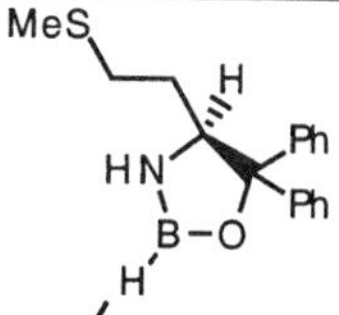 79	18	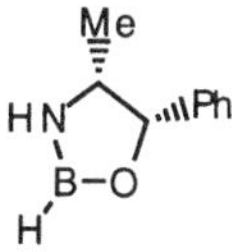	84	20
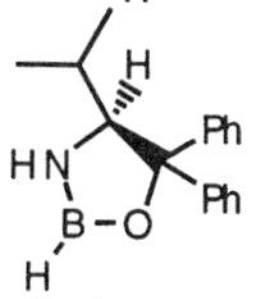 94 (1 eq.)	6	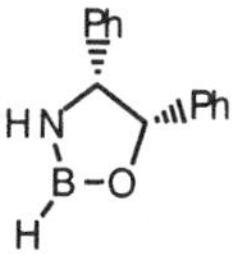	94	20
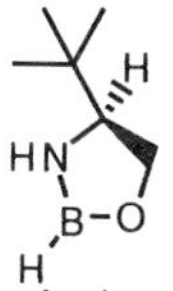 80	19	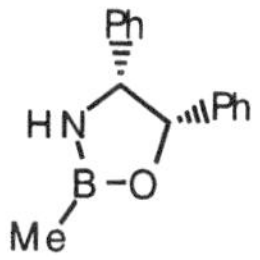	92	9
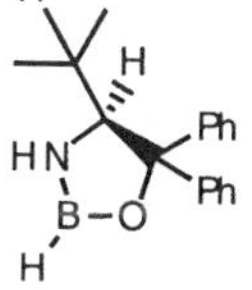 89	14	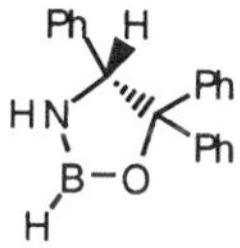	96	21

side, the discovery of this technique prompted many groups to investigate its versatility. At Merck, a group of chemists led by Mathre investigated the use of oxazaborolidines to prepare various optically pure pharmaceutical intermediates. In parallel with Corey's work, they studied the mechanistic aspects of this reaction and clearly demonstrated the pivotal role of the borane adduct CBS-B [10]. Their research culminated in the isolation of this stable complex described as a "free-flowing crystalline solid." They were the first to report its single-crystal x-ray structure.

11.3.2. Preparation of the Catalyst

Two different approaches can be followed to prepare and use the catalyst. The first is to prepare it in situ by mixing (R)- or (S)-diphenylprolinol (DPP) (**4**) and a borane complex (Scheme 2). This route is advantageous because there is no need to use boronic acids (or boroxines) and to remove water to form the catalyst. Another possible way is to use preformed catalysts, some of which are commercially available from suppliers such as Callery.

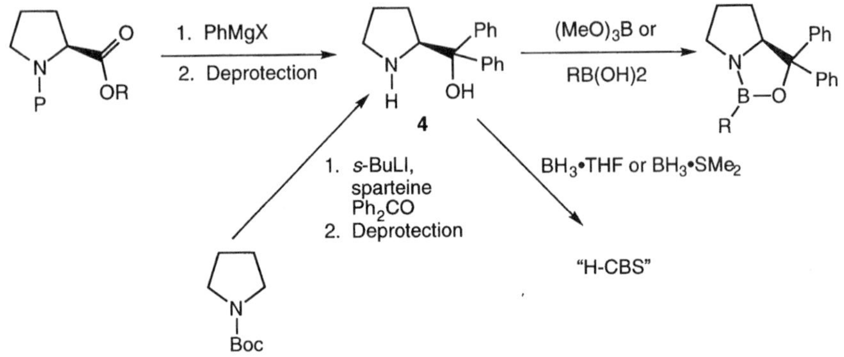

SCHEME 2.

Diphenylprolinol (**4**) itself is now commercially available at scale, or it can be prepared several ways (Scheme 2): direct addition of an aryl Grignard reagent to a proline ester leads to the diarylprolinol with a low yield in the range of 20–25% [22]. A more efficient route is based on an earlier publication from Corey. Suitably protected proline is reacted with phenylmagnesium halide, and after deprotection, diphenylprolinol is obtained [23a–c]. An alternative two-step enantioselective synthesis of diphenylprolinol from proline, based on the addition of proline-*N*-carboxyanhydride to three equivalents of phenylmagnesium chloride has been recently reported by the Merck group [24a,b]. An elegant asymmetric preparation of **4** has been reported by Beak et al. [25]. The enantioselective deprotonation of Boc-pyrrolidine is achieved with *sec*-butyllithium in the presence of sparteine as chiral inducer. Subsequent quench with benzophenone gives *N*-Boc-DPP in 70% yield and 99% ee after one recrystallization. To cope with the rather high price of (*R*)-proline, Corey developed an alternative route to this ligand from racemic pyroglutamic acid [26].

The oxazaborolidine is finally made by heating diphenylprolinol (**4**) under reflux with a suitable alkyl(aryl)boronic acid or, even better, with the corresponding boroxine in toluene in the presence of molecular sieves—the water can also be removed by azeotropic distillation. According to the literature, methyl-oxazaborolidine (Me-CBS) can be either distilled or recrystallized. The key point is that the catalyst must be free from any trace of water or alkyl(aryl)boronic acid because those impurities decrease enantioselection.

In the case of H-CBS, this catalyst could be prepared by mixing diphenylprolinol with borane-tetrahydrofurane (THF) or borane dimethylsulfide. Despite numerous efforts in many groups to isolate or even characterize H-CBS by nuclear magnetic resonance (NMR) spectroscopy, all attempts have been unsuccessful. H-CBS is used in situ and good results can be obtained in many cases.

11.3.3. Reducing Agent

Borane complexes, borane–THF and borane dimethylsulfide, are generally the appropriate reducing agents in this reaction. Borane thioxane has been used but it suffers from the inconvenience that thioxane is expensive. Diborane gas has been tried by Callery and has proven to be very efficient. The use of this reagent could open new industrial developments for this technology. Catecholborane has shown some advantages at low temperature when selectivity is needed.

11.3.4. Main Parameters of the Reaction

The best reaction conditions depend both on the nature of the substrate and the catalyst. A wide range of conditions have been studied by Mathre at Merck [27], by Quallich et al. at Pfizer [20,28], by Stone at Sandoz [29], and our group at Sipsy. The reaction must be conducted under anhydrous conditions. Traces of moisture have a dramatic effect on enantiomeric excesses [27a]. The catalyst loading is typically 1–20% depending on the nature of the ketone to be reduced. Diphenylprolinol can be recovered at the end of the reaction and recycled in some cases. The temperature is the most important parameter of the reaction. Although a clear effect has been shown by Stone, no good mechanistic explanations have been presented. The effect of temperature could be interpreted by assessing the accumulation of an oxazadiborane intermediate at low temperature, which would reduce the ketone with a low enantioselectivity [30].

When enantiomeric excess versus temperature is plotted, all reductions show the same shape of curve. At a low temperature, lower enantiomeric excesses are obtained. When the temperature is increased, the enantiomeric excess reaches a maximum value that depends on the reducing agent and the structure of the amino alcohol, of the catalyst itself (methyl, butyl, and phenyl oxazaborolidines do not give the same result), and the substrate. As the temperature increases further, the enantiomeric excess decreases once again. This general behavior was experienced in the reduction of p-chloro-chloroacetophenone (5), where the range of optimal reaction temperature is broad and very practical. Indeed, a very high enantiomeric excess could be obtained at 20°C (Scheme 3).

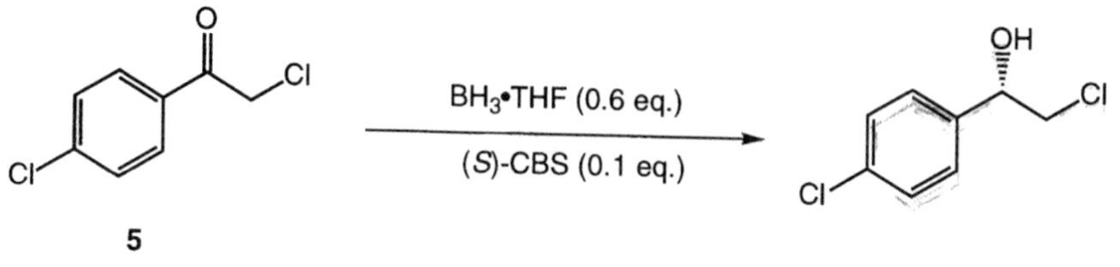

SCHEME 3.

218   BULLIARD

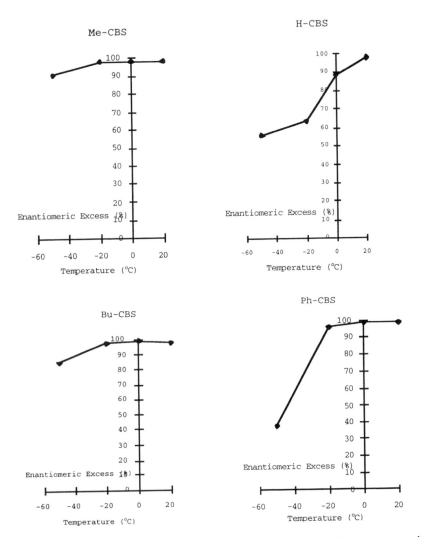

FIGURE I   The effect of temperature on ee for the reduction of p-chlorochloroacetophenone (5).

More interestingly, reduction of complex ketones could be dramatically improved by optimization of the reaction temperature. In the case of phenoxyphenyl-vinyl methyl ketone (6), a 5-lipoxygenase inhibitor synthesis intermediate, we were able to improve the enantiomeric excess from the 80–85% range up to 96% by selection of the optimal temperature for the reduction. In this case,

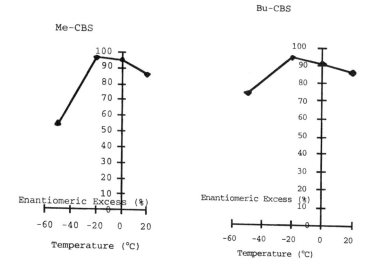

FIGURE 2   The reduction of the ketone (6).

the temperature range for an acceptable enantiomeric excess is very narrow. Generally, Me-CBS is the best catalyst, with borane complexes as the reducing agent.

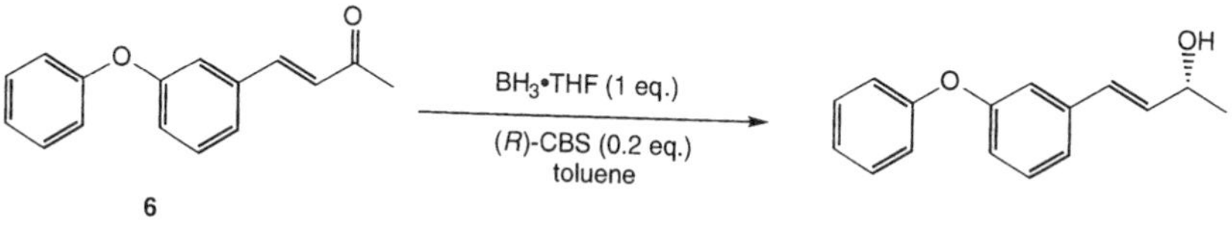

$$\text{6} \quad \xrightarrow[\substack{(R)\text{-CBS (0.2 eq.)}\\ \text{toluene}}]{\text{BH}_3\bullet\text{THF (1 eq.)}} \quad$$

SCHEME 4.

11.3.5. Versatility of the Reaction

The enantioselectivity versus the nature of the substrate is highly predictable. The best results are obtained with aryl ketones, where the difference between the large and the small substituent as well as the basicity of the ketone are favorable.

Beside the preparation of aryl alkyl carbinol, many other secondary alcohols have been prepared in good to excellent optical purity through the use of this method, although improvements of the catalytic system have to be found to obtain good results with all types of ketones, especially dialkyl ketones. Numerous complex target molecules have been prepared where the asymmetric reduction is a key step. The Corey reduction is now a standard reaction for laboratory-scale preparation.

TABLE 3 Versatility of the Reduction

Starting material	Product	ee or de (%)	Reference
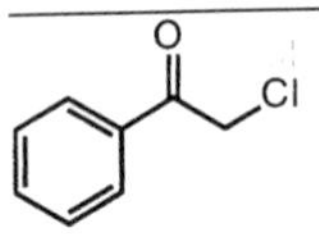 (phenyl α-chloro ketone, O, Cl)	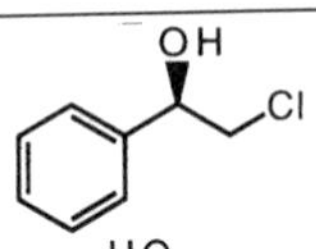 (phenyl α-chloro alcohol, OH, Cl)	96	31
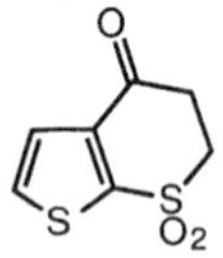 (thieno-fused ketone, O, S, O$_2$)	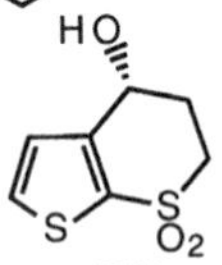 (thieno-fused alcohol, HO, S, O$_2$)	98	27a
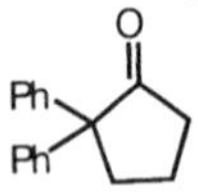 (Ph, Ph cyclopentanone, O)	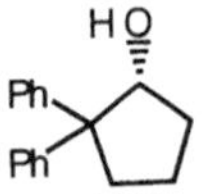 (Ph, Ph cyclopentanol, HO)	97	32
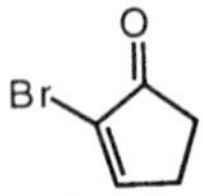 (Br cyclopentenone, O)	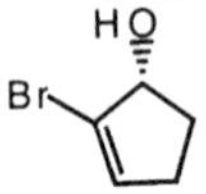 (Br cyclopentenol, HO)	90	33
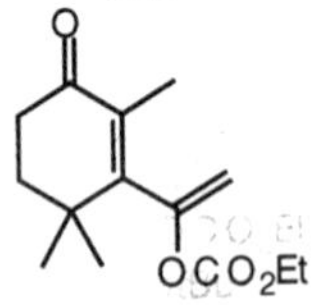 (cyclohexenone, O, OCO$_2$Et)	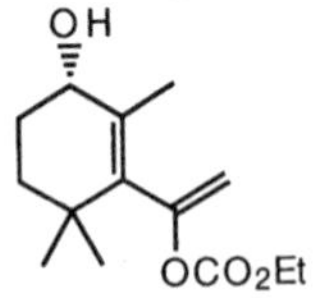 (cyclohexenol, OH, OCO$_2$Et)	93	34
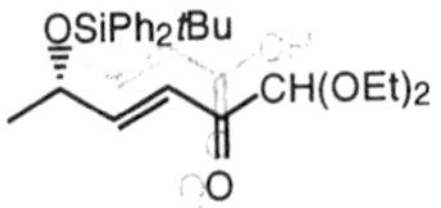 OSiPh$_2$tBu ... CH(OEt)$_2$, O	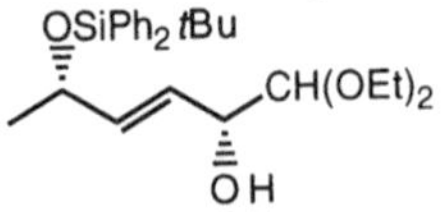 OSiPh$_2$ tBu ... CH(OEt)$_2$, OH	94	35
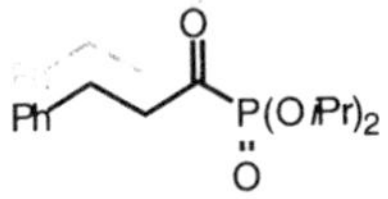 Ph ... P(OiPr)$_2$, O, O	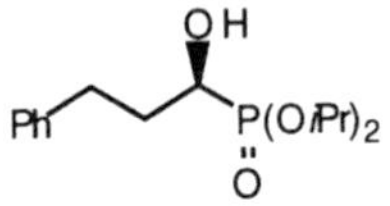 Ph ... P(OiPr)$_2$, OH, O	95	36

11.4. INDUSTRIAL APPLICATION IN THE SYNTHESIS OF PHARMACEUTICALS

The first investigator to become involved in the preparation of pharmaceutical molecules using this technology was Corey, who reported several preparations of important drugs. His first success was the synthesis of a pure enantiomer of fluoxetine, the Lilly antidepressant introduced under the brand name Prozac [37]. Shortly after, he published the preparation of both enantiomers of isoproterenol, a β-adrenoreceptor agonist and denopamine, a useful drug for congestive heart failure [38]. From a pharmaceutical point of view, the structure of isoproterenol is remarkable since its basic structure is encountered in a variety of drugs such as those that act at the adrenoreceptor site. Therapeutic indications are broad, ranging from asthma (β_2-agonist), to obesity and diabetes (β_3-agonist). Within this frame, halo aryl ketones have been shown to be valuable intermediates, giving rise after reduction and basic cyclization to the corresponding epoxystyrenes (Scheme 5). These are important intermediates for the preparation of β_3-agonists.

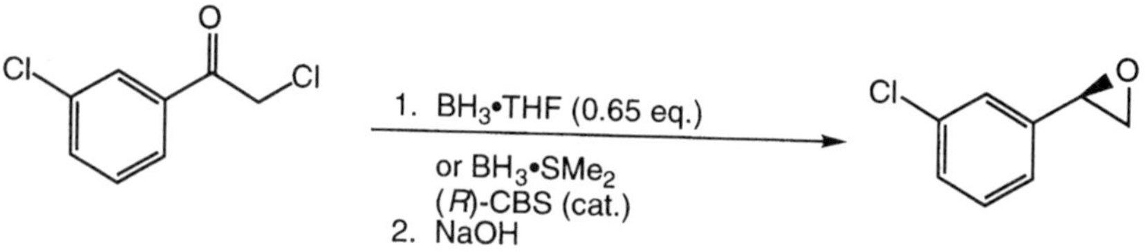

1. BH$_3$•THF (0.65 eq.)
   or BH$_3$•SMe$_2$
   (R)-CBS (cat.)
2. NaOH

SCHEME 5.

The regiospecific ring opening of these epoxides is achieved with the appropriate amine to give advanced intermediates possessing the right configuration at the carbinol stereocenter. Sipsy has demonstrated the industrial feasibility of this approach and produced hundreds of kilos of related molecules.

Along the same lines, several groups of chemists at Merck successfully produced substantial quantities of various important pharmaceutical intermediates used in the preparation of drugs under development. Among these are MK-0417 (7), a soluble carbonic anhydrase inhibitor used to treat glaucoma [27a], the LTD4 antagonist L-699392 (8) [27b], and the antiarrhythmic MK-499 (9) [27c,d].

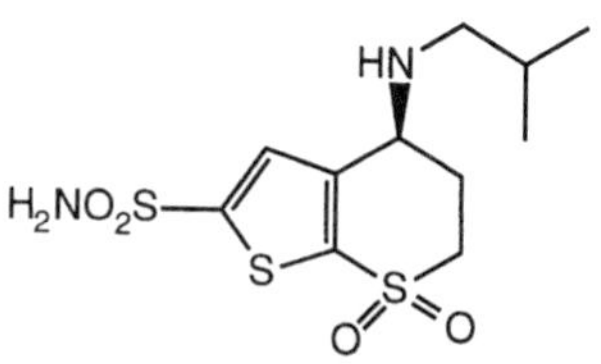

7

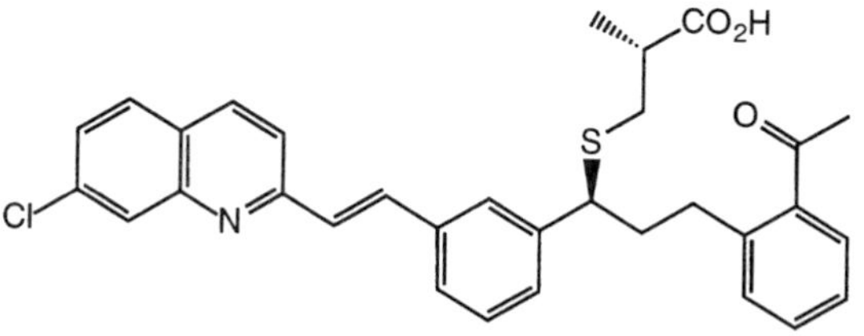

8

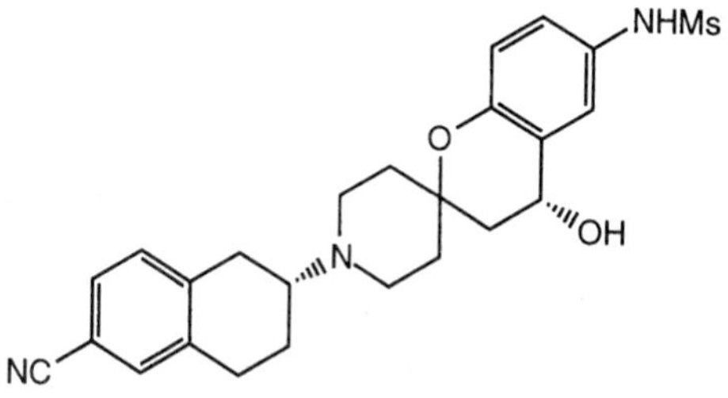

9

Equally successful has been the Quallich's group at Pfizer [28a], who reported the preparation of a substituted dihydronaphthalenone, a pivotal intermediate in the preparation of the antidepressant sertraline (Lustral) (10) (Scheme 6). In this synthesis, the new chiral center created subsequently determines the absolute stereochemistry of the final product.

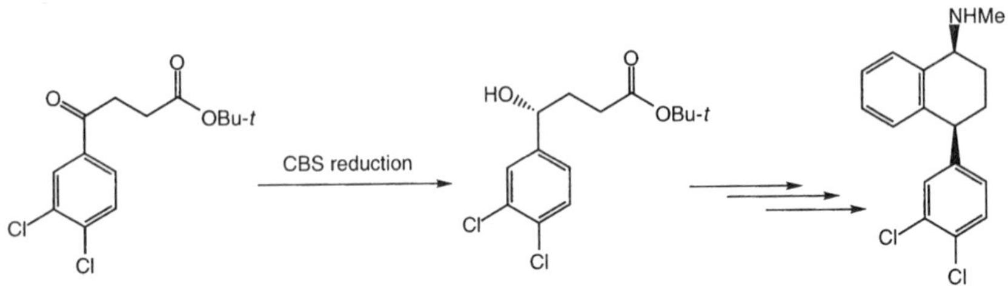

SCHEME 6.

Corey reduction is a valuable tool to introduce the chirality in a complex molecule because of its high compability with other reactive centers. Quallich demonstrated that high enantiomeric excesses were also obtained when other heteroatoms were present in the starting material. Specific chelation/protection of nitrogen functions can occur in addition to the reduction without affecting it (Scheme 7) [28b].

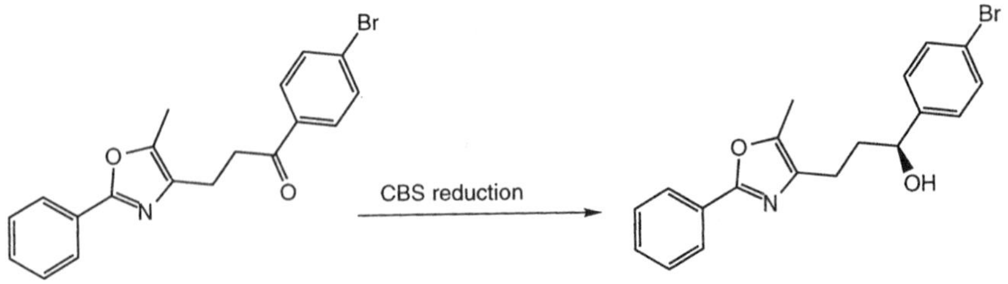

SCHEME 7.

Moreover, reduction of alkyl aryl ketones can be used to access optically pure secondary aryl alkyl amines, as illustrated in an enantioselective synthesis of SDZ-ENA-713 (11) [39], and as we have demonstrated in a related process (Scheme 8).

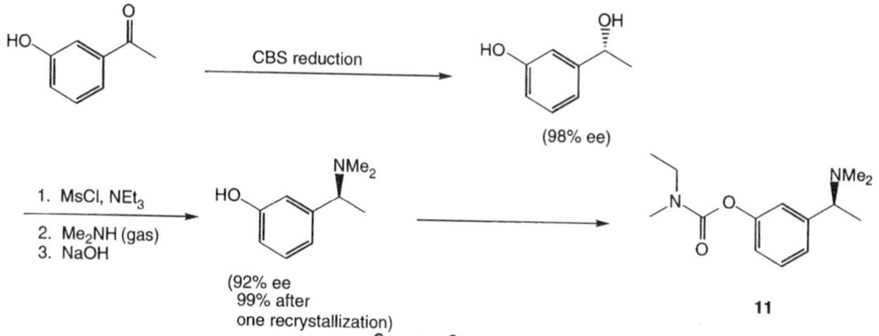

SCHEME 8.

In the synthesis of another carbonic anhydrase inhibitor, we have successfully performed the following reduction with a very high enantioselectivity.

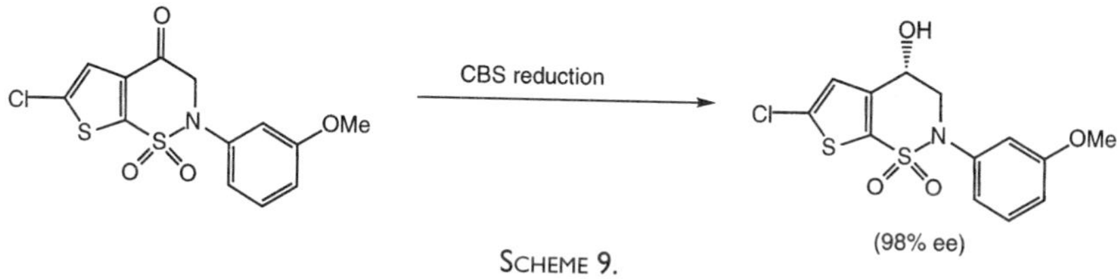

SCHEME 9.

Finally, Corey reduction was used to solve the particular problem of the diastereoselective reduction of the ketone in C-15 in a prostaglandin synthesis. Corey reported a selectivity of 90:10 [26]. We have investigated the reduction of a modified prostaglandin and despite considerable efforts, we were not able to reach excesses greater than 72% (Scheme 10). The yield of stereochemically pure product was 60% after one recrystallization. This yield is higher than that obtained by reduction with sodium borohydride, followed by chromatography and recycle of the undesired isomer by oxidation. Even in difficult cases in which stereoselectivity is good but not excellent, Corey reduction can still be used to advantage.

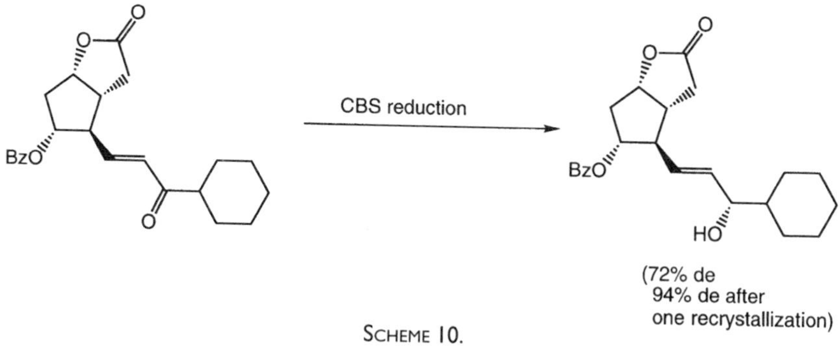

SCHEME 10.

11.5. CONCLUSION

The asymmetric reduction of ketones by borane catalyzed by oxazaborolidines has been widely studied since the beginning of the 1980s. Despite the use of borane complexes, which are hazardous chemicals, this reaction is an excellent tool to introduce the chirality in a synthesis and has demonstrated its usefulness in industrial preparation of chiral pharmaceutical intermediates. Because of its performance, versatility, predictability, and scale-up features, this method is particularly suitable for the rapid preparation of quantities of complex chiral molecules for clinical trials.

ACKNOWLEDGMENTS

I would like to thank everyone at Callery and Sipsy, and especially Dr. J. M. Barendt, Dr. J.-C. Caille, Dr. B. Laboue, and S. Frein for their contribution to the development of CBS technology.

REFERENCES

1. Singh, V. K. *Synthesis* 1992, 605.
2. Weigel, L. O. *Tetrahedron Lett.* 1990, 7101.
3. Noyori, R., Tomino, I., Tanimoto, Y., Nishizawa, M. *J. Am. Chem. Soc.* 1984, *106*, 6709.
4. Brown, H. C. In *Organic Synthesis via Boranes.* Wiley-Interscience: New York, 1975.
5. Brown, H. C., Park, W. S., Cho, B. T., Ramachandran, P. V. *J. Org. Chem.* 1987, *52*, 5406.
6a. Itsuno, S., Hirao, A., Nakahama, S., Yamazaki, N. *J. Chem. Soc. Perkin Trans. I* 1983, 1673.
6b. Itsuno, S., Nakano, M., Miyazaki, K., Masuda, H., Ito, K., Mhizao, A., Nakahama, S. *J. Chem. Soc. Perkin Trans. I* 1985, 2039.
7. Kraatz, U. DE3609152 A1.
8a. Corey, E. J., Bakshi, R. K., Shibata, S. *J. Am. Chem. Soc.* 1987, *109*, 5551.
8b. Corey, E. J. *US Patent* 4,943,635.
9. Quallich, G. J., Woodall, T. M. *Tetrahedron Lett.* 1993, *34*, 4145.
10. Thompson, A. S. Douglas, A. W., Hoogsteen, K., Carroll, J. D., Corley, E. G., Grabowski, E. J. J., Mathre, D. J. *J. Org. Chem.* 1993, *58*, 2880.
11. Willems, J. G. H., Dommerholt, F. J., Hamminck, J. B., Vaarhorst, A. M., Thÿss, L., Zwanenburg, B. *Tetrahedron Lett* 1995, *36*, 603.
12. Mehler, T., Behnen, W., Wilken, J., Martens, J. *Tetrahedron: Asymmetry* 1994, *5*, 185.
13. Rao, A. V. R., Gurjar, M. K., Kaiwar, V. *Tetrahedron: Asymmetry* 1992, *3*, 859.
14. Behnen, W., Dauelsberg, C., Wallbaum, S., Martens, J. *Synth. Comm.* 1992, *22*, 2143.

15. Rao, A. V. R., Gurjar, M. K., Sharma, P. A., Kaiwar, V. *Tetrahedron Lett.* 1990, *31*, 2341.
16. Huong, Y., Gao, Y., Nie, X., Zepp, C. M. *Tetrahedron Lett.* 1994, *35*, 6631.
17. Didier, E., Loubinoux, B., Tombo, G. M. R., Rihs, G. *Tetrahedron* 1991, *47*, 4941.
18. Mehler, T.; Martens, J. *Tetrahedron: Asymmetry* 1993, *4*, 1983.
19. Wallbaum, S., Martens, J. *Tetrahedron: Asymmetry* 1992, *3*, 1475.
20. Quallish, G. J. Woodall, T. M. *Synlett* 1993, 929.
21. Berenguer, R., Garcia, J., Vilarrasa, J. *Tetrahedron: Asymmetry* 1994, *5*, 165.
22. EP 0237902 A2.
23a. Corey, US 4,943,635.
23b. Maertens, DE 4,341,605 C2.
23c. Klinger, EP 0682013 A1.
24a. Mathre, US 5,039,802.
24b. Mathre, D. J., Jones, T. K., Xavier, L. C., Blacklock, T. J., Reamer, R. A., Mohan, J. J., Turner-Jones, E. T., Hoogsteen, K., Baum, M. W., Grabowski, E. J. J. *J. Org. Chem.* 1991, *56*, 751.
25. Beak, P., Kerrick, S. T., Wu, S., Chu, J. *J. Am. Chem. Soc.* 1994, *116*, 3231.
26. Corey, E. J., Bakshi, R. K., Shibata, S., Chn C. P., Singh, V. K. *J. Am. Chem. Soc.* 1987, *109*, 7925.
27a. Jones, T. K., Mohan, J. J., Xavier, L. C., Blacklock, T. J., Mathre, D. J., Sohar, P., Turner-Jones, E. T., Reaner, R. A., Robers, F. E., Grabowski, E. J. J. *J. Org. Chem.* 1991, *56*, 763.
27b. King, A. O., Corley, E. G., Anderson, R. K., Larsen, R. D., Verhoeven, T. R., Reider, P. J. *J. Org. Chem.* 1993, *58*, 3731.
27c. Shi, Y. J., Cai, D., Dolling, U. M., Douglas, A. W., Tschaen, D., Veroheven, T. R. *Tetrahedron Lett.* 1994, *35*, 6409.
27d. Cai, D., Tschaen, D., Shi, Y. J., Veroheven, T. R., Reamer, R. A., Douglas, A. W. *Tetrahedron Lett.* 1993, *34*, 3243.
27e. Tschaen, D., Abramson, M. L., Cai, D., Desmond, R., Dolling, U. M., Frey, L., Karady, S., Shi, Y. J., Veroheven, T. R. *J. Org. Chem.* 1995, *60*, 4324.
28a. Quallich, G. J., Woodall, T. M. *Tetrahedron* 1992, *48*, 10239.
28b. Quallich, G. J., Woodall, T. M. *Tetrahedron Lett.* 1993, *34*, 785.
28c. Quallich, G. J., Woodall, T. M. *Synlett* 1993, 929.
29. Stone, G. B. *Tetrahedron: Asymmetry* 1994, *5*, 465.
30. Mathre, D. J. *Proceedings of the Chiral USA '96 Symposium*, Spring Innovation Ltd, Boston, MA 1996, p. 69.
31. Corey, E. J., Shibata, S., Bakshi, R. K. *J. Org. Chem.* 1988, *53*, 2861.
32. Denmark, S. E., Schnute, M. E., Marcin, L. R., Thorarensen, A. *J. Org. Chem.* 1995, *60*, 3205.
33. Corey, E. J., Rao, K. S. *Tetrahedron Lett.* 1991, *32*, 4623.
34. Corey, E. J., Jardine, P. D. S. *Tetrahedron Lett.* 1989, *30*, 7297.
35. Dumartin, H., Lefloc'h, Y., Gree, R. *Tetrahedron Lett.* 1994, *35*, 6681.
36. Meier, C., Laux, W. H. G. *Tetrahedron: Asymmetry* 1995, *61*, 1089.
37. Corey, E. J. Reichard, G. A. *J. Org. Chem.* 1989, *30*, 5207.
38. Corey, E. J. Link, J. O. *J. Org. Chem.* 1991, *56*, 442.
39. Chen, C.-P., Prasad, K., Repic, O. *Tetrahedron Lett.* 1991, *32*, 7175.

12
Asymmetric Oxidations

DAVID J. AGER and DAVID R. ALLEN
NSC Technologies, Mount Prospect, Illinois

12.1. INTRODUCTION

Considerable advances have been made in catalytic methodologies to perform asymmetric oxidations. Although no large-scale processes for commercial pharmaceuticals currently use the technology, the methods are relatively new compared with catalytic asymmetric hydrogenation. However, the approach is now in the synthetic arsenal and it is surely just a matter of time before it comes to fruition.

Asymmetric oxidations have followed the usual development pathway in which face selectivity was observed through the use of chiral auxiliaries and templates. The breakthrough came with the Sharpless asymmetric epoxidation method, which, although stoichiometric, allowed for a wide range of substrates and the stereochemistry of the product to be controlled in a predictable manner [1]. The need for a catalytic reaction was very apparent, but this was developed and now the Sharpless epoxidation is a viable process at scale, although subject to the usual economic problems of a cost-effective route to the substrate (see later) [2]. The Sharpless epoxidation has now been joined by other methods and a wide range of products are now available. The power of these oxidations is augmented by the synthetic utility of the resultant epoxides or diols that can be used for further transformations, especially those that use a substitution reaction (see Chapter 7) [1].

12.2. SHARPLESS EPOXIDATION

One of the major advantages of the Sharpless [3,4] titanium asymmetric epoxidation is the simple method by which the stereochemical outcome of the reaction can be predicted [5]. The other powerful feature is the ability to change this selectivity to the other isomer by simple means (Figure 1) [6–8].

227

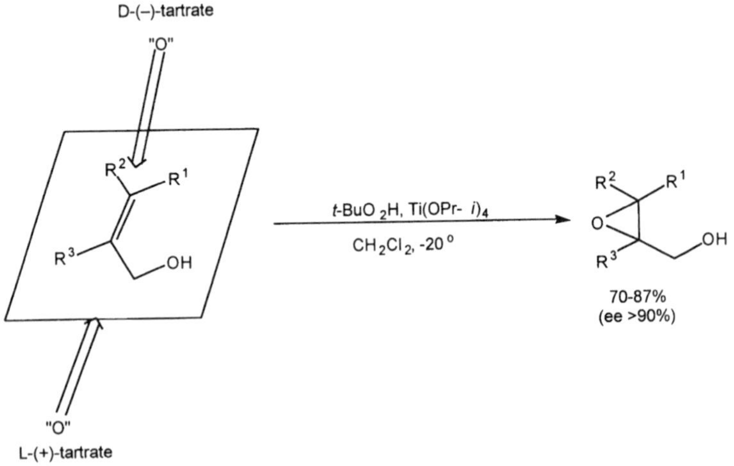

D-(–)-tartrate

"O"

R^2 R^1

R^3 OH

t-BuO$_2$H, Ti(OPr-i)$_4$

CH$_2$Cl$_2$, -20 °

R^2 R^1

O

R^3 OH

70-87%
(ee >90%)

"O"

L-(+)-tartrate

FIGURE I   Face selectivity for the Sharpless epoxidation of allyl alcohols.

Although the original Sharpless epoxidation method was stoichiometric, the development of a catalytic method has allowed the reaction to be amenable to scale up. The addition of molecular sieves for the removal of trace amounts of water is important in the catalytic procedure [2,9–11].

In addition to the fact that the reagent's ingredients are commercially available, the reaction is promiscuous and proceeds in good chemical yield with excellent enantiomeric excesses. The reaction, however, does suffer when bulky substituents are cis to the hydroxymethyl functionality (R^1 in Figure 1). For prochiral alcohols, the absolute stereochemistry of the transformation is predictable, whereas for a chiral alcohol, the diastereofacial selectivity of the reagent is often sufficient to override those preferences inherent in the substrate. When the chiral atom is in the E-β-position of the allyl alcohol (R^2), then the epoxidation can be controlled to access either diastereoface of the alkene. In contrast, when the chirality is at either the α- or Z-β-positions (R^1 or R^3), the process is likely to give selective access of the reagent from only one of the two diastereotopic faces [6,12]. Many examples of substrates for the epoxidation protocol are known [1,13,14].

An improved work-up procedure increases the yield for allyl alcohols containing a small number of carbon atoms [15–17]. Futhermore, less reactive substrates provide the epoxide readily [18].

The structure of the titanium–tartrate derivatives has been determined [13,14,19–25], and based on these observations together with the reaction selectivity, a mechanistic explanation has been proposed (Scheme 1) [26]. The complex 1 contains a chiral titanium atom through the appendant tartrate ligands. The intramolecular hydrogen bond ensures that internal epoxidation is only favored at one face of the allyl alcohol. This explanation is in accord with the experimental observations that substrates with an α-substituent (b = alkyl; a = alkyl or hydrogen) react much slower than when this position is not substituted (b = hydrogen).

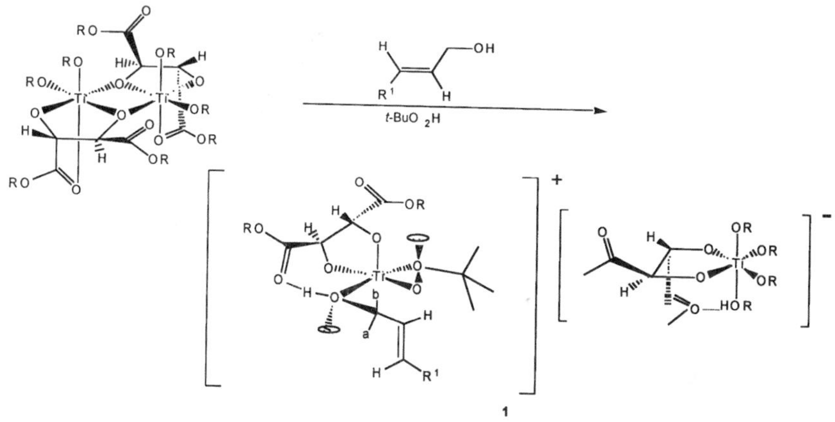

SCHEME 1.

The reaction time has been reduced dramatically by the addition of calcium hydride, silica gel, or montmorillonite catalysts [18,27–29].

The power of the Sharpless epoxidation method is augmented by the versatility of the resultant 2,3-epoxy alcohols [1] and the development of the catalytic variation [2].

Although glycidols are important chemical intermediates, it is unfortunate that the Sharpless approach gives lower enantiomeric purities in this simple case. However, if the arylsulphonates are prepared from the oxidation products, enantioenrichment during crystallization is observed [30]. There is an additional advantage as the simple glycidyl epoxide undergoes attack at C-1 and C-3, but the presence of the sulphonate leaving group gives rise to preponderance of C-1 attack (see later).

12.2.1. Kinetic Resolutions

The ability of the Sharpless epoxidation catalyst to differentiate between the two enantiomers of an asymmetric allyl alcohol affords a powerful synthetic tool to obtain optically pure materials through kinetic resolution [31]. Because the procedure relies on one enantiomer of a secondary allyl alcohol undergoing epoxidation at a much faster rate than its antipode, reactions are usually run to 50–55% completion [10]. In this way, resolution can often be impressive [6,13,14,32–

34]. An increase in steric bulk at the olefin terminus increases the rate of reaction [34,35].

This resolution method has been used to resolve furfuryl alcohols—the furan acts the alkene portion of the allyl alcohol (Scheme 2) [36–38].

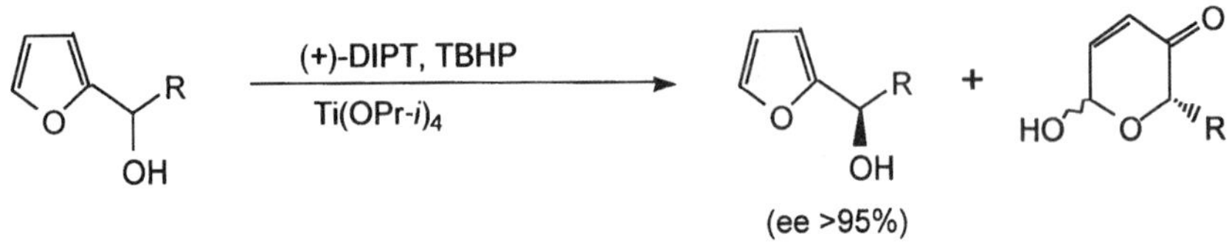

SCHEME 2.

12.2.2. Reactions of 2,3-Epoxy Alcohols

As previously noted, one reason for the powerful nature of the Sharpless epoxidation is the ability of the resultant epoxy alcohols to undergo regio- and stereoselective reaction with nucleophiles [1,6–8,39–43]. Often, the regiochemistry is determined by the functional group within the substrate [44–71].

A key reaction of 2,3-epoxy alcohols is the Payne rearrangement, an isomerization that produces an equilibrium mixture. This rearrangement then allows for the selective reaction with a nucleophile at the most reactive, primary position (Scheme 3) [6,72].

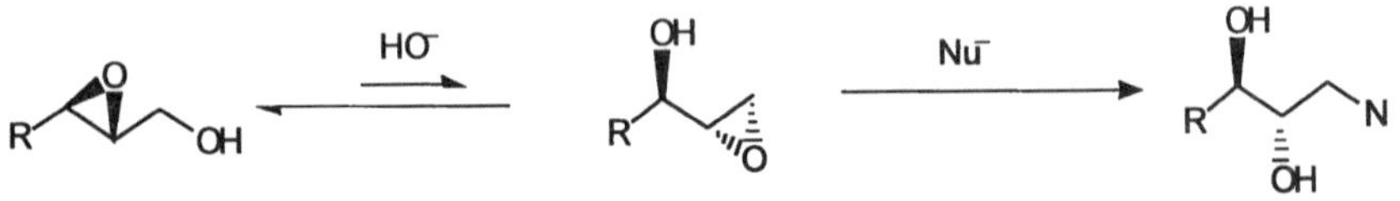

SCHEME 3.

Under Payne rearrangement conditions, sodium t-butylthiolate provides 1-t-butylthio-2,3-diols with very high regioselectivity. However, the selectivity is affected by many factors, including reaction temperature, base concentration, and the rate of addition of the thiol. These sulphides can then be converted to the 1,2-epoxy-3-alcohols, which in turn react with a wide variety of nucleophiles specifically at the 1-position (Scheme 4). This methodology circumvents the problems associated with the instability of many nucleophiles under Payne conditions [73].

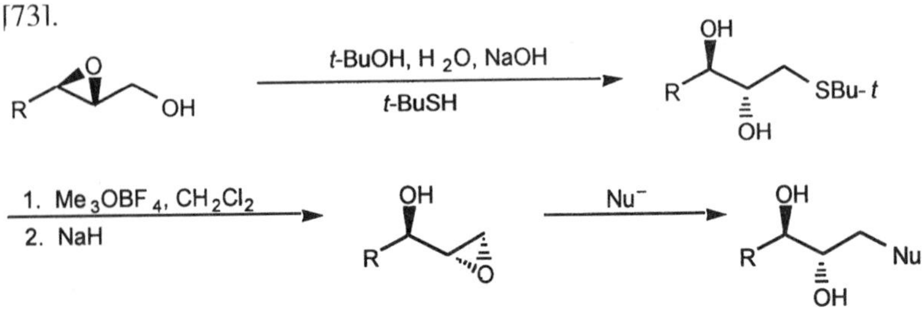

SCHEME 4.

With good nucleophiles, under relatively mild conditions, 2,3-epoxy alcohols will undergo epoxide ring opening at C-2 or C-3. In simple cases, nucleophilic attack at C-3 is the preferred mode of reaction. However, as the steric congestion at C-3 is increased, or if substituents play a significant electronic role, attack at C-2 can predominate [74].

12.2.3. Commercial Applications

Commercial applications include the synthesis of both isomers of glycidols (ARCO), oct-3-en-1-ol (Upjohn) and the synthesis of disparulene (**2**) (Scheme 5) [9,75–79].

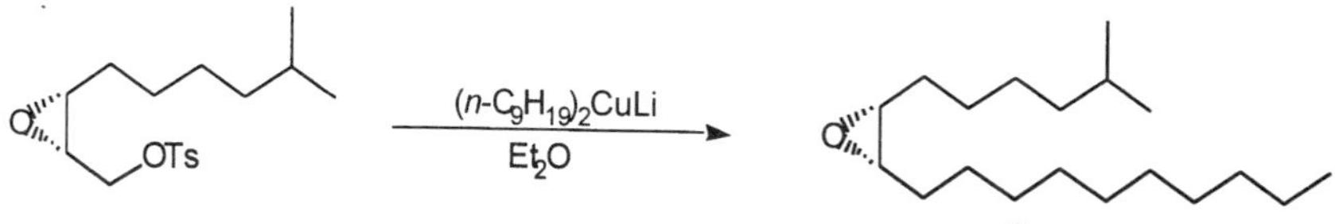

SCHEME 5.

Although no specific synthesis uses a carbohydrate-type synthon that has been prepared by an asymmetric oxidation, the methodology offers many advantages over a synthesis starting from a carbohydrate (see Chapter 5). The useful building block, **3**, is available by a Sharpless protocol (Scheme 6) [80].

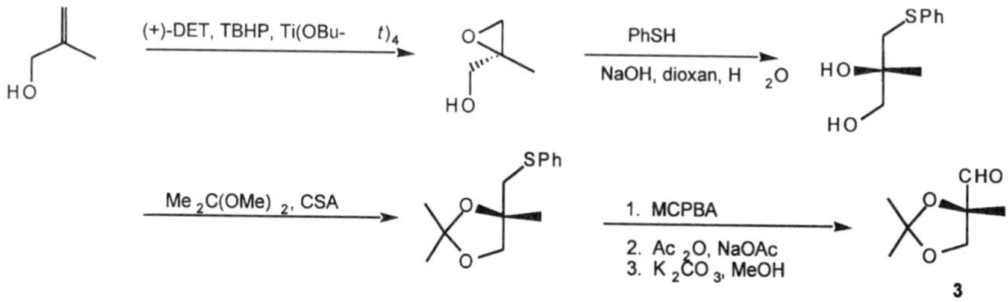

SCHEME 6.

This methodology has been used for the preparation of L-threitol (**4**) (Scheme 7) and erythritol derivatives [81]. This simple iterative process has not only been used for the simple alditols [81,82], deoxyalditols [83], and aldoses [81], but all of the L-hexoses as well [84,85].

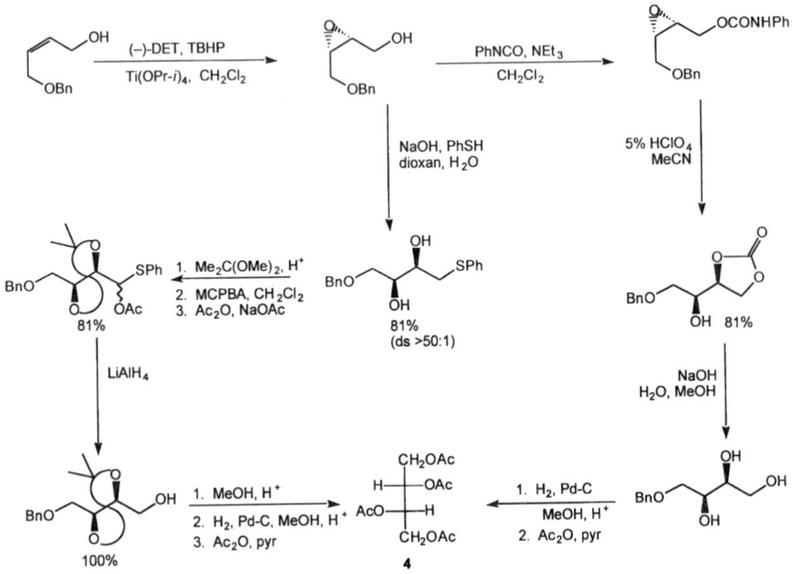

SCHEME 7.

12.3. ASYMMETRIC DIHYDROXYLATION

This asymmetric dihydroxylation problem was first solved by the use of cinchona alkaloid esters (**5** and **6**; R = p-ClC$_6$H$_4$) together with a catalytic amount of osmium tetroxide [86]. The alkaloid esters act as pseudoenantiomeric ligands (Scheme 8) [87–91]. They can also be supported on a polymer [92,93].

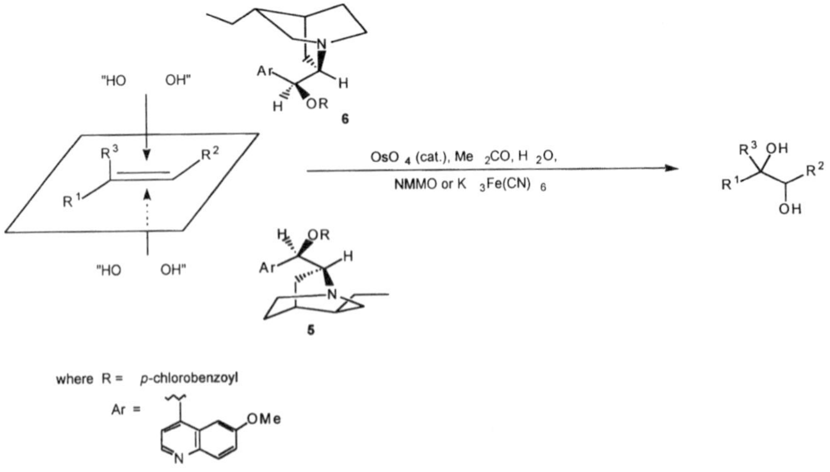

where R = p-chlorobenzoyl

Ar =

SCHEME 8.

The original procedure has been modified by the use of a slow addition of the alkene to afford the diol in higher optical purity, and ironically, this modification results in a faster reaction. This behavior can be rationalized by consideration of two catalytic cycles operating for the alkene (Scheme 9); the use of low alkene concentrations effectively removes the second, low enantioselective cycle [88,94]. The use of potassium ferricyanide in place of *N*-methylmorpholine-*N*-oxide (NMMO) as oxidant also improves the level of asymmetric induction [95,96].

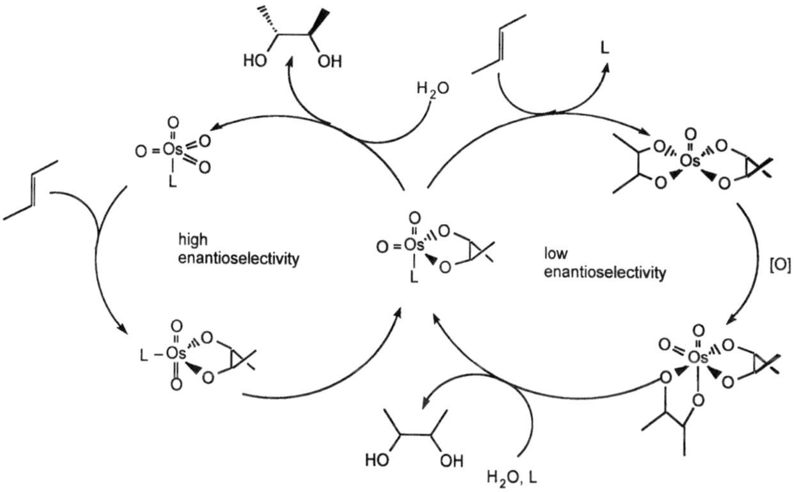

SCHEME 9.

Although the use of cinchona alkaloids as chiral ligands does provide high asymmetric induction with a number of types of alkene, the search for better systems has resulted in better catalysts (Scheme 10) [97–101].

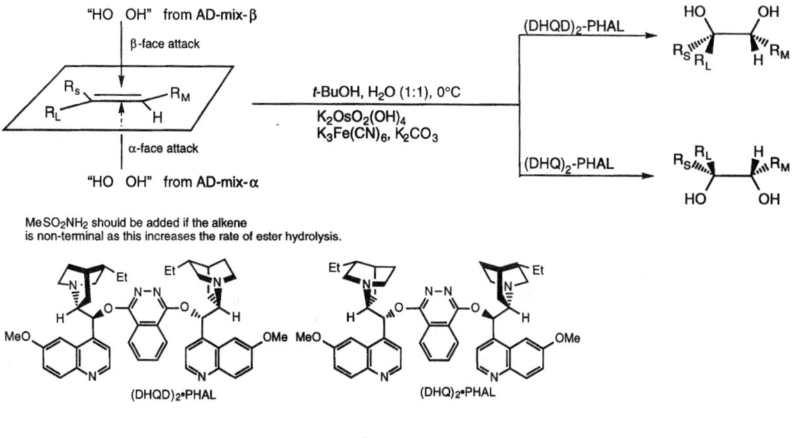

SCHEME 10.

Kinetic resolutions can be achieved by the dihydroxylation approach [102]. The Sharpless dihydroxylation method has been run at scale by Pharmacia-Upjohn with *o*-isopropoxy-*m*-methoxysytrene as substrate and NMMO as oxidant [103]. A variation on the Sharpless dihydroxylation methodology allows for the preparation of amino alcohols. As the groups are introduced in a *syn* manner, this compliments an epoxide opening or a similar reaction that uses substitution at one center (see Chapter 7). Higher selectivity is observed when one of the alkene substituents is electron withdrawing, as in an ester group (Scheme 11) [104–106].

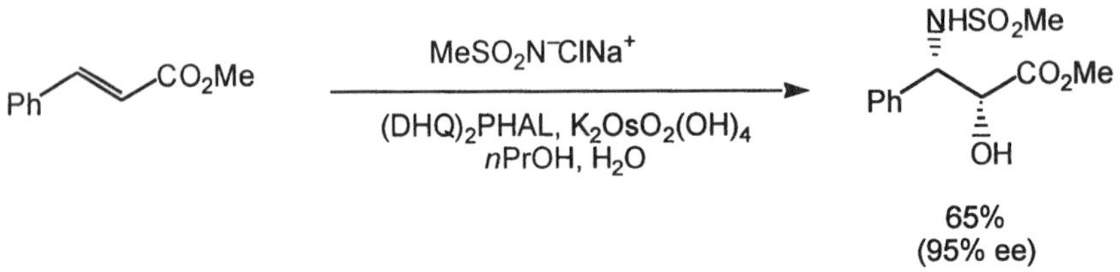

SCHEME 11.

The methodology has been extended to provide α-arylglycinols from styrenes [107].

12.3.1. Reactions of Diols

The 1,2-diols formed by the asymmetric oxidation can be used as substrates in a wide variety of transformations. Conversion of the hydroxy groups to *p*-toluenesulphonates then allows nucleophilic displacement by azide at both centers with inversion of configuration (Scheme 12) [108].

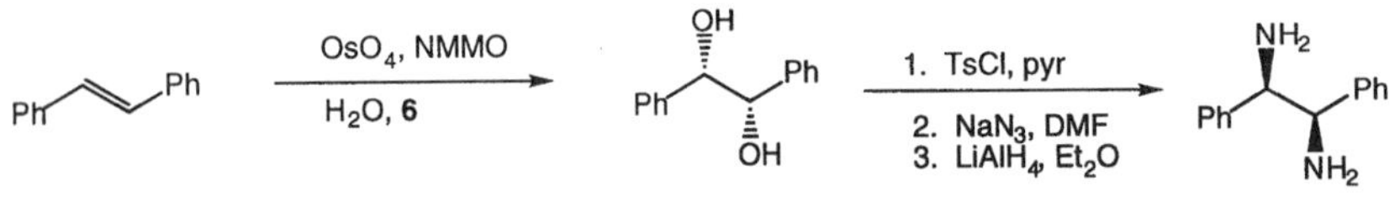

SCHEME 12.

Cyclic sulphates provide a useful alternative to epoxides now that it is viable to produce a chiral diol from an alkene. These cyclic compounds are prepared by reaction of the diol with thionyl chloride, followed by ruthenium catalyzed oxidation of the sulphur (Scheme 13) [109]. This oxidation has an advantage over previous procedures because it only uses a small amount of the transition metal catalyst [110,111].

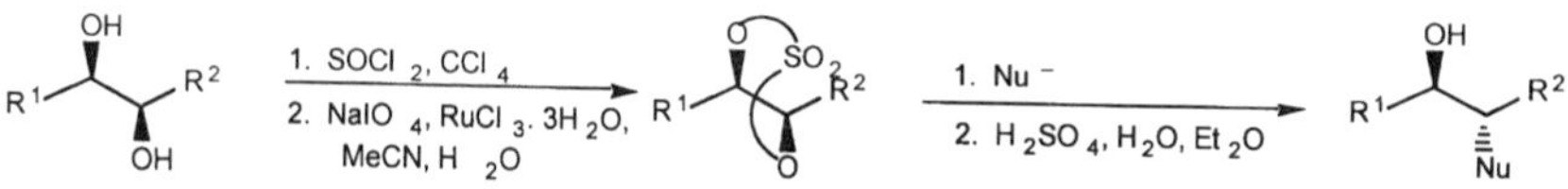

SCHEME 13.

The cyclic sulphates undergo ring opening with a wide variety of nucleophiles such as hydride, azide, fluoride, benzoate, amines, and Grignard reagents. The reaction of an amidine with a cyclic sulphate provides an expeditious entry to chiral imidazolines (**7**) and 1,2-diamines (Scheme 14) [112].

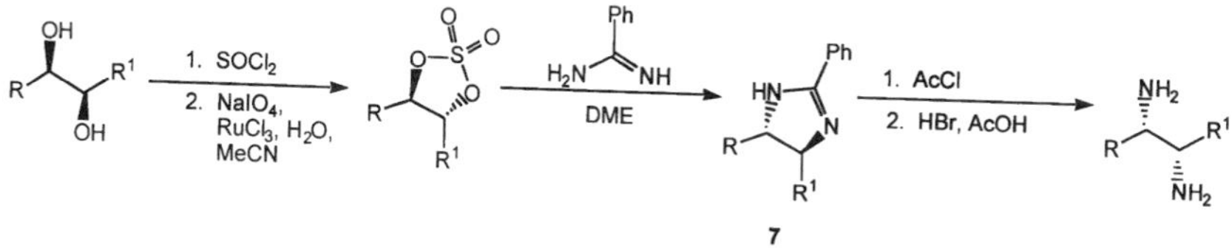

SCHEME 14.

In the case of an ester (R^2 = CO_2Me), the addition occurs exclusively at C-2 (Scheme 15); the analogous epoxide does not demonstrate such selectivity [109,113–116]. Terminal cyclic sulphates (see Chapter 10, Scheme 25; R^2 = H) open in a manner completely analogous to the corresponding epoxide [113].

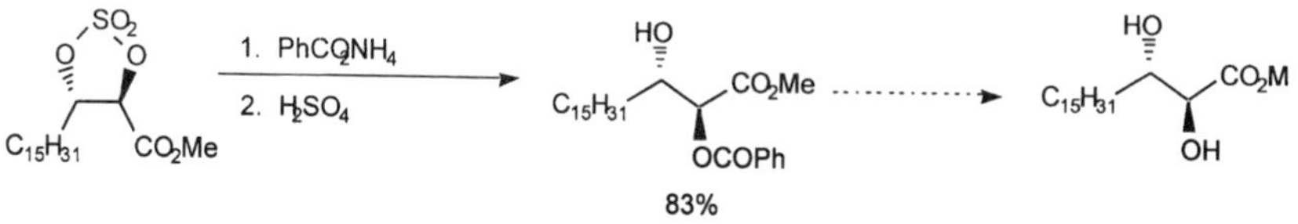

SCHEME 15.

The resultant sulphate ester can be converted to the alcohol by acid hydrolysis. If an acid-sensitive group is present, this hydrolysis is still successful through use of a catalytic amount of sulphuric acid in the presence of 0.5–1.0 equivalents of water with tetrahydrofuran as solvent. The use of base in the formation of the cyclic sulphates themselves can also alleviate problems associated with acid-sensitive groups [117,118].

An alternative to the use of cyclic sulphates is the use of cyclic carbonates as illustrated for the synthesis of 1,2-amino alcohols (Scheme 16) [119].

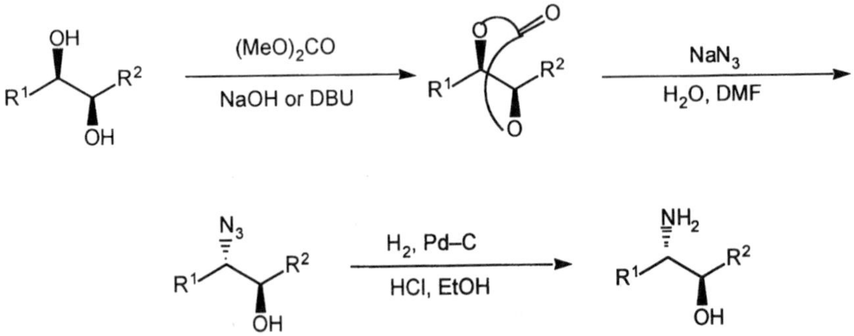

SCHEME 16.

12.4. JACOBSEN EPOXIDATION

The facial selectivity required for an asymmetric epoxidation can be achieved with manganese complexes to provide sufficient induction for synthetic utility (Scheme 17) [120–126]. This manganese (III) salen complex 8 can also use bleach as the oxidant rather than an iodosylarene [127,128]. The best selectivities are seen with *cis*-alkenes.

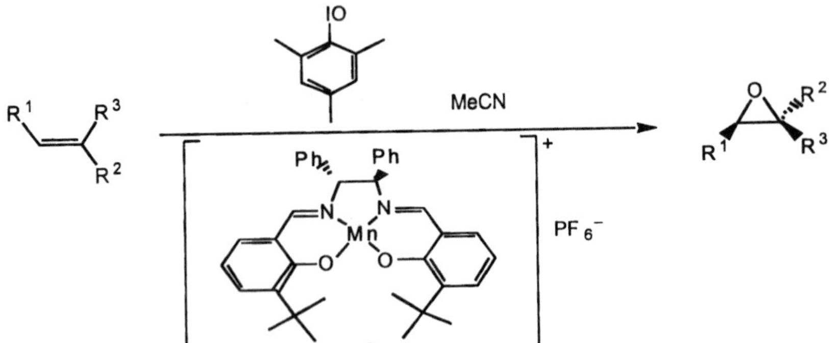

SCHEME 17.

In effect, the metal is planar, but the ligands are slightly buckled and direct the approach of the alkene to the metal reaction site in a particular orientation. The enantioselectivity is highest for Z-alkenes and correlates directly with the electronic properties of the ligand substituents (Scheme 17) [122,129,130]. The use of blocking groups other than *t*-butyl provides similar selectivities [121,127, 131,132].

12.4.1. Resolutions

Although terminal epoxides are not yet accessible by an asymmetric oxidation methodology, a resolution method has been developed based on the catalytic opening with water (Scheme 18) (see Chapter 8) [133,134].

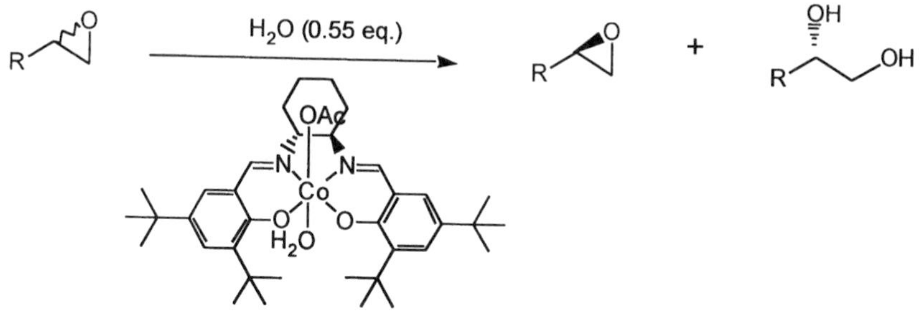

SCHEME 18.

Although this is a resolution approach, racemic epoxides are readily available, and other than the catalysts, the substrate and water are the only materials present.

A similar approach uses a chromium–salen complex to open an epoxide with trimethylsilyl azide, as illustrated by the synthesis of the antihypertensive agent, (S)-propranolol (9) (Scheme 19) [135,136].

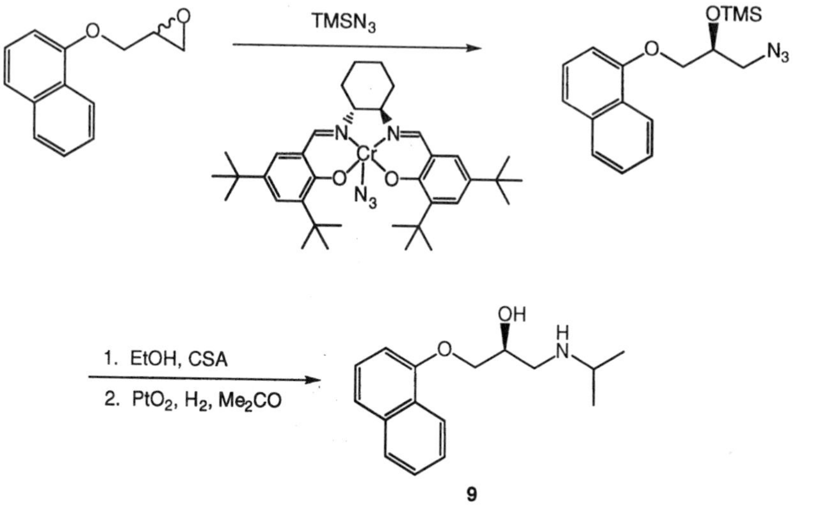

SCHEME 19.

The methodology, including the use of other nucleophiles, can be used to produce chiral compounds from *meso*-epoxides in high yield [137–139].

12.5. HALOHYDROXYLATIONS

The conversion of an alkene to a halohydrin can also be considered as an epoxidation because this can be achieved by a simple ring closure [140]. Although no reagent is yet available to perform an asymmetric conversion of an isolated alkene to a halohydrin, the reaction can be controlled through diastereoselection. One such case is the halolactonization of γ,δ unsaturated carboxylic acids, *N,N*-dialkylamides [141–145].

One of the few industrial examples of the asymmetric synthesis of halohydrins is in a process to the human immunodeficiency virus protease inhibitor, Indinavir [146]. The γ,δ unsaturated carboxamide **10** is smoothly converted into iodohydrin **11** (92%, 94% de) (Scheme 20) [147].

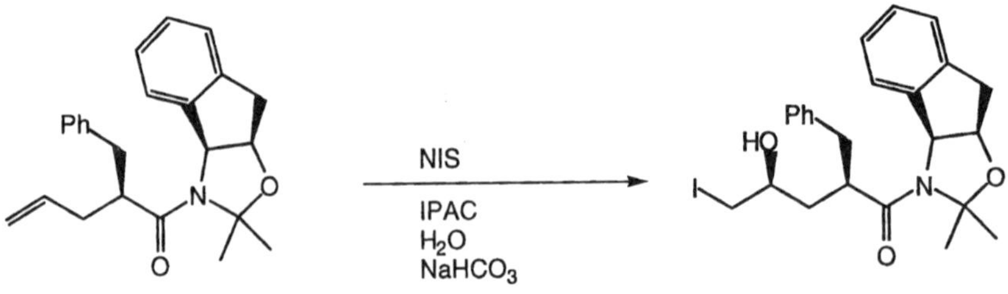

10   NIS / IPAC / H$_2$O / NaHCO$_3$ **11**

SCHEME 20.

The reaction presumably proceeds by attack of the amide carbonyl on iodonium ion, followed by collapse of tetrahedral intermediate (Scheme 21) [148].

N-Iodosuccinimide, iodine, diiodohydantoin [149], or electrochemical means [150] can be used as the oxidant. Other *N,N*-dialkyl-2-substituted-4-enamides work equally as well [147,149].

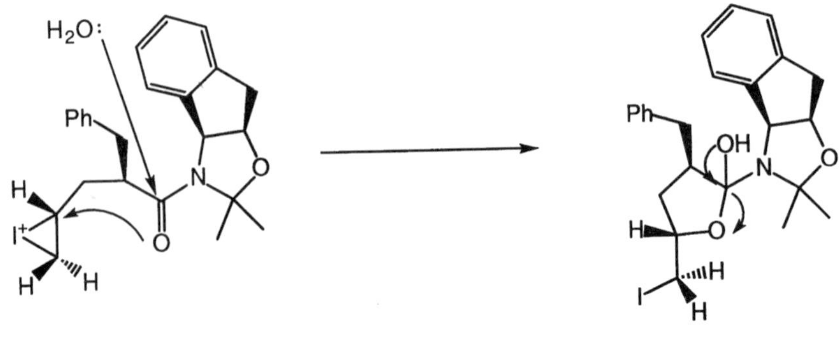

SCHEME 21.

12.6. ENZYMATIC METHODS

The oxidation of an aromatic substrate has provided a synthesis of 2,3-isopropyli-
dene L-ribonic γ-lactone (12) (Scheme 22) [151]. Indeed, this dihydroxylation of
aryl substrates provides a wealth of synthetic intermediates and reaction pathways
[151–164].

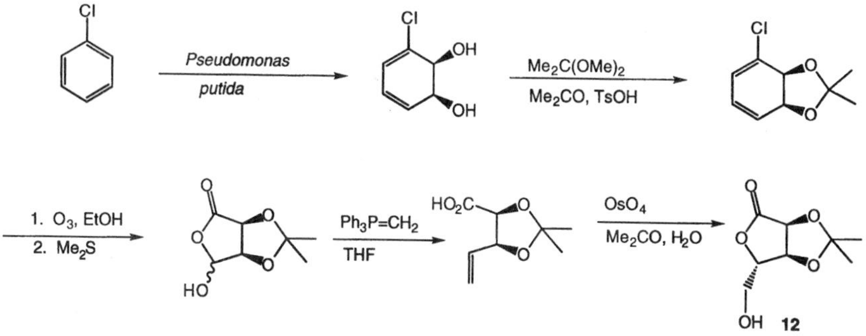

SCHEME 22.

Other oxidations catalyzed by enzymes are known (see Chapter 13). As
our understanding of the processes coupled with schemes to alleviate the need
for cofactors develop, we will no doubt see more examples of the use of enzymes
to achieve oxidations.

12.7. SUMMARY

Methodology has been found that allows for the asymmetric oxidation of alkenes
and allyl alcohol to the corresponding epoxides or diols. Although the methodologies have found widespread application at a laboratory scale for the preparation
of a wide variety of compounds, examples of their use at scale are rare. With
the potential of the approach, this will surely change.

REFERENCES

1. Ager, D. J., East, M. B. *Asymmetric Synthetic Methodology*; CRC Press: Boca Raton, 1995.
2. Shum, W. P., Cannarsa, M. J. In *Chirality in Industry II: Developments in the
   Commercial Manufacture and Applications of Optically Active Compounds*; Collins, A. N., Sheldrake, G. N., Crosby, J., Eds.; Wiley: Chichester, 1997; p. 363.
3. Katsuki, T. *J. Syn. Org. Chem. Jpn* 1987, *45*, 90.

4. Jorgensen, K. A. *Chem. Rev.* 1989, *89*, 431.
5. Katsuki, T., Sharpless, K. B. *J. Am. Chem. Soc.* 1980, *103*, 5974.
6. Sharpless, K. B., Behrens, C. H., Katsuki, T., Lee, A. W. M., Martin, V. S., Takatani, M., Viti, S. M., Walker, F. J., Woodard, S. S. *Pure Appl. Chem.* 1983, *55*, 589.
7. Pfenninger, A. *Synthesis* 1986, 89.
8. Sato, F., Kobayashi, Y. *Synlett* 1992, 849.
9. Sheldon, R. A. *Chriotechnology: Industrial Synthesis of Optically Active Compounds*; Marcel Dekker: New York, 1993.
10. Gao, Y., Hanson, R. M., Klunder, J. M., Ko, S. Y., Masamune, H., Sharpless, K. B. *J. Am. Chem. Soc.* 1987, *109*, 5765.
11. Hansen, R. M., Sharpless, K. B. *J. Org. Chem.* 1986, *51*, 1922.
12. Kende, A. S., Rizzi, J. P. *J. Am. Chem. Soc.* 1981, *103*, 4247.
13. Finn, M. F., Sharpless, K. B. in *Asymmetric Synthesis*; Morrison, J. D., Ed.; Academic Press: Orlando, 1986; *Vol. 5*; p. 247.
14. Rossiter, B. E. in *Asymmetric Synthesis*; Morrison, J. D., Ed.; Academic Press: Orlando, 1985; *Vol. 5*; p. 193.
15. Mulzer, J., Angermann, A., Munch, W., Schlichthorl, G., Hentzschel, A. *Liebigs Ann. Chem.* 1987, 7.
16. Kang, J., Park, M., Shin, H. T., Kim, J. K. *Bull. Korean Chem. Soc.* 1985, *6*, 376.
17. Rossiter, B. E., Katsuki, T., Sharpless, K. B. *J. Am. Chem. Soc.* 1981, *103*, 464.
18. Wang, Z.-M., Zhou, W.-S. *Tetrahedron* 1987, *43*, 2935.
19. Pedersen, S. F., Dewan, J. C., Eckman, R. R., Sharpless, K. B. *J. Am. Chem. Soc.* 1987, *109*, 1279.
20. Burns, C. J., Martin, C. A., Sharpless, K. B. *J. Org. Chem.* 1989, *54*, 2826.
21. Carlier, P. R., Sharpless, K. B. *J. Org. Chem.* 1989, *54*, 4016.
22. Woodard, S. S., Finn, M. G., Sharpless, K. B. *J. Am. Chem. Soc.* 1991, *113*, 106.
23. Finn, M. G., Sharpless. *J. Am. Chem. Soc.* 1991, *113*, 113.
24. Hawkins, J. M., Sharpless, K. B. *Tetrahedron Lett.* 1987, *28*, 2825.
25. Potvin, P. G., Bianchet, S. *J. Org. Chem.* 1992, *57*, 6629.
26. Corey, E. J. *J. Org. Chem.* 1990, *55*, 1693.
27. Wang, Z.-M., Zhou, W.-S., Lin, C. Q. *Tetrahedron Lett.* 1985, *26*, 6221.
28. Choudary, B. M., Valli, V. L. K., Prasad, A. D. *J. Chem. Soc. Chem. Commun.* 1990, 1186.
29. Whang, Z.-M., Zhou, W.-S *Synth. Commun.* 1989, *19*, 2627.
30. Klunder, J. M., Onami, T., Sharpless, K. B. *J. Org. Chem.* 1989, *54*, 1295.
31. Brown, J. M. *Chem. Ind. (London)* 1988, *612*.
32. Schweiter, M. J., Sharpless, K. B. *Tetrahedron Lett.* 1985, *26*, 2543.
33. Martin, V. S., Woodard, S. S., Katsuki, T., Yamada, Y., Ikeda, M., Sharpless, K. B. *J. Am. Chem. Soc.* 1981, *103*, 6237.
34. Carlier, P. R., Mungall, W. S., Schroder, G., Sharpless, K. B. *J. Am. Chem. Soc.* 1988, *110*, 2978.
35. Kitano, Y., Matsumoto, T., Wakasa, T., Okamoto, S., Shimazaki, T., Kobayashi, Y., Sato, F., Miyaji, K., Arai, K. *Tetrahedron Lett.* 1987, *28*, 6351.
36. Kobayashi, Y., Kusakabe, M., Kitano, Y., Sato, F. *J. Org. Chem.* 1988, *53*, 1586.

37. Kusakabe, M., Kitano, Y., Kobayashi, Y., Sato, F. *J. Org. Chem.* 1989, *54*, 2085.
38. Martin, S. F., Zinke, P. W. *J. Am. Chem. Soc.* 1989, *111*, 2311.
39. Behrens, C. H., Sharpless, K. B. *Aldrichimica Acta* 1983, *16*, 67.
40. Hanson, R. M. *Chem. Rev.* 1991, *91*, 437.
41. Ward, R. S. *Chem. Soc. Rev.* 1990, *19*, 1.
42. Ager, D. J., East, M. B. *Tetrahedron* 1992, *48*, 2803.
43. Ager, D. J., East, M. B. *Tetrahedron* 1993, *49*, 5683.
44. Marshall, J. A., Trometer, J. D. *Tetrahedron Lett.* 1987, *28*, 4985.
45. Marshall, J. A., Trometer, J. D., Blough, B. E., Crute, T. D. *Tetrahedron Lett.* 1988, *29*, 913.
46. Chakraborty, T. K., Joshi, S. P. *Tetrahedron Lett.* 1990, *31*, 2043.
47. Meyers, A. I., Comins, D. L., Roland, D. M., Henning, R., Shimizu, K. *J. Am. Chem. Soc.* 1979, *101*, 7104.
48. Trost, B. M., Angle, S. R. *J. Am. Chem. Soc.* 1985, *107*, 6124.
49. Hungerbuhler, E., Deebach, D., Wasmuth, D. *Angew. Chem. Int. Ed. Engl.* 1979, *18*, 958.
50. Askin, D., Volante, R. P., Ryan, K. M., Reamer, R. A., Shinkai, I. *Tetrahedron Lett.* 1988, *29*, 4245.
51. Howe, G. P., Wang, S., Procter, G. *Tetrahedron Lett.* 1987, *28*, 2629.
52. Bernet, B., Vasella, A. *Tetrahedron Lett.* 1983, *24*, 5491.
53. Roush, W. R., Adam, M. A., Peseckis, S. M. *Tetrahedron Lett.* 1983, *24*, 1377.
54. Isobe, M. I., Kitamura, M., Mio, S., Goto, T. *Tetrahedron Lett.* 1982, *23*, 221.
55. Ohfune, Y., Kurokawa, N. *Tetrahedron Lett.* 1984, *25*, 1587.
56. Solladie, G., Hamouchi, C., Vicente, M. *Tetrahedron Lett.* 1988, *29*, 5929.
57. Page, P. C. B., Rayner, C. M., Sutherland, I. O. *J. Chem. Soc. Perkin Trans. I* 1990, 1375.
58. Mulzer, J., Schollhorn, B. *Angew. Chem. Int. Ed. Engl.* 1990, *29*, 1476.
59. Rao, A. V. R., Bose, D. S., Gurjar, M. K., Ravindranathan, T. *Tetrahedron* 1989, *45*, 7031.
60. Marshall, J. A., Blough, B. E. *J. Org. Chem.* 1991, *56*, 2225.
61. White, J. D., Bolton, G. L. *J. Am. Chem. Soc.* 1990, *112*, 1626.
62. Rao, A. V. R., Dhar, T. G. M., Bose, D. S., Chakraborty, T. K., Gurjar, M. K. *Tetrahedron* 1989, *45*, 7361.
63. Jung, M. E., Jung, Y. H. *Tetrahedron Lett.* 1989, *30*, 6637.
64. Wang, Z., Schreiber, S. L. *Tetrahedron Lett.* 1990, *31*, 31.
65. Mori, Y., Kuhara, M., Takeuchi, A., Suzuki, M. *Tetrahedron Lett.* 1988, *29*, 5419.
66. Murphy, P. J., Procter, G. *Tetrahedron Lett.* 1990, *31*, 1059.
67. Hummer, W., Gracza, T., Jager, V. *Tetrahedron Lett.* 1989, *30*, 1517.
68. Mori, Y., Takeuchi, A., Kageyama, H., Suzuki, M. *Tetrahedron Lett.* 1988, *29*, 5423.
69. Mori, Y., Suzuki, M. *J. Chem. Soc. Perkin Trans. I* 1990, 1809.
70. Mori, Y., Kohchi, Y., Ota, T., Suzuki, M. *Tetrahedron Lett.* 1990, *31*, 2915.
71. Williams, N. R. *Adv. Carbohydr. Chem. Biochem.* 1970, *25*, 109.
72. Payne, G. B. *J. Org. Chem.* 1962, *27*, 3819.
73. Behrens, C. H., Ko, S. Y., Sharpless, K. B., Walker, F. J. *J. Org. Chem.* 1985, *50*, 5687.

74. Behrens, C. H., Sharpless, K. B. *J. Org. Chem.* 1985, *50*, 5696.
75. Mori, K., Kbata, T. *Tetrahedron* 1986, *42*, 3471.
76. Bell, T. W., Clacclo, J. A. *Tetrahedron Lett.* 1988, *29*, 865.
77. Kurth, M. J., Abreo, M. A. *Tetrahedron* 1990, *46*, 5085.
78. Marczak, S., Masnyk, M., Wicha, J. *Tetrahedron Lett.* 1989, *30*, 2845.
79. Kang, S.-K., Kim, Y.-S., Lim, J.-S., Kim, K.-S., Kim, S.-G. *Tetrahedron Lett.* 1991, *32*, 363.
80. Dung, J.-S., Armstrong, R. W., Anderson, O. P., Williams, R. M. *J. Org. Chem.* 1983, *48*, 3592.
81. Katsuki, T., Lee, A. W. M., Ma, P., Martin, V. S., Masamune, S., Sharpless, K. B., Tuddenham, D., Walker, F. J. *J. Org. Chem.* 1982, *7*, 1373.
82. Lee, A. W. M., Martin, V. S., Masamune, S., Sharpless, K. B., Walker, F. J. *J. Am. Chem. Soc.* 1982, *104*, 3515.
83. Ma, P., Martin, V. S., Masamune, S., Sharpless, K. B., Viti, S. M. *J. Org. Chem.* 1982, *47*, 1378.
84. Ko, S. Y., Lee, A. W. M., Masamune, S., Reed, L. A., Sharpless, K. B., Walker, F. J. *Science* 1983, *220*, 949.
85. Ko, S. Y., Lee, A. W. M., Masamune, S., Reed, L. A., Sharpless, K. B., Walker, F. J. *Tetrahedron* 1990, *46*, 245.
86. Kolb, H. C., VanNieuwenhze, M. S., Sharpless, K. B. *Chem. Rev.* 1994, *94*, 2483.
87. Jacobsen, E. N., Markó, I., Mungall, W. S., Schröder, G., Sharpless, K. B. *J. Am. Chem. Soc.* 1988, *110*, 1968.
88. Jacobsen, E. N., Marko, I., France, M. B., Svendsen, J. S., Sharpless, K. B. *J. Am. Chem. Soc.* 1989, *111*, 737.
89. Lohray, B. B., Kalantar, T. H., Kim, B. M., Park, C. Y., Shibata, T., Wai, J. S. M., Sharpless, K. B. *Tetrahedron Lett.* 1989, *30*, 2041.
90. Shibata, T., Gilheany, D. G., Blackburn, B. K., Sharpless, K. B. *Tetrahedron Lett.* 1990, *31*, 3817.
91. Jorgensen, K. A. *Tetrahedron Lett.* 1990, *31*, 6417.
92. Kim, B. M., Sharpless, K. B. *Tetrahedron Lett.* 1990, *31*, 3003.
93. Han, H., Janda, K. D. *J. Am. Chem. Soc.* 1996, *118*, 7632.
94. Wai, J. S. M., Marko, I., Svendsen, J. S., Finn, M. G., Jacobsen, E. N., Sharpless, K. B. *J. Am. Chem. Soc.* 1989, *111*, 1123.
95. Kwong, H.-L., Sorato, C., Ogino, Y., Chen, H., Sharpless, K. B. *Tetrahedron Lett.* 1990, *31*, 2999.
96. Minato, M., Yamamoto, K., Tsuji, J. *J. Org. Chem.* 1990, *55*, 766.
97. Sharpless, K. B., Amberg, W., Bennani, Y. L., Crispino, G. A., Hartung, J., Jeong, K.-S., Kwong, H.-L., Morikawa, K., Wang, Z.-M., Xu, D., Zhang, X.-L. *J. Org. Chem.* 1992, *57*, 2768.
98. Vidari, G., Giori, A., Dapiaggi, A., Lanfranchi G. *Tetrahedron Lett.* 1993, *34*, 6925.
99. Arrington, M. P., Bennani, Y. L., Göbel, T., Walsh, P., Zhao, S.-H., Sharpless, K. B. *Tetrahedron Lett.* 1993, *34*, 7375.
100. Morikawa, K., Park, J., Andersson, P. G., Hashiyama, T., Sharpless, K. B. *J. Am. Chem. Soc.* 1993, *115*, 8463.
101. Wang, L., Sharpless, K. B. *J. Am. Chem. Soc.* 1992, *114*, 7568.

102. VanNieuwenhze, M. S., Sharpless, K. B. *J. Am. Chem. Soc.* 1993, *115*, 7864.
103. Ahrgren, L., Sutin, L. *Org. Proc. Res. Dev.* 1997, *1*, 425.
104. Rudolph, J., Sennhenn, P. C., Vlaar, C. P., Sharpless, K. B. *Angew. Chem. Int. Ed. Engl.* 1996, *35*, 2810.
105. Li, G., Chang, H.-T., Sharpless, K. B. *Angew. Chem. Int. Ed. Engl.* 1996, *35*, 451.
106. Rubin, A. E., Sharpless, K. B. *Angew. Chem. Int. Ed. Engl.* 1997, *36*, 2637.
107. Reddy, K. L., Sharpless, K. B. *J. Am. Chem. Soc.* 1998, *120*, 1207.
108. Pini, D., Iuliano, A., Rosini, C., Salvadori, P. *Synthesis* 1990, 1023.
109. Gao, Y., Sharpless, K. B. *J. Am. Chem. Soc.* 1988, *110*, 7538.
110. Denmark, S. E. *J. Org. Chem.* 1981, *46*, 3144.
111. Lowe, G., Salamone, S. J. *J. Chem. Soc. Chem. Commun.* 1983, 1392.
112. Oi, R., Sharpless, K. B. *Tetrahedron Lett.* 1991, *32*, 999.
113. Gao, Y. PhD Thesis, Massachusetts Institute of Technology, 1988.
114. Lohray, B. B., Gao, Y., Sharpless, K. B. *Tetrahedron Lett.* 1989, *30*, 2623.
115. Berridge, M. S., Franceschini, M. P., Rosenfeld, E., Tewson, T. J. *J. Org. Chem.* 1990, *55*, 1211.
116. Shao, H., Goodman, M. *J. Org. Chem.* 1996, *61*, 2582.
117. Kim, B. M., Sharpless, K. B. *Tetrahedron Lett.* 1989, *30*, 655.
118. Pearlstein, R. M., Blackburn, B. K., Davis, W. M., Sharpless, K. B. *Angew. Chem. Int. Ed. Engl.* 1990, *29*, 639.
119. Chang, H.-T., Sharpless, K. B. *Tetrahedron Lett.* 1996, *37*, 3219.
120. Zhang, W., Loebach, J. L., Wilson, S. R., Jacobsen, E. N. *J. Am. Chem. Soc.* 1990, *112*, 2801.
121. Irie, R., Noda, K., Ito, Y., Matsumoto, N., Katsuki, T. *Tetrahedron Lett.* 1990, *31*, 7345.
122. Zhang, W., Loebach, J. L., Wilson, S. R., Jacobsen, E. J. *J. Am. Chem. Soc.* 1990, *112*, 2801.
123. Van Draanen, N. A., Arseniyadis, S., Crimmins, M. T., Heathcock, C. H. *J. Org. Chem.* 1991, *56*, 2499.
124. Okamoto, Y., Still, W. C. *Tetrahedron Lett.* 1988, *29*, 971.
125. Schwenkreis, T., Berkessel, A. *Tetrahedron Lett.* 1993, *34*, 4785.
126. Quan, R. W., Li, Z., Jacobsen, E. N. *J. Am. Chem. Soc.* 1996, *118*, 8156.
127. Zhang, W., Jacobsen, E. N. *J. Org. Chem.* 1991, *56*, 2296.
128. Deng, L., Jacobsen, E. N. *J. Org. Chem.* 1992, *57*, 4320.
129. Palucki, M., Finney, N. S., Pospisil, P. J., Güler, M. L., Ishida, T., Jacobsen, E. N. *J. Am. Chem. Soc.* 1998, *120*, 948.
130. Finney, N. S., Pospisil, P. J., Chang, S., Palucki, M., Konsler, R. G., Hansen, K. B., Jacobsen E. N. *Angew. Chem. Int. Ed. Engl.* 1997, *36*, 1720.
131. Irie, R., Noda, K., Ito, Y., Katsuki, T. *Tetrahedron Lett.* 1991, *32*, 1055.
132. O'Connor, K. J., Wey, S.-J., Burrows, C. J. *Tetrahedron Lett.* 1992, *33*, 1001.
133. Tokunaga, M., Larrow, J. F., Kakiuchi, F., Jacobsen, E. N. *Science* 1997, *277*, 936.
134. Brandes, B. D., Jacobsen, E. N. *Tetrahedron: Asymmetry* 1997, *8*, 3927.
135. Larrow, J. F., Schaus, S. E., Jacobsen, E. N. *J. Am. Chem. Soc.* 1996, *118*, 7420.
136. Schaus, S. E., Jacobsen, E. N. *Tetrahedron Lett.* 1996, *37*, 7937.
137. Hansen, K. B., Leighton, J. L., Jacobsen, E. N. *J. Am. Chem. Soc.* 1996, *118*, 10924.

138. Jacobsen, E. N., Kakiuchi, F., Konsler, R. G., Larrow, J. F., Tokunaga, M. *Tetrahedron Lett.* 1997, *38*, 773.

139. Schaus, S. E., Larrow, J. F., Jacobsen, E. N. *J. Org. Chem.* 1997, *62*, 4197.

140. Ng, J. S., Przybyla, C. A., Liu, C., Yen, J. C., Muellner, F. W., Weyker, C. L. *Tetrahedron* 1995, *51*, 6397.

141. Fuji, K., Node, M., Naniwa, Y., Kawabata, T. *Tetrahedron Lett.* 1990, *31*, 3175.

142. Hart, D. J., Huang, H. C., Krishnamurthy, R., Schawrtz, T. *J. Am. Chem. Soc.* 1989, *111*, 7507.

143. Moon, H.-S., Eisenberg, S. W. E., Wilson, M. E., Schore, N. E., Kurth, M. J. *J. Org. Chem.* 1994, *59*, 6504.

144. Tamaru, Y., Mizutani, M., Furukawa, Y., Kawamura, S., Yoshida, Z., Yanagi, K., Minobe, M. *J. Am. Chem. Soc.* 1984, *106*, 1079.

145. Kurth, M. J., Brown, E. G. *J. Am. Chem. Soc.* 1987, *109*, 6844.

146. Dorsey, B. D., Levin, R. B., McDaniel, J. P., Vacca, J. P., Guare, J. P., Darke, P. L., Zugay, J. A., Emini, E. A., Schleif, W. A., Quintero, J. C., Lin. J. H., Chen, I.-W., Holloway, M. K., Fitzgerald, P. M. D., Axel, M. G., Ostovic, D., Anderson, P. S., Huff, J. R. *J. Med. Chem.* 1994, *37*, 3443.

147. Maligres, P. E., Upadhyay, V., Rossen, K., Cianciosi, S. J., Purick, R. M., Eng, K. K., Reamer, R. A., Askin, D., Volante, R. P., Reider, P. J. *Tetrahedron Lett.* 1995, *36*, 2195.

148. Rossen, K., Reamer, R. A., Volante, R. P., Reider, P. J. *Tetrahedron Lett.* 1996, *38*, 6843.

149. Maligres, P. E., Weissman, S. A., Upadhyay, V., Cianciosi, S. J., Reamer, R. A., Purick, R. M., Sager, J., Rossen, K., Eng, K. K., Askin, D., Volante, R. P., Reider, P. J. *Tetrahedron* 1996, *52*, 3327.

150. Rossen, K., Volante, R. P., Reider, P. J. *Tetrahedron Lett.* 1997, *38*, 777.

151. Hudlicky, T., Price, J. D. *Synlett* 1990, 159.

152. Hudlicky, T., Entwistle, D. A., Pitzer, K. K., Thorpe, A. J. *Chem. Rev.* 1996, *96*, 1195.

153. Ley, S. V., Sternfeld, F., Taylor, S. *Tetrahedron Lett.* 1987, *28*, 225.

154. Hudlicky, T., Luna, H., Price, J. D., Rulin, F. *Tetrahedron Lett.* 1989, *30*, 4053.

155. Ley, S. V., Sternfeld, F. *Tetrahedron* 1989, *45*, 3463.

156. Amici, M. D., Micheli, C. D., Carrea, G., Spezia, S. *J. Org. Chem.* 1989, *54*, 2646.

157. Ley, S. V., Sternfeld, F. *Tetrahedron Lett.* 1988, *29*, 5305.

158. Carless, H. A. J. *Tetrahedron: Asymmetry* 1992, *3*, 795.

159. Ley, S. V., Yeung, L. L. *Synlett* 1992, 997.

160. Dumortier, L., Liu, P., Dobbelaere, S., Van der Eycken, J., Vandewalle, M. *Synlett* 1992, 243.

161. Ley, S. V., Yeung, L. L. *Synlett* 1992, 291.

162. Königsberger, K., Hudlicky, T. *Tetrahedron: Asymmetry* 1993, *4*, 2469.

163. Hudlicky, T., Boros, E. E., Boros, C. H. *Tetrahedron: Asymmetry* 1993, *4*, 1365.

164. Hudlicky, T., Seoane, G., Pettus, T. *J. Org. Chem.* 1989, *54*, 4239.

13

Biotransformations: "Green" Processes for the Synthesis of Chiral Fine Chemicals

DAVID P. PANTALEONE
NSC Technologies, Mount Prospect, Illinois

13.1. INTRODUCTION

The need for chiral fine chemical intermediates has become commonplace today in light of the trend toward single enantiomer drugs manufactured by pharmaceutical companies [1–5]. In addition, as waste disposal costs skyrocket, processes developed today must emphasize environmental safety. Although, at present, chemical approaches account for the majority of processes to produce fine chemicals, biotransformation steps are making more of an impact and are being integrated into chemical process sequences, especially where the chirality of the target compound must be maintained [6,7]. New biocatalysts are being isolated and marketed by such companies as Diversa, formerly Recombinant BioCatalysis, (San Diego, CA), ThermoGen (Chicago, IL), and Altus Biologics (Cambridge, MA) [8–10]. Many of these new types of biocatalysts have altered substrate specificity and enhanced thermal stability with greater organic solvent tolerance. In addition, commercial enzyme suppliers such as Amano Enzyme Co. (Nagoya, Japan), Novo Nordisk Bioindustrials, Inc. (Danbury, CT), and Boehringer Mannheim Biochemicals (Indianapolis, IN) are promoting their enzymes for specific biotransformations. Regardless of the source of these enzymes, because they are chiral catalysts and operate under mild reaction conditions, their utility to produce chiral molecules is being exploited more and more.

Evidence of biocatalyst utility can be seen from the number of scientific journals discussing biotransformations, including *Biocatalysis and Biotransformation* (Harwood Academic), *Tetrahedron: Asymmetry* (Elsevier Science), *Ap-*

plied Biochemistry and Biotechnology (Humana Press), *Applied Microbiology and Biotechnology* (Springer-Verlag), and *Applied and Environmental Microbiology* (American Society for Microbiology). Along with these journals, numerous textbooks, review articles, and symposia proceedings have been published that exemplify this rapidly developing field [11–18]. Additionally, two searchable databases have been commercialized on CD-ROM to allow easy access for the determination of a suitable biocatalyst for one's needs. They are Bio*Catalysis* by Synopsis Scientific Systems (Leeds, UK) developed in collaboration with Professors Bryan Jones (University of Toronto) and Herbert Holland (Brock University), and Biotransformations by Chapman and Hall using data from Professor Klaus Kieslich (GBF, Braunschweig, Germany) and the Warwick Biotransformation Club (Coventry, UK).

The term "biotransformation" means many things to many people. Throughout this chapter, it is defined as the use of a biological catalyst from a bacterial, fungal, plant, or mammalian source to carry out a desired transformation; the discussion, however, is limited to the formation of a single enantiomer or diastereoisomer. This biocatalyst could be a commercial enzyme (often an impure mixture of many different activities), whole cell, or cell lysate. These biocatalysts are sometimes immobilized onto solid supports that can provide added benefits such as greater thermal stability or reusability, despite an increase in catalyst cost. Enzymatic resolutions are a specific type of biotransformation; they will not be discussed here because they do not result in the creation of an asymmetric center, but rather convert one isomer of a mixture to a compound that will generally allow easy separation of the two isomers. Because this field is extremely large, only selected examples of some industrially important biotransformations are discussed along with certain biotransformations that offer potential to become significant routes to chiral fine chemicals.

13.2. BIOCATALYST CLASSIFICATIONS

Enzymes are classified by a four-digit number referred to as the Enzyme Commission number, or EC number, which is assigned according to the enzyme's function [19]. There are six general groups into which enzymes are classified, and the first digit of the EC number corresponds to the following general categories: (1) oxidoreductases; (2) transferases; (3) hydrolases; (4) lyases; (5) isomerases; and (6) ligases. The other three digits are then assigned based on further specifics of the type of reaction catalyzed. The EC number is used here to conveniently categorize the biocatalysts of industrial importance that are discussed in this chapter. Only the first five classes are discussed here since the ligase class currently does not represent any significant number of commercial bioprocesses. Although this number is assigned to a specific enzyme, reference is also made to microorganisms catalyzing the same reaction as the purified enzyme. A publication of sources of enzymes has been compiled that allows one to locate a commercial supplier of a particular enzyme using the EC number [20].

13.2.1.   Oxidoreductases (EC 1.x.x.x)

This class of biocatalysts is one of the most common of all biological reactions, comprising dehydrogenases, oxidases, and reductases. All of these enzymes act on substrates through the transfer of electrons with various cofactors or coenzymes serving as acceptor molecules. Only a select group of reactions are discussed here due to space limitations; therefore, the reader is referred to other texts for more in-depth discussions of other oxidation/reduction reactions [21–24].

13.2.1.1.   Dehydrogenases

There are a number of different types of dehydrogenases that catalyze redox reactions, that have been used to synthesize chiral molecules [25]. Some are cofactor dependent, generally requiring nicotinamide adenine dinucleotide/reduced nicotinamide adenine dinucleotide (NAD^+/NADH) or nicotinamide adenine dinucleotide phosphate/reduced nicotinamide adenine dinucleotide phosphate ($NADP^+$/NADPH), such as the alcohol dehydrogenases (AdHs), and some are cofactor independent, such as the alcohol dehydrogenase from *Gluconobacter suboxydans* [26]. The dependent enzymes use the cofactor to supply a hydride ion stereospecifically to reduce the substrate, whereas the independent dehydrogenases are often membrane bound and thus associated with the cytochromes and quinoproteins that facilitate hydrogen removal and ultimately reduce oxygen to water. For AdHs, the hydride can be delivered to either the *si* or *re* face, depending on how the substrate binds in the enzyme active site based on the steric bulk of the R^1 and R^2 substituents (Scheme 1). Enzymes such as yeast (YAdH), horse liver (HLAdH), and *Thermoanaerobium brockii* AdHs deliver the *pro-R* hydrogen to the *re* face, thus forming the (*S*)-alcohol. Another enzyme from *Mucor javanicus* delivers the *pro-S* hydrogen to the *si* face to yield the (*R*)-alcohol. An enzyme isolated from *Pseudomonas* sp, strain PED (ATCC 49794) has also been shown to form (*R*)-alcohols from a wide variety of substrates [27]. Further mechanistic details, discussion of Prelog's rule to predict the stereochemical outcome of a particular reduction, and listings of other enzymes with different stereospecificities are discussed elsewhere [23]. Some specific examples of dehydrogenases and their use to produce chiral alcohols and amino acids are given in the next section.

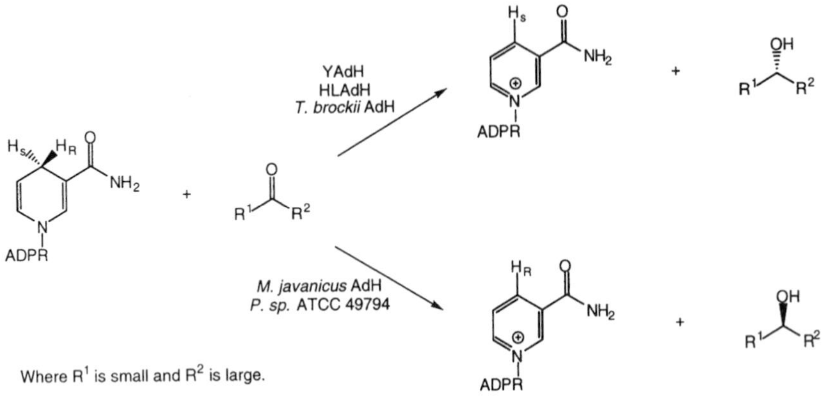

SCHEME 1.

13.2.1.1.1. REDUCTIONS OF KETONES.

There are two lactate dehydrogenases (LdHs) with different stereospecificities that have been very useful for the preparation of chiral alcohols [28–31]. L-LdH (EC 1.1.1.27) and D-LdH (EC 1.1.1.28) are obtained from various microbial sources (Table 1). Along with hydroxyisocaproate dehydrogenase (HicdH), these enzymes have been used in conjunction with a cofactor regeneration system, usually the formate dehydrogenase (FdH) [EC 1.2.1.2] system from *Candida boidinii*, to generate important chiral intermediates and synthons (Scheme 2 and Table 1) [32] that illustrate the broad range of substrates that these enzymes work on. In all cases, the enantiomeric excess and chemical yields of the products were

TABLE I Selected Substrates and Stereochemical Product Configuration with Lactate (L) and Hydroxyisocaproate (Hic) Dehydrogenases (dH) from Various Sources that Catalyzes the Reaction Shown in Scheme 2

R	Enzyme	Source	Product configuration	Reference
CH_3	L-LdH	Rabbit muscle (iso M)	S	28
CH_3	D-LdH	*L. mesenteroides*	R	28
$(CH_3)_3C$	L-2-HicdH	*L. confusus*	S	33
$(CH_3)_3C$	D-2-HicdH	*L. casei*	R	33
$C_6H_5(CH_2)_2$	D-LdH	*L. mesenteroides*	R	30
$C_6H_5(CH_2)_2$	D-LdH	*S. epidermis*	R	31
$ZHN(CH_2)_4$	L-2-HicdH	*S. epidermis*	R	34
$E\text{-}CH_3(CH)_2$	L-LdH	*B. stearothermophilus*	S	35
$E\text{-}CH_3(CH)_2$	D-LdH	*S. epidermis*	R	35
$c\text{-}(CH_2)CH$	D-LdH	*S. epidermis*	R	29

high. The reader is referred to the specific references in Table 1 for additional substrates tested with the respective enzymes.

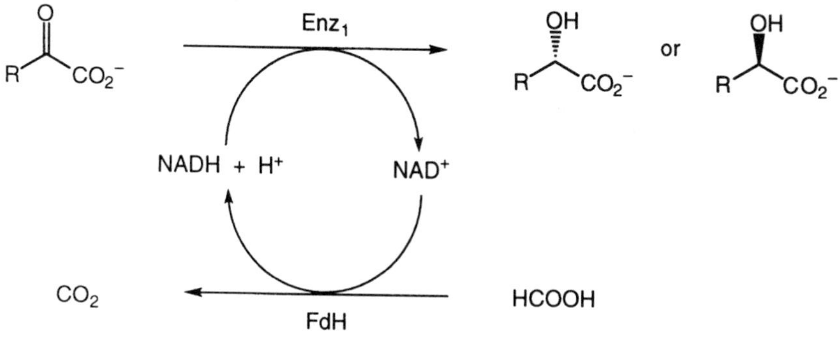

For Enz$_1$ see Table 1

SCHEME 2.

13.2.1.1.2. *REDUCTIVE AMINATION WITH AMINO ACID DEHYDROGENASES.*

Natural and unnatural amino acids have been prepared through reductive amination reactions catalyzed by several amino acid dehydrogenases (dHs) [EC 1.4.1.x] (Scheme 3) along with the formate dH cofactor recycling system [25,36]. Another cofactor recycling system that is sometimes used requires glucose dehydrogenase for the oxidation of glucose to gluconate and produces NADH. Those enzymes, whose natural substrates are alanine (R = CH$_3$) (x = 1), leucine (R = (CH$_3$)$_2$ CHCH$_2$) (x = 9), and phenylalanine (R = C$_6$H$_5$CH$_2$) (x = 20), are the most well studied, although others exist. For example, cloned, thermostable alanine dH has recently been used with a coupling enzyme system to prepare D-amino acids, and the alanine dH gene has been incorporated into selected bacterial strains to enhance L-alanine production [37,38]. In addition, halogenated derivatives of L-alanine have also been prepared using alanine dH [39,40].

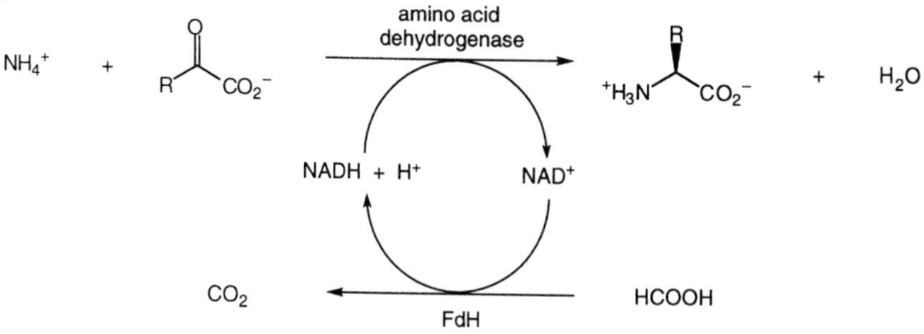

SCHEME 3.

In the mid to late 1980s, many research groups focused on methods and processes to prepare L-phenylalanine (see Chapter 4, this book by Fotheringham). This was a direct result of the demand for the synthetic, artificial sweetener aspartame. One of the many routes studied was the use of phenylalanine dH (Scheme 3, R = $C_6H_5CH_2$) with phenylpyruvate (PPA) as substrate [41,42]. This enzyme from *Bacillus sphaericus* shows a broad substrate specificity and, thus, has been used to prepare a number of derivatives of L-phenylalanine [43]. A phenylalanine dH isolated from a *Rhodococcus* strain M4 has been used to make L-homophenylalanine [(S)-2-amino-4-phenylbutanoic acid], a key chiral component in many angiotensin-converting enzyme (ACE) inhibitors [30]. More recently, that same phenylalanine dH has been used to synthesize a number of other unnatural amino acids that do not contain an aromatic side chain [33].

Leucine dH is the enzyme used as the biocatalyst in the process commercialized by Degussa AG (Hanau, Germany) to produce L-*tert*-leucine (L-Tle) [24]. This unnatural amino acid has found widespread use in peptidomimetic drugs in development, and the demand for this unique amino acid continues to increase [44]. This process, which has been the subject of much study, requires a cofactor recycling system (Scheme 3, R = $(CH_3)_3C$) [45,46]. Similar to phenylalanine dH, leucine dH has been used to prepare numerous unnatural amino acids because of its broad substrate specificity [33,47,48].

13.2.1.2.   Reductions

13.2.1.2.1.   YEAST REDUCTIONS.

Yeast reductions have shown great versatility by the synthetic organic chemist to prepare chiral alcohols from prochiral ketones, most of which have used *Saccharomyces cerevisiae* (Baker's yeast). In addition to prochiral ketone reductions, hundreds of other examples of yeast reductions have been cited in the literature using a variety of substrates such as β-keto esters, β-diketones, and analogs such as sulfur and nitrogen-containing compounds [49–52]. Examples of yeast reductions used to prepare some important chiral intermediates as reported by several pharmaceutical companies are given in this section.

On the synthetic route to a novel class of orally active 2,3-benzodiazepines, scientists at Lilly recently described a yeast reduction using *Zygosaccharomyces rouxii* (ATCC 14462) (Scheme 4) [53–55]. This intermediate is used in the synthesis of a novel class of benzodiazepines such as LY-300164 (**1**). The choice of this yeast strain came from a screen of numerous microbes. The product was isolated in > 95% yield and > 99.9% ee by use of an adsorbent polymeric resin [53–55].

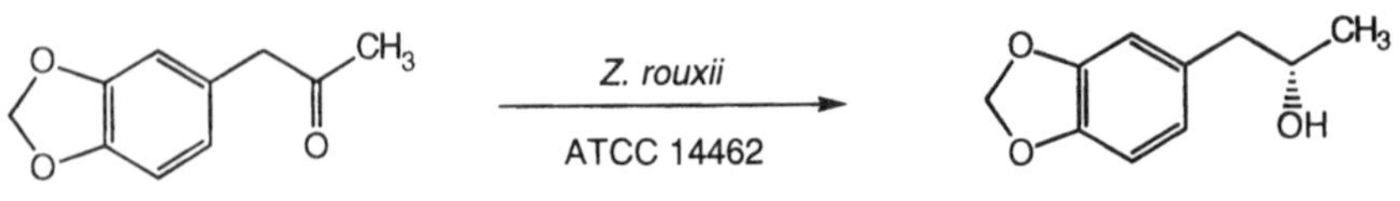

SCHEME 4.

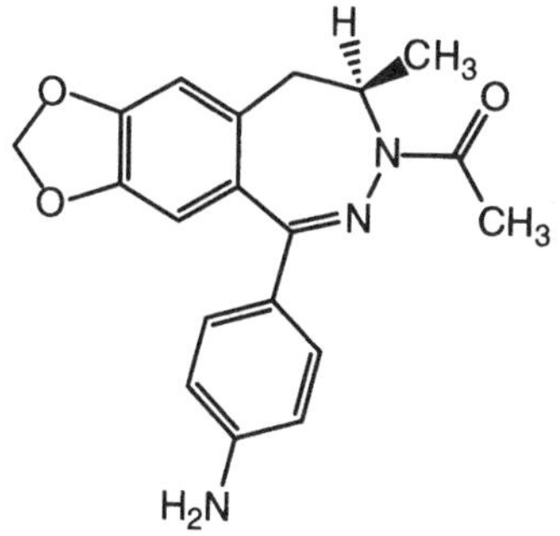

1

Scientists at Merck have reported a number of recent biocatalytic routes derived from screening various microorganisms targeted to produce key intermediates which are then combined with chemical reactions to prepare the target molecule. The biocatalytic step was often conducted by necessity as a result of poor chemical yield, low optical purity, or both. 6-Bromo-β-tetralone (**2**) is reduced to (*S*)-6-bromo-β-tetralol (**3**) by the yeast *Trichosporon capitatum* MY 1890 (Scheme 5) [56]. The tetralol **3** is a key intermediate for the synthesis of MK-0499 (**4**), a potassium channel blocker. The (*S*)-β-tetralol **3** was produced in gram quantities with an ee of > 99% to support further development of MK-0499. Baker's yeast was tested for its ability to carry out this reduction, but showed insignificant product formation.

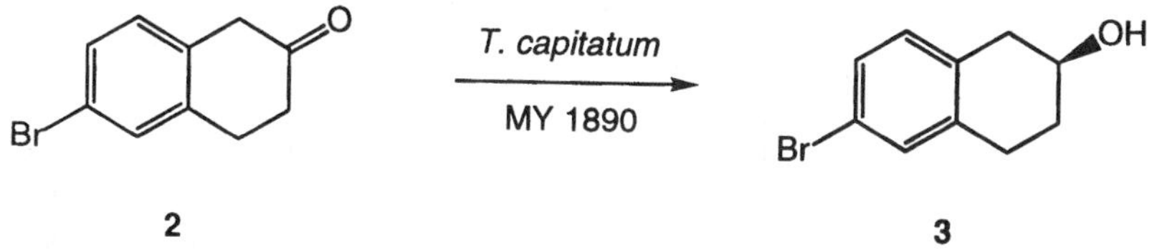

2   **3**

SCHEME 5.

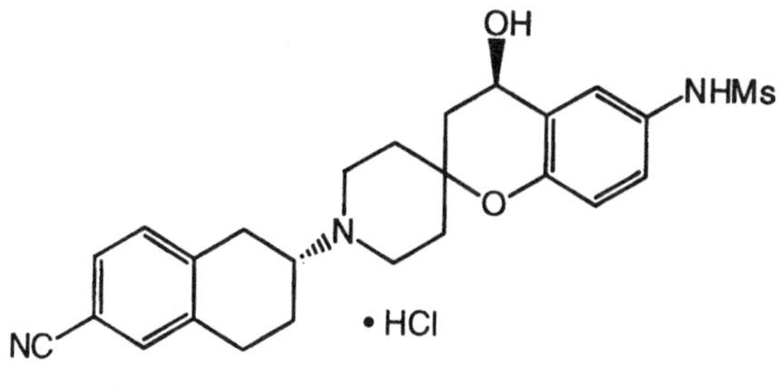

4

Benzylacetoacetate (**5**) is reduced to benzyl-(*S*)-(+)-3-hydroxybutyrate (**6**) by the yeast *Candida schatavii* MY 1831 (Scheme 6), which is considered a possible intermediate in the synthesis of L-734,217 (**7**), an experimental fibrinogen receptor antagonist [57]. Initially, eight microorganisms were tested for their reduction ability using **5** as substrate. The best culture, *C. schatavii* MY 1831, yielded product with an ee of 93% and an estimated conversion yield of > 95% (by thin-layer chromatography (TLC)).

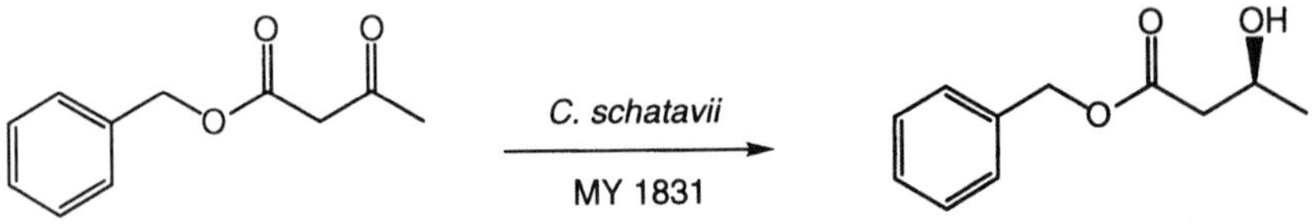

SCHEME 6.

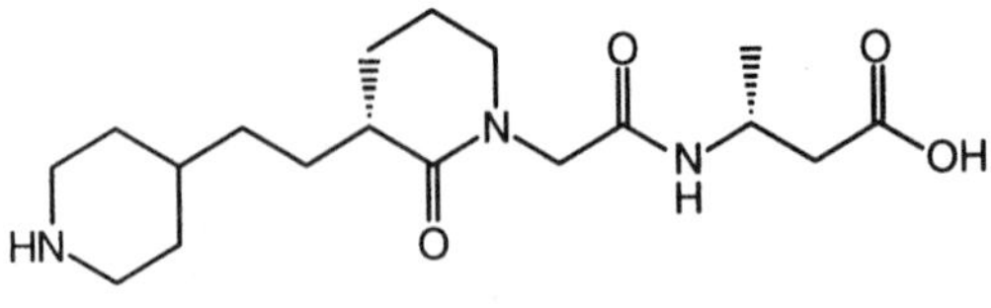

7

Another yeast, *Rhodotorula rubra* MY 2169, has been shown to reduce a ketosulfone **8** to the corresponding *trans*-hydroxysulfone **9** (Scheme 7). This hydroxysulfone is an intermediate in the drug candidate L-685,393 (**10**), a carbonic anhydrase inhibitor [58]. Results of this biotransformation yielded gram quantities of product with a de of > 96%. Studies by Zeneca discuss additional screening experiments aimed at finding microorganisms to reduce a similar ketosulfone [59].

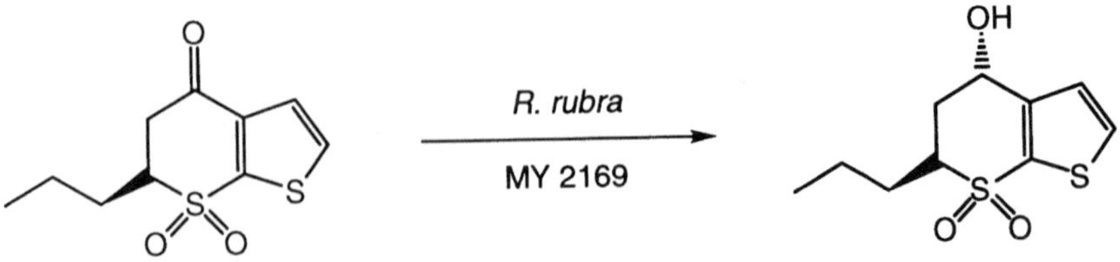

SCHEME 7.

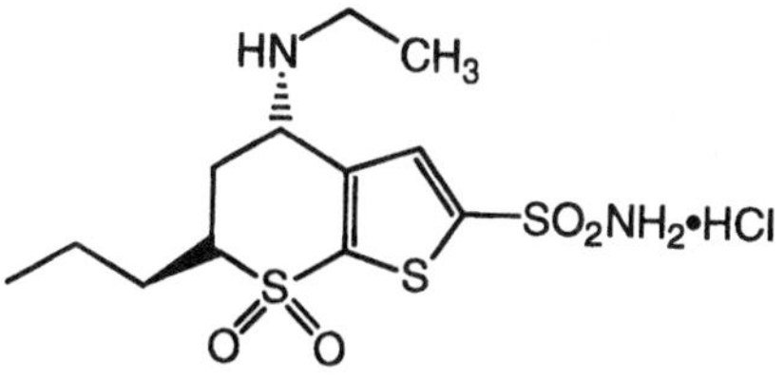

10

Another example of a β-keto ester reduction was studied by scientists at Zeneca with collaborators from the University of Warwick (Coventry, UK). They have reported on the reduction of ethyl 4-chloroacetoacetate (**11**) after screening a number of yeast strains [60,61]. The reduction of this substrate has been studied previously, as it has been realized for a number of years to be a potential intermediate for a number of pharmaceuticals, including the lipid carrier molecule, carnitine [62–64]. Interestingly, the Zeneca group and colleagues found that a cosubstrate (e.g., glucose, 2-propanol, xylose, glycerol), could dramatically effect the chirality of the stereogenic center of the product. Two different yeast strains can catalyze the reduction, but result in opposite stereochemistry, depending on which cosubstrate was added (Scheme 8). The authors [60] suggested that these cosubstrates inhibit certain enzymes possibly involved in the NADH regeneration rather than the (R)-specific enzymes themselves. Nevertheless, this reversal in enantioselectivity depending on cosubstrate indicates the complexity of yeast reductions.

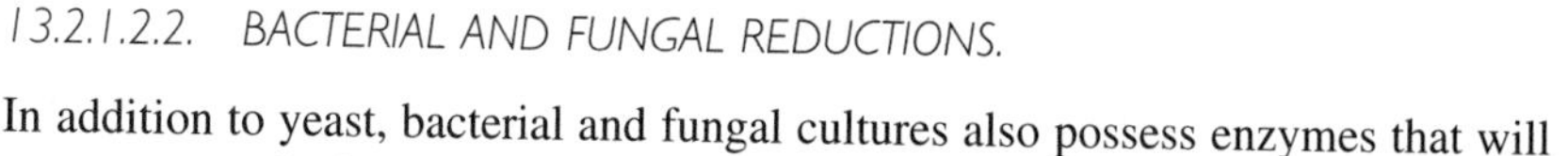

11

SCHEME 8.

13.2.1.2.2. BACTERIAL AND FUNGAL REDUCTIONS.

In addition to yeast, bacterial and fungal cultures also possess enzymes that will carry out synthetically useful biotransformations en route to a particular chiral intermediate. Some recent examples of these types of bioconversions are given here.

In the route to BO-2727 (**12**), a broad-spectrum β-methyl carbapenem being developed by Merck, a bioreduction catalyzed by the fungus *Mortierella alpina* MF 5534 is used to form a precursor (R)-β-hydroxy ester **13** (Scheme 9) [65]. This fungal culture was a result of screening some 260 strains of microorganisms and resulted in the production of gram quantities of product with a de of > 98%.

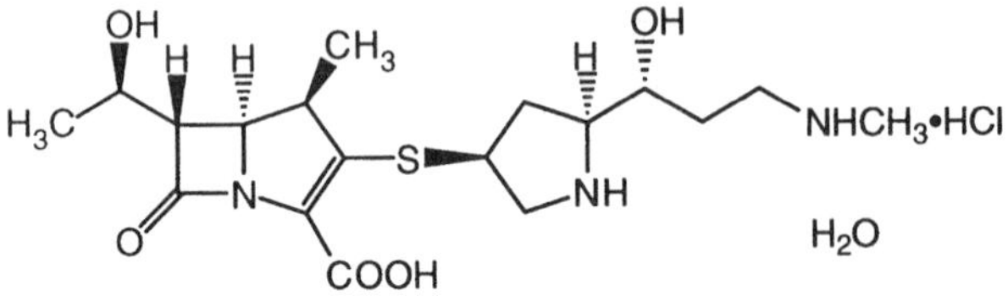

12

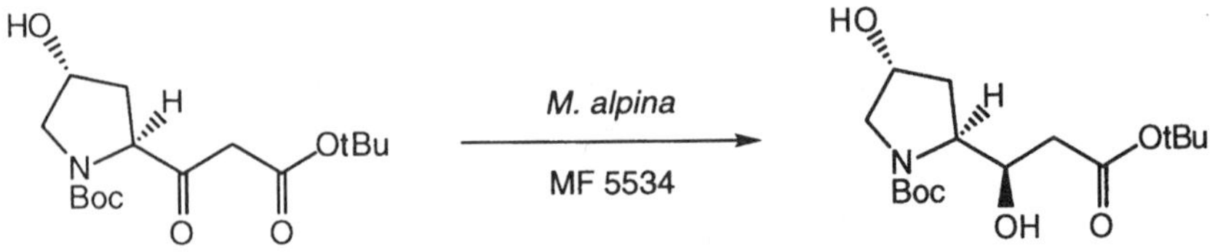

SCHEME 9.

After screening various microorganisms, scientists at Bristol-Myers Squibb selected a bacterial strain of *Acinetobacter calcoaceticus* SC 13876 to reduce a 3,5-dioxo ester **14** to the dihydroxy ester **15** (Scheme 10) [66]. The diol **15** is a key intermediate in the synthesis of an 3-hydroxy-3-methylglutanyl coenzyme A (HMG-CoA) reductase inhibitor (**16**). In a one-liter batch reaction, a yield of 92% was obtained with an optical purity of 99%.

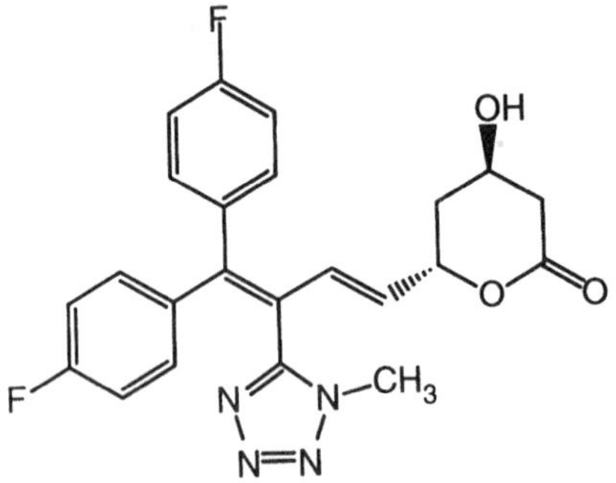

16

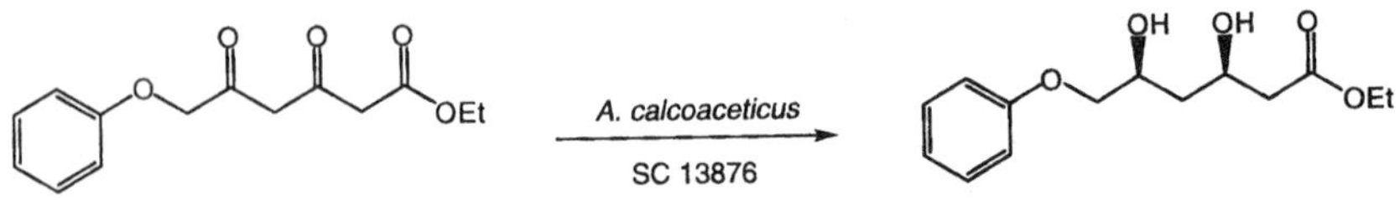

SCHEME 10.

In another study, screening was performed for reduction of substituted benzazepin-2,3-dione **17** to a 3-hydroxy derivative **18** (Scheme 11). This was accomplished by a bacterial strain of *Rhodococcus fascians* ATCC 12975 (*Norcardia salmonicolor* SC 6310) with a conversion of 97% and an optical purity of > 99.9%. This reaction product **18** is a key intermediate in the synthesis of the calcium antagonist SQ 31765 (**19**) [67,68]. The Bristol-Myers Squibb group have also shown the selective reduction of the β-keto ester, methyl-4-chloro-3-oxobutanoate, by the fungus *Geotrichum candidum* SC 5469 to the corresponding (*S*)-hydroxy ester [69].

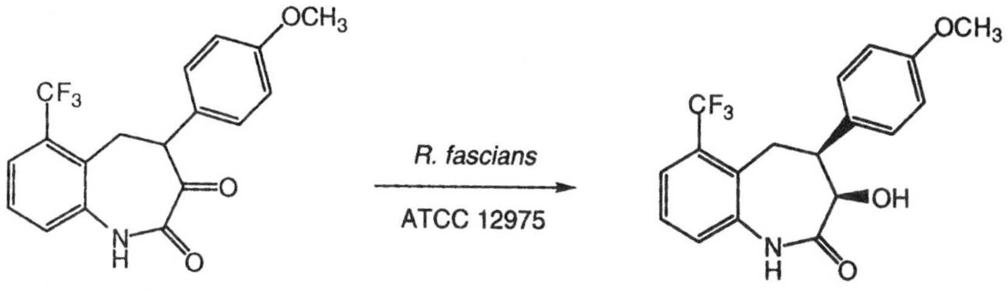

	R. fascians	
	⟶	
	ATCC 12975	
17	SCHEME II.	**18**

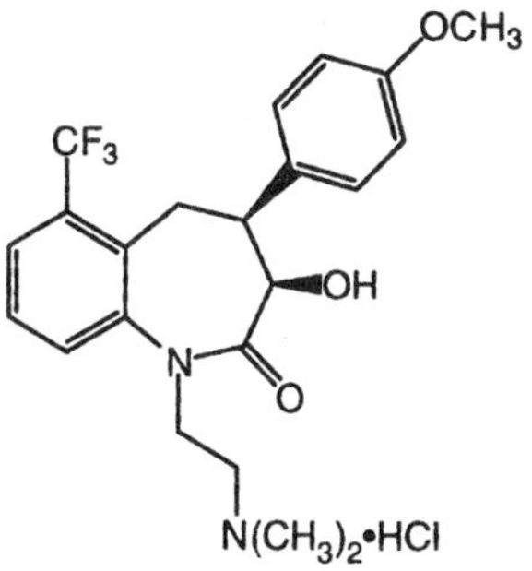

19

The reactions shown here illustrate the versatility of yeast, bacterial, and fungal reductions. In all of these cases, screening studies were usually conducted first, followed by refinement of the bioreduction using the best microorganism. Other examples in which reductions are used to produce pharmaceutical intermediates can be found in a recent review by Patel [70].

13.2.2.   Transferases (EC 2.x.x.x)

This class of biocatalysts is one of the most common of all biological reactions, catalyzing the general reaction shown in Scheme 12. This type of reaction uses a wide variety of substrates such as amino acids, keto acids, nucleotides, and carbohydrates, to name a few. In general, the carbohydrate transferases do not result in the generation of a new stereocenter, but have been used most extensively to generate novel saccharides [71].

$$X-Y \quad + \quad Z \quad \xrightleftharpoons{\text{transferase}} \quad X \quad + \quad Z-Y$$

SCHEME 12.

A transferase that also has aldolase activity and has been used to prepare a number of chiral compounds is the enzyme serine hydroxymethyltransferase (SHMT) (EC 2.1.2.1). This enzyme, also known as threonine aldolase, catalyzes the physiological reaction of the interconversion of serine and glycine with pyridoxal phosphate (PLP) and tetrahydrofolate (FH_4) as the shuttling cofactor of the C-1 unit. It also catalyzes a number of other reactions, some of which are independent of PLP and FH_4 [72]. The SHMT-catalyzed aldolase reaction generates two stereocenters, which it does stereospecifically at the α-carbon, whereas it is less strict at the β-carbon (Scheme 13). Nevertheless, this enzyme from porcine liver, *Escherichia coli* and *Candida humicola* (threonine aldolase) has been used to prepare a number of β-hydroxy-α-amino acids [73–76].

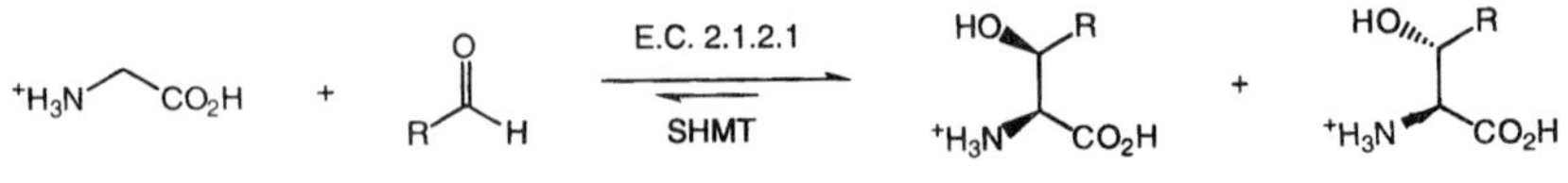

SCHEME 13.

The enzyme has also been used in the production of several natural amino acids such as L-serine from glycine and formaldehyde and L-tryptophan from glycine, formaldehyde, and indole [77–79]. In addition, SHMT has also been used for the production of a precursor, **20**, to the artificial sweetener aspartame (**21**) through a non–phenylalanine-requiring route (Scheme 14) [80–83]. Glycine methyl ester (**22**) is condensed with benzaldehyde under kinetically controlled conditions to form L-*erythro*-β-phenylserine (**23**). This is then coupled enzymatically using thermolysin with Z-aspartic acid (**24**) to form N-carbobenzyloxy-L-α-aspartyl-L-*erythro*-β-phenylserine (**20**), and affords aspartame upon catalytic hydrogenation.

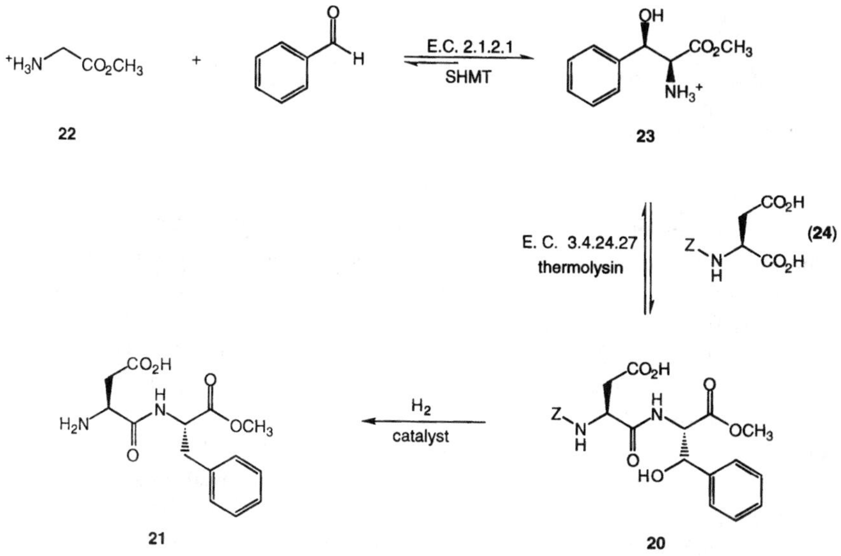

SCHEME 14.

13.2.2.1. Aminotransferases

The aminotransferase, or transaminase class of enzymes, are ubiquitous, PLP-requiring enzymes that have been used extensively to prepare natural L-amino acids [84,85]. They catalyze the general reaction shown in Scheme 15, where an amino group from one L-amino acid is transferred to an α-keto acid to produce a new L-amino acid and the respective α-keto acid. Those enzymes most commonly used have been cloned, overexpressed, and generally used as whole cell or immobilized preparations. These include the following: branched chain aminotransferase (BCAT) (EC 2.6.1.42), aspartate aminotransferase (AAT) (EC 2.6.1.1), and tyrosine aminotransferase (TAT) (EC 2.6.1.5). A transaminase patented by Celgene Corporation (Warren, NJ), called an ω-aminotransferase, does not require an α-amino acid as amino donor and hence is used to produce chiral amines [86,87]. Another useful transaminase, D-amino acid transaminase (DAT) (EC 2.6.1.21), has been the subject of much study [37,88,89]. This enzyme catalyzes the reaction using a D-amino acid donor, either alanine or aspartate (Scheme 16).

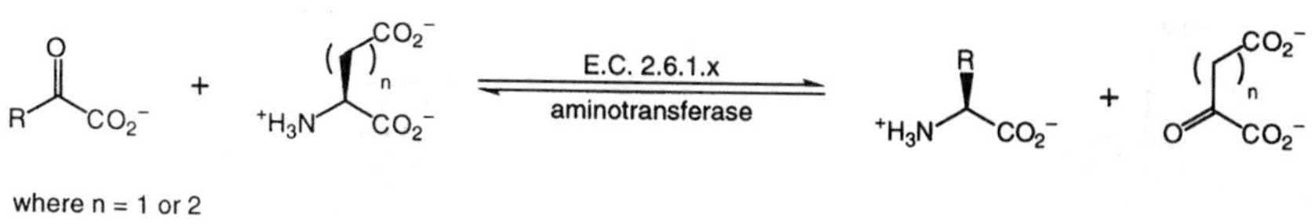

where n = 1 or 2

SCHEME 15.

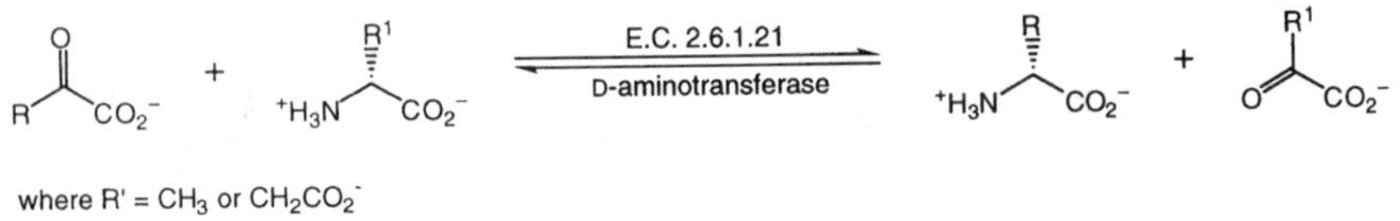

where R' = CH$_3$ or CH$_2$CO$_2^-$

SCHEME 16.

Although the utility of transaminases has been widely examined, one such limitation is the fact that the equilibrium constant for the reaction is near unity. Therefore, a shift in this equilibrium is necessary for the reaction to be synthetically useful. A number of approaches to shift the equilibrium can be found in the literature [37,85,90]. Another method to shift the equilibrium is a modification of that previously described. Aspartate is used as the amino donor, which is converted into oxaloacetate (25) upon reaction (Scheme 17). Since 25 is unstable, it decomposes to pyruvate (26) and thus favors product formation. However, since pyruvate is itself an α-keto acid, it must be removed, or else it will be converted into alanine, which could cause downstream processing problems. This is accomplished by inclusion of the *alsS* gene encoding for the enzyme acetolac

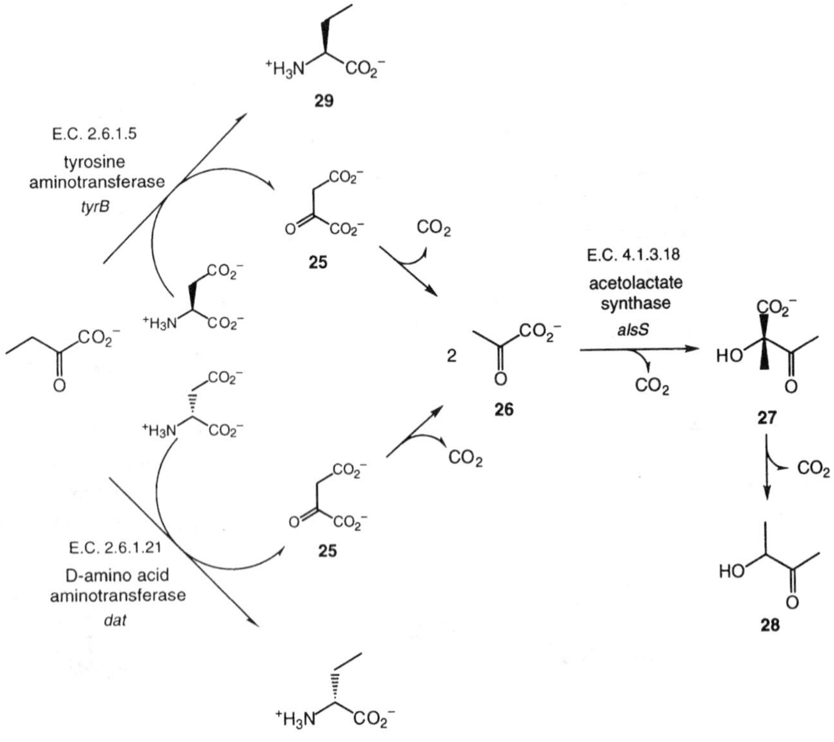

SCHEME 17.

tate synthase (EC 4.1.3.18) in the bacteria, which condenses two moles of pyruvate to form acetolactate (27). The acetolactate undergoes decarboxylation to the final by-product, acetoin (28), which is metabolically inert. This process, for example, can be used for the production of both L- and D-2-aminobutyrate (29 and 30) (Scheme 17) [6,91].

As well as unnatural amino acids such as L- and D-2-aminobutyrate (29 and 30, respectively), other examples of the use of transaminases to synthesize unnatural amino acids have also been described in the literature, including L-Tle, phosphinothricin, and L-thienylalanines [6,92–95]. These unnatural amino acids are becoming increasely important for their use in peptidomimetics, especially as pharmaceutical companies turn toward rational drug design.

13.2.3. Hydrolases (EC 3.x.x.x)

The hydrolase class of enzymes is the largest, representing approximately 75% of all the industrial enzymes, the use of which ranges from the hydrolysis of polysaccharides (carbohydrases) and nitriles to proteins (proteases) [96]. Most of these industrial enzymes are used in processing-type reactions to reduce protein, carbohydrate (i.e., viscosity), and lipid content in the detergent and food industries. Some of the more common reactions catalyzed by hydrolases are shown in Figure 1. The glycosidases (Figure 1; reaction 1) will not be discussed here because they do not result in the preservation or formation of a chiral center.

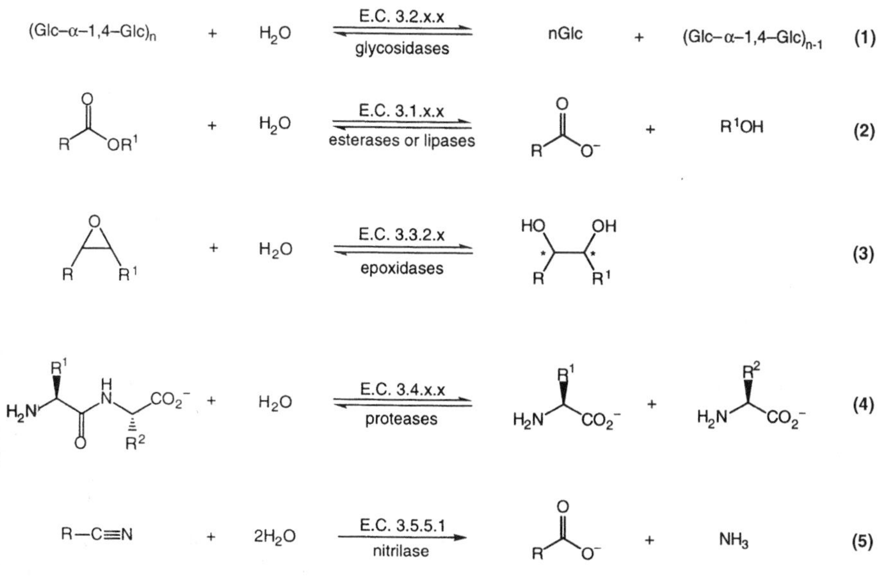

$$(\text{Glc-}\alpha\text{-1,4-Glc})_n \quad + \quad H_2O \quad \underset{\text{glycosidases}}{\overset{\text{E.C. 3.2.x.x}}{\rightleftharpoons}} \quad n\text{Glc} \quad + \quad (\text{Glc-}\alpha\text{-1,4-Glc})_{n-1} \quad (1)$$

$$(2)$$

$$(3)$$

$$(4)$$

$$R-C{\equiv}N \quad + \quad 2H_2O \quad \underset{\text{nitrilase}}{\overset{\text{E.C. 3.5.5.1}}{\rightleftharpoons}} \quad \text{(acid)} \quad + \quad NH_3 \quad (5)$$

FIGURE I   General reactions of hydrolase enzymes.

Nevertheless, their industrial use in the corn wet milling industry for starch conversion is of utmost significance [97]. Reactions catalyzed by esterases and lipases, epoxidases and proteases (Figure 1; reactions 2–4, respectively), have most frequently been used to resolve substrates into their desired enantiomers.

The degradation of nitriles by nitrilases (EC 3.5.5.1) has been the subject of intense study, especially as it relates to the preparation of the commodity chemical acrylamide. Nitrilases catalyze the hydrolysis of nitriles to the corresponding acid plus ammonia (Figure 1; reaction 5), whereas nitrile hydratases (EC 4.2.1.84) add water to form the amide. Strains such as *Rhodococcus rhodochrous* J1, *Brevibacterium* sp., and *Pseudomonas chlororaphis* have been used to prepare acrylamide from acrylonitrile, which contain the hydratase and not nitrilase activity [12]. A comparison of these strains has been discussed elsewhere [98]. Other uses of nitrilases, however, have primarily been directed at resolution processes to stereoselectively hydrolyze one enantiomer over another or regioselectively hydrolyze dinitriles [99–101].

In the same group as nitrilase enzymes are the amidases. This includes amino acid amidase (EC 3.5.1.4), used to prepare amino acids, usually through resolution, and also penicillin G acylase (penicillin G amidohydrolase) (EC 3.5.1.11), used in the manufacture of semisynthetic penicillins [102,103]. Immobilized penicillin G acylase has most recently been used to catalyze the formation of *N*-α-phenylacetyl amino acids, which can then be used in peptide coupling reactions (see Section 13.2.3.2) [104].

The esterases, lipases, and proteases are often used to prepare chiral intermediates when the reactions are conducted in the synthetic mode. Selected examples of these enzymatic biotransformations are discussed in the following respective sections. The reader is directed to the following reviews and textbooks for additional information, especially regarding resolutions [12,22,52,105–107].

13.2.3.1. Esterases and Lipases

One of the reactions catalyzed by esterases and lipases is the reversible hydrolysis of esters (Figure 1; reaction 2). These enzymes also catalyze transesterifications and the asymmetrization of *meso*-substrates (Section 13.2.3.1.1). Many esterases and lipases are commercially available, making them easy to use for screening desired biotransformations without the need for culture collections and/or fermentation capabilities. As more and more research has been conducted with these enzymes, a less empirical approach is being taken due to the different substrate profiles amassed for various enzymes. These profiles have been used to construct active site models for such enzymes as pig liver esterase (PLE) (EC 3.1.1.1) and the microbial lipases (EC 3.1.1.3) *Pseudomonas cepacia* lipase (PCL), formerly *P. fluorescens* lipase, *Candida rugosa* lipase (CRL), formerly *C. cylindracea* lipase, lipase SAM-2 from *Pseudomonas* sp., and *Rhizopus oryzae* lipase (ROL) [108–116]. In addition, x-ray crystal structure information on PCL and CRL has been most helpful in predicting substrate activities and isomer preferences [117–119].

Any discussion of lipases and esterases would be remiss without specifically mentioning PLE. Because of its extreme biocatalytic versatility, this enzyme has been the subject of numerous reviews and has been screened for stereoselective hydrolysis using hundreds of substrates [120,121]. PLE has been the workhorse of enzymes for many organic chemists catalyzing both enantio- and regioselective reactions.

A number of major pharmaceutical companies have used biocatalytic approaches based on esterases and lipases to prepare target drugs or intermediates [70,122–126]. Most of these approaches involve resolutions that start with a racemic ester or amide, and as such, yields of $\leq 50\%$ can only be realized. Recent examples of resolutions applied to pharmaceutical intermediates such as the paclitaxel (Taxol) side chain and R-(+)-BMY-14802, an antipsychotic agent, have been described by the Bristol-Myers group [70]. The following sections discuss selected examples of the use of esterases and lipases to hydrolyze prochiral or *meso*-substrates, where theoretical yields of 100% can be realized, followed by a brief discussion of dynamic kinetic resolution where reaction yields of 100% can also be achieved.

13.2.3.1.1. PROCHIRAL AND MESO-SUBSTRATES.

Selected prochiral and *meso*-substrates have been used with various esterases and lipases (Figure 2) and illustrate the wide variety of substrates that can be used with these enzymes. A complete listing of these types of substrates can be found in other sources [12,105,107,120,121]. It should be noted that the use of organic solvent can have a profound effect as to what product isomer is formed [127]. The use of porcine pancreatic lipase (PPL) to carry out an asymmetric hydrolysis of a *meso*-diacetate for the production of an intermediate in pheromone synthesis was recently reported [128]. Two specific examples are discussed here that used this approach directed at key chiral pharmaceutical intermediates.

Dodds et al. at Schering-Plough have described the acylation of a prochiral

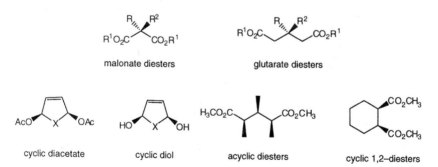

FIGURE 2   Examples of prochiral and *meso*-substrates (hydrolytic and synthetic) for esterases and lipases.

diol **31** on the synthetic route to a potential anitfungal drug, SCH 56592 (**32**) (Scheme 18). Numerous commercial enzymes were screened for their ability to carry out the stereoselective acylation of the diol using many different conditions of solvents, temperature and enzyme, substrate, and acylating agent concentrations. Some of the better enzymes for monoacylation were Amano CE (*Humicola lanuginosa*) and Biocatalyst (*Pseudomonas fluorescens*). Yields using both enzymes were 100% with ee's of 97 and 99, respectively. The enzyme Novo SP435 (*Candida antarctica*) gave ~70% yield of (*S*)-mono-acetate **33** with good ee (> 99), but with the remaining product (~30%) being the diacetate [122,124,126].

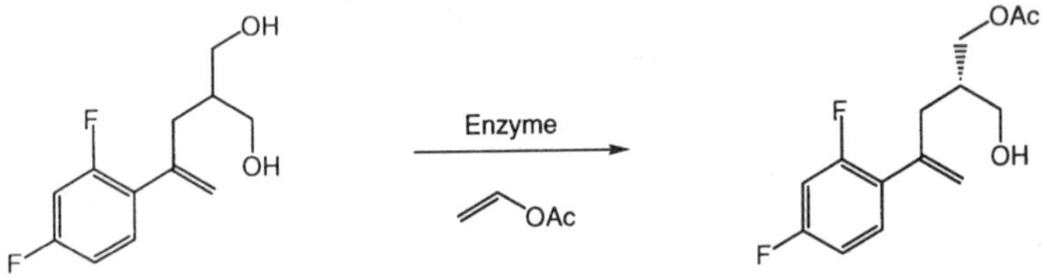

SCHEME 18.

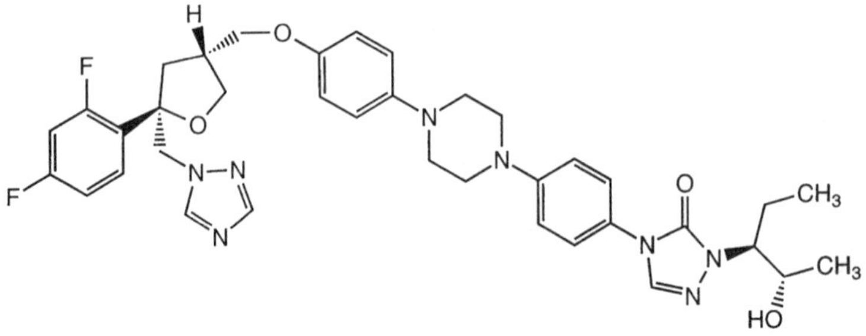

32

Another example using a prochiral acetate and asymmetic hydrolysis was described by Bristol-Myers for an intermediate in the synthesis of Fusinopril (**34**), an ACE inhibitor (Scheme 19). The prochiral substrate **35** is hydrolyzed both when R = phenyl or cyclohexyl to the corresponding (*S*)-(−)-monoacetate **36**. The reaction was conducted in a 10% toluene biphasic system with either PPL or *Chromobacterium viscosum* lipase. The cyclohexyl monoacetate was obtained in 90% yield with an optical purity of 99.8% [70,129].

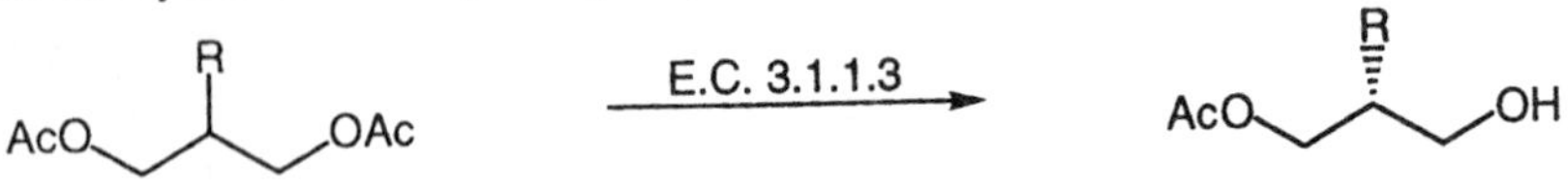

R = C_6H_5 or C_6H_{11}

SCHEME 19.

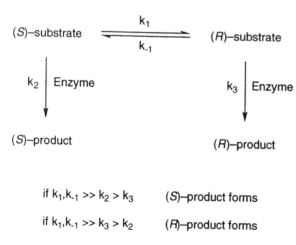

34

13.2.3.1.2. DYNAMIC KINETIC RESOLUTION.

Another method that has been used to approach 100% theoretical yield in asymmetric syntheses is dynamic kinetic resolution (DKR), which has been reviewed [130–133] and has been applied to just chemical or a combination of chemical and biochemical reactions. Only those examples in which biochemical transformations are included in the approach are presented here. If the interconversion between the two enantiomeric substrates is rapid and the product is relatively stable, and, thus, irreversibly formed, then the magnitude of the rate constants, k_2 and k_3, will dictate which product isomer is formed (Figure 3). This interconversion between the isomers is sometimes catalyzed by metal ions, silica, or ion exchange resin, or it could be due to the lability of the stereogenic center.

$$(S)\text{–substrate} \underset{k_{-1}}{\overset{k_1}{\rightleftharpoons}} (R)\text{–substrate}$$

k_2 | Enzyme k_3 | Enzyme

(S)–product (R)–product

if $k_1, k_{-1} \gg k_2 > k_3$ (S)–product forms

if $k_1, k_{-1} \gg k_3 > k_2$ (R)–product forms

FIGURE 3 Dynamic kinetic resolution (DKR).

An example is described for the unnatural amino acid (S)-*tert*-leucine (**37**) (Scheme 20), developed by Chiroscience Ltd. (now ChiroTech) [134]. It uses the commercially available Lipozyme (*Mucor miehei*) from Novo to hydrolyze the racemic 2-phenyl-4-*tert*-butyl-oxazolin-5-(4H)-one (**38**) to the (S)-N-benzoyl-*tert*-leucine butyl ester (**39**), followed by Alcalase (*Bacillus licheniformis*) treatment to hydrolyze the butyl ester, which upon debenzoylation, yields (S)-*tert*-leucine (**37**) with an ee of 99.5% and a yield of 74%.

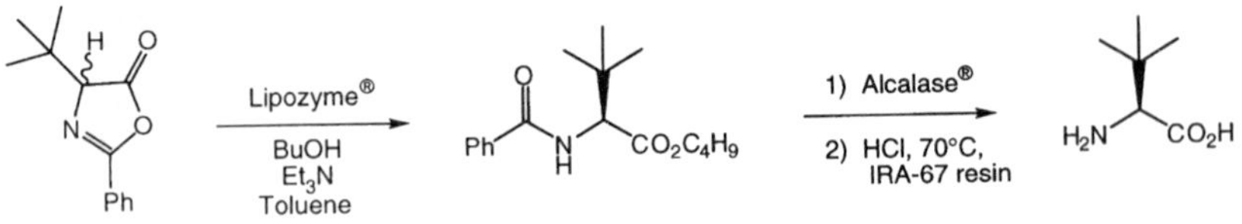

SCHEME 20.

An example in which a transition metal catalyst is used in combination with an enzyme has been described (Scheme 21) [135]. The racemic alcohol **40** is converted to the (R)-acetate **41**, using a 2% ruthenium catalyst along with Novozym 435, 3 equivalents of p-chlorophenylacetate in t-BuOH, and 1 equivalent of 1-indanone. The reaction yield was 81% with an optical purity of > 99.5% ee.

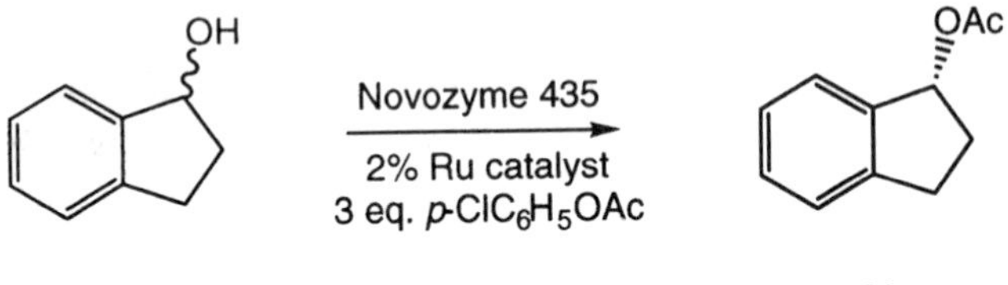

SCHEME 21.

These are just two examples of many DKR reports that have been published [136,137]. This approach to asymmetric synthesis, especially combining chemical and biochemical regimes, is going to be a powerful tool in synthesizing asymmetric molecules with high yields and optical purities.

13.2.3.2. Proteases

As previously mentioned, proteases have been extensively studied to conduct resolutions both of natural and unnatural amino acids. These useful reactions provide chiral compounds that very often cannot be obtained by chemical means. The use of proteases to conduct enantio- and regioselective coupling reactions, especially incorporating the use of organic solvents, has provided numerous examples of synthetic and semisynthetic peptides [138–143].

Regarding the use of proteases in the synthetic mode, the reaction can be conducted using a kinetic or thermodynamic approach. The kinetic approach requires proteases that form an acyl-enzyme intermediate, such as trypsin, chymotrypsin, subtilisin, or papain, and the amino donor substrate must be "activated" as the ester (Scheme 22). Here, the nucleophile R^3-NH_2 competes with water to form the peptide bond. Reaction conditions such as pH, temperature, and organic solvent modifiers are manipulated to maximize synthesis. Examples of this approach using carboxypeptidase Y from baker's yeast have been described [144].

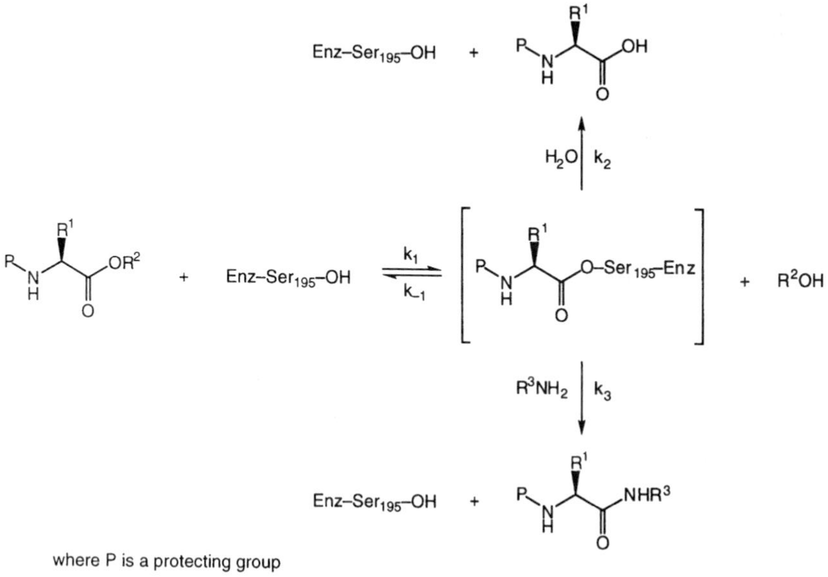

where P is a protecting group

SCHEME 22.

The thermodynamic approach is shown whereby the uncharged amino acids are coupled to form the peptide bond (Scheme 23). This requires suppression of the substrate charges, which again can be accomplished by altering reaction conditions similar to that described for the kinetic approach. A prominent example of peptide coupling is used for the production of the synthetic artificial sweetener aspartame (α-APM) (**21**). The chemical coupling method yields an approximately 80:20 mixture of the α- and β-isomers, whereas the regioselectivity of the enzymatic method produces 100% α-APM, the desired isomer. The thermodynamic approach is the basis of the Tosoh process for α-APM synthesis, which is practiced by the Holland Sweetener Company. Carbobenzyloxy-L-aspartic acid (**24**) is coupled with L-phenylalanine methyl ester (**42**) using the enzyme thermolysin from *Bacillus thermoproteolyticus* to form Z-α-APM (**43**), which then, upon catalytic hydrogenation, forms α-APM (**Scheme 24**). Because of the enantioselectivity of enzymes, D/L-PM can also be used with only the L-isomer being coupled. The D-isomer does not react and forms a salt with Z-α-APM, which can then be recycled upon acidification. Many variations of this process have been explored for the enzymatic preparation of α-APM [145–147].

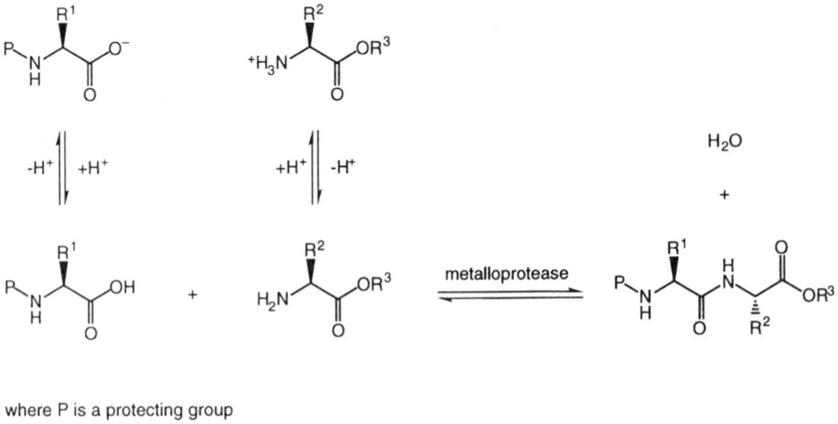

where P is a protecting group

SCHEME 23.

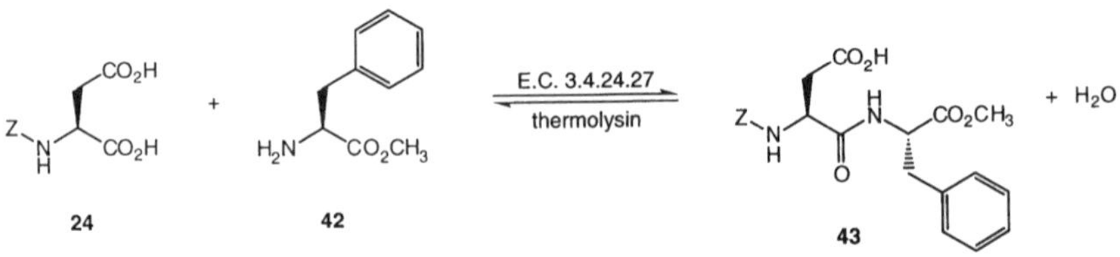

SCHEME 24.

With the increased efforts in using peptidomimetics as selective enzyme inhibitors in a number of disease states such as hypertension (renin), cancer (matrix metalloproteases), and acquired immunodeficiency syndrome (human immunodeficiency virus protease), the incorporation of unnatural amino acids into peptides is becoming necessary. The use of proteases such as subtilisin (EC 3.4.21.62), trypsin (EC 3.4.21.4), α-chymotrypsin (EC 3.4.21.1), and thermolysin (EC 3.4.24.27) have been used with or without organic solvents to synthesize peptides containing unnatural amino acids. In addition, protein engineering has redesigned certain proteases to effect thermostability and organic solvent tolerance [148]. As shown in Table 2, N-protecting groups such as Z (benzyloxycarbonyl), Boc (t-butyloxycarbonyl), and Moz [(p-methoxybenzyl)oxy]carbonyl have been used with a variety of solvent conditions and enzymes to prepare the final di- or tripeptide, in most cases, with high yield. A number of different acyl donors and acceptors containing unnatural amino acids are shown to function very well in the synthesis of coupled peptides. These include D-amino acids (Leu, Phe, Ala, Trp), an α, α-disubstituted amino acid (Aib, α-aminobutyrate), a statine-type isostere, and two noncoded amino acids, norvaline (Nvl) and homophenylalanine

TABLE 2

Product[a]	Enzyme	Solvent	Yield (%)	Reference
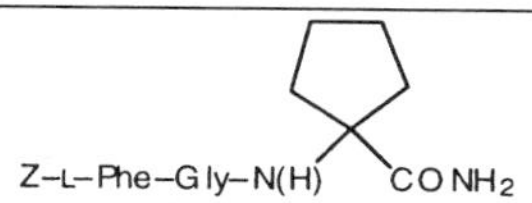 Z-L-Phe-Gly-N(H) — cyclopentyl-CONH₂	Subtilisin BPN'	DMF/H₂O (8/3)	89	150
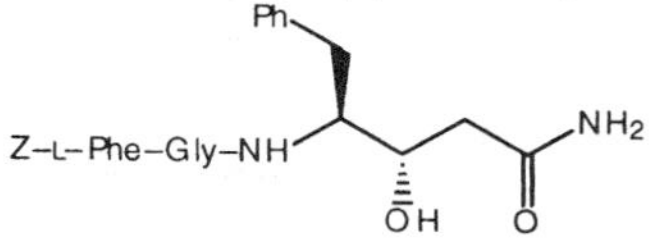 Z-L-Phe-Gly-NH (structure with NH₂, O, OH)	Subtilisin BPN'	DMF/H₂O (8/3)	70	150
Z-L-Phe-D-Leu-NH₂	Subtilisin CLEC	CH₃CN	96	151
Z-L-Phe-D-Phe-NH₂	Subtilisin CLEC	CH₃CN	95	151
Boc-Aib-L-Ala-pNA	Trypsin (*S. griseus*)	DMSO/buffer (1/1)	96	152
Boc-L-Phe-D-Ala-L-Ala-pNA	Trypsin (bovine pancreas)	DMSO/buffer (1/1)	80	153
Z-L-Val-L-Tyr-D-Phe-NH₂	α-Chymotrypsin (bovine pancreas)	Isooctane/THF (7/3)	68	154
Moz-L-Phe-D-Ala-NH₂	Subtilisin (Alcalase) (*B. licheniformis*)	*t*-Butanol	87	155
Boc-L-Phe-D-TrpOMe	Subtilisin (Alcalase) (*B. licheniformis*)	EtOH	60	156
Z-D-Phe-L-LysOMe	Subtilisin (Alcalase) (*B. licheniformis*)	EtOH	53	156
Z-L-Phe-L-Nvl-NH₂	Thermolysin (*B. thermoproteolyticus*)	Buffer	92	157
Z-L-hPhe-L-Leu-NH₂	Thermolysin (*B. themoproteolyticus*)	Buffer	91	157

[a] The arrow indicates the peptide bond that was synthesized.

(hPhe). Halogenated phenylalanines can be used in peptide-coupling reactions with thermolysin, but the coupling was conducted with a racemic derivative [149]. Nevertheless, this approach using proteases to synthesize peptides containing unnatural amino acids is likely to become a key technology as additional peptidomimetics become approved as drugs.

13.2.4. Lyases (EC 4.x.x.x)

Lyases are an attractive group of enzymes from a commercial perspective as demonstrated by their use in many industrial processes [158]. They catalyze the cleavage of C–C, C–N, C–O, and other bonds by means other than hydrolysis, often forming double bonds. For example, two well-studied ammonia lyases, aspartate ammonia lyase (aspartase) (EC 4.3.1.1) (see Chapter 16, this book by P. Taylor) and phenylalanine ammonia lyase (PAL) (EC 4.3.1.5) catalyze the *trans*-elimination of ammonia from the amino acids, L-aspartate and L-phenylalanine, respectively. Most commonly used in the synthetic mode, the reverse reaction has been used to prepare the L-amino acids at the ton scale (Schemes 25 and 26) [158–160]. These reactions are conducted at very high substrate concentrations such that the equilibrium is shifted, which results in very high conversion to the amino acid products.

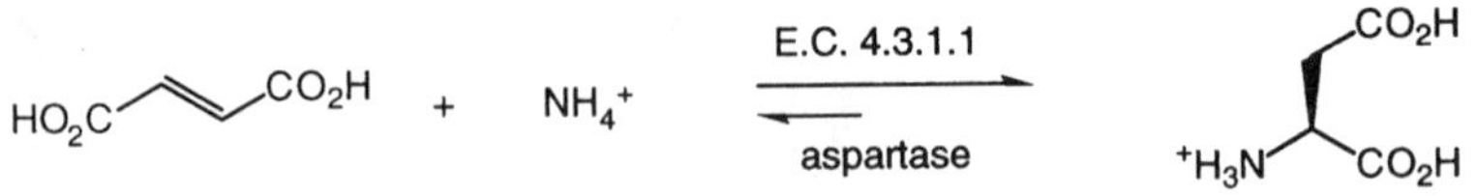

SCHEME 25.

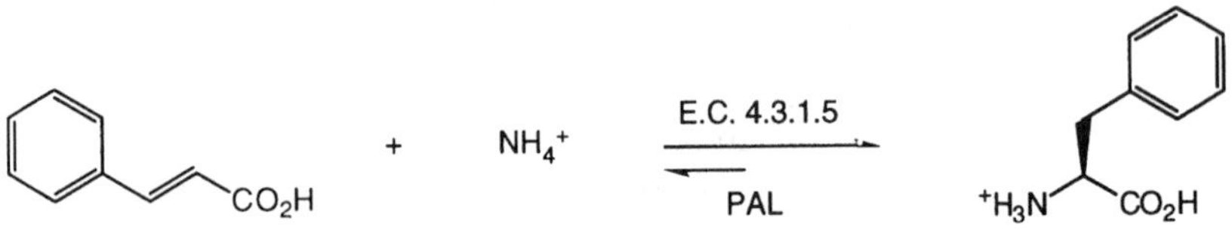

SCHEME 26.

Although both aspartase and PAL have shown great utility for the preparation of their respective amino acids, they have an incredibly strict substrate specificity; therefore, in general, they have not found much utility for the preparation of derivatives. There have been several reports in the literature, however, in which the yeast *Rhodotorula glutinis* and an *E. coli* strain containing a cloned PAL gene were used to prepare a number of L-phenylalanine analogs [161,162]. Future work in this area will likely include protein engineering, such that modifications to active site residues will allow substrates with altered structures to be accepted in the substrate binding site more readily.

13.2.4.1. Aldolases

A specific type of lyase, the aldolase class of enzymes, catalyzes the formation of an asymmetric C–C bond, which is a most useful reaction to the synthetic organic chemist [163,164]. These enzymes are most often used in the synthesis of novel saccharides, such as aminosugars, thiosugars, and disaccharide mimetics, since they use these types of substrates in nature and their substrate specificity is quite broad [165–167]. Aldolases are classified as type I or type II enzymes depending on their source and mechanism of action [168,169]. Type I aldolases are from higher plants and mammalian sources, require no metal ion cofactor, and use a Schiff-base intermediate in catalysis. Type II enzymes are from bacterial and fungal sources and require a metal ion, usually Zn^{2+}, for catalysis. In addition to different mechanisms, depending on the specific aldolase, the stereochemistry of the two stereogenic centers formed can be controlled (Scheme 27) [12,16]. These types of reactions in which stereochemistry at two contiguous carbon centers can be controlled is of extreme interest to synthetic organic chemists, especially when the synthesis of carbohydrates and their derivatives is desired [170,171].

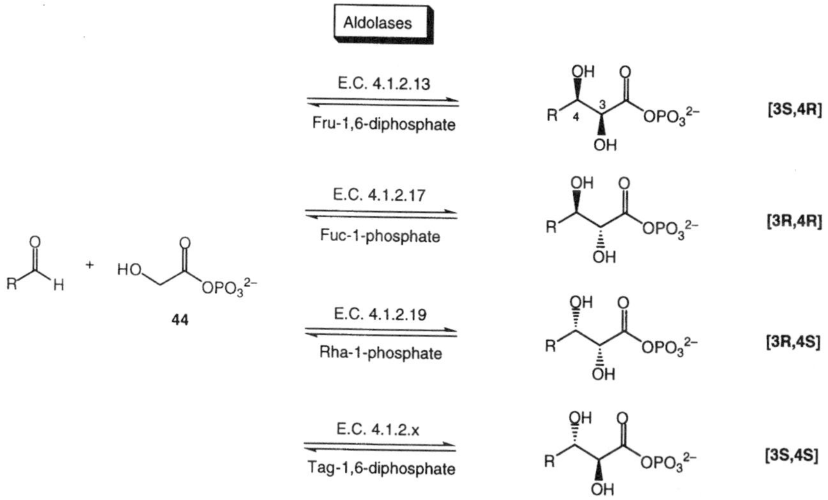

SCHEME 27.

Although many aldolases have been characterized for research purposes, these enzymes have not been developed commercially to any significant extent. This is likely due to the availability of the various biocatalysts and the need for dihydroxyacetone phosphate (DHAP) (44), the expensive donor substrate required in nearly all aldolase reactions. A number of chemical and enzymatic routes have been described for DHAP synthesis, which could alleviate these concerns [12]. In terms of the enzyme supply issue, this may change with the introduction of products from Boehringer Mannheim and their Chirazyme Aldol reaction kit. They have three kits, each containing a different aldolase: fructose-1,6-diphosphate (*FruA*) (EC 4.1.2.13), L-rhamnulose-1-phosphate (*RhuA*) (EC 4.1.2.19), and L-fuculose-1-phosphate (*FucA*) (EC 4.1.2.17). As more screening is performed with kits of this type and as more aldolases are cloned and become available at a reasonable cost, this class of enzymes will very likely be developed for use in an industrial process in the near future. Specific reference to cloning of selected aldolases has been described [172].

13.2.4.2. Decarboxylases

The decarboxylase family of enzymes catalyzes the PLP-dependent removal of CO_2 from substrates. This reaction has found great industrial utility with one specific enzyme in particular, L-aspartate-β-decarboxylase (EC 4.1.1.12) from *Pseudomonas dacunhae*. This enzyme, most often used as immobilized whole cells, has been used to synthesize L-alanine on an industrial scale (multitons) since the mid-1960s by Tanabe (Scheme 28) [159,173]. Another use for this biocatalyst has been the resolution of racemic aspartic acid to produce L-alanine and D-aspartic acid (Scheme 29).

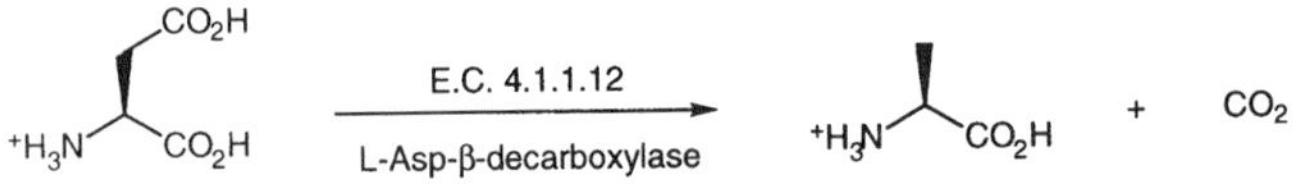

SCHEME 28.

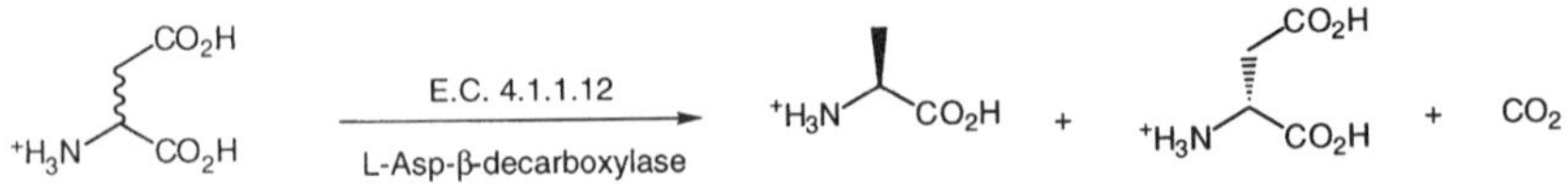

SCHEME 29.

Numerous other decarboxylases have been isolated and characterized, and much interest has been shown due to the irreversible nature of the reaction with the release of CO_2 as the thermodynamic driving force. Although these enzymes have narrow substrate specificity profiles, their utility has been widely demonstrated. Additional industrial processes will continue to be developed once other decarboxylases become available. Such biocatalysts would include the aromatic amino acid (EC 4.1.1.28), phenylalanine (EC 4.1.1.53), and tyrosine (EC 4.1.1.25) decarboxylases, which likely could be used to produce derivatives of their respective substrates. These derivatives are finding increasing uses in the development of peptidomimetic drugs and as possible positron emission tomography imaging agents [174].

13.2.5. Isomerases (EC 5.x.x.x)

The isomerase class of biocatalysts represents a small number of enzymes that are mainly composed of the racemases, epimerases, and mutases. In the case of racemases and epimerases, the stereochemistry of at least one carbon center is changed, while mutases catalyze the transfer of a functional group, such as an amino function, to an adjacent carbon, forming a new stereocenter. One of the most industrially important biotransformations involves an enzyme from this class, D-xylose isomerase (EC 5.3.1.5), also known as glucose isomerase. It catalyzes the conversion of D-glucose (**44**) to D-fructose (**45**), which is necessary to produce high-fructose corn syrup (HFCS), a sucrose substitute used by the food and beverage industries (Scheme 30). It is one of the highest tonnage value enzymes produced in the world and is used in nearly all processes as a cell-free or whole-cell immobilized preparation [4,160,175,176].

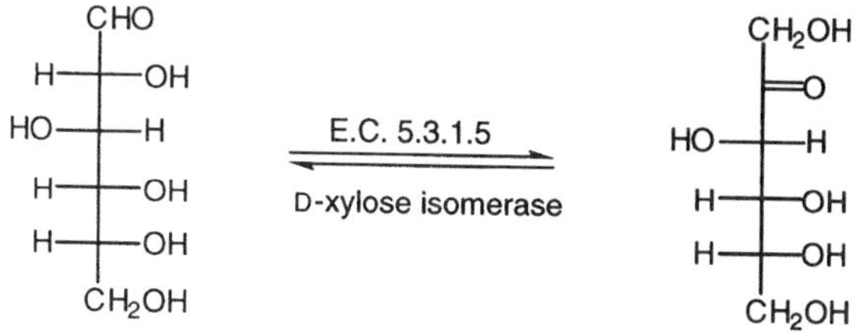

SCHEME 30.

Because this biocatalyst is of such industrial significance, efforts to redesign it with altered properties could have a profound economic effect on the cost of HFCS. With the advances in molecular biology and prediction of protein structure–function relationships, these studies have been underway for a number of years, and include thermal stabilization, alteration in pH and temperature optima, and alteration in substrate specificity [176–179].

13.2.5.1. Mutases

Although this class of biocatalyst is currently not used commercially, the potential for these types of enzymes, such as the aminomutases (EC 5.4.3.x), has yet to be realized. The general reaction is the conversion of an α-L-amino acid to its corresponding β-L-amino acid with the migration of the 2-amino group to carbon-3 (Scheme 31). Some of these biocatalysts require coenzymes such as PLP and are activated by S-adenosylmethionine or use vitamin B_{12} [180,181]. The amino acid aminomutases using substrates such as L-lysine, L-leucine, and L-tyrosine have been described [181]. An aminomutase with L-phenylalanine as substrate has been identified in the biosynthesis of the paclitaxel (Taxol) side chain [182]. These β-amino acids are finding utility in pharmaceuticals and in the speciality chemical arena and are not easy to produce chemically [183,184].

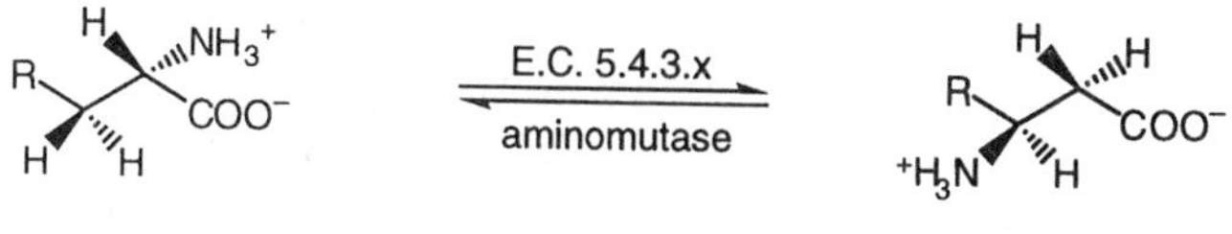

$R = -(CH_2)_3NH_3^+, -C_6H_4OH, - C_6H_5, or -CH(CH_3)_2$

SCHEME 31.

13.3. METABOLIC PATHWAY ENGINEERING

Metabolic pathway engineering is an approach to more complex types of bio-transformations. This is accomplished by the manipulation of several, usually sequential, biotransformation reactions, often through the use of recombinant DNA techniques to overexpress the selected enzymes of interest. This approach has been used to improve production of metabolites [185–187] and to manipulate metabolic pathways for the synthesis of aromatic compounds (see 13.3.1). The techniques of metabolic pathway engineering have been applied to many commodity and specialty chemicals such as amino acids, polyketides, and poly-β-hydroxyalkanoates. Much of this work is driven to ultimately use renewable substrates as opposed to those derived from peteroleum-based feedstocks. Reviews have been published on amino acids, general cellular, and metabolic engineering [188,189]. A few pertinent examples of this exciting technology are discussed in more detail in the following sections.

13.3.1. Aromatic Amino Acid Pathway

With the increasing use of L-phenylalanine, especially due to the artificial sweetener aspartame (21), this amino acid, along with L-tryptophan, a feed additive, have become major compounds of focused research. The L-phenylalanine pathway has been and continues to be the subject of intense research, details of which can be found in Chapter 4. Early work focused on the aromatic amino acid pathway itself and the deregulation of the specific genes leading to L-phenylalanine synthesis. More recent work has been concerned with other pathways that produce precursors to direct carbon flow to the common aromatic pathway, such as phosphoenolpyruvate (PEP) and erythrose-4-phosphate (E4P) [190]. This has been demonstrated whereby other genes encoding enzymes such as transketolase (*tktA*) and PEP synthase (*pps*) have been introduced into selected bacterial strains and striking effects have been observed for L-phenylalanine synthesis [191,192]. Another gene (*tal*) encoding the enzyme transaldolase was cloned into *E. coli* and showed a significant increase in the production of 3-deoxy-D-arabinoheptulosonate-7-phosphate (DAHP) from glucose [193]. DAHP is the first common intermediate in the aromatic amino acid pathway (see Chapter 4, this book by Fotheringham).

The work of Frost et al. has focused on the aromatic pathway with respect to the biosynthesis of quinoid organics such as benzoquinone and hydroquinone, as well as adipic acid production [194–197]. The same principles have been used whereby a cloned gene is overexpressed, which allows the recombinant organism to funnel carbon down a specific, selected metabolic pathway. Although currently these compounds are synthesized from petroleum-based feed stocks primarily for economic reasons, environmental concerns and the shortage of these starting materials in the future will necessitate other approaches such as these for the production of such commodity chemicals [197].

13.3.2. Poly-β-Hydroxyalkanoates (PHAs)

Another example where metabolic pathway engineering has made a dramatic impact is in the biodegradable polymer field. The polymer of this family most widely studied is poly-β-hydroxybutyrate (PHB) (**46**). Another member of the PHA family commercialized by Imperial Chemical Industries (ICI), which later became Zeneca under the trade name Biopol, is a copolymer consisting of β-hydroxybutyric acid and β-hydroxyvaleric acid. This biodegradable polymer was first used in plastic shampoo bottles by the Wella Corporation [198]. In the early part of 1996, the Biopol product line, was purchased from Zeneca by the Monsanto Company.

The *phbA, phbB*, and *phbC* genes from *Alcaligenes eutrophus* have been cloned into *E. coli* (Scheme 32). These genes encode for the biosynthetic enzymes β-ketothiolase, acetoacetyl-CoA reductase (NADPH-dependent), and PHB synthase, respectively [199–202]. These genes have been also cloned into plants, since the cost of the bacterially produced PHB was higher than other biomaterials such as starches or lipids [203,204].

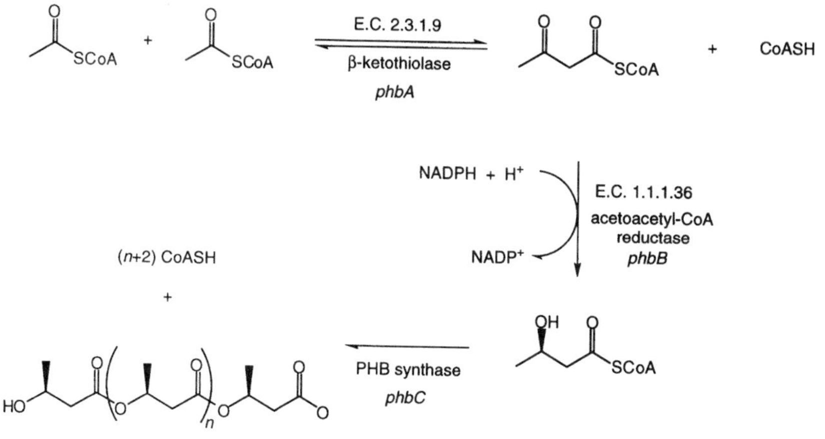

SCHEME 32.

PHB has been shown to accumulate to levels approaching 90% of the bacterial dry cell weight when these pathway enzymes have been overexpressed [205]. With these efficiencies, it is only a matter of time before the production of these biopolymers can compete economically with the synthetically prepared material [206].

13.3.3. Polyketides and Analogs

A final example of pathway engineering is based on polyketide biosynthesis. This is a very active area of research and has been recently reviewed [207,208]. Polyketides are complex natural products with numerous chiral centers, which are of substantial economic benefit as pharmaceuticals. The complexity of polyketide structures is illustrated by erythromycin A (**47**), FK506 (**48**), and avermectin A1a (**49**). These natural products function as antibacterial, antifungal, and antiparasitic agents, and some exhibit antitumor and immunosuppresive activities. In addition, polyketide derivatives are being sought as possible sources for new antibacterial agents or other pharmaceutically important compounds [209–211].

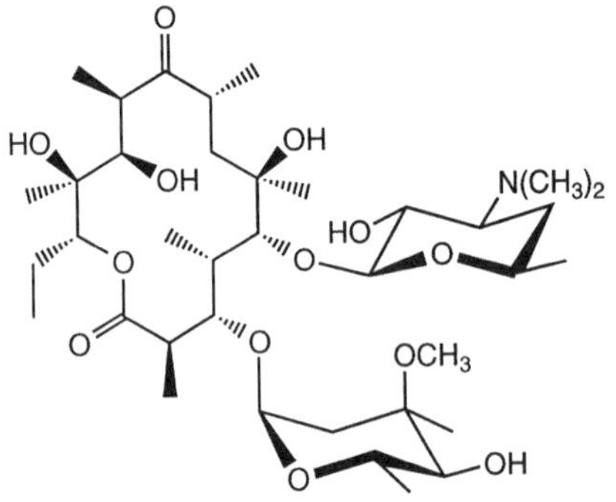

47

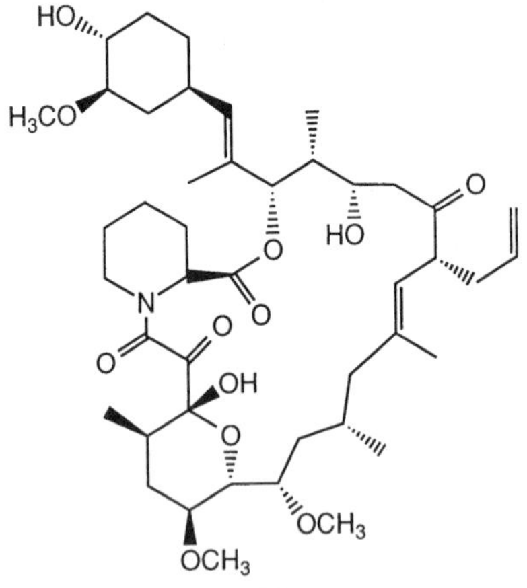

48

49

The biosynthesis of polyketides is analogous to the formation of long-chain fatty acids catalyzed by the enzyme fatty acid synthase (FAS). These FASs are multienzyme complexes that contain numerous enzyme activities. The complexes condense coenzyme A thioesters (usually acetyl, propionyl, or malonyl) followed by a ketoreduction, dehydration, and enoylreduction of the β-keto moiety of the elongated carbon chain to form specific fatty acid products. Polyketide synthases (PKSs) possess similar multienzyme activities which are assembled in modular fashion. A model system that has been studied thoroughly and used to develop this modular organization of enzymes has been the PKS deoxyerythronolide B synthase (DEBS) from *Saccharopolyspora erythraea* [212]. This enzyme consists of six modules encoding the various enzymes for condensation and further processing activities to produce 6-deoxyerythronolide B (6-dEB) (50), the aglycon of erythromycin. The modular arrangement of enzymes is shown in Figure 4. More recent studies have shown that through manipulation of these enzyme modules, numerous derivatives of polyketides can be synthesized. Through the reprogramming of genes in their respective modules by site-directed mutagenesis, by deletion of certain modules, or by introduction of an unnatural enzymatic activity, the synthesis of polyketide derivatives has been demonstrated [213–217]. Selected derivative structures that have been confirmed are shown by structures 51–53 [218]. In addition, by incubating certain precursors with strains containing designed, mutant DEBSs, various analogs of erythromycin have been synthesized [219]. Purification and characterization of various DEBS mutants have also recently been published, allowing further insight into the design of polyketide derivatives [220].

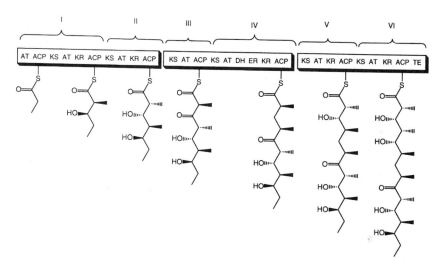

FIGURE 4 Modular organization of the six modules (I–VI) of 6-deoxyerythronolide B synthase (DEBS) enzyme as derived from *Saccharopolyspora erythraea*. Enzyme activities are acyltransferases (AT), acyl carrier proteins (ACP), β-ketoacyl-ACP synthases (KS), β-ketoreductases (KR), dehytratases (DH), enoyl reductases (ER), and thioesterases (TE). The TE-catalyzed release of the polyketide chain results in the formation of 6-dEB (50) [212,216,217].

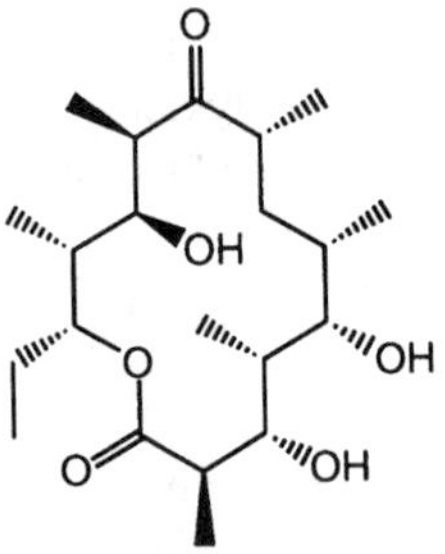

50

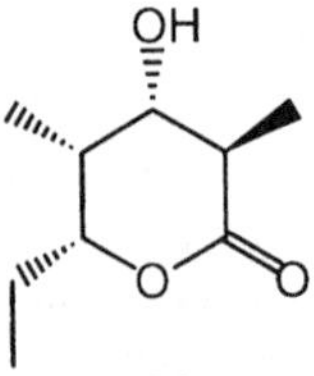

51

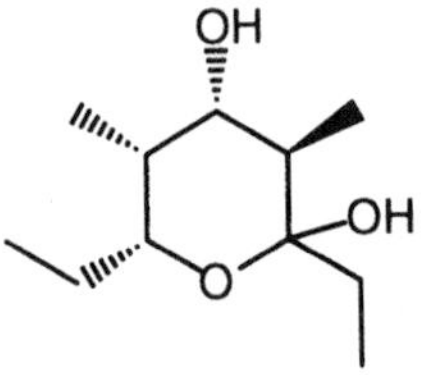

52

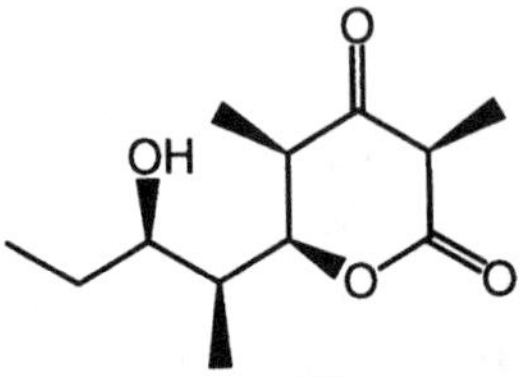

53

Much work has been conducted to elucidate the genes and characterize the enzymes and reaction products involved in the polyketide biosynthetic pathway. This example of metabolic pathway engineering goes one step further than typically using native enzymes to funnel carbon down a particular pathway to produce a selected metabolite. Here, mutagenesis is being used to create enzymes with altered activities that can lead to novel reaction products, which, upon screening, might be identified to have pharmaceutical applications.

13.4. SCREENING FOR BIOCATALYSTS

If a biocatalyst is not known to exist for a certain biotransformation, then screening for that activity is generally undertaken, especially when the chemical route is either too difficult or costly [221]. As previously mentioned, numerous companies are marketing kits aimed at organic chemists, designed to allow rapid screening of selected commercial enzymes for a particular biotransformation. These kits are designed to screen multiple enzymes under a number of reactions conditions. The major conditions of choice would be either the use of a water-miscible or -immiscible solvent. The use of organic solvents is often necessary since many substrates are usually not soluble in 100% aqueous solution. Investigation of commercial enzymes would generally be the first step of a screening program before embarking on the more labor-intensive testing of fungal or bacterial cultures [222].

In addition, screening for biotransformation activity is important when a "natural" claim is needed, which occurs most notably in the flavor industry [223]. Screening for novel biocatalysts to perform specific biotransformations can require extensive resources, depending on the details of the screening protocol. Therefore, the choice of when to screen for a biocatalyst can only be justified when the resultant product exhibits significant commercial potential. A number of recent articles and patents have appeared that describe screening using enzymes and/or cultures for microbial reductions and oxidations, ester synthesis, acylation of a diol, or chiral epoxide hydrolysis [122,224–228]. These are but a few examples in which biotransformations involving chiral compounds have been successfully performed.

An example of a project of this nature is the screening for microorganisms to produce aspartame (**21**). Microorganisms were screened for their ability to catalyze the *trans*-addition of ammonia across the double bond of *N*-fumaryl-L-phenylalanine methyl ester (FumPM) (**54**) (Scheme 33) [229]. This is essentially the reaction of a mutant aspartase, since the native enzyme has such strict substrate specificity (see Section 13.2.4). Although the literature touts this as a successful screening effort, this process has not been practiced commercially, as the yields are extremely low [230,231].

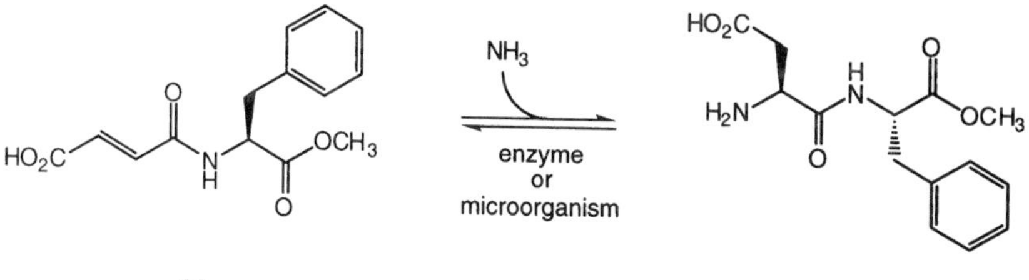

SCHEME 33.

Another example of screening was for the hydantoinase enzymes used for the production of either L- or D-amino acids [232,233]. These studies were conducted in the mid to late 1970s, and 5-monosubstituted hydantoins **55** were used as the sole carbon and nitrogen sources with a variety of soil microorganisms. The microorganisms that grew were then futher characterized as having the desired activity. This has resulted in the isolation and characterization of a number of biocatalysts from various microorganisms that catalyze the enantioselective cleavage of hydantoins. Both L- and D-hydantoinase enzymes have been described in the literature that can be further coupled to either a chemical or enzymatic hydrolysis of the *N*-carbamoyl-amino acid (Scheme 34). In addition, many different substituent R-groups at carbon 5 have been studied for each hydantoinase [232,233]. For industrial purposes, R = *p*-hydroxyphenyl and phenyl have found most utility, leading to the production of D-*p*-hydroxyphenylglycine and D-phenylglycine, respectively, which are used as the unnatural side chains in a number of well-known β-lactam antibiotics. Most recently, two new hydantoinases have been isolated from thermophilic microorganisms and are being sold commercially in an immobilized form by Boehringer Mannheim for the production of D-amino acids [234]. These purified hydantoinases, along with three selected bacterial strains, *Agrobacterium tumefaciens* CIP 67.1, *Pseudomonas fluorescens* CIP 69.13, and *Pseudomonas* sp. ATCC 43648, were tested with a number of ring-substituted D-phenylglycines and shown to be effective at producing these derivatives [235].

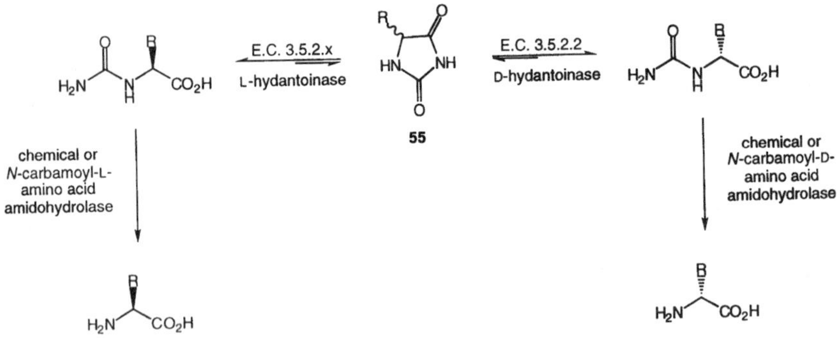

SCHEME 34.

13.5. SUMMARY

The use of microorganisms or commercially available enzymes to perform biotransformations is well established. Numerous routes have been developed to prepare chiral fine chemicals with enzymes from all classes, often in combination with chemical methods, and it is clear that many of the major pharmaceutical companies are investing heavily in this type of research and development. With approaches such as dynamic kinetic resolution, pathway engineering, and screening for biocatalysts showing such promise, many new biotransformations will undoubtedly be discovered to develop novel, "green" bioprocesses in the future.

REFERENCES

1. Stinson, S. C. *Chem. Eng. News*, 1995, *73*(41), 44.
2. Stinson, S. C. *Chem. Eng. News*, 1996, *74*(29), 35.
3. Martineau, W. in *Chiral Chemicals*; The Freedonia Group: Cleveland, OH, 1996.

4. Rotheim, P. in *The Enzyme Industry: Specialty and Medical Applications*; Business Communications Company: Norwalk, CT, 1994.
5. Rotheim, P. in *Opportunities in Chiral Technology: Emphasizing Enabling Products*; Business Communications Company Norwalk, CT, 1995.
6. Stinson, S. C. *Chem. Eng. News*, 1997, *75*(20), 34.
7. Stinson, S. C. *Chem. Eng. News*, 1997, *75*(28), 37.
8. Robertson, D. E., Mathur, E. J., Swanson, R. V., Marrs, B. L., Short, J. M. *SIM News* 1996, *46*, 3.
9. Margolin, A. L. *TIBTECH* 1996, *14*, 223.
10. Zelinski, T., Waldmann, H. *Angew. Chem. Int. Ed. Engl.* 1997, *36*, 722.
11. West, S. in *Industrial Enzymology*; Godfrey, T., West, S., Eds; Macmillan Press Ltd: London, 1996; p. 155.
12. Faber, K. in *Biotransformations in Organic Chemistry*, 2nd ed; Springer-Verlag: Berlin, 1995.
13. Blanch, H. W., Clark, D. S. Eds; in *Applied Biocatalysis*; Marcel Dekker: New York, 1991; *Vol. 1.*
14. *Enzymes in Organic Synthesis*; Ciba Foundation Symposium 111; Pitman: London, 1985.
15. *Biocatalysis in Organic Media*; Laane, C., Tramper, J., Lilly, M. D., Eds; in series *Studies in Organic Chemistry*; Elsevier: Amsterdam, 1987; *Vol. 29.*
16. Roberts, S. M., Wiggins, K., Casy, G., Phythian, S. in *Preparative Biotransformations: Whole Cell and Isolated Enzymes in Organic Synthesis*; Wiley: Chichester, 1992.
17. Santaniello, E., Ferraboschi, P., Grisenti, P., Manzocchi, A. *Chem. Rev.* 1992, *92*, 1071.
18. Wong, C.-H., Whitesides, G. M. in *Enzymes in Synthetic Organic Chemistry*; Elsevier: New York, 1994.
19. Webb, Edwin C. *Enzyme Nomenclature*, Academic Press: New York, 1992.
20. White, J. S., White, D. C. in *Source Book of Enzymes*; CRC Press: Boca Raton, FL, 1997.
21. Chapter 2. Reduction Reactions in *Preparative Biotransformations: Whole Cell and Isolated Enzymes in Organic Synthesis*; Roberts, S. M., Wiggins, K., Casy, G., Phythian, S., Eds; Wiley: Chichester, 1992.
22. Sheldon, R. A. in *Chirotechnology: Industrial Synthesis of Optically Acive Compounds*; Marcel Dekker: New York, 1993; p. 205.
23. Wong, C.-H., Whitesides, G. M. in *Enzymes in Synthetic Organic Chemistry*; Elsevier: New York, 1994; p. 131.
24. Turner, N. J. in *Advanced Asymmetric Synthesis*; Stephenson, G. R., Ed; Chapman & Hall: London, 1996; p. 260.
25. Hummel, W., Kula, M.-R. *Eur. J. Biochem.* 1989, *184*, 1.
26. Matsushita, K., Yakushi, T., Takaki, Y., Toyama, H., Adachi, O. *J. Bacteriol.* 1995, *177*, 6552.
27. Bradshaw, C. W., Wong, C.-H., Shen, G.-J. *US Patent*, 1995, 5,385,833.
28. Simon, E. S., Plante, R., Whitesides, G. M. *Appl. Biochem. Biotechnol.* 1989, *22*, 169.
29. Kim, M.-J., Kim, J. Y. *J. Chem. Soc. Chem. Commun.* 1991, 326.

30. Bradshaw, C. W., Wong, C.-H., Hummel, W., Kula, M.-R. *Bioorg. Chem.* 1991, *19*, 29.
31. Schmidt, E., Ghisalba, O., Gygax, D., Sedelmeier, G. *J. Biotechnol.* 1992, *24*,315.
32. Kallwass, H. K. W. *Enzyme Microb. Technol.* 1992, *14*, 28.
33. Krix, G., Bommarius, A. S., Drauz, K., Kottenhahn, M., Schwarm, M., Kula, M.-R. *J. Biotechnol.* 1997, *53*, 29.
34. Hanson, R. L., Bembenek, K. S., Patel, R. N., Szarka, L. J. *Appl. Microbiol. Biotechnol.* 1992, *37*, 599.
35. Casy, G., Lee, T. V. *World Patent*, 1993, WO 93/13215.
36. Ohshima, T., Soda, K. *TIBTECH* 1989, *7*, 210.
37. Galkin, A., Kulakova, L., Yamamoto, H., Tanizawa, K., Tanaka, H., Esaki, N., Soda, K. *J. Ferment. Bioeng.* 1997, *83*, 299.
38. Katsumata, R., Hashimoto, S. *US Patent*, 1996, 5,559,016.
39. Ohshima T., Wandrey, C., Conrad, D. *Biotechnol. Bioeng.* 1989, *34*, 394.
40. Kato, Y., Fukumoto, K., Asano, Y. *Appl. Microb. Biotechnol.* 1993, *39*, 301.
41. Hummel, W., Schmidt, E., Wandrey, C., Kula, M.-R. *Appl. Microb. Biotechnol.* 1986, *25*, 175.
42. Asano, Y., Nakazawa, A. *Agric. Biol. Chem.* 1987, *51*, 2035.
43. Asano, Y., Yamada, A., Kato, Y., Yamaguchi, K., Hibino, Y., Hirai, K., Kondo, K. *J. Org. Chem.* 1990, *55*, 5567.
44. Bommarius, A. S., Schwarm, M., Stingl, K., Kottenhahn, M., Huthmacher, K., Drauz, K. *Tetrahedron: Asymmetry* 1995, *6*, 2851.
45. Wichmann, R., Wandrey, C., Bückmann, A. F., Kula, M.-R. *Biotechnol Bioeng.* 1981, *23*, 2789.
46. Kragl, U., Kruse, W., Hummel, W., Wandrey, C. *Biotechnol. Bioeng.* 1996, *52*, 309.
47. Hanson, R. L., Singh, J., Kissick, T. P., Patel, R. N., Szarka, L. J., Mueller, R. H. *Bioorg. Chem.* 1990, *18*, 116.
48. Sutherland, A., Willis, C. L. *Tetrahedron Lett.* 1997, *38*, 1837.
49. Ward, O. P., Young, C. S. *Enzyme Microb. Technol.* 1990, *12*, 482.
50. Servi, S. *Synthesis* 1990, 1.
51. Csuk, R., Glänzer, B. I. *Chem. Rev.* 1991, *91*, 49.
52. Ager, D. J., East, M. B. in *Asymmetric Synthetic Methodology*; CRC Press: Boca Raton, FL, 1996; p. 391.
53. Anderson, B. A., Hansen, M. M., Harkness, A. R., Henry, C. L., Vicenzi, J. T., Zmijewski, M. J. *J. Am. Chem. Soc.* 1995, *117*, 12358.
54. Vicenzi, J. T., Zmijewski, M. J., Reinhard, M. R., Landen, B. E., Muth, W. L., Marler, P. G. *Enzyme Microb. Technol.* 1997, *20*, 494.
55. Zmijewski, M. J., Vicenzi, J., Landen, B. E., Muth, W., Marler, P., Anderson, B. *Appl. Microb. Biotechnol.* 1997, *47*, 162.
56. Reddy, J., Tschaeh, D., Shi, Y.-J., Pecore, V., Katz, L., Greasham, R., Chartrain, M. *J. Ferment. Bioeng.* 1996, *81*, 304.
57. Chartrain, M., McNamara, J., Greasham, R. *J. Ferment. Bioeng.* 1996, *82*, 507.
58. Katz, L., King, S., Greasham, R., Chartrain, M. *Enzyme Microb. Technol.* 1996, *19*, 250.
59. Blacker, A. J., Holt, R. A. in *Chirality in Industry II*; Collins, A. N., Sheldrake, G. N., Crosby, J., Eds; Wiley: Chichester, 1997; p. 245.

60. Hunt, J. R., Carter, A. S., Murrell, J. C., Dalton, H., Hallinan, K. O., Crout, D. H. G., Holt, R. A., Crosby, J. *Biocatalysis Biotransform.* 1995, *12*, 159.
61. Hallinan, K. O., Crout, D. H. G., Hunt, J. R., Carter, A. S., Dalton, H., Murrell, J. C., Holt, R. A., Crosby, J. *Biocatalysis Biotransform.* 1995, *12*, 179.
62. Sih, C. J., Zhou, B.-N., Gopalan, A. S., Shieh, W.-R., Chen, C.-S., Girdaukas, G., Vanmiddlesworth, F. *Ann. N. Y. Acad. Sci.* 1984, *434*, 186.
63. Sih, C. J. *US Patent*, 1987, 4,710,468.
64. Bare, G., Jacques, P., Hubert, J. B., Rikir, R., Thonart, P. *Appl. Biochem. Biotechnol.* 1991, *28/29*, 445.
65. Chartrain, M., Armstrong, J., Katz, L., Keller, J., Mathre, D., Greasham, R. *J. Ferment. Bioeng.* 1995, *80*, 176.
66. Patel, R. N., Banerjee, A., McNamee, C. G., Brzozowski, D., Hanson, R. L., Szarka, L. J. *Enzyme Microb. Technol.* 1993, *15*, 1014.
67. Patel, R. N., Robinson, R. S., Szarka, L. J., Kloss, J., Thottathil, J. K., Mueller, R. H. *Enzyme Microb. Technol.* 1991, *13*, 906.
68. Patel, R. N., Szarka, L. J., Mueller, R. H. *US Patent*, 1996, 5,559,017.
69. Patel, R. N., McNamee, C. G., Banerjee, A., Howell, J. M., Robinson, R. S., Szarka, L. J. *Enzyme Microb. Technol.* 1992, *14*,731.
70. Patel, R. N. *Adv. Appl. Microbiol.* 1997, *43*, 91.
71. Wong, C.-H., Whitesides, G. M. in *Enzymes in Synthetic Organic Chemistry*; Elsevier: New York, 1994; p. 252.
72. Schirch, L. *Adv. Enzymol. Relat. Areas Mol. Biol.* 1982, *53*, 83.
73. Miller, M. J. *Eur. Patent*, 1991, EP 460,883 A2.
74. Saeed, A., Young, D. W. *Tetrahedron* 1992, *48*, 2507.
75. Dotzlaf, J. E., Gazak, R. J., Kreuzman, A. J., Kroeff, E. P., Queener, S. W., Vicenzi, J. T., Yeh, J. M., Zock, J. M. *Eur. Patent*, 1994, EP 628,638 A2.
76. Vassilev, V. P., Uchiyama, T., Kajimoto, T., Wong, C.-H. *Tetrahedron Lett.* 1995, *36*, 4081.
77. Ura, D., Hashimukai, T., Matsumoto, T., Fukuhara, N. *Eur. Patent*, 1990, EP 421,477 A1.
78. Nikumaru, S., Fukuhara, N. *Jap. Patent*, 1996, JP 08084593.
79. Hatakeyama, K., Goto, M., Terasawa, M., Yugawa, H. *Jap. Patent*, 1997, JP 09028391.
80. Chmurny, A. B., Gross, A. T., Kupper, R. J., Roberts, R. L. *US Patent*, 1987, 4,710,583.
81. Chmurny, A. B., Gross, A. T., Kupper, R. J., Roberts, R. L. *US Patent*, 1989, 4,873,359.
82. Gross, A. T. *US Patent*, 1991, 5,002,872.
83. Bull, C., Durham, D. R., Gross, A., Kupper, R. J., Walter, J. F. in *Biocatalytic Production of Amino Acids and Derivatives*; Rozzell, J. D., Wagner, F., Eds; Carl Hanser Verlag: Munich, 1992; p. 241.
84. Christen, P., Metzler, D. E. in *Transaminases*; in series *Biochemistry*; Meister, A., Ed.; Wiley: New York, 1985; *Vol. 2*.
85. Crump, S. P., Rozzell, J. D. in *Biocatalytic Production of Amino Acids and Derivatives*; Rozzell, J. D., Wagner, F., Eds; Carl Hanser Verlag: Munich, 1992; p. 43.
86. Stirling, D. I., Zeitlin, A. L., Matcham, G. W. *US Patent*, 1990, 4,950,606.

87. Stirling, D. I. in *Chirality in Industry*; Collins, A. N., Sheldrake, G. N., Crosby, J., Eds; Wiley: Chichester 1992; p. 209.

88. Tanizawa, K., Asano, S., Masu, Y., Kuramitsu, S., Kagamiyama, H., Tanaka, H., Soda, K. *J. Biol. Chem.* 1989, *264*, 2450.

89. Taylor, P. P., Fotheringham, I. G. *Biochim. Biophys. Acta* 1997, *1350*, 38.

90. Rozzell, J. D. *US Patent,*1989, 4,826,766.

91. Ager, D. J., Fotheringham, I. G., Laneman, S. A., Pantaleone, D. P., Taylor, P. P. *Chim. Oggi* 1997, *15* (3/4), 11.

92. Then, J., Bartsch, K., Deger, H.-M., Grabley, S., Marquardt, R. *Eur. Patent*, 1987, EP 248,357.

93. Bartsch, K., Schneider, R., Schulz, A. *Appl. Environ. Microbiol.* 1996, *62*, 3794.

94. Meiwes, J., Schudok, M., Kretzschmar, G. *Tetrahedron: Asymmetry* 1997, *8*, 527.

95. Kretzchmar, G., Meines, J., Schudok, M., Hamann, P., Lerch, U., Grabley, S. *US Patent*, 1996, 5,480,786.

96. Godfrey, T., West, S. I. in *Industrial Enzymology*; Godfrey, T., West, S., Eds; Macmillan Press Ltd: London, 1996; p. 3.

97. Bentley, I. S., Williams, E. C. in *Industrial Enzymology*; Godfrey, T., West, S., Eds; Macmillan Press Ltd: London, 1996; p. 339.

98. Nagasawa, T., Yamada, H. *Pure Appl. Chem.* 1995, *67*, 1241.

99. Fallon, R. D., Stieglitz, B., Turner, I. *Appl. Microb. Biotechnol.* 1997, *47*, 156.

100. Crosby, J., Moilliet, J., Parratt, J. S., Turner, N. J. *J. Chem. Soc. Perkin Trans. I* 1994, 1679.

101. Meth-Cohn, O., Wang, M.-X. *J. Chem. Soc. Chem. Commun.* 1997, 1041.

102. Kamphuis, J., Boesten, W. H. J., Kaptein, B., Hermes, H. F. M., Sonke, T., Broxterman, Q. B., van den Tweel, W. J. J., Schoemaker, H. E. in *Chirality in Industry*; Collins, A. N., Sheldrake, G. N., Crosby, J., Eds; Wiley: Chichester, 1992; p. 187.

103. Powell, L. W. in *Industrial Enzymology*; Godfrey, T., West, S., Eds; Macmillan Press Ltd: London, 1996; p. 265.

104. Fité, M., Capellas, M., Benaiges, M. D., Caminal, G., Clapés, P., Alvaro, G. *Biocatalysis Biotransform.* 1997, *14*, 317.

105. Wong, C.-H., Whitesides, G. M. in *Enzymes in Synthetic Organic Chemistry*; Elsevier: New York, 1994; p. 41.

106. Collins, A. N., Sheldrake, G. N., Crosby, J. Eds. *Chirality in Industry II*; Wiley: Chichester, 1997.

107. Theil, F. *Chem. Rev.* 1995, *95*, 2203.

108. Lemke, K., Lemke, M., Theil, F. *J. Org. Chem.* 1997, *62*, 6268.

109. Kazlauskas, R. J., Weissfloch, A. N. E., Rappaport, A. T., Cuccia, L. A. *J. Org. Chem.* 1991, *56*, 2656.

110. Ader, U., Andersch, P., Berger, M., Goergens, U., Seemayer, R., Schneider, M. *Pure Appl. Chem.* 1992, *64*, 1165.

111. Jones, J. B. *Aldrichimica Acta* 1993, *26*, 105.

112. Provencher, L., Wynn, H., Jones, J. B., Krawczyk, A. R. *Tetrahedron: Asymmetery* 1993, *4*, 2025.

113. Cygler, M., Grochulski, P., Kazlauskas, R. J., Schrag, J. D., Bouthillier, F., Rubin, B., Serreqi, A. N., Gupta, A. K. *J. Am. Chem. Soc.* 1994, *116*, 3180.

114. Holzwarth, H.-C., Pleiss, J., Schmid, R. D. *J. Mol. Cat. B: Enzymatic* 1997, *3*, 73.

115. Nishizawa, K., Ohgami, Y., Matsuo, N., Kisida, H., Hirohara, H. *J. Chem. Soc. Perkin Trans. II* 1997, 1293.
116. Toone, E. J., Jones, J. B. *Tetrahedron: Asymmetry* 1991, *2*, 1041.
117. Kazlauskas, R. J. *TIBTECH* 1994, *12*, 464.
118. Kazlauskas, R. J., Weissfloch, A. N. E. *J. Mol. Cat. B: Enzymatic* 1997, *3*, 65.
119. Zuegg, J., Hönig, H., Schrag, J. D., Cygler, M. *J. Mol. Cat. B: Enzymatic* 1997, *3*, 83.
120. Ohno, M., Otsuka, M. *Org. Reactions* 1989, *37*, 1.
121. Tamm, C. *Pure Appl. Chem.* 1992, *64*, 1187.
122. Saksena, A. K., Girijavallabhan, V. M., Pike, R. E., Wang, H., Lovey, R. G., Liu, Y.-T., Ganguly, A. K., Morgan, W. B., Zaks, A. *US Patent*, 1995, 5,403,937.
123. Patel, R. N., Banerjee, A., Szarka, L. J. *J. Am. Oil Chem. Soc.* 1996, *73*, 1363.
124. Dodds, D. R., Heinzelman, C., Homann, M., Morgan, W. B., Previte, E., Roehl, R. A., Vail, R., Zaks, A. *Chim. Oggi* 1996, *14*, 9.
125. Roberge, C., Cvetovich, R. J., Amato, J. S., Pecore, V., Hartner, F. W., Greasham, R., Chartrain, M. *J. Ferment. Bioeng.* 1997, *83*, 48.
126. Morgan, B., Dodds, D. R., Zaks, A., Andrews, D. R., Klesse, R. *J. Org. Chem.* 1997, *62*, 7736.
127. Klibanov, A. M. *Acc. Chem. Res.* 1990, *23*, 114.
128. Mori, K. *Synlett* 1995, *11*, 1097.
129. Patel, R. N., Robison, R. S., Szarka, L. J. *Appl. Microb. Biotechnol* 1990, *34*, 10.
130. Ward, R. S. *Tetrahedron: Asymmetry* 1995, *6*, 1475.
131. Noyori, R., Tokunaga, M., Kitamura, M. *Bull. Chem. Soc. Jpn* 1995, *68*, 36.
132. Caddick, S., Jenkins, K. *Chem. Soc. Rev.* 1996, *25*, 447.
133. Stecher, H., Faber, K. *Synthesis* 1997, 1.
134. Turner, N. J., Winterman, J. R., McCague, R., Parratt, J. S., Taylor, S. J. C. *Tetrahedron Lett.* 1995, *36*, 1113.
135. Larsson, A. L. E., Persson, B. A., Bäckvall, J.-E. *Angew. Chem. Int. Ed. Engl.* 1997, *36*, 1211.
136. Thuring, J. W. J. F., Klunder, A. J. H., Nefkens, G. H. L., Wegman, M. A., Zwanenburg, B. *Tetrahedron Lett.* 1996, *37*, 4759.
137. Stürmer, R. *Angew. Chem. Int. Ed. Engl.* 1997, *36*, 1173.
138. Fruton, J. S. *Adv. Enzymol. Relat. Areas Mol. Biol.* 1982, *53*, 239.
139. Chaiken, I. M., Komoriya, A., Ohno, M., Widmer, F. *Appl. Biochem. Biotechnol.* 1982, *7*, 385.
140. Jakubke, H.-D., Kuhl, P., Könnecke, A. *Angew. Chem. Int. Ed. Engl* 1985, *24*, 85.
141. Kullmann, W. in *Enzymatic Peptide Synthesis*; CRC Press: Boca Raton, FL, 1987.
142. Heiduschka, P., Dittrich, J., Barth, A. *Pharmazie* 1990, *45*, 164.
143. Schellenberger, V., Jakubke, H.-D. *Angew. Chem. Int. Ed. Engl.* 1991, *30*, 1437.
144. Breddam, K. *Carlsberg Res. Commun.* 1986, *51*, 83.
145. Oyama, K. in *Biocatalysis in Organic Media*; Laane, C., Tramper, J., Lilly, M. D., Eds.; in series *Studies in Organic Chemistry*, Elsevier: Amsterdam, 1987; *Vol 29*, p. 209.
146. Pantaleone, D. P., Dikeman, R. N. in *Science for the Food Industry of the 21st Century*; Yalpani, M. Ed.; ATL Press: Mt. Prospect, IL, 1993; *Vol. 1*; p. 173.
147. Ager, D. J., Pantaleone, D. P., Henderson, S. A., Katritzky, A. R., Prakash, I., Walters, D. E. *Angew. Chem. Int. Ed. Engl.* 1998, *37*.

148. Sears, P., Wong, C.-H. *Biotechnol. Prog.* 1996, *12*, 423.
149. Imaoka, Y., Kawamoto, T., Ueda, M., Tanaka, A. *Appl. Microbiol. Biotechnol.* 1994, *40*, 653.
150. Moree, W. J., Sears, P., Kawashiro, K., Witte, K., Wong, C.-H. *J. Am. Chem. Soc.* 1997, *119*, 3942.
151. Wang, Y.-G., Yakovlevsky, K., Zhang, B., Margolin, A. L. *J. Org. Chem.* 1997, *62*, 3488.
152. Sekizaki, H., Itoh, K., Toyota, E., Tanizawa, K. *Tetrahedron Lett.* 1997, *38*, 1777.
153. Sekizaki, H., Itoh, K., Toyota, E., Tanizawa, K. *Chem. Pharm. Bull.* 1996, *44*, 1585.
154. Sergeeva, M. V., Paradkar, V. M., Dordick, J. S. *Enzyme Microb. Technol.* 1997, *20*, 623.
155. Chen, S.-T., Chen, S.-Y., Wang, K.-T. *J. Org. Chem.* 1992, *57*, 6960.
156. Zhang, X.-Z., Wang, X., Chen, S., Fu, X., Wu, X., Li, C. *Enzyme Microb. Technol.* 1996, *19*, 538.
157. Eichhorn, U., Bommarius, A. S., Drauz, K., Jakubke, H.-D. *J. Peptide Sci.* 1997, *3*, 245.
158. Werf, M. J. Y. D., van den Tweel, W. J. J., Kamphuis, J., Hartmans, S., de Bont, J. A. M. *TIBETCH* 1994, *12*, 95.
159. Collins, A. N., Sheldrake, G. N., Crosby, J. Eds; in *Chirality in Industry*; Wiley: Chichester, 1992.
160. Gerhartz, W. Ed. in *Enzymes in Industry*; VCH: Weinheim, 1990.
161. Renard, G., Guilleux, J.-C., Bore, C., Malta-Valette, V., Lerner, D. A. *Biotechnol. Lett.* 1992, *14*, 673.
162. Yanaka, M., Ura, D., Takahashi, A., Fukahara, N. *Jap. Patent*, 1994, JP 06113870 A2.
163. Wong, C.-H. *Science* 1989, *244*, 1145.
164. Wong, C.-H., Whitesides, G. M. in *Enzymes in Synthetic Organic Chemistry*; Elsevier: New York, 1994; p. 195.
165. Bednarski, M. D. in *Applied Biocatalysis*; Blanch, H. W., Clark, D. S., Eds; Marcel Dekker: New York, 1991; *Vol. 1*; p. 87.
166. Gefflaut, T., Blonski, C., Perie, J., Willson, M. *Prog. Biophys. Mol. Biol.* 1995, *63*, 301.
167. Wang, P. G., Fitz, W., Wong, C.-H. *CHEMTECH* 1995, *25*, 22.
168. Abeles, R. H., Frey, P. A., Jencks, W. P. in *Biochemistry*; Jones and Bartlett: Boston, 1992; p. 45.
169. Hupe, D. J. in *Enzyme Mechanisms*; Page, M. I., Williams, A., Eds; The Royal Society of Chemistry: London, 1987; p. 317.
170. Wong, C.-H., Liu, K. C. *World Patent*, 1992, WO 9221655.
171. Wong, C.-H., Sharpless, K. B. *World Patent*, 1995, WO 9516049.
172. Garcia-Junceda, E., Shen, G.-J., Sugai, T., Wong, C.-H. *Bioorg. Med. Chem.* 1995, *3*, 945.
173. Calton, G. J. in *Biocatalytic Production of Amino Acids and Derivatives*; Rozzell, J. D., Wagner, F., Eds; Carl Hanser Verlag: Munich, 1992; p. 59.
174. Ritter, S. K. *Chem. Eng. News*, 1995, *73*(9), 39.
175. Pedersen, S. *Bioprocess Technol.* 1993, *16*, 185.

176. Bhosale, S. H., Rao, M. B., Deshpande, V. V. *Microbiol. Rev.* 1996, *60*, 280.
177. Starnes, R. L., Kelly, R. M., Brown, S. H. *US Patent*, 1993, 5,219,751.
178. Quax, W. J., Luiten, R. G. M., Schuurhuizen, P. W., Mrabet, N. *US Patent*, 1994, 5,376,536.
179. Liao, W.-X., Earnest, L., Kok, S. L., Jeyaseelan, K. *Biochem. Mol. Biol. Int.* 1995, *36*, 401.
180. Frey, P. A. *Chem. Rev.* 1990, *90*, 1343.
181. Frey, P. A., Reed, G. H. *Adv. Enzymol. Relat. Areas Mol. Biol.* 1992, *66*, 1.
182. Fleming, P. E., Mocek, U., Floss, H. G. *J. Am. Chem. Soc.* 1993, *115*, 805.
183. Cardillo, G., Tomasini, C. *Chem. Soc. Rev.* 1996, *25*, 117.
184. Borman, S. *Chem. Eng. News* 1997, *75*(24), 32.
185. Bailey, J. E. *Science* 1991, *252*, 1668.
186. Stephanopoulos, G., Vallino, J. J. *Science* 1991, *252*, 167.
187. Stephanopoulos, G. *Curr. Opinion Biotechnol.* 1994, *5*, 196.
188. Krämer, R. *J. Biotechnol.* 1996, *45*, 1.
189. Cameron, D. C., Tong, I.-T. *Appl. Biochem. Biotechnol.* 1993, *38*, 105.
190. Flores, N., Xiao, J., Berry, A., Bolivar, F., Valle, F. *Nature Biotechnol.* 1996, *14*, 620.
191. Patnaik, R., Liao, J. C. *Appl. Environ. Microbiol.* 1994, *60*, 3903.
192. Patnaik, R., Spitzer, R. G., Liao, J. C. *Biotechnol. Bioeng.* 1995, *46*, 361.
193. Lu, J.-l., Liao, J. C. *Biotechnol. Bioeng.* 1997, *53*, 132.
194. Draths, K. M., Pompliano, D. L., Conley, D. L., Conley, D. L., Frost, J. W., Berry, A., Disbrow, G. L., Staversky, R. J., Lievense, J. C. *J. Am. Chem. Soc.* 1992, *114*, 3956.
195. Draths, K. M., Ward, T. L., Frost, J. W. *J. Am. Chem. Soc.* 1992, *114*, 9725.
196. Draths, K. M., Frost, J. W. *J. Am. Chem. Soc.* 1994, *116*, 399.
197. Frost, J. W., Lievense, J. *New J. Chem.* 1994, *18*, 341.
198. Keeler, R. *R&D Magazine*, 1991, *33*, 46.
199. Slater, S. C., Voige, W. H., Dennis, D. E. *J. Bacteriol.* 1988, *170*, 4431.
200. Schubert, P., Steinbüchel, A., Schlegel, H. G. *J. Bacteriol.* 1988, *170*, 5837.
201. Peoples, O. P., Sinskey, A. J. *J. Biol. Chem.* 1989, *264*, 15293.
202. Peoples, O. P., Sinskey, A. J. *J. Biol. Chem.* 1989, *264*, 15298.
203. Poirier, Y., Dennis, D. E., Klomparens, K., Somerville, C. *Science* 1992, *256*, 520.
204. John, M. E., Keller, G. *Proc. Natl. Acad. Sci. USA* 1996, *93*, 12768.
205. Steinbüchel, A., Schlegel, H. G. *Mol. Microbiol.* 1991, *5*, 535.
206. Lee, S. Y. *TIBTECH* 1996, *14*, 431.
207. Katz, L., Donadio, S. *Ann. Rev. Microbiol.* 1993, *47*, 875.
208. Hutchinson, C. R., Fujii, I. *Ann. Rev. Microbiol.* 1995, *49*, 201.
209. Chu, D. T. W., Plattner, J. J., Katz, L. *J. Med. Chem.* 1996, *39*, 3858.
210. Khosla, C., Caren, R., Kao, C. M., McDaniel, R., Wang, S.-W. *Biotechnol. Bioeng.* 1996, *52*, 122.
211. Khosla, C., Zawada, R. J. X. *TIBTECH* 1996, *14*, 335.
212. Donadio, S., Staver, M. J., McAlpine, J. B., Swanson, S. J., Katz, L. *Science* 1991, *252*, 675.
213. Donadio, S., McAlpine, J. B., Sheldon, P. J., Jackson, M., Katz, L. *Proc. Natl. Acad. Sci. USA* 1993, *90*, 7119.

214. Kao, C. M., Luo, G., Katz, L., Cane, D. E., Khosla, C. *J. Am. Chem. Soc.* 1995, *117*, 9105.

215. Pieper, R., Ebert-Khosla, S., Cane, D., Khosla, C. *Biochemistry* 1996, *35*, 2054.

216. Tsukamoto, N., Chuck, J.-A., Luo, G., Kao, C. M., Khosla, C., Cane, D. E. *Biochemistry* 1996, *35*, 15244.

217. McDaniel, R., Kao, C. M., Fu, H., Hevezi, P., Gustafsson, C., Betlach, M., Ashley, G., Cane, D. E., Khosla, C. *J. Am. Chem. Soc.* 1997, *119*, 4309.

218. Kao, C. M., Luo, G., Katz, L., Cane, D. E., Khosla, C. *J. Am. Chem. Soc.* 1996, *118*, 9184.

219. Jacobsen, J. R., Hutchinson, C. R., Cane, D. E., Khosla, C. *Science* 1997, *277*, 367.

220. Pieper, R., Gokhale, R. S., Luo, G. Cane, D. E., Khosla, C. *Biochemistry* 1997, *36*, 1846.

221. Kieslich, K., van der Beek, C. P., de Bont, J. A. M., van den Tweel, W. J. J., Eds. *New Frontiers in Screening for Microbial Biocatalysts*; Elsevier: Amsterdam, 1997.

222. Cheetham, P. S. J. *Enzyme Microb. Technol.* 1987, *9*, 194.

223. Cheetham, P. S. J. *TIBTECH* 1993, *11*, 478.

224. Lorraine, K., King, S., Greasham, R., Chartrain, M. *Enzyme Microb. Technol.* 1996, *19*, 250.

225. Patel, R. N., Banerjee, A., Szarka, L. J. *J. Am. Oil Chem. Soc.* 1995, *72*, 1247.

226. Takahashi, E., Nakamichi, K., Furui, M. *J. Ferment. Bioeng*, 1995, *80*, 247.

227. Linko, Y.-Y., Lamsa, M., Huhtala, A., Rantanen, O. *J. Am. Oil Chem. Soc.* 1995, *72*, 1293.

228. Zhang, J., Reddy, J., Roberge, C., Senanayake, C., Greasham, R., Chartrain, M. *J. Ferment. Bioeng.* 1995, *80*, 244.

229. Chung, W., Goo, Y. M. *Arch. Pharm. Res.* 1988, *11*, 139.

230. Xu, J., Wutong, J., Jin, S., Yao, W. *Zhongguo Yaoke Daxue Xuebao* 1994, *25*, 53.

231. Nakayama, A., Torigoe, Y. *Jap. Patent,* 1987, JP 63185395.

232. Syldatk, C., Muller, R., Siemann, M., Krohn, K., Wagner, F. in *Biocatalytic Production of Amino Acids and Derivatives*; Rozzell, J. D., Wagner, F., Eds; Carl Hanser Verlag: Munich, 1992: p. 75.

233. Syldatk, C., Muller, R., Pietzsch, M., Wagner, F. in *Biocatalytic Production of Amino Acids and Derivatives*; Rozzell, J. D., Wagner, F., Eds; Carl Hanser Verlag: Munich, 1992; p. 129.

234. Keil, O., Schneider, M. P., Rasor, J. P. *Tetrahedron: Asymmetry* 1995, *6*, 1257.

235. Garcia, M. J., Azerad, R. *Tetrahedron: Asymmetry* 1997, *8*, 85.

14

Industrial Applications of Chiral Auxiliaries

NSC Technologies, Mount Prospect, Illinois

14.1. INTRODUCTION

Chiral auxiliaries have traditionally been used in the academic and small-scale synthetic arenas. This, in part, has been due to the infancy of chiral auxiliaries themselves in organic synthesis. Many of these chiral substrates have been developed only in the last 10–20 years, and their synthetic utility is only now being realized. Additionally, many chiral auxiliaries are difficult to prepare and handle at large scale. Only recently has their been considerable effort to produce large quantities of the auxiliaries. Another important consideration is the cost of the auxiliary compared with alternate methods such as resolution. The experimental reaction conditions for some of the more common chiral auxiliaries often call for specialized equipment and/or extreme temperature conditions. These factors, to date, have combined to limit the use of chiral auxiliaries on an industrial scale.

This chapter summarizes the application of chiral auxiliaries at a scale important to the chiral fine chemical business. In addition, and as important, there are numerous cases in which a chiral auxiliary has been used in the early stages of pharmaceutical development and the potential exists for this approach to be used in future work at larger scale. Lastly, there exists a limited number of cases in which the auxiliary exists as part of an active drug. The known industrial applications, the potential applications, and the presence of auxiliaries in active drugs are all addressed.

14.1.1. General Chiral Auxiliary Review

By definition, a chiral auxiliary differs from a chiral template in that the auxiliary is capable of being recycled after the desired asymmetric reaction. Hence, chiral templates are not included in this chapter. The uses of a chiral moiety as a ligand for a reagent have also been excluded.

A number of reviews exist on the formation and uses of chiral auxiliaries [1, 3–7]. Some of these auxiliaries, such as oxazolines, can be used on their own or incorporated into chiral ligands for asymmetric transition metal catalyzed synthesis [7].

Many chiral auxiliaries are derived from 1,2-amino alcohols [7], including oxazolidinones (1) [7–9], oxazolines (2) [10,11], bis-oxazolines (3) [12,13], oxazinones (4) [14], and oxazaborolidines (5) [15–17]. Even the 1,2-amino alcohol itself can be used as a chiral auxiliary [18–22]. Other chiral auxiliary examples include camphorsultams (6) [23], piperazinediones (7) [24], (S)-1-amino-2-(methoxymethyl)-pyrrolidine (SAMP) (8) and (R)-1-amino-2-(methoxymethyl)-pyrrolidine (RAMP) [25], chiral boranes such as isopinocampheylborane (9) [26], and tartaric acid esters (10). Some of these auxiliaries have been used as ligands in reagents, such as **3** and **5,** whereas others have only been used at laboratory scale (e.g., **6** and **7**).

14.2. CHIRAL AUXILIARY STRUCTURES IN PHARMACEUTICALS

Chiral auxiliaries can serve not only as a chiral template, but the structural motif can be incorporated into an active pharmaceutical. Although the chemistry of the chiral unit can be exploited, this is not a true application of an auxiliary as the unit becomes part of the final molecule.

The oxazolidinone-based compounds are the most common (Figure 1), and examples include antibacterial [27,28], adhesion receptor antagonists [29,30], tumor rectosis factor tumor necrosis factor (TNF) inhibitors [31], platelet aggregation inhibitors [32], antimigraine drugs [33], and monoamine oxidase inhibitors [34,35]. In one case, an oxazolidinone (SR-38) has been used as a transdermal drug delivery agent [36].

Oxazoles and isooxazoles are also found in many active drugs; although not as prevalent compared with the oxazolidinones, examples include antiinflammatories [37,38] and nitric oxide–associated diseases [39].

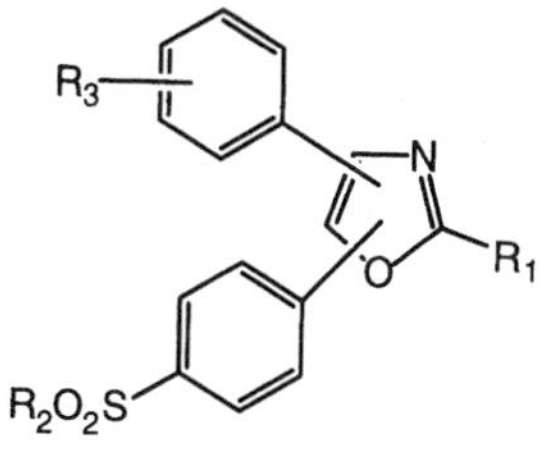

Searle's antiinflammatory

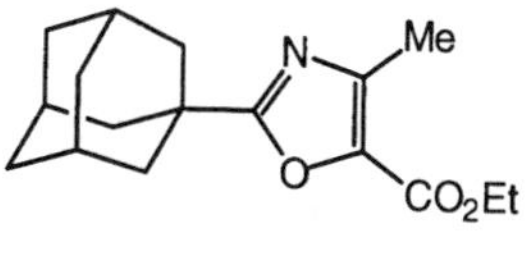

Adir et Cie's antiinflammatory

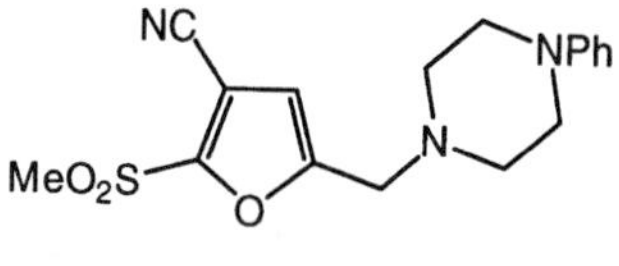

Takeda NO production inhibitor

Although this application area is intriguing, the focus of this chapter remains on the broader application of chiral auxiliaries for asymmetric synthesis.

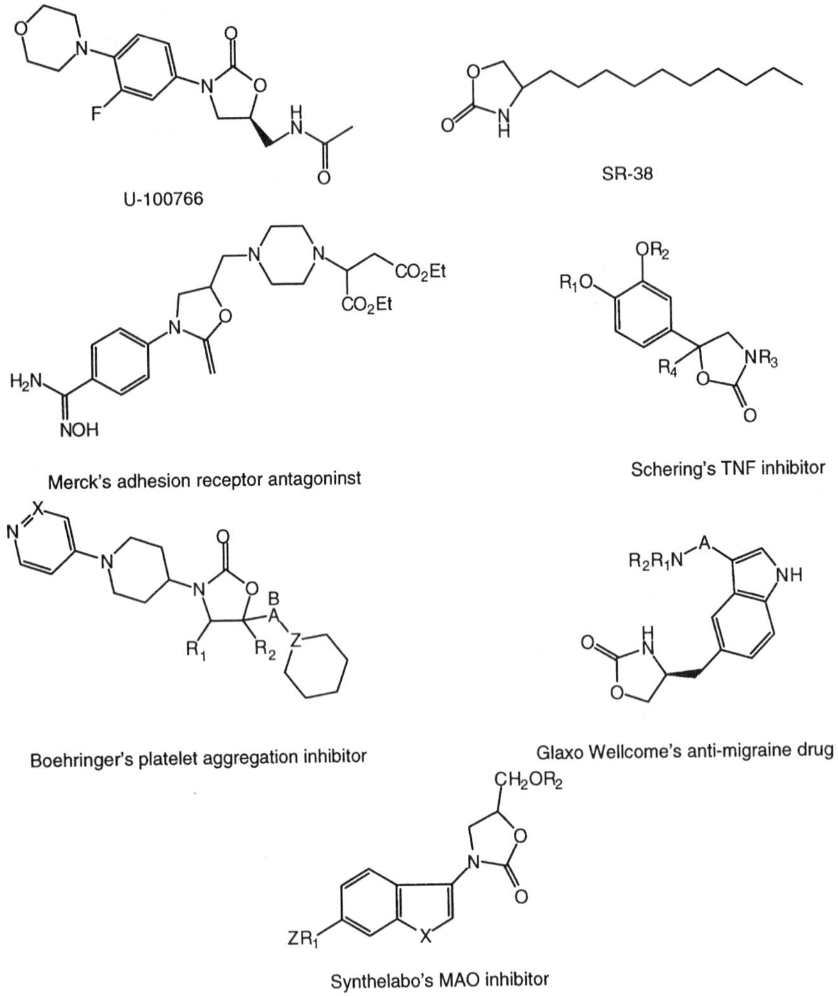

U-100766

SR-38

Merck's adhesion receptor antagoninst

Schering's TNF inhibitor

Boehringer's platelet aggregation inhibitor

Glaxo Wellcome's anti-migraine drug

Synthelabo's MAO inhibitor

FIGURE 1   Examples of oxazolidinones as part of the structures of active pharmaceuticals.

14.3.   APPLICATION OF CHIRAL AUXILIARIES IN INDUSTRY

Despite the relatively recent advent of chiral auxiliaries, they have some limited use at multikilogram industrial scale.

14.3.1. Zambon Process

Tartaric acid has been used as a chiral auxiliary in a patented route to (S)-na-proxen (**11**) (Scheme 1) [40–44]. In initial studies, the acetal **12** was used to allow stereoselective bromination that resulted in a 91:9 ratio of the (RRS)- and (RRR)-bromo derivatives (**13** and **14**). The bromo acetal diesters could be completely separated. Debromination of **13** followed by acid hydrolysis led to formation of (S)-naproxen in 80% yield, > 99% enantiomeric excess (ee), and recovery of the auxiliary. Conversely, debromination and hydrolysis of **14** gave only 12% yield of (R)-naproxen and 86% ee. In this case, the hydroxy acetal **15** was the major product (68%). However, the auxiliary was recovered in enantiomerically pure form.

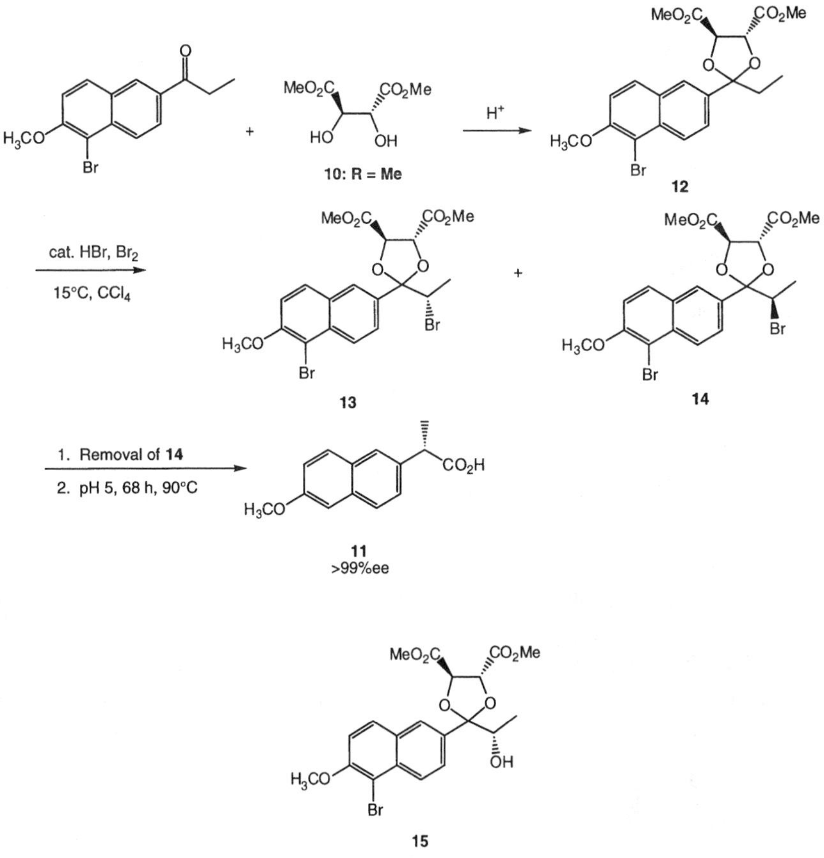

SCHEME 1.

14.3.2. Lilly Route to CI-1008

An anticonvulsant, CI-1008,(16), was recently developed using the Evans' oxa-zolidinone auxiliary 17 derived from (1S,2R)-(−)-norephedrine [45]. The route was scaled up to produce several hundred kilogram quantities at production and pilot plant scale and provided early clinical trial material (Scheme 2).

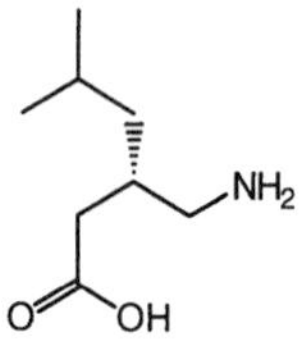

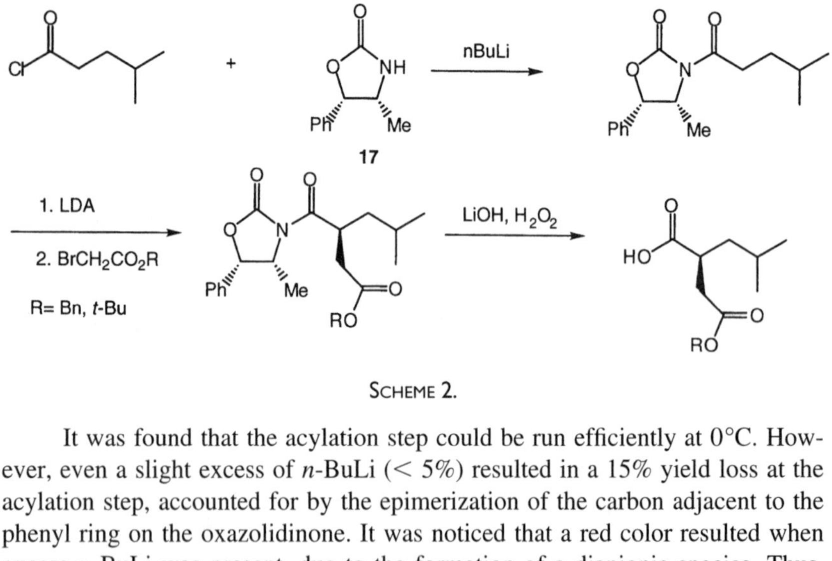

SCHEME 2.

It was found that the acylation step could be run efficiently at 0°C. How-ever, even a slight excess of n-BuLi (< 5%) resulted in a 15% yield loss at the acylation step, accounted for by the epimerization of the carbon adjacent to the phenyl ring on the oxazolidinone. It was noticed that a red color resulted when excess n-BuLi was present, due to the formation of a dianionic species. Thus, stopping the addition when only the characteristic yellow color was present main-tained the normally high yields (95%) for this acylation.

The alkylation step could be run as high as −35°C, but above this tempera-ture a side product formed as a result of decomposition of the lithium enolate through a ketene that underwent alkylation with tert-butyl bromoacetate (Scheme 3).

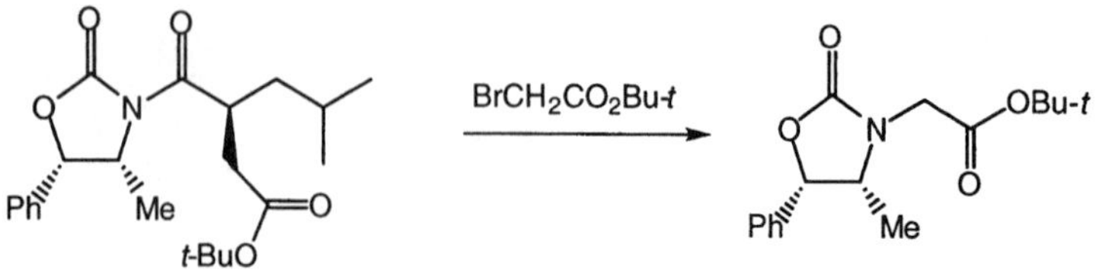

SCHEME 3.

Overall recovery of the auxiliary was 64%. Upon cost estimates for scale-up work, it was found that the aforementioned chiral auxiliary route exceeded the desired cost by a factor of six, primarily because of the cost of the chiral oxazolidinone and a yield of 33% over 10 steps. The route was ultimately replaced with a classical resolution protocol using mandelic acid.

14.4. POTENTIAL APPLICATIONS OF CHIRAL AUXILIARIES

There are numerous examples in which the potential for the application of chiral auxiliaries, once again primarily oxazolidinones, exists for an industrial setting.

The synthesis of calcium channel blockers of the diltiazem group, effective in lowering blood pressure, has been reported by Hoffmann-LaRoche (Scheme 4) [46]. A screening of several auxiliaries yielded acceptable ee, but it was found that the use of (1R,2S)-2-phenylcyclohexanol (18) gave the required diastereoisomer 19 as the major isomer [47–49]. The intermediate could be isolated readily and the auxiliary recycled simply by base hydrolysis. Multikilogram quantities of enantiomerically pure naltiazem (20) and diltiazem (21) were produced by this method.

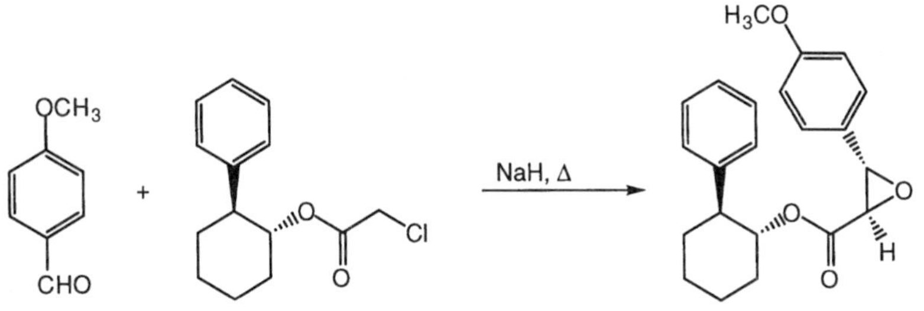

18   **19**

SCHEME 4.

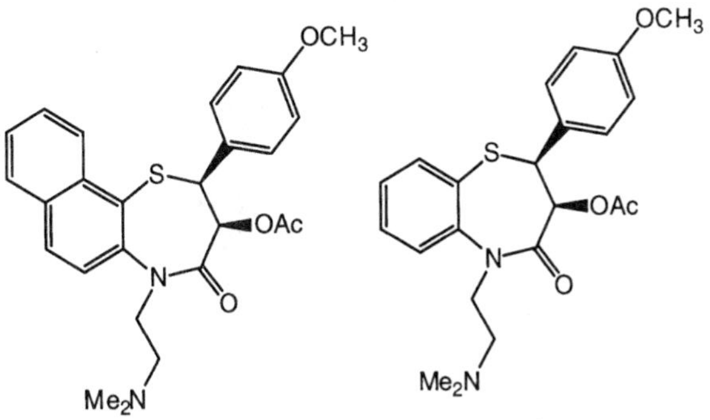

20   **21**

Alternatively, diltiazem (21) has been prepared using the Evans auxiliary 22 derived from L-valine (Scheme 5) [50]. After dehydration of the adduct from the condensation of 22 with anisaldehyde through the mesylate, the enol ether was formed with a $Z:E$ ratio of 4:1. This imide was then treated with 2-aminothiophenol in the presence of 0.1 equivalents of 2-aminothiophenoxide with no change in the isomer ratio. The auxiliary was removed with trimethylaluminum, with concomitant formation of the lactam. After separation by crystallization, the correct diastereoisomer was converted to diltiazem in > 99% ee.

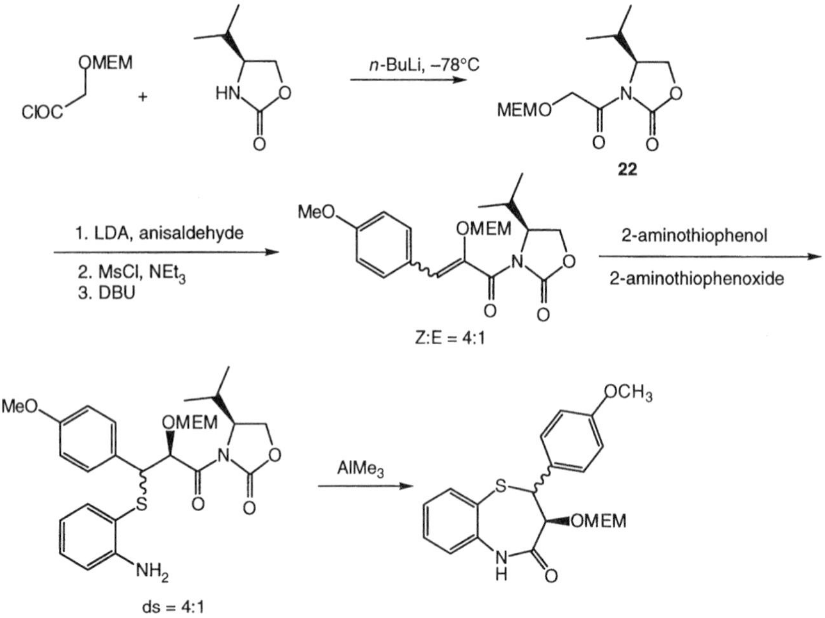

SCHEME 5.

Eli Lilly & Co. has used an Evans auxiliary in its synthesis of LY309887 (23), a dideazafolate antitumor agent (Scheme 6) [2]. The diastereomeric excess (de) of the key step, reaction of the titanium-derived enolate, was > 98%. It was noted that the auxiliary could be recycled after cleavage with lithium borohydride.

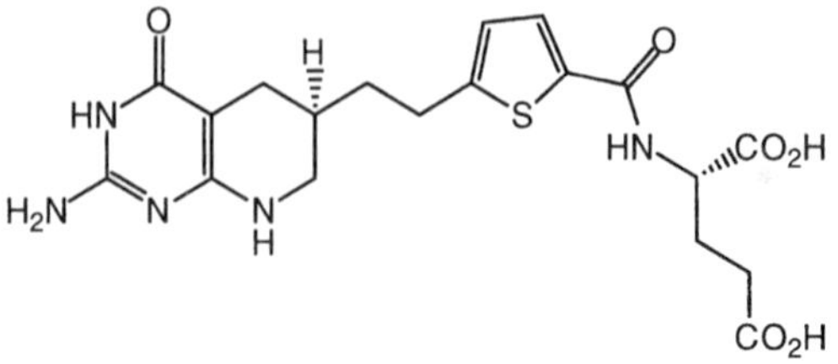

23

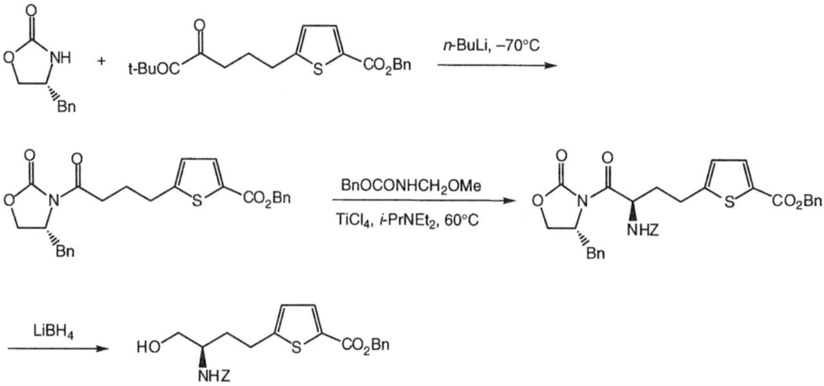

SCHEME 6.

Schering-Plough Corp. has used an oxazolidinone in its synthesis of the cholesterol adsorption inhibitor (+)-SCH 54016 (**24**) (Scheme 7) [51]. Condensation of an acid chloride intermediate with the (*R*)-phenylglycine-derived Evans auxiliary followed by reaction of the titanium enolate with *N*-(4-methoxybenzylidene)aniline gave the intermediate **25**. This was silylated and treated with TBAF, resulting in removal of the auxiliary and cyclization to form the 2-azetidinone ring. The stereochemistry was exclusively *trans*. The azetidinone was then converted to the bromide, followed by intramolecular alkylation to form SCH 50416 in > 99% ee.

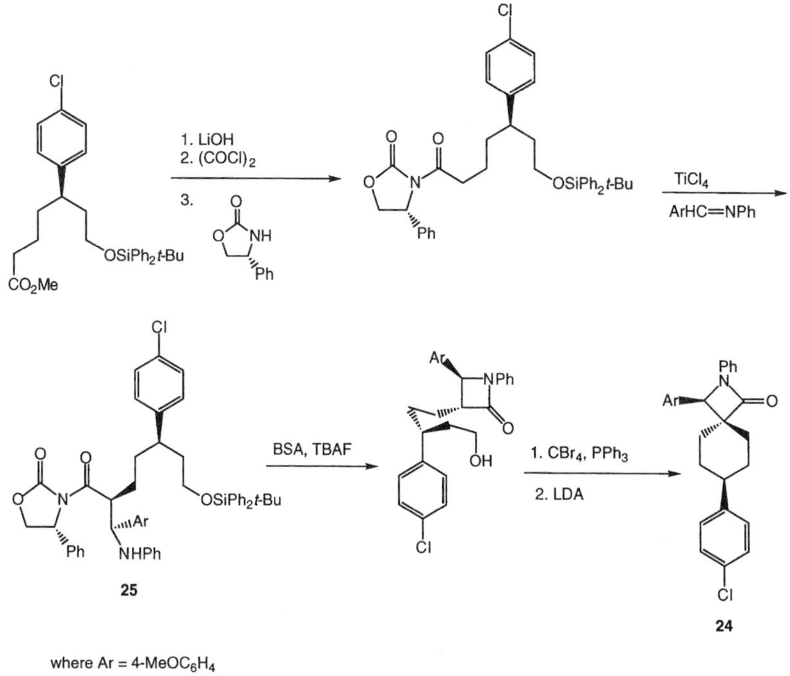

where Ar = 4-MeOC₆H₄

SCHEME 7.

Bio-Mega/Boehringer Ingelheim has used an oxazolidinone auxiliary in the synthesis of renin inhibitors (e.g., **26**) for the treatment of hypertension and congestive heart failure [52]. A multi-step derivitization of the oxazolidinone **27** from reaction of (S)-4-(1-methylethyl)-2-oxazolidinone and 4-bromo-4-pentenoic acid yielded the desired compound.

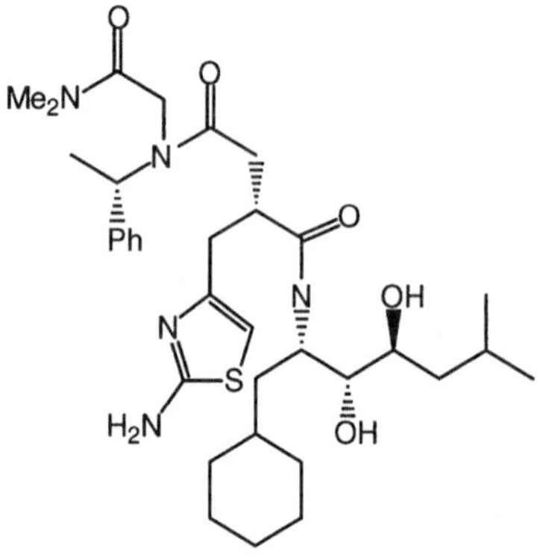

26 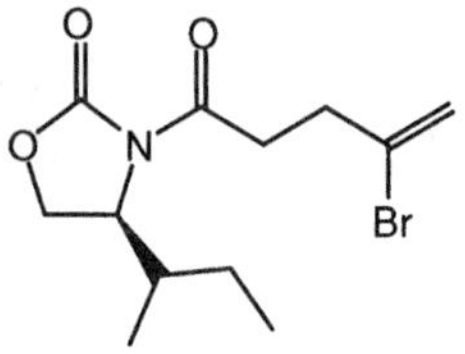   27

The optically active alkaloid Bao Gong Teng A (**28**), used to treat glaucoma, has been prepared with methyl (S)-lactate as the auxiliary. The key step was the asymmetric 1,3-dipolar cycloaddition of a 3-hydroxypyridinium betaine with the acrylate **29**, where the auxiliary is attached directly to the dienophile (Scheme 8). The auxiliary was removed after several steps, yielding the desired product in optically pure form, but in only 9% overall yield [53,54].

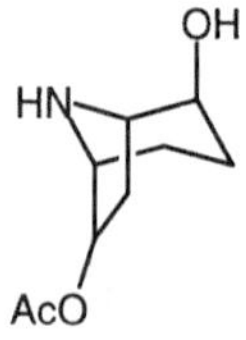

27

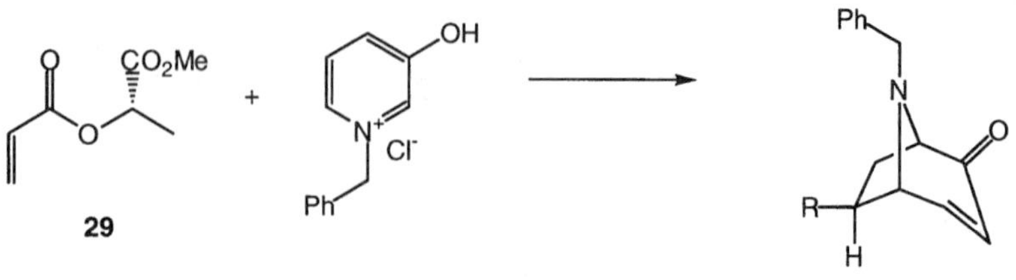

29

SCHEME 8.

(S)-Proline has been used in the synthesis of (R)-etomoxir (**30**), a powerful hypoglycemic agent (Scheme 9) [55]. The proline was used for the diastereoselective bromolactonization step. After hydrolysis of the auxiliary and further functional group transformations, the desired product was obtained in an overall yield of 45%.

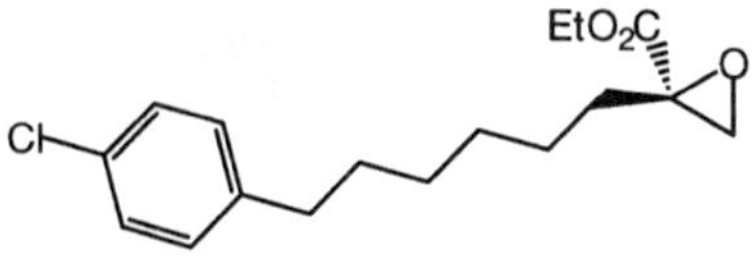

30

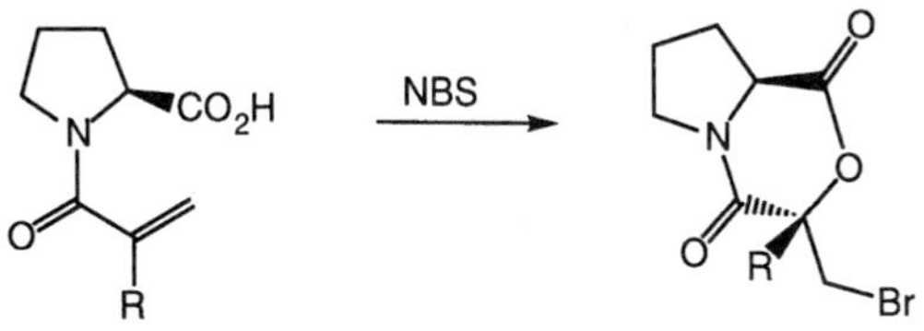

R= -(CH$_2$)$_6$OBn

Scheme 9.

Merck & Co., Inc., has used the L-phenylalanine-derived Evans oxazolidinone to make matrix metalloproteinase inhibitors such as **31** (Scheme 10) [56]. The mixed anhydride of 4-butyric acid was reacted with the lithium anion of the oxazolidinone. This was enolized with the standard titanium reagents. An enantioselective Michael addition was then performed by the addition of *t*-butyl acrylate at low temperature. The auxiliary was removed with LiOH/peroxides to give the acid, which was further derivatized over multiple steps to yield the desired drug.

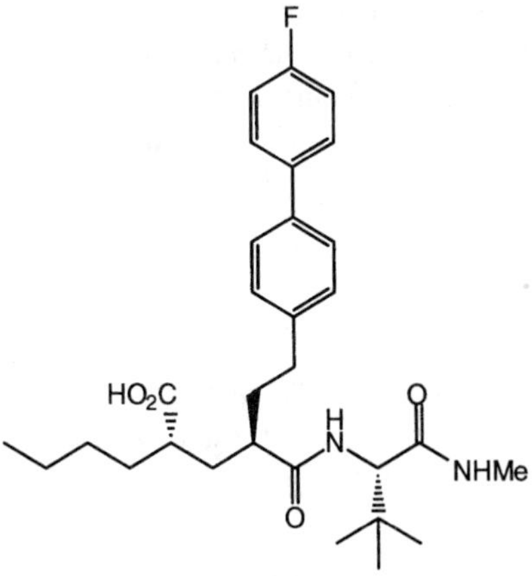

31

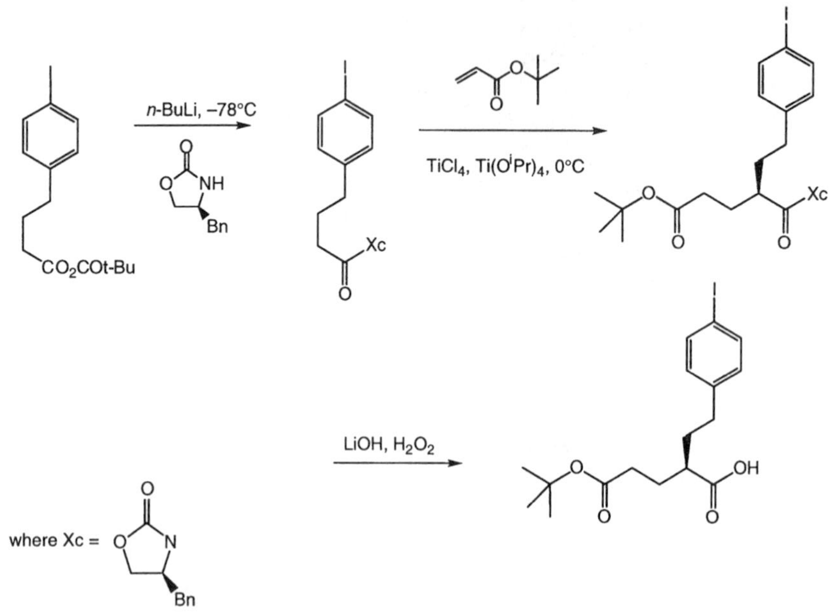

SCHEME 10.

14.5. CONCLUSIONS

The use of chiral auxiliaries has and will continue to see an ever-increasing use in an industrial setting. Heretofore, the use of such auxiliaries was limited by their price, availability, and novelty to large-scale production. The synthetic utility of many of the auxiliaries, particulary the oxazolidinones, is just now being realized. Several of these auxiliaries are just now becoming commercially available, which in turn should make them more competitive with alternate production methods currently in practice. With the demand for producing single enantiomer pharmaceuticals continuing to increase, the future industrial use of chiral auxiliaries appears promising.

REFERENCES

1. Gao, Y., Boschetti, E., Guerrier, L. *Ann. Pharm. Rev.* 1994, *52*, 184.
2. Barnett, C. J., Wilson, T. M., Evans, D. A., Somers, T. C. *Tetrahedron Lett.* 1997, *38*, 735.
3. Blaser, H.-U. *Chem. Rev.* 1992, *92*, 935.

4. Meyers, A. I. in *Stereocontrolled Organic Synthesis;* Trost, B. M., Ed.; Blackwell: Oxford, 1994; p. 144.

5. Foskolos, G. *Pharmakeutike* 1995, *8,* 155.

6. Marchand-Brynaert, J. *Belg. Chim. Nouv.* 1991, *8,* 155.

7. Ager, D. J., Prakash, I., Schaad, D. R. *Chem. Rev.* 1996, *96,* 835.

8. Evans, D. A. *Aldrichimica Acta* 1982, *15,* 23.

9. Swern, D., Dyen, M. E. *Chem. Rev.* 1967, *67,* 197.

10. Gant, T. G., Meyers, A. I. *Tetrahedron* 1994, *50,* 2297.

11. Reuman, M., Meyers, A. I. *Tetrahedron* 1985, *41,* 837.

12. Bolm, C. *Angew. Chem. Int. Ed. Engl.* 1991, *30,* 542.

13. Pfaltz, A. *Acc. Chem. Res.* 1993, *26,* 339.

14. Williams, R. M. *Aldrichimica Acta* 1992, *25,* 11.

15. Corey, E. J. *US Patent,* 1990, 4,943,635.

16. Itsuno, S., Sakurai, Y., Ito, K., Hirao, A., Nakahama, S. *Bull. Chem. Soc. Jpn* 1987, *60,* 395.

17. Itsuno, S., Nakano, M., Miyazaki, K., Masuda, H., Ito, K., Hirao, A., Nakahama, S. *J. Chem. Soc. Perkin Trans. I* 1985, 2039.

18. Meyers, A. G., Gleason, J. L., Yoon, T. *J. Am. Chem. Soc.* 1995, *117,* 8488.

19. Meyers, A. G., Yang, B. H., Chen, H., McKinistry, L., Kopecky, D., Gleason, J. L. *J. Am. Chem. Soc.* 1997, *119,* 6496.

20. Meyers, A. G., Yang, B. H., Chen, H., Gleason, G. L. *J. Am. Chem. Soc.* 1994, *116,* 9361.

21. Larcheveque, M., Ignatova, E., Cuvigny, T. *Tetrahedron Lett.* 1978, *19,* 3961.

22. Meyers, A. G., Gleason, J. L., Yoon, T., Kung, D. W. *J. Am. Chem. Soc.* 1997, *119,* 656.

23. Oppolzer, W., Moretti, R., Thomi, S. *Tetrahedron Lett.* 1989, *30,* 5603.

24. Schoellkopf, U. *Pure Appl. Chem.* 1983, *55,* 1799.

25. Enders, D., Papadopoulos, K., Rendenbach, B. E. M. *Tetrahedron Lett.* 1986, *27,* 3491.

26. Brown, H. C., Singaram, B. *J. Am. Chem. Soc.* 1984, *106,* 1797.

27. Brickner, S. J., Hutchinson, D. K., Barbachyn, M. R., Manninen, P. R., Ulanowicz, D. A., Garmon, S. A., Grega, K. C., Hendges, S. K., Toops, D. S., Ford, C. W., Zurenko, G. E. *J. Med. Chem.* 1996, *39,* 673.

28. Barbachyn, M. R., Hutchinson, D. K., Brickner, S. J., Cynamon, M. H., Kilburn, J. O., Klemens, S. P., Glickman, S. E., Grega, K. C., Hendges, S. K., Toops, D. S., Ford, C. W., Zurenko, G. E. *J. Med. Chem.* 1996, *39,* 680.

29. Gante, J., Juraszyk, H., Raddatz, P., Wurziger, H., Bernotat-Danielowski, S., Melzer, G. *Eur. Patent,* 1996, EP 96-101746.

30. Gante, J., Juraszyk, H., Raddatz, P., Wurziger, H., Bernotat-Danielowski, S., Melzer, G. *Eur. Patent,* 1996, EP 96-106423.

31. Graf, H., Wachtel, H., Schneider, H., Faulds, D., Perez, D., Dinter, H. *World Patent,* 1996, WO 96-DE257.

32. Tsaklakidis, C., Schaefer, W., Doerge, L., Esswein, A., Friebe, W.-G. *Ger. Offen.* 1995, DE 95-19524765.

33. Patel, R. *World Patent,* 1996, WO 96-GB1886.

34. Jegham, S., Puech, F., Burnier, P., Berthon, D. *World Patent*, 1996, WO 96-FR1732.
35. Jegham, S., Puech, F., Burnier, P., Berthon, D., Leclerc, O. *World Patent*, 1996, WO 96-FR1511.
36. Rajadhyaksha, V. J., Pfister, W. R. *Drug Cosmetic Ind.* 1996, *157,* 36.
37. de Nanteuil, G., Vincetn, M., Lila, C., Bonnet, J., Fradin, A. *Eur. Patent*, 1993, EP 93-402967.
38. Talley, J. J., Bertenshaw, S., Rogier, D. J., Graneto, M., Brown, D. L., Devadas, D., Lu, H.-F., Sikorski, J. A. *World Patent*, 1996, WO 96-US6992.
39. Sugihara, Y., Uchibayashi, N., Matsumura, K., Nozaki, Y., Ichimori, Y. *World Patent*, 1996, WO 96-JP3857.
40. Castaldi, G., Giordano, C. *Synthesis* 1987, 1039.
41. Crosby, J. *Tetrahedron* 1991, *47,* 4789.
42. Giordano, C., Castaldi, G., Cavicchioli, S. *US Patent*, 1989, 4,888,433.
43. Giordano, C., Cataldi, G., Cavicchioli, S., Villa, M. G. *Tetrahedron* 1989, *45,* 4243.
44. Giordano, C., Castaldi, C., Uggeri, F., Calvicchioli, S. *US Patent*, 1989, 4,855,464.
45. Hoekstra, M. S., Sobieray, D. M., Schwindt, M. A., Mulhern, T. A., Grote, T. M., Huckabee, B. K., Hendrickson, V. S., Franklin, L. C., Granger, E. J., Karrick, G. L. *Org. Process R. & D.* 1997, *I,* 26.
46. Schwartz, A., Madan, P. B., Mohacsi, E., O'Brien, J. P., Todaro, L. J., Coffen, D. L. *J. Org. Chem.* 1992, *57,* 851.
47. Oppolzer, W., Chapuis, C., Kelly, M. J. *Helv. Chim. Acta.* 1983, *66,* 2358.
48. Helmchen, G., Selim, A., Dorsch, D., Taufer, I. *Tetrahedron Lett.* 1983, *24,* 3213.
49. Whitesell, J. K., Chen, H. C., Lawrence, R. M. *J. Org. Chem.* 1986, *51,* 4663.
50. Miyata, O., Shinada, T., Ninomiya, N., Naito, T. *Tetrahedron* 1997, *53,* 2421.
51. Chen, L.-Y., Zaks, A., Chackalamannil, S., Dugar, S. *J. Org. Chem.* 1996, *61,* 8341.
52. Lavallee, P., Simoneau, B. *US Patent*, 1993, 5,541,163.
53. Pham, V. C., Charlton, J. L. *J. Org. Chem.* 1995, *60,* 8051.
54. Jung, M. E., Longmei, Z., Tangsheng, P., Huiyan, Z., Yan, L., Jingyu, S. *J. Org. Chem.* 1992, *57,* 3528.
55. Jew, S., Kim, H., Jeong, B., Park, H. *Tetrahedron: Asymmetry* 1997, *8,* 1187.
56. Esser, C. K., Bugianesi, R. L., Caldwell, C. G., Chapman, K. T., Durette, P. L., Girotra, N. N., Kopka, I. E., Lanza, T. J., Levorse, D. A., MacCoss, M., Owens, K. A., Ponpipom, M. M., Simeone, J. P., Harrison, R. K., Niedzwiecki, L., Becker, J. W., I., M. A., Axel, M. G., Christen, A. J., McDonnell, J., Moore, V. L., Olszewski, J. M., Saphos, C., Visco, D. M., Shen, F., Colletti, A., Krieter, P. A., Hagmann, W. K. *J. Med. Chem.* 1997, *40,* 1026.

15

Synthesis of Unnatural Amino Acids: Expansion of the Chiral Pool

DAVID J. AGER, DAVID R. ALLEN, MICHAEL B. EAST, IAN G. FOTHERINGHAM, SCOTT A. LANEMAN, WEIGUO LIU, DAVID P. PANTALEONE, DAVID R. SCHAAD, and PAUL P. TAYLOR
NSC Technologies, Mount Prospect, Illinois

15.1. INTRODUCTION

Although nature has been the primary source of chiral materials through extraction processes, which may also include a bioprocess, modern asymmetric synthetic methodology is now in the position to provide cheap, pure, chiral materials at scale. Thus, the chiral pool can be expanded by synthesis of the stereogenic center. Rather than duplicate sections of this book, we have chosen to discuss the problems associated with the synthesis of unnatural amino acids at various scales. This illustrates that a single, cheap method need not fulfill all of the criteria to provide a chiral pool material to a potential customer—a number of approaches are required (see also Chapter 2).

15.1.1. Unnatural Amino Acid Synthesis

A variety of methods have been described in this book for the preparation of amino acids. They range from the use of chiral auxiliaries (see Chapter 14), to the use of enzymes (see Chapters 13 and 16) and asymmetric hydrogenation (see Chapter 9), to total fermentation (see Chapter 4). The aim of this chapter is to put into perspective the choice of the appropriate method. We have chosen the synthesis of unnatural amino acids for illustration, but the philosophy could be applied to any class of compound.

The majority of chiral fine chemicals produced at present are used in pharmaceutical candidates and products. During the development of a compound as a drug candidate, the factors that govern the needs for intermediates change (see Chapter 2). Methodology must be available to make material at a relatively small scale, but which can be implemented rapidly. In addition, there must be a low probability of failure because the change to a second method would be very time consuming. As the material needs increase, then cost becomes more of a factor, but the time scale lengthens, allowing for the development of cheaper processes. However, there is still a reasonable chance that the drug candidate will not pass through the testing phase, and investment into method development must be made prudently. When commercialization of the compound is close and requirements become large, then cost is the overriding factor, and the manufacturing processes discussed elsewhere in this book and in other sources are paramount [1,2].

Fine chemical companies are providing the resources for pharmaceutical companies to "outsource" chiral intermediates. Many of the strengths of these fine chemical companies are highlighted in this book. Some have concentrated on specific chiral methodologies, whereas others have developed portfolios of methods. Here we discuss our experience and methods that allow NSC Technologies to provide a wide range of unnatural amino acids at a variety of scales, although our emphasis is on large-volume materials.

15.2. The Choice of Approach

Unnatural amino acids are becoming more important as a starting material for synthesis because of their expanded usage in pharmaceutical drug candidates. This increase has been spurred by a number of factors. Rational drug design has allowed natural substrates of an enzyme to be structurally modified without the need for a large screening program. Because many of these substrates are polypeptides, substitution of unnatural amino acids for the native component is a relatively small step. In contrast, many combinatorial approaches rely on the production of a large number of potential compounds that are then screened. With a wide range of reactions available, the history of chemistry performed on solid supports, and unnatural amino acids extending the library of amino acids to almost boundless limits, these building blocks are being used in a wide range of applications.

Thus, the need has arisen for larger amounts of unnatural amino acids—those that contain unusual side chains or are in the D-series. As α-amino acids contain an epimerizable center, D-amino acids are usually accessible through epimerization of the natural isomer, followed by a resolution. Of course, the racemic mixture can also be accessed by synthesis. Because resolution can be either wasteful—if the undesired isomer is discarded—or clumsy, when the other isomer is recycled through an epimerization protocol, many large-scale methods now rely on a dynamic resolution in which all of the starting material is converted to the desired isomer (see Chapter 8). With the advent of asymmetric reactions that can be performed at large scale, a substrate can now be converted to the required stereoisomer without the need for any extra steps associated with a resolution approach.

The provision of intermediates to the pharmaceutical, agrochemical, and other industries requires a number of robust synthetic methods to be available for the preparation of a single class of compounds. The choice of which method to use is determined by the scale of the reaction, economics for a specific run or campaign, ease of operation, time frame, and the availability of equipment. There are a large number of methodologies available in the literature for the preparation of "unnatural" amino acids [3,4].

Chiral auxiliaries play a key role in the scale up of initial samples of materials and for small quantities. In addition, this method of approach can be modified to allow for the preparation of closely related materials that are invariably required for toxicological testing during the development of a pharmaceutical. There are a number of advantages associated with the use of an established auxiliary: the scope and limitations of the system are well defined; it is simple to switch to the other enantiomeric series (as long as mismatched pairs do not occur); and concurrent protection of sensitive functionality can be achieved. This information can result in a short development time. The cost of the auxiliary has the potential to be limited through recycles. On the other hand, the need to place and remove the auxiliary unit adds two extra steps to a synthetic sequence, which will reduce the overall yield. Most auxiliaries are not cheap, which must be considered carefully when large amounts of material are needed. Finally, because the auxiliary must be used on a stoichiometric scale, a by-product—recovered auxiliary—will be formed somewhere in the sequence. This by-product must be separated from the desired product, which is not always a trivial task.

Chiral auxiliaries modify the structure of the precursor to allow induction at a new stereogenic center. An alternative approach uses a modifier group—a template—in a similar manner to an auxiliary, but this group is destroyed during removal. Because the modifier group cannot be recovered, the cost of the stereogenic center is incorporated into the product molecule. This type of approach is becoming less common on the grounds of simple economics.

Chiral reagents allow an external chiral influence to act on an achiral precursor. Chiral reagents usually use chiral ligands that modify a more familiar reagent. Often, the ligand has to be used in stoichiometric quantities, although it usually can be recovered at the end of the reaction. In contrast, a chiral catalyst allows many stereogenic centers to be derived from one catalyst molecule. Thus, the cost of the catalyst can be shared over a larger quantity of product, even when the potential for recycle is minimal, as in small production runs. The catalysts can be chemical or biological. The development of these catalyst systems is time consuming, and we look for general applicability so that a wide range of compounds can be prepared from a single system. Our preference is to use a chiral catalyst as early as possible.

Each of these approaches is discussed in detail in the following sections.

15.3.   SMALL-SCALE APPROACHES

As previously noted, the usual purpose of these methods is to provide relatively small amounts of material, in a rapid manner, with little risk of the chemistry failing. Cost is not of primary concern. One of the methods cited in the following sections for larger-scale approaches may be applicable for the product candidate, but experience with these methodologies has to be such that success can be attained in a synthesis campaign with little work involved in the transfer of technology to this project.

One of the most general approaches available is to use a chiral auxiliary, because many examples are available in the literature and the scope and limitations of the various structures available as auxiliary units has been documented (see also Chapter 14) [4–6].

15.3.1.   Chiral Templates

Dehydromorpholinones (**1**) can be used to prepare unnatural amino acids and α-substituted amino acids (Scheme 1) [7]. The phenylglycinol unit is cleaved by hydrogenolysis.

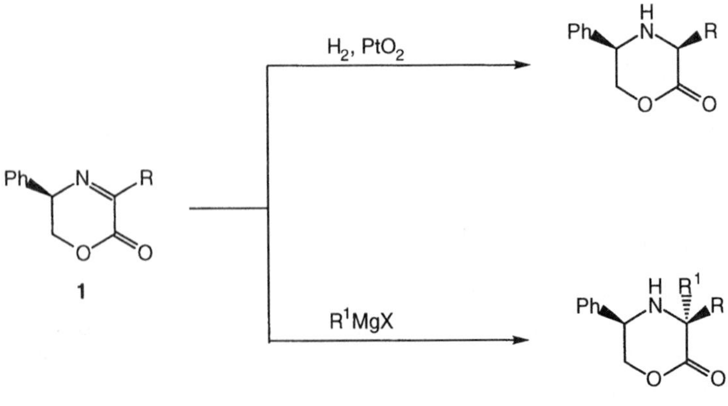

SCHEME 1.

This class of compounds serves as a chiral template rather than a chiral auxiliary, as the original stereogenic center is destroyed in the reaction sequence.

Dehydromorpholinones have been prepared from the amino alcohol by a rather extended reaction sequence that uses expensive reagents (Scheme 2) [7].

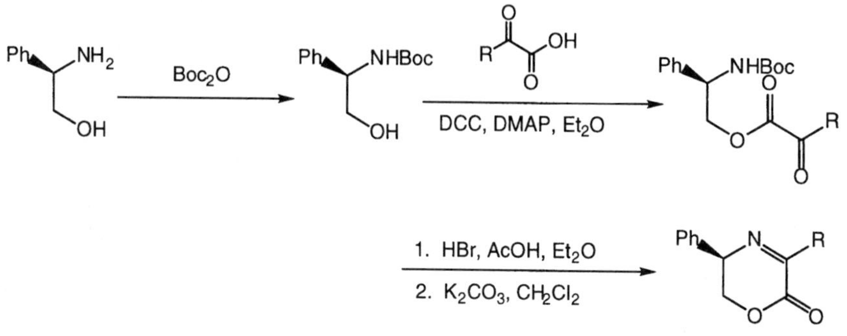

SCHEME 2.

It has been shown that the heterocycles can be made simply and cheaply in a one-pot reaction (Scheme 3).

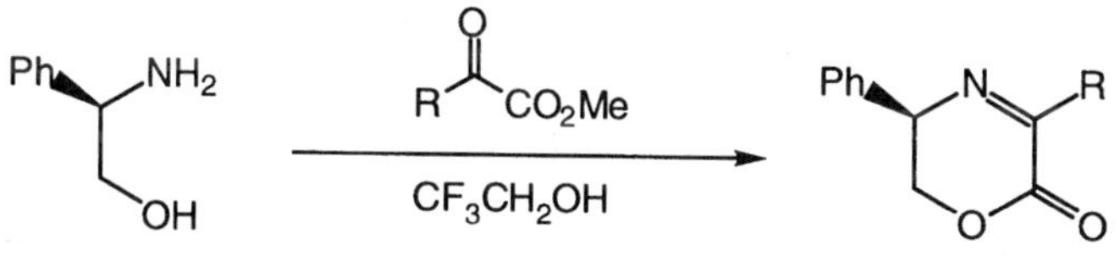

SCHEME 3.

The template can be used to control pericyclic reactions (Scheme 4) [8,9].

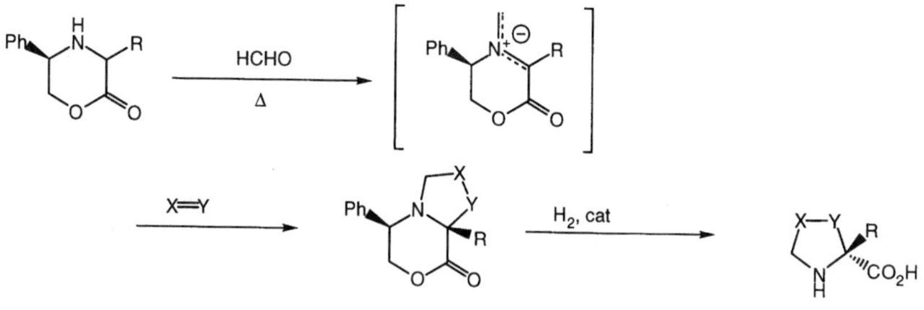

SCHEME 4.

A specific example is provided by the synthesis of the substituted proline derivative **2** (Scheme 5) [9].

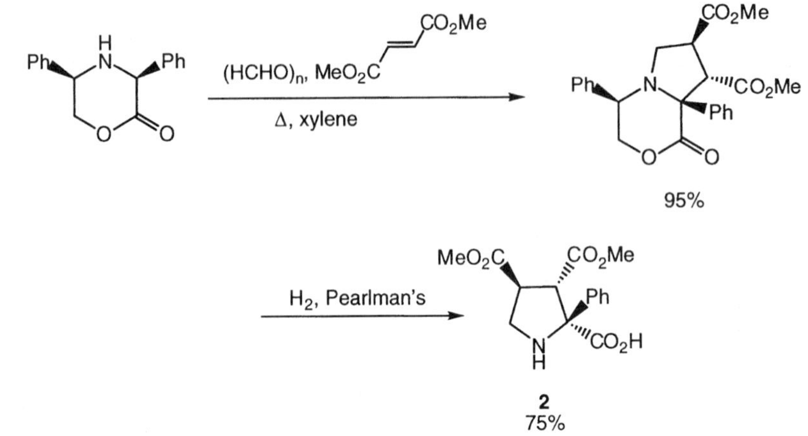

95%

2
75%

SCHEME 5.

The reaction can also be performed in an intramolecular manner [10].

15.3.2.   Chiral Auxiliaries

The preferred auxiliaries in chemistry at NSC Technologies are oxazolidinones. Oxazolidinones have found widespread use as chiral auxiliaries. A wide range of reactions is available and well documented [5], including aldol; alkylations; α-substitution with a heteroatom such as halogenation, aminations, hydroxylations, and sulphenylations; Diels-Alder cycloadditions; and conjugate additions.

A method has been developed for a one-pot procedure from an amino acid to an oxazolidinone (Scheme 6). The sequence can be run at scale.

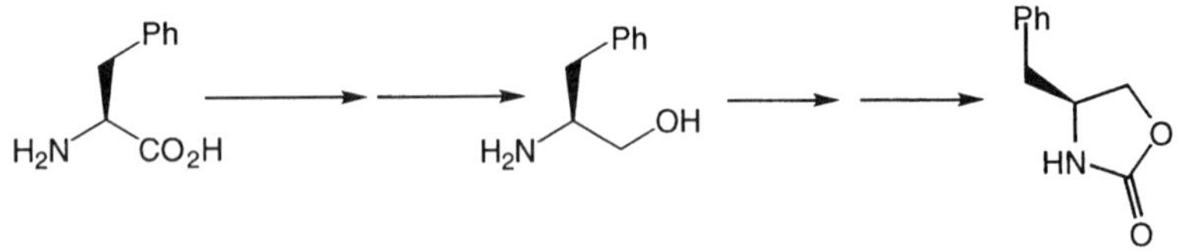

SCHEME 6.

In addition to the availability issues of the oxazolidinone unit, there has been some reluctance to scale up the reactions because formation of the N-acyl derivatives usually employs n-butyl lithium as a base. We have found a simple acylation procedure that avoids the use of this strong base and copper salts [11,12]. Our procedure uses an acid chloride or anhydride (symmetrical or mixed) with triethylamine as the base in the presence of a catalytic amount of DMAP. The reaction is general and provides good yields even with α,β-unsaturated acid derivatives (Scheme 7) [13].

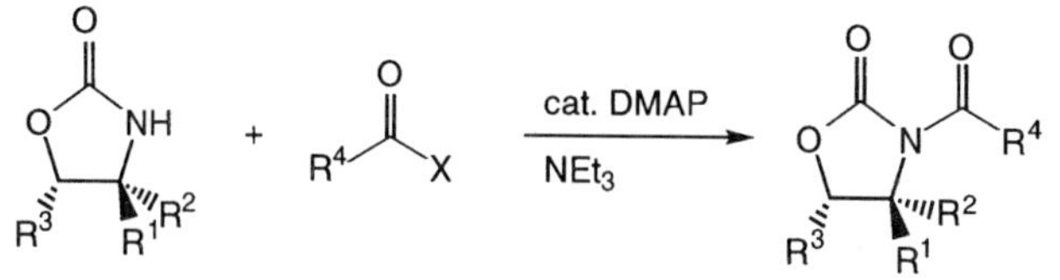

where X = Cl, R^4CO_2, or R^5CO_2

SCHEME 7.

N-Acyl-oxazolidinones are, therefore, readily available to us. The use of this class of compounds for the preparation of unnatural amino acids has been well documented [5]. Di-*tert*-butyl azodicarboxylate (DBAD) (**3**) reacts readily with the lithium enolates of *N*-acyloxazolidinones to provide hydrazides **4** in excellent yield and high diastereomeric ratio (Scheme 8); these adducts, **4**, can be converted to amino acids [14,15].

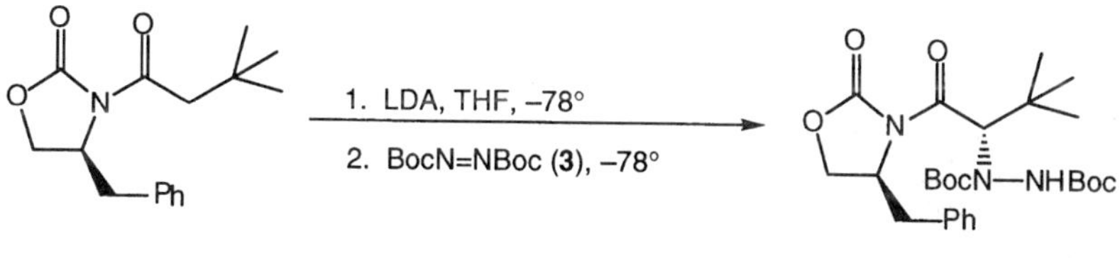

1. LDA, THF, –78°

2. BocN=NBoc (**3**), –78°

4
(ds >99:1)

SCHEME 8.

The electrophilic introduction of azide with chiral imide enolates has also been used to prepare α-amino acids with high diastereoselection (Scheme 9). The reaction can be performed with either the enolate directly [16–19] or through a halo intermediate [20]. The resultant azide can be reduced to an amine [21].

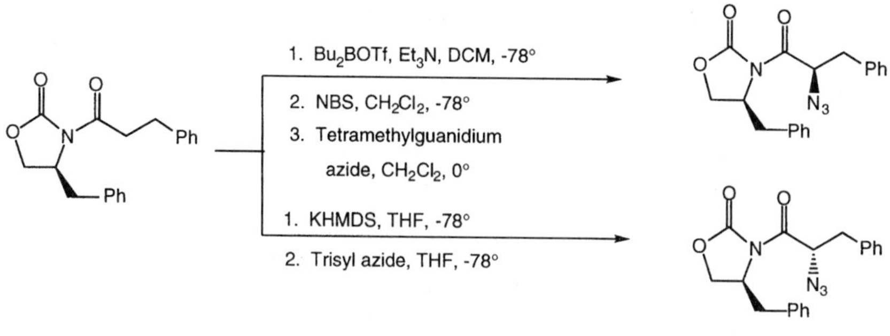

1. Bu_2BOTf, Et_3N, DCM, -78°

2. NBS, CH_2Cl_2, -78°

3. Tetramethylguanidium azide, CH_2Cl_2, 0°

1. KHMDS, THF, -78°

2. Trisyl azide, THF, -78°

SCHEME 9.

15.3.3.   Resolutions

Although our general philosophy is to undertake an asymmetric synthesis, there are occasions when a resolution method can be useful. On the chemical side, this approach usually comes into play when small amounts of material are required and alternative methodology is under development. In contrast, biological approaches can be efficient if a dynamic resolution can be achieved or if the starting materials are very cheap and readily available (see later).

Although the asymmetric hydrogenation methodology used at NSC Technologies does not allow access to cyclic amino acid derivatives at present, it is an area that is being researched. To obtain D-proline, we have performed a dynamic resolution with L-proline as the starting material. The presence of butyraldehyde allows racemization of the L-proline in solution, whereas the desired D-isomer is removed as a salt with D-tartaric acid [22,23].

Another example is provided by D-*tert*-leucine (**4**) (Scheme 10), where an asymmetric approach cannot be used. This unnatural amino acid is one of the few that cannot be made by our current D-amino acid transaminase. While this enzymatic method is being researched, small amounts of material have been prepared by a resolution method [24].

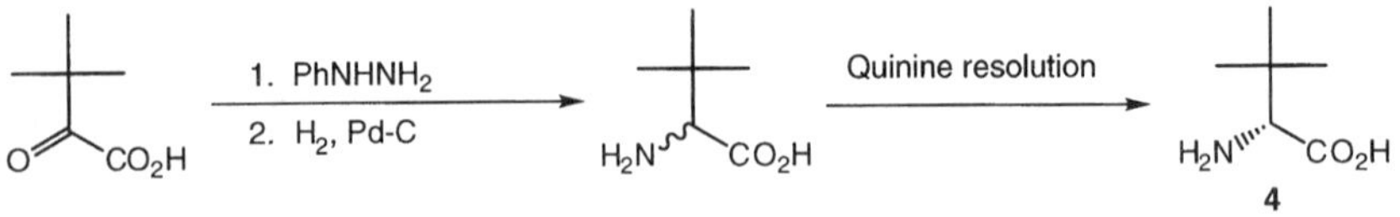

SCHEME 10.

15.4.   INTERMEDIATE-SCALE APPROACHES

These reactions take the scale of reactions above laboratory and kilo laboratory scale to pilot plant. It is usually at this stage that cost becomes a factor, although the speed by which a material can be obtained may still be important. In certain cases, the intermediate may only be needed at this multikilo level even when the pharmaceutical end product is commercialized.

Although the trend is changing, the use of chiral auxiliaries, such as oxazolidinones, is often thought of as too expensive. There are examples in which this approach has been taken to large scale (see Chapter 14). To circumvent some of the problems associated with the use of strong bases and other hazardous materials, we usually entertain the introduction of the methods that we will use for large-scale manufacture, namely asymmetric hydrogenation or bioprocesses.

15.4.1. Chiral Auxiliaries

One stoichiometric method that avoids the use of an expensive chiral auxiliary and allows for the use of nonpyrophoric bases is based on diketopiperazine chemistry. The use of this system as a chiral auxiliary is associated with a method that was developed for the preparation of the sweetener aspartame. At the same time, we were looking at the alkylation reactions of amino acid derivatives and dipeptides. These studies showed that high degrees of asymmetric induction were not simple, were limited to expensive moieties as the chiral units, and required the use of large amounts of lithium [25,26]. The cyclic system of the diketopiperazine has been used successfully by other investigators [27,28], and we also chose to exploit the face selectivity of this unit. L-Aspartic acid was chosen as the auxiliary unit because it is readily available and cheap. All of the studies were performed with sodium as the counterion because it is a more cost-effective metal at scale. Finally, we concentrated in the use of aldehydes rather than alkyl halides to allow for a general approach and so as not to limit the reaction to reactive alkyl halides.

The general approach was a base catalyzed condensation of the diketopiperazine with an aldehyde, followed by hydrogenation and hydrolysis. The glycine-L-aspartic acid diketopiperazine ester (5) can be prepared by standard peptide coupling procedures, but our preferred method at scale is to prepare the glycine unit from chloroacetyl chloride and ammonia (Scheme 11).

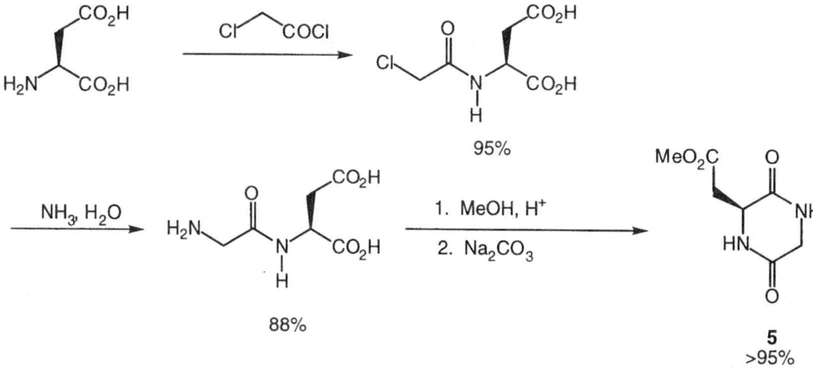

SCHEME 11.

Despite numerous attempts, reaction of the diketopiperazine with base and an electrophile resulted in very low conversion yields. To circumvent this problem, N-acylation was performed. The condensation step now proceeded smoothly and resulted in a stereoselective alkene synthesis. A sodium alkoxide was used as the base. The transfer of the proximal N-acetyl group was found to be rapid.

Hydrogenation of the unsaturation with a standard catalyst resulted in high facial selectivity. Subsequent hydrolysis cleaved the diketopiperazine to afford L-aspartic acid and the new L-amino acid (Scheme 12).

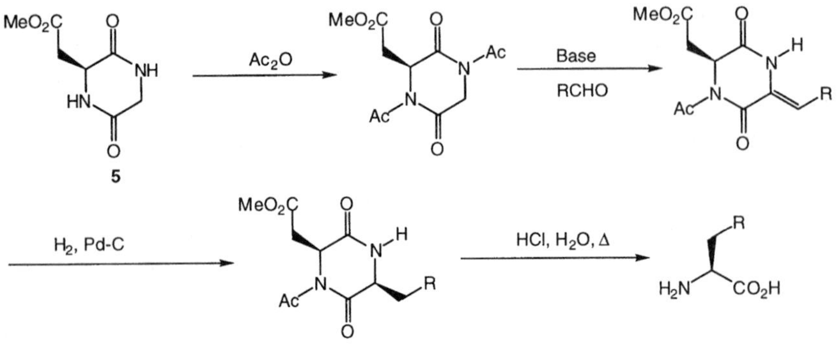

SCHEME 12.

15.5. LARGE-SCALE METHODS

15.5.1. Asymmetric Hydrogenations

The main catalyst used by NSC Technologies was developed at the dawn of asymmetric synthetic methodology. Knowles and Horner independently developed the ligand system, but Monsanto has been successful in the scale up to produce L-Dopa commercially (see Chapter 9) [29,30]. The chemistry of the Rh(dipamp) system has been reviewed [31–34]. Although the catalyst system is more than 25 years old and, in certain cases, has been superseded by other systems that can provide higher ee's with certain substrates, it is still extremely useful. The turnover numbers are often high, the scope and limitations of the system are well defined, and the recycle of the catalyst has been practiced at scale (as part of the Dopa process) [35]. The mechanism has been elucidated; it was found that the major product is derived from the minor, but less stable, organometallic intermediate, and results from the latter's higher reactivity to hydrogen [36].

The substrate requirements have been rationalized. Thus, for an examine, they can be divided into four specific regions (Figure 1) [34].

The hydrogen region will only tolerate hydrogen and still give a reasonable rate. When other groups fall into this region, the stereospecificity decreases (see later). The aryl region will tolerate a wide range of groups, including heterocycles, alkyl groups, and hydrogen. In a similar manner, the amide region will allow a wide variety of amides to be used not only in terms of size, but also in electronic features so that protected amino acids can be synthesized directly. The acid region is also forgiving as long as it contains an electronic withdrawing group [35]. Some of these conclusions are illustrated in Table 1.

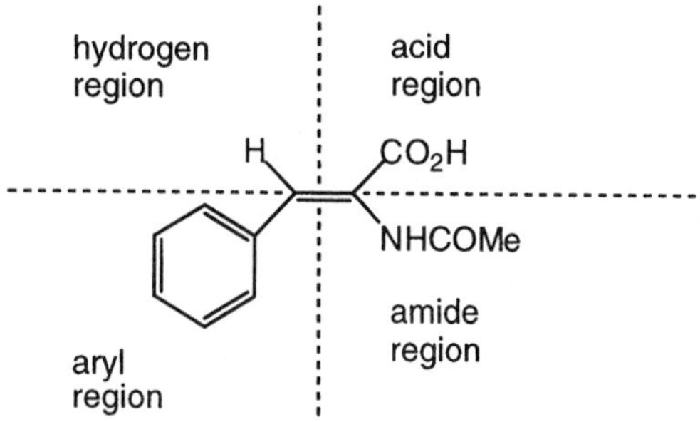

FIGURE 1 Regions of the enamide substrate for asymmetric hydrogenations.

Fortunately, the general method for the preparation of the enamides is stereoselective to the Z-isomer when an aryl aldehyde is used (Scheme 13) [37].

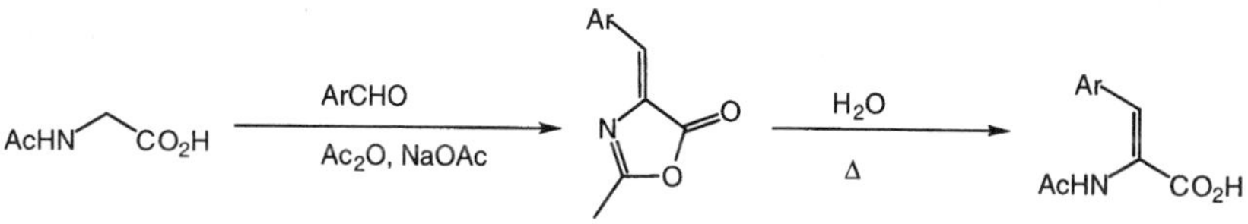

SCHEME 13.

15.5.2. Resolutions

These are very rarely performed at large scale because of the economic disadvantages associated with recycle of the off-isomer and resolution agent. However, when the substrate is very cheap, a biological resolution can be used.

TABLE 1 Selectivity of Rh(dipamp) with the Enamide 6

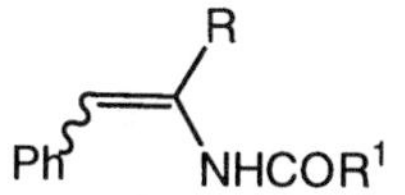

Substrate	Isomer	Substrate/cat	Relative rate	ee
R = CO$_2$H, R^1 = Me	Z	1000	100	94
	E	200	1	47
R = CO$_2$Me, R^1 = Me	Z	900	78	96
	E	1200	5	23
R = CO$_2$H, R^1 = Ph	Z	800	75	93
R = CN, R^1 = Ph	Z	200		89

15.5.2.1. Decarboxylases

Amino acid decarboxylases can be used to catalyze the resolution of several amino acids, and for the most part, their utility has been underestimated, especially with respect to the irreversible reaction equilibrium. The decarboxylases are ideally suited to large-scale industrial application because of their robust nature. D-Aspartate and D-glutamate can be made economically by use of the corresponding L-amino acids as starting material through an enzymatic racemization before the resolution with decarboxylase (Scheme 14). The process for D-glutamate consists of racemization of L-glutamate at pH 8.5 with the cloned glutamate racemase enzyme from *Lactobacillus brevis*, expressed in *Escherichia coli* followed by adjustment of the pH to 4.2 to allow *E. coli* cells with elevated decarboxylase activity to decarboxylate the L-glutamate [38]. This reaction can be greatly improved by combination and overexpression of the glutamate racemase and glutamate decarboxylase in one bacterial strain. The combination of both activities is possible because glutamate decarboxylase is inactive at pH 8.5 during the racemization step, but regains full activity when the pH is 4.2. Conversely, glutamate racemase is active at pH 8.5, but is completely inhibited at pH 4.2.

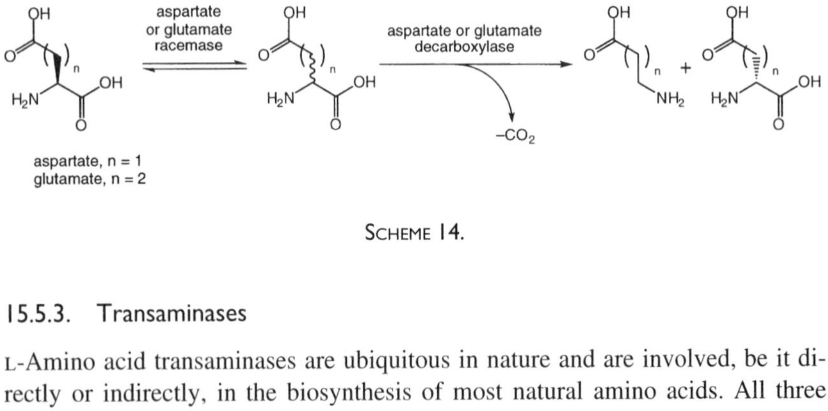

aspartate, n = 1
glutamate, n = 2

SCHEME 14.

15.5.3. Transaminases

L-Amino acid transaminases are ubiquitous in nature and are involved, be it directly or indirectly, in the biosynthesis of most natural amino acids. All three common types of the enzyme, aspartate, aromatic, and branched chain transaminases require pyridoxal 5'-phosphate as cofactor, covalently bound to the enzyme through the formation of a Schiff base with the ε-amino group of a lysine side chain. The reaction mechanism is well understood, with the enzyme shuttling between pyridoxal and pyridoxamine forms [39]. With broad substrate specificity and no requirement for external cofactor regeneration, transaminases have appropriate characteristics to function as commercial biocatalysts. The overall transformation is comprised of the transfer of an amino group from a donor, usually aspartic or glutamic acids, to an α-keto acid (Scheme 15). In most cases, the equilibrium constant is approximately 1.

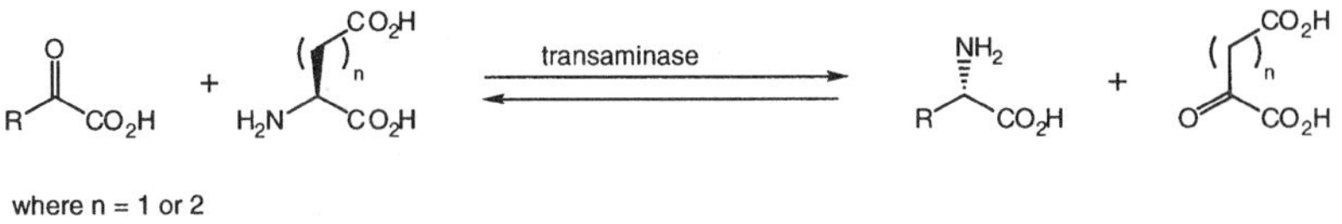

where n = 1 or 2

SCHEME 15.

Examples of unnatural L-amino acids that can be accessed by transaminase methodology are L-2-aminobutyric acid, L-homophenylalanine, and L-*tert*-leucine; the latter is not accessible by an asymmetric hydrogenation approach because of the absence of a β-hydrogen. We have previously described the isolation and cloning of genes encoding microbial L-amino acid transaminases [40], and the use of site-directed mutagenesis to enhance enzyme function [41]. The reversibility of the reaction has frequently been considered a drawback of transaminase commercial application because it results in reduced yields and complicates product isolation. Through molecular cloning we have constructed bacterial strains that effectively eliminate this concern by removal of the keto acid by-product, and thereby significantly increase the reaction yield.

This removal of the reaction by-product has been achieved through the use of aspartic acid as the amino donor (Scheme 16). The amine group transfer results in the formation of oxaloacetate (**7**), an unstable compound that decarboxylates under the reaction conditions to afford pyruvate (**8**). As **8** is still an α-keto acid, and is a substrate for a transaminase reaction that results in the production of alanine, another enzyme is used to dimerize the pyruvate. The product of this reaction is acetolactate (**9**), which, in turn, spontaneously undergoes decarboxylation to result in the overall formation of acetoin (**10**) as the final by-product. Acetoin is simple to remove and does not participate in any further reactions. Thus, the equilibrium is driven to provide the desired unnatural amino acid that makes the isolation straightforward.

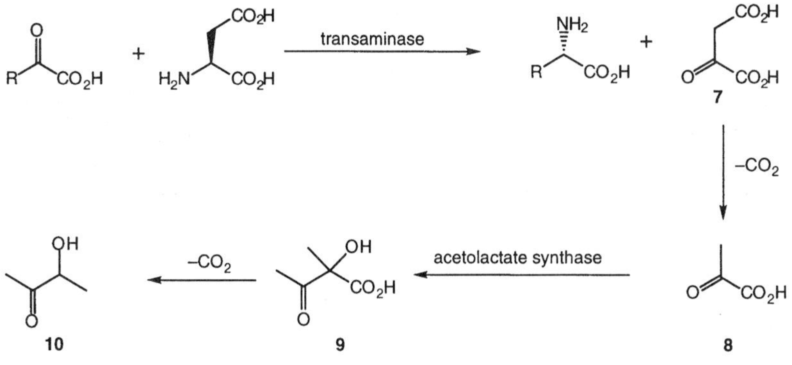

SCHEME 16.

D-Amino acid transaminases have been less well characterized but proceed by a similar catalytic mechanism and show similar potential as effective biocatalysts. Further strain development has incorporated amino acid racemases, enabling complete utilization of racemic amino donors. Thus, L-aspartic acid can be used as the D-amino acid donor through use of aspartate racemase within the system.

The α-keto acid substrates are available by chemical synthesis. They are also accessible by enzymatic methods, such as amino acid deaminase, that generate these substrates from inexpensive L-amino acids.

The approach has also been extended to enable direct synthesis of D-amino acids from metabolic intermediates during fermentation of the organism. This has the advantage of eliminating the need to provide external substrates and reduces processing steps.

15.5.4. Chiral Pool Approaches

Unnatural amino acids are available from more readily available amino acids. A wide variety of reactions are described in the literature. We use these coupled with our background in amino acid chemistry. An example is provided by the synthesis of L-homoserine (**11**) from methionine (Scheme 17) [42]. Of course, the problems associated with the formation of one equivalent of dimethyl sulphide must be overcome.

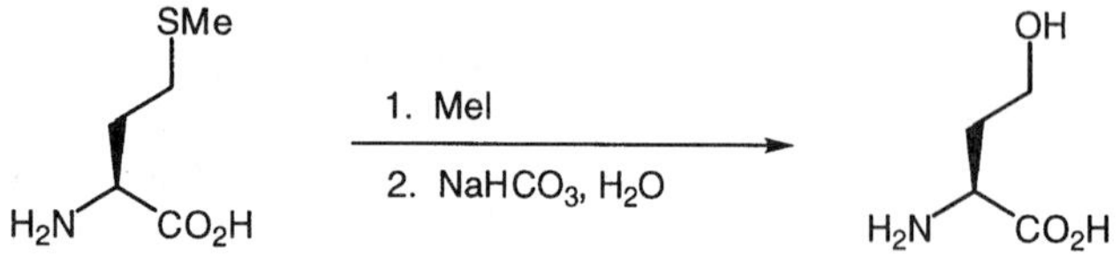

SCHEME 17.

A wide variety of simple transformations on chiral pool materials can lead to unnatural amino acid derivatives. This is illustrated by the Pictet-Spengler reaction of L-phenylalanine, followed by amide formation and reduction of the aromatic ring (Scheme 18) [43]. The resultant amide **12** is an intermediate in a number of commercial human immunodeficiency virus protease inhibitors.

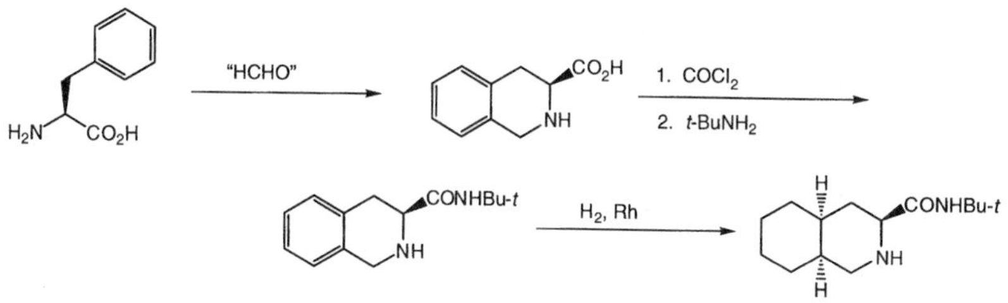

SCHEME 18.

15.6. SUMMARY

The two large-scale methodologies outlined here, asymmetric hydrogenation and enzymatic transamination, allow for the large-scale production of unnatural amino acids in a cost-effective manner. The two methodologies are somewhat complementary and can compete, on an economical basis, with resolution methodologies. In both the chemical and biological processes, the high yields and simple by-products allow for a simple isolation procedure of the desired product. Transformations of natural, readily available amino acids can also lead to useful unnatural amino acids. Decarboxylases allow access to a limited number of functionalized D-amino acids. For smaller-scale reactions, chiral auxiliaries such as oxazolidinones or aspartic acid can be used.

REFERENCES

1. Collins, A. N., Sheldrake, G. N., Crosby, J., Eds. *Chirality in Industry: The Commercial Manufacture and Applications of Optically Active Compounds;* Wiley: Chichester, 1992.
2. Collins, A. N., Sheldrake, G. N., Crosby, J., Eds. *Chirality in Industry II: Developments in the Commercial Manufacture and Applications of Optically Active Compounds;* Wiley: Chichester, 1997.
3. Williams, R. M. *Synthesis of Optically Active α-Amino Acids;* Pergamon Press: Oxford, 1989.
4. Ager, D. J., East, M. B. *Asymmetric Synthetic Methodology;* CRC Press: Boca Raton, 1995.
5. Ager, D. J., Prakash, I., Schaad, D. R. *Chem. Rev.* 1996, *96*, 835.
6. Ager, D. J., Prakash, I., Schaad, D. R. *Aldrichimica Acta* 1997, *30*, 3.
7. Cox, G. G., Harwood, L. M. *Tetrahedron: Asymmetry* 1994, *5*, 1669.
8. Harwood, L. M., Lilley, I. A. *Tetrahedron: Asymmetry* 1995, *6*, 1557.
9. Anslow, A. S., Harwood, L. M., Lilley, I. A. *Tetrahedron: Asymmetry* 1995, *6*, 2465.
10. Harwood, L. M., Kitchen, L. C. *Tetrahedron Lett.* 1993, *34*, 6603.
11. Lee, J. Y., Chung, Y. J., Kim, B. H. *Synlett* 1994, 197.
12. Thom, C., Kocienski, P. *Synthesis* 1992, 582.
13. Ager, D. J., Allen, D. R., Schaad, D. R. *Synthesis* 1996, 1283.

14. Evans, D. A., Britton, T. C., Dorow, R. L., Dellaria, J. F. *J. Am. Chem. Soc.* 1986, *108*, 6395.
15. Trimble, L. A., Vederas, C. J. *J. Am. Chem. Soc.* 1986, *108*, 6397.
16. Evans, D. A., Britton, T. C., Ellman, J. A., Dorow, R. L. *J. Am. Chem. Soc.* 1990, *112*, 4011.
17. Evans, D. A., Britton, T. C. *J. Am. Chem. Soc.* 1987, *109*, 6881.
18. Doyle, M. P., Dorow, R. L., Terpstra, J. W., Rodenhouse, R. A. *J. Org. Chem.* 1985, *50*, 1663.
19. Evans, D. A., Lundy, K. M. *J. Am. Chem. Soc.* 1992, *114*, 1495.
20. Evans, D. A., Ellman, J. A., Dorow, R. L. *Tetrahedron Lett.* 1987, *28*, 1123.
21. Evans, D. A., Everard, D. A., Rychnovsky, S. D., Fruh, T., Whittington, W. G., DeVries, K. M. *Tetrahedron Lett.* 1992, *33*, 1189.
22. Shiraiwa, T., Shinjo, K., Kurokawa, H. *Chem. Lett.* 1989, 1413.
23. Shiraiwa, T., Shinjo, K., Kurokawa, H. *Bull. Chem. Soc. Jpn.* 1991, *64*, 3251.
24. Miyazawa, T., Takashima, K., Mitsuda, Y., Yamada, T., Kuwota, S., Watanabe, H. *Bull. Chem. Soc. Jpn.* 1979, *52*, 1539.
25. McIntosh, J. M., Thangarasa, R., Ager, D. J., Zhi, B. *Tetrahedron* 1992, *48*, 6219.
26. McIntosh, J. M., Thangarasa, R., Foley, N., Ager, D. J., Froen, D. E., Klix, R. C. *Tetrahedron* 1994, *50*, 1967.
27. Schollkopf, U. *Tetrahedron* 1983, *39*, 2085.
28. Schollkopf, U. *Top. Curr. Chem.* 1983, *109*, 65.
29. Knowles, W. S., Sabacky, M. J. *J. Chem. Soc. Chem. Commun.* 1968, 1445.
30. Horner, L., Siegel, H., Buthe, H. *Angew. Chem. Int. Ed. Engl.* 1968, *7*, 942.
31. Knowles, W. S. *Acc. Chem. Res.* 1983, *16*, 106.
32. Koenig, K. E., Sabacky, M. J., Bachman, G. L., Christopfel, W. C., Barnstorff, H. D., Friedman, R. B., Knowles, W. S., Stults, B. R., Vineyard, B. D., Weinkauff, D. *J. Ann. N. Y. Acad. Sci.* 1980, *33*, 16.
33. Koenig, K. E. in *Catalysis of Organic Reactions;* Kosak, J. R., Ed.; Marcel Dekker: New York, 1984; p. 63.
34. Koenig, K. E. in *Asymmetric Synthesis;* Morrison, J. D., Ed.; Academic: Orlando, 1985; *Vol. 5;* p. 71.
35. Vineyard, B. D., Knowles, W. S., Sabacky, M. J., Bachman, G. L., Weinkauff, D. *J. Am. Chem. Soc.* 1977, *99*, 5946.
36. Landis, C. R., Halpern, J. *J. Am. Chem. Soc.* 1987, *109*, 1746.
37. Herbst, R. M., Shemin, D. *Org. Synth.* 1943, *Coll. Vol. II*, 1.
38. Yagasaki, M., Azuma, M., Ishino, S., Ozaki, A. *J. Ferm. Bioeng.* 1995, *79*, 70.
39. Kirsch, J. F., Eichele, G., Ford, G. C., Vincent, M. G., Jansonius, J. N. *J. Mol. Biol.* 1983, *174*, 497.
40. Fotheringham, I. G., Dacey, S. A., Taylor, P. P., Smith, T. J., Hunter, M. G., Finlay, M. E., Primrose, S. B., Parker, D. M., Edwards, R. M. *Biochem. J.* 1986, *234*, 593.
41. Kohler, E., Seville, M., Jager, J., Fotheringham, I., Hunter, M., Edwards, M., Jansonius, J. N., Kirschner, K. *Biochemistry* 1994, *33*, 90.
42. Boyle, P. H., Davis, A. P., Dempsey, K. J., Hoskin, G. D. *Tetrahedron: Asymmetry* 1995, *6*, 2819.
43. Allen, D. R., Jenkins, S., Klein, L., Erickson, R., Froen, D. *US Patent*, 1996, 5,587,481.

16
Synthesis of L-Aspartic Acid

PAUL P. TAYLOR
NSC Technologies, Mount Prospect, Illinois

16.1. INTRODUCTION

L-Aspartic acid is an industrially important, large-volume, chiral compound. Worldwide production of L-aspartic acid was approximately 11,000 metric tons in 1993 and is estimated to grow to 13–15 thousand tons in 1998 [1]. More than half of all the aspartic acid produced is consumed by the United States market. The primary use of aspartic acid is in the production of aspartame (*N*-L-α-aspartyl-L-phenylalanine, methyl ester), a high-potency sweetener. Other uses of L-aspartate include dietary supplements, pharmaceuticals, production of alanine, antibacterial agents, and lubricating compounds.

There have been a number of excellent reviews on L-aspartate production [2–4] that have focused on the immobilization of the key enzyme, aspartase, and productivity of the immobilized biocatalyst. The aspartase enzyme has only been given cursory consideration with regard to its importance to the overall efficiency of the L-aspartic acid process. This chapter reviews some of the ways in which better aspartase biocatalysts have been produced.

16.2. COMMERCIAL PRODUCTION

Historically, L-aspartic acid was produced by hydrolysis of asparagine, isolated from protein hydrolysates, or by the resolution of chemically synthesized D,L-aspartate. With the discovery of aspartase (L-aspartate ammonia lyase, Enzyme Commission [EC] 4.3.1.1) [5] fermentation routes to L-aspartic acid quickly superseded the initial chemical methods. After further characterization, enzymatic routes to the production aspartic acid from ammonium fumarate using aspartase

317

were developed. These processes were far more cost effective than the fermentation routes, and aspartate is now made exclusively by enzymatic methods that use variations of the general scheme outlined in Scheme 1.

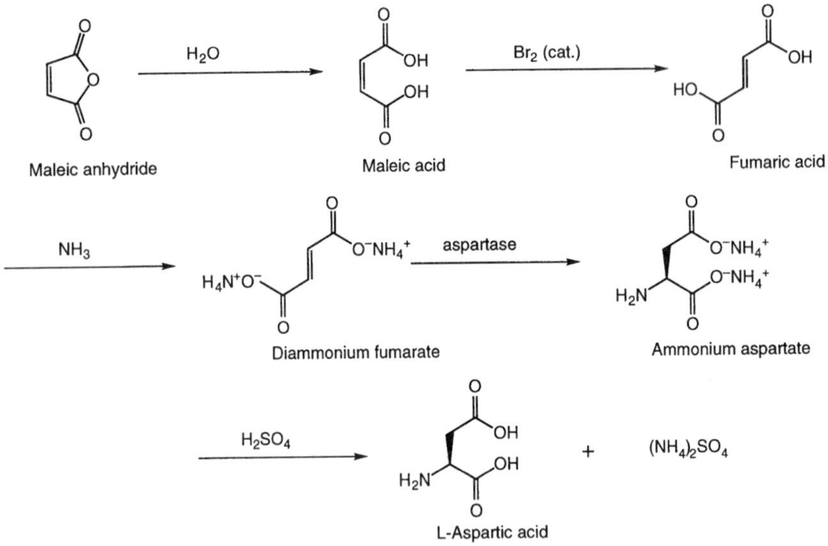

SCHEME 1.

The basic enzymatic procedure usually involves immobilization of bacterial cells containing aspartase activity or of semipurified enzyme, in either a fed batch or a continuous, packed column process. The starting material is maleic anhydride, an inexpensive, high-volume, bulk chemical. The process is initiated by the hydration of maleic anhydride to maleic acid, followed by bromine catalyzed isomerization to fumaric acid. This is subsequently converted to diammonium fumarate by the addition of ammonia. The aspartase catalyst is then added, and diammonium fumarate is converted to ammonium aspartate. The final step is the addition of sulfuric acid to precipitate the L-aspartic acid with removal of the ammonia as its sulphate salt. Conversion rates of maleic anhydride to L-aspartic acid are generally very high (95–99%). An alternative method that uses maleic acid to precipitate the final product has also been described [6]. The latter process allows recycle of the mother liquor after crystallization because it does not produce ammonium sulphate, but has only a 74% yield at the bromine catalyzed isomerization step. By reduction of the pH to 1.5 with sulphuric acid, the efficiency of isomerization step can be increased to 96%, but large amounts of ammounium sulphate waste are then formed, making recycling of the mother liquor uneconomical.

16.3. GENERAL PROPERTIES OF ASPARTASE

Aspartase is ubiquitous in nature and has been studied in many bacterial species with the specific aim of using the enzyme to make L-aspartic acid [5,7–14]. It catalyzes the reversible deamination of aspartic acid to fumaric acid (Scheme 2). This in vivo reaction procedes exclusively in the direction of fumarate, but in vitro it is readily reversible and can go to 100% completion. Much of the early characterization of aspartase was conducted on enzyme purified from strains of *Escherichia coli* B and W [15,16] and, for this reason, many of the early commercial processes were developed using the *E. coli* aspartase.

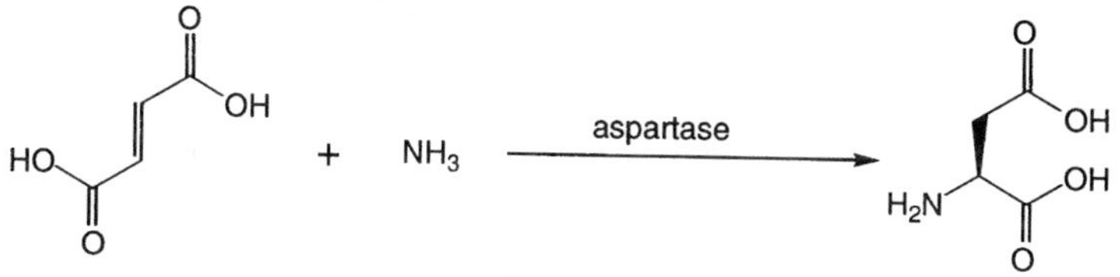

SCHEME 2.

The *E. coli* aspartase is a tetrameric enzyme that is composed of four identical subunits, each with a molecular weight of 52,224, a pH optima of 8.7 [16], and absolute specificity for its natural substrate [17,18]. It is activated by the product of its own reaction, L-aspartic acid [19] and by divalent cations at alkaline pH [15]. The enzyme can also be activated in by a variety of physical and chemical methods, which will be described in more detail in the following sections.

16.4. BIOCATALYST DEVELOPMENT

Because the starting substrates for the aspartase reaction are relatively inexpensive, the most effective way to develop a rapid, cost-efficient L-aspartate production process is to improve the overall productivity of the biocatalyst. One way to achieve this is to increase the half-life or stability of the system, thus reducing the up-front biocatalyst production costs. Another strategy is to increase the amount of enzyme in the process, thus getting greater through-put per unit of biocatalyst. Perhaps the most elegant way to produce an efficient biocatalyst is the application of protein engineering to yield an enzyme of higher specific activity than the native enzyme. Some of the key strategies that have been used to improve the overall efficiency of the aspartase biocatalyst are outlined in the following sections.

16.4.1. Aspartase Hyperproducing Strains

Initially, conventional selection methods were used to select aspartase overproducing strains. For example, Nishimura and Kisumi isolated aspartase hyperproducing mutants of *E. coli B* ATCC11303 (a strain with naturally high aspartase activity) by selection for growth on aspartate as the sole nitrogen source in the presence of glucose. Mutants selected in this way were no longer subject to catabolite repression, and one such mutant produced sevenfold more aspartase compared with the wild-type strain [20]. This strain was successfully used in an immobilized industrial process to produce aspartic acid [21,22]. Two classes of mutants were selected by this method. The first class were linked to the *aspA* gene and led to overproduction of aspartase. The second class of mutants were linked to the adenylate cyclase gene (*cyaA*) and had elevated levels of cyclic adenosine monophosphate (cAMP). Because catabolite repression depends on the level of cAMP, the later strain overproduced aspartase due to its high adenylate cyclase activity [23]. The two mutations were combined in one strain by transduction, and the resultant strain AT202 was found to be completely resistant to catabolite repression. Overproduction of aspartase was 18-fold compared with the wild-type strain or threefold compared with either of the parental strains [23].

16.4.1.1.   Recombinant Escherichia coli Aspartase Strains

Using recombinant DNA technology, John Guest cloned the aspartase gene (*aspA*) from *E. coli* K12 [24]. Subsequently, the *aspA* gene has been cloned from a variety of other bacterial sources: *E. coli* W [13], *Psuedomonas fluorescens* [25], *Serratia marcescens* [26], *Bacillus subtilis* [27], and *Brevibacterium flavum* [28]. For the commercial production of aspartate, however, the *E. coli* aspartase has been the main enzyme of choice, and a number of groups have developed processes using the cloned gene.

The cloning of the *aspA* gene onto the multicopy plasmid pBR322 [29] led to a modest fourfold to sixfold overproduction of aspartase compared with host strain without the clone [24]. The aspartase activity of the clone was further enhanced by deregulation of the gene using the lactose inducible, *tac* promoter [30,31]. The resultant plasmid (pPT174), when fully induced, produced 190-fold more aspartase activity than the *E. coli* K12 parental strain without the plasmid [31].

Another approach, which combines both recombinant DNA technology and conventional strain development, was used by Nishimura et al. In parallel to Guest, Komatsubara et al. also cloned the *aspA* gene from *E. coli* K12 into plasmid pBR322 [32]. This clone appeared to be already somewhat deregulated because it produced 30-fold more aspartase than the parental strain, whereas a similar clone isolated by Guest et al. only exhibited 4–6-fold overproduction. The cloned aspartase (pYT471) was, however, highly unstable in the absence of antibiotic selection, making it unsuitable for large-scale commercial production of aspartate [32]. This problem was subsequently alleviated by the addition of the partition locus (*par*) from plasmid pSC101 to pYT471—the par locus is known to confer stable inheritance of plasmid DNA [33]. The resultant plasmid, pNK101,was almost completely stablised by *par*. The cells harboring this plasmid produced 30-fold more aspartase activity than pYT471,and this high activity was maintained throughout at least 30 cell generations. The aspartase activity was further boosted to 80-fold of the wild-type levels by transferring pNK101 into the aspartase catabolite repression resistant mutant AT202 (described in Section 16.4.1), thus effectively relieving catabolite repression of the plasmidborne aspartase [34]. Again the high aspartase activity of this strain was also maintained through at least 30 cell generations.

16.4.1.2. Aspartate Production Using Brevibacterium flavum

The Mitsubishi Chemical Company has described a process for the commercial production of L-aspartate using an α-amino-*n*-butyric acid resistant mutant of *B. flavum* [11]. The enzyme is moderately thermal resistant, allowing the process to be run at 45°C. The process is run using immobilized cells in a fed batch system in which the biocatalyst is recycled [4]. An initial problem was the conversion of fumarate to malic acid by an intracellular fumarase activity, which led to low L-aspartic acid yields during the first cycle. This problem was circumvented by preheating the biocatalyst for 1 hour at 45°C, which completely destroyed the fumarase activity [4,11]. Recently, the aspartase gene from *B. flavum* has been cloned [28] and has presumably been used to improve the efficiency of this process.

16.4.2. Activation by Truncation

Early studies of the *E. coli* B aspartase showed that highly purified enzyme could be activated threefold by limited digestion with trypsin [35–37]. It was subsequently shown that this activation was caused by the release of one or more carboxy-terminal peptides in the range of 7–17 residues [38]. Further refinement of this protease-mediated activation using subtilisin BPN' strongly suggested that proteolytic cleavage within a limited range around the eighth residue from the carboxy-terminal residue was responsible for the protease-induced activation [39]. To test this hypothesis, an aspartase gene was engineered with a carboxy-terminus that could easily be altered by cassette mutagenesis [31]. The wild-type *tac* deregulated *aspA* clone (pPT174,described in Section 16.4.1.1) was mutagenized by inserting a truncation adapter that truncated the *aspA* gene by 15 amino acid residues (aa). A DNA linker that truncated the *aspA* gene by eight amino acid residues was inserted between unique restriction endonuclease sites of the truncation adapter (all DNA changes were conservative with respect to the protein sequence). As a control, a linker recreating the natural carboxy-terminus was also cloned. One of the most noticeable effects of *C*-terminal engineering of the *E.*

TABLE I Activity of Genetically Engineered Aspartase Plasmids

Plasmid[a]	Total activity[b]	Protein expression[c]	Specific activity[d]
pPT174 (w.t. *tac* aspartase)	100	100	1.00
pPT175 (15 aa truncation)	0	0	0
pPT177 (8 aa truncation)	110	42	2.61
pPT188 (natural C-terminus)	65	62	1.12
pPT190 (8 aa truncation + terminator)	312	99	3.15

[a] Host strain was *E. coli* K12 strain HW1111 (lacI^q L8 tyrA Δ).
[b] Total activity normalized to wild type aspartase expressed using the *tac* promoter.
[c] Protein expression was determined using a Joyce Loebl Chromosdan 3 gel scanner, numbers reflect % of wild type *tac* aspartase expression.
[d] Specific activity of aspartase constructs normalized with respect to protein expression.

coli K12 aspartase was the effect on gene expression. Expression of the 15 aa truncate was undetectible by polyacrylamide gel electrophoresis, the 8 aa truncate was reduced to 42% of wild type (w.t.) levels, and the recreated natural C-terminus clone was reduced to 62% of w.t. levels (Table 1). Addition of a transcriptional terminator downstream of the 8 aa truncate returned expression to w.t. levels. This strongly suggests that the effects on protein expression are probably due to increased turnover of the aspartase messenger ribonucleic acid (mRNA) of the C-terminal contructs. The relative aspartase activity of the various constructs is shown in Table 1. From these data, it was concluded that the eight amino acid deletion did, in fact, yield an enzyme with approximately threefold more aspartase activity, confirming Yumoto's data [39]. As might be expected, the 15–amino acid truncation produced an enzyme that had no detectable activity. The final clone (pPT190), when fully induced in an *E. coli* K12 host, is capable of producing approximately 600-fold more aspartase activity than the w.t. strain without the plasmid [31] and is an excellent biocatalyst for the commercial production of L--aspartic acid.

16.4.3. Activation by Site-Directed Mutagenesis

Replacement of Cys-430 of the *E. coli* W aspartase gene with a tryptophan residue has been shown to enhance aspartase activity threefold at pH 6.0, but only slightly at pH 8, the optimum for the wild-type enzyme [40]. Limited proteolysis of the mutated protein completely inactivates the enzyme, showing that truncation activation and the Cys-430 mutation are not additive. Substitution of Lys-126 to an arginine residue of the *E. coli* AS. 881 aspartase causes a fivefold enhancement in specific activity at pH 8.5 with a concomitant increase in the thermostability of the enzyme [41]. It is unknown if this mutant can be further activated by truncation or chemical activation, but it would appear to be an excellent candidate for further optimization and use in the commercial production of L-aspartic acid.

16.4.4.   Chemical Activation

Although of more academic than commercial importance, aspartase can also be chemically activated by acetylation using acetic anhydride or N-hydroxy-succinimide [42]. Acetylated enzyme exhibits a twofold activation at pH 7.0, but at pH 8.5, the activity is lower than the native enzyme and an increased requirement for divalent cations. Limited proteolysis of acetylated enzyme does not further enhance the aspartase activity.

Glycerol, ethylene glycol, propylene glycol, dimethylsulfoxide, and dioxane have also been shown to activate aspartase, but only at low substrate concentrations [43]. At high substrate concentrations, glycerol was found to be inhibitory. Thus, this type of activation has little commercial application, except for the observation that glycerol also protects the enzyme against heat inactivation [16].

16.4.5.   Thermostable Aspartase

There has been only one report on the use of a thermostable aspartase to produce aspartate in the literature. Suzuki et al. described the use of the *Bacillus stearothermophilis* aspartase to produce L-aspartate in a whole-cell bioconversion [9]. To date, there have been substantiated reports of aspartase genes cloned from thermophilic organisms (excluding the moderately thermal resistant enzyme from *B. flavum*, previously described). This lack of interest in thermophilic aspartases may well be testament to the commercial success of the mesophilic aspartase processes or may simply reflect difficulty in obtaining clones of thermophilic aspartase genes. Increased through-put, greater enzyme stability, and higher substrate concentrations are still compelling reasons for the development of a thermophilic aspartate process.

16.5.   FUTURE PERSPECTIVES

It is clear that recombinant DNA technology has had a major impact on commercial production of L-aspartic acid. With the exception of a thermophilic process, there would appear little need for further development of a better enzyme, as the currently available biocatalysts are extremely cost efficient [44,45]. Future development will be aimed at L-aspartic acid processes that produce less waste and use cheaper starting materials. Current L-aspartic acid processes require bromine as the catalyst for the isomerization of maleic acid to fumaric acid, while large amounts of ammonium sulphate are produced during conversion of ammonium aspartate to the final crystallized product, both of which cause problems of waste disposal. The following sections describe two approaches that are under development that could produce a more cost-efficient and more environmentally friendly L-aspartic acid process.

16.5.1. Fermentative Production of Fumaric Acid

A recent report from Cao et al. outlined an ingenious method for the fermentative production of fumaric acid from glucose [46]. *Rhizopus oryzae* was grown in a nitrogen-rich medium in a special fermentation tank containing plastic disks, called a rotory biofilm contactor (RBC). Under these conditions, *R. oryzae* forms a thin film on the plastic discs of the contactor, thus effecting self immobilization of the fungal mycelia. Under conditions of nitrogen starvation, *R. oryzae* overproduces fumaric acid [47,48]. Unfortunately, high concentrations of fumaric acid are detrimental to cell viablity and limit the amount of fumaric acid that can be accumulated. The continuous removal by absortion of the fumaric acid to polyvinyl pyridine resin was found to alleviate this problem and improve both the fermentation rate and cell viability. By this method, a yield of 85 g/L fumaric acid from 100 g/L glucose was obtained by repetitive fed-batch fermentations at a productivity rate of 4.25 g/L/hr [46]. Assuming that the bioreactor can be used continuously for several months, this procedure could be a cost-effective, "green" alternative for the production of fumaric acid.

16.5.2. Aspartate from Maleic Acid Using Maleate Isomerase

The existence of maleate isomerase in *Pseudomonas* sp., an enzyme that catalyzes the isomerization of maleic acid to fumaric acid, was first described by Ken'ichi Otsuka [49]. Subsequently, the enzyme was also found in *Alcaligenes faecalis* [50] and *Pseudomanas fluorescens* [51]. The first demonstration of aspartate production from maleic acid using maleate isomerase was in 1966 [50], but it was not until recently that this process has realized its full potential with the development of a semicontinuous process that allows recycling of the ammonia in the mother liquor, which eliminates the ammonium sulphate waste at the final L-aspartic acid crystallization step [52–54]. The basic reaction scheme is shown in Scheme 3, in which maleic anhydride is converted to ammonium aspartate by the combined actions of maleate isomerase and aspartase. The whole process can be made semicontinuous, as shown in Scheme 4. After removal of the biocatyst, the aspartic acid is precipitated using maleic acid, the mother liquor is adjusted to optimal maleic acid and ammonia concentrations and is recycled to produce more aspartic acid [52,54]. Initial problems with enzyme production and stability should be alleviated by the recent cloning of maleate isomerase genes from *Alcaligenes faecalis, Bacillus stearothermophilis,* and *Serratia marcescens* [55] to allow the development of an extremely efficient, virtually waste-free, L-aspartic acid process.

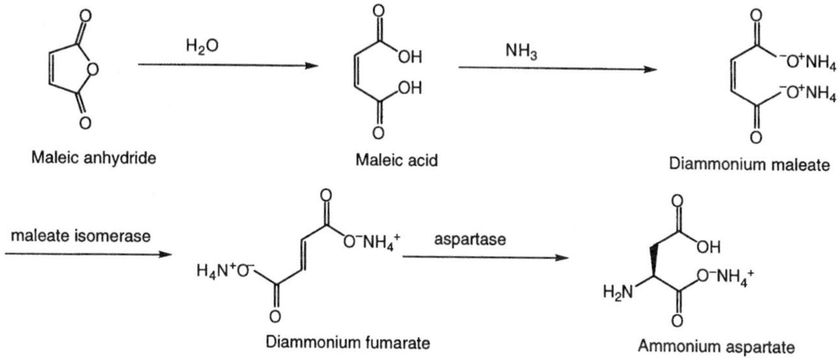

Maleic anhydride H_2O Maleic acid NH_3 Diammonium maleate

maleate isomerase   Diammonium fumarate   aspartase   Ammonium aspartate

SCHEME 3.

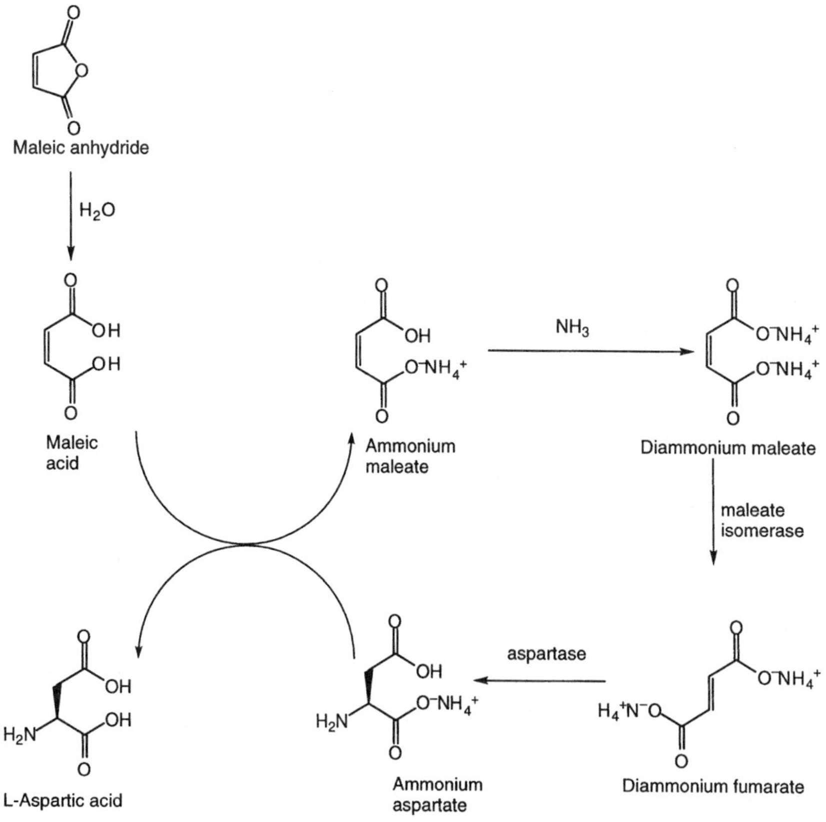

Maleic anhydride

H_2O

Maleic acid   Ammonium maleate NH_3 Diammonium maleate

maleate isomerase

L-Aspartic acid   Ammonium aspartate   aspartase   Diammonium fumarate

SCHEME 4.

REFERENCES

1. Borgeson, N. S., Bizzari, S. N., Janshekar, H., Takei, N. CEH Chemical Marketing Research Report: Amino Acids. In *Chemical Economics Handbook;* SRI International, 1995; Vol. 502.5000 A; pp. 502.5004 C–502.5004 F.
2. Chibata, I., Tosa, T., Sato, T. Aspartic Acid. In *Progress in Industrial Microbiology;* Aida, K., Chibata, I., Nakayama, K., Takinami, K., Yamada, H., Eds.; Elsevier: New York, 1986; Vol. 24; pp. 144.
3. Carlton, G. J. The Enzymatic Production of L-Aspartic Acid. In *Biocatalytic Production of Amino Acids and Derivatives;* Rozzell, J. D., Wagner, F., Eds.; Carl Hanser Verlag, Publishers: Munich, 1992; pp. 4.
4. Terasawa, M., Yukawa, H. *Bioprocess Technology* 1993, *16*, 37.
5. Quastel, J. H., Wolf, B. *J. Biochem.* 1926, *20*, 545.
6. Sherwin, M. B., Blouin, J. *US Patent,* 1985, 4560653.
7. Plachy, J., Sikyta, B. *Folia Microbiol.* 1977, *22*, 410.
8. Laanbroek, H. J., Lambers, J. T., Vos, W. M. D., Veldkamp, H. *Arch. Microbiol.* 1978, *117*, 109.
9. Suzuki, Y., Yasui, T., Mino, Y., Abe, S. *Eur. J. Appl. Microbiol. Biotech.* 1980, *11*, 23.
10. Shiio, I., Ozaki, H., Ujigawa-Takeda, K. *Agric. Biol. Chem.* 1982, *46*, 101.
11. Terasawa, M., Yukawa, H., Takayama, Y. *Process Biochem.* 1985, *20*, 124.
12. Takagi, T., Kisumi, M. *J. Bacteriol.* 1985, *161*, 1.
13. Takagi, J. S., Ida, N., Tokushige, M., Sakamoto, H., Shimura, Y. *Nucl. Acids Res.* 1985, *13*, 2063.
14. Costa-Ferreira, M., Duarte, J. C. *Biotech. Lett.* 1992, *14*, 1025.
15. Rudolph, F. B., Fromm, H. J. *Arch. Biochem. Biophys.* 1971, *147*, 92.
16. Suzuki, S., Yamaguchi, J., Tokushige, M. *Biochim. Biophys. Acta* 1973, *321*, 369.
17. Elfolk, N. *Acta Chem. Scand.* 1954, *8*, 151.
18. Falzone, C. J., Karsten, W. E., Conley, J. D., Viola, R. E. *Biochem.* 1988, *27*, 9089.
19. Ida, N., Ushige, M. *J. Biochem.* 1985, *98*, 35.
20. Nishimura, N., Kisumi, M. *Appl. Environ. Microbiol.* 1984, *48*, 1072.
21. Umemura, I., Takamatsu, S., Sato, T., Tosa, T., Chibata, I. *Appl. Microbiol. Biotech.* 1984, *20*, 291.
22. Chibata, I., Tosa, T., Sato, T. *Appl. Biochem. Biotech.* 1986, *13*, 231.
23. Nishimura, N., Kisumi, M. *J. Biotech.* 1988, *7*, 11.
24. Guest, J. R., Roberts, R. E., Wilde, R. J. *J. Gen. Micriobiol.* 1984, *130*, 1271.
25. Takagi, J. S., Tokushige, M., Shimura, Y. *J. Biochem.* 1986, *100*, 697.
26. Omori, K., Akatsuka, H., Komatsubara, S. *Plasmid* 1994, *32*, 233.
27. Sun, D., Setlow, P. *J. Bacteriol.* 1991, *173*, 3831.
28. Asai, Y., Inui, M., Vertes, A., Kobayashi, M., Yukawa, H. *Gene* 1995, *158*, 97.
29. Bolivar, F., Rodriguez, R. L., Greene, P. J., Betlach, M. C., Heynecker, H. L., Boyer, H. W., Crosa, J. H., Falkow, S. *Gene* 1977, *2*, 95.
30. Russell, D. R., Bennett, G. N. *Gene* 1982, *20*, 231.
31. Taylor, P. P. *Unpublished results.* 1986.
32. Komatsubara, S., Taniguchi, T., Kisumi, M. *J. Biotech.* 1986, *3*, 281.
33. Meacock, P. A., Cohen, S. N. *Cell* 1980, *20*, 529.

34. Nishimura, N., Taniguchi, T., Komatsubara, S. *J. Ferment. Bioeng.* 1989, *67,* 107.
35. Mizuta, K., Tokushige, M. *Biochim. Biophys. Acta* 1975, *403,* 221.
36. Mizuta, K., Tokushige, M. *Biochem. Biophys. Res. Commun.* 1975, *67,* 741.
37. Mizuta, K., Tokushige, M. *Biochim. Biophys. Acta* 1976, *452,* 253.
38. Yumoto, N., Tokushige, M., Hayashi, R. *Biochim. Biophys. Acta* 1980, *616,* 319.
39. Yumoto, N., Mizuta, K., Tokushige, M., Hayashi, R. *Physiol. Chem. Physics* 1982, *14,* 391.
40. Murase, S., Takagi, J. S., Higashi, Y., Imaishi, H., Yumoto, N., Tokushige, M. *Biochem. Biophys. Res. Commun.* 1991, *177,* 414.
41. Zhang, H. V., Zhang, J., Lin, L., Du, W. V., Lu, J. *Biochem. Biophys. Res. Commun.* 1993, *192,* 15.
42. Yumoto, N., Tokushige, M. *Biochim. Biophys. Acta* 1983, *749,* 101.
43. Tokushige, M., Mizuta, K. *biochem. Biophys. Res. Commun.* 1976, *68,* 1082.
44. Jayasekera, M. M. K., Shi, W., Farber, G. K., Viola, R. E. *Biochemistry* 1997, *36,* 9145.
45. Shi, W., Dunbar, J., Jayasekera, M. M. K., Viola, R. E., Farber, G. K. *Biochem.* 1997, *36,* 9136.
46. Cao, N., Du, J., Gong, C. S., Tsao, G. T. *Appl. Environ. Microbiol.* 1996, *62,* 2926.
47. Gangl, I. C., Weigand, W. A., Keller, F. A. *Appl. Biochem. Biotech.* 1990, *24/25,* 663.
48. Gangl, I. C., Weigand, W. A., Keller, F. A. *Appl. Biochem. Biotech.* 1991, *28/29,* 471.
49. Otsuka, K. *J. Agric. Biol. Chem.* 1961, *25,* 726.
50. Takamura, Y., Kitamura, I., Iikura, M., Kono, K., Ozaki, A. *Agric. Biol. Chem.* 1966, *30,* 338.
51. Scher, W., Jakoby, W. B. *J. Biol. Chem.* 1969, *244,* 1873.
52. Goto, M., Nara, T., Tokumaru, I., Fugono, N., Uchida, Y., Terawasa, M., Yukawa, H. *Eur. Patent,* 1995, EP 0 693 557 A2.
53. Sakano, K., Hayashi, T., Mukouyama, M. *Eur. Patent,* 1995, EP 683231 A1.
54. Kato, N., Mori, Y., Mine, N., Watanabe, N., Fujii, S. *Japan Patent,* 1996, Wo 9617950 A1.
55. Hatakeyama, K., Asai, Y., Kato, Y., Uchida, Y., Kobayashi, M., Yukawa, H. *Japan Patent,* 1996, WO 9619571 A1.

17

Synthesis of Homochiral Compounds: A Small Company's Role

BASIL J. WAKEFIELD
Ultrafine Chemicals, UFC Pharma, Manchester, England

17.1. INTRODUCTION

A small chemical company may occupy a small, specialized niche, or it may provide services covering a wide area. Ultrafine Chemicals falls into the latter category, and in many ways may be compared with the chemical research and development section of a large company. Because it is so much smaller, emphasis must be placed on versatility as well as high scientific expertise.

Many of the custom synthesis and research contracts tackled at Ultrafine Chemicals have required the preparation of homochiral compounds; therefore, it has been necessary to build up expertise in the whole range of available methodology. Some of the most interesting examples remain confidential, but the selection that follows illustrates how some of the challenges were tackled.

17.2. CLASSICAL RESOLUTION

17.2.1. 2-Chloropropionic Acid

A number of herbicides are 2-aryloxypropionic acids. These herbicides were originally used as racemates, but in the agrochemical field, as in the pharmaceutical field, there has been an increasing requirement for homochiral active enantiomers. We were approached by a manufacturer of a key intermediate for these herbicides, 2-chloropropionic acid (**1**). Because the manufacturing process was already in place, a resolution rather than an asymmetric synthesis was needed. The process should give as high a yield as possible of the required enantiomer, the resolving agent should be cheap and/or recyclable, and it should be possible to recover, racemize and then recycle the "unwanted" 2-chloropropionic acid.

The (S)-(−)-isomer (**1b**) was the one required for synthesis of the herbicides. Reaction of racemic 2-chloropropionic acid with one equivalent of (R)-(+)-α-phenylethylamine (**2**) gave a quantitative yield of the mixture of diastereoisomeric salts, which after four recrystallizations gave the required diastereoisomer in 20% yield [i.e., 40% based on the (S)-(−)-isomer present] with diastereomeric excess (de) 88% (determined by nuclear magnetic resonance [NMR]). The free acid (**1b**) was obtained quantitatively and without racemization on acidification of the salt, and the (R)-(+)-α-phenylethylamine was also recoverable without racemization in 89% yield.

Some preliminary experiments were performed on the reaction of racemic 2-chloropropionic acid with half an equivalent of (R)-(+)-α-phenylethylamine (**2**). These gave the required diastereoisomeric salt with de approximately 80% in 39% yield based on **2** after only one recrystallization.

Our methodology was regarded by the client as well suited to scale up. However, although the resolution was then, 15 years ago, regarded as an efficient example of its kind, it is noteworthy that it is now conducted on a large scale by an enzymatic method [1].

17.2.2. Anatoxin-a

Anatoxin-a (**3**) is a powerful neurotoxin (inhibitor of acetylcholinesterase) found in freshwater blue-green algae. The compound was required as an analytical standard and also for development of an immunoassay. The compound was synthesized by the eight-step sequence summarized in Scheme 1. The key intermediate (**4**) was resolved by separation of the dibenzoyl tartrates, and the remaining steps then gave both (+)- and (−)-anatoxin-a [2]. The efficiency of the resolution was monitored by formation of the "homemade Mosher's amide," Boc-Ala-(+)-anatoxin and Boc-Ala-(−)-anatoxin being distinguishable by NMR; an enantiomeric excess (ee) of ≥ 98% was achieved even before recrystallization of the salts. It is also noteworthy that the published absolute stereochemistry of the intermediate (**5**) [3] was shown to be in error by x-ray crystallography.

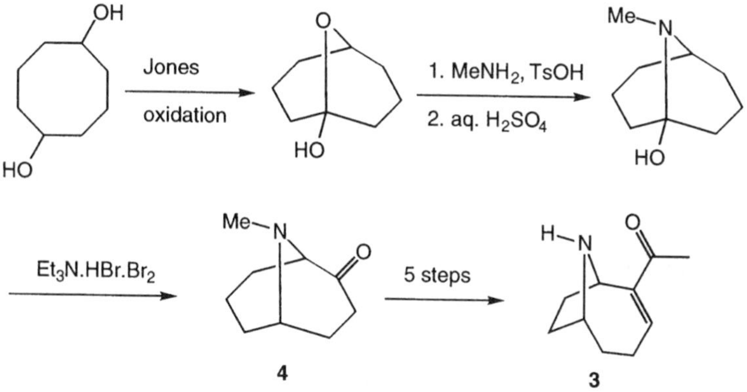

SCHEME 1.

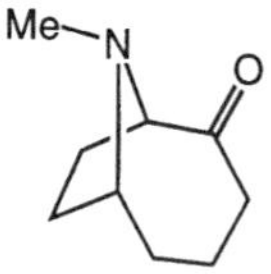

5

17.3. THE CHIRAL POOL

17.3.1. A Key Leukotriene Intermediate

The predecessor of Ultrafine Chemicals was the Fine Chemicals Unit of Salford University Industrial Centre, which was set up partly to exploit a new route to prostaglandins [4]. When Ultrafine Chemicals was founded, it made sense to offer other eicosanoids. A key intermediate for the synthesis of leukotriene A_4 (LTA_4) (**6**), and thence LTC_4, LTD_4, and LTE_4, was the epoxyalcohol (**8**), whose synthesis from 2-deoxy-D-ribose (**7**) (Scheme 2) had been reported [5].

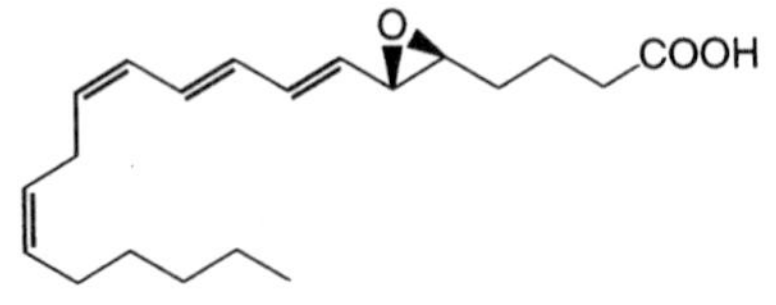

6

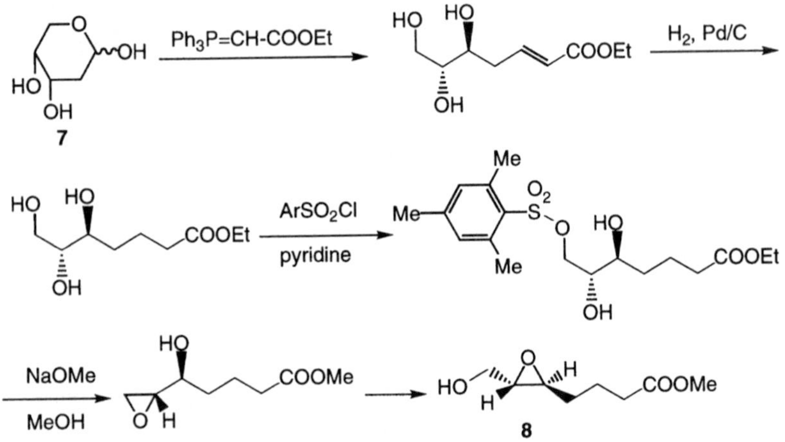

SCHEME 2.

This route was successfully used for the synthesis of the desired leukotrienes, but was later modified and improved as outlined in Scheme 3; this modified route was also used for the synthesis of [8,9,10,11-^{13}C$_4$]-leukotriene A$_4$ methyl ester.

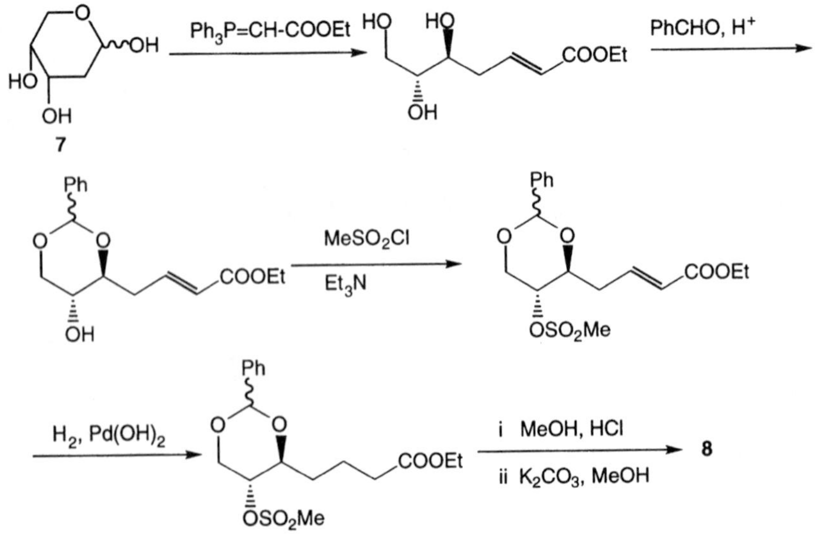

SCHEME 3.

17.4. ENZYME-CATALYZED KINETIC RESOLUTION

17.4.1. Synthesis of LTB₄

One way in which a small company can keep abreast of new developments is to collaborate with universities. Moreover, such collaboration also helps to motivate staff (and publication of results provides advertising). One fruit of such a collaboration was a synthesis of LTB₄, in which chiral moieties of the molecule are derived from the enantiomers of a common intermediate (9) [6]. Several routes have been devised for enzyme-catalyzed kinetic resolution of bicyclo[3.2.0]heptenones [7]; an efficient one that was used for the synthesis of LTB₄ is shown in Scheme 4.

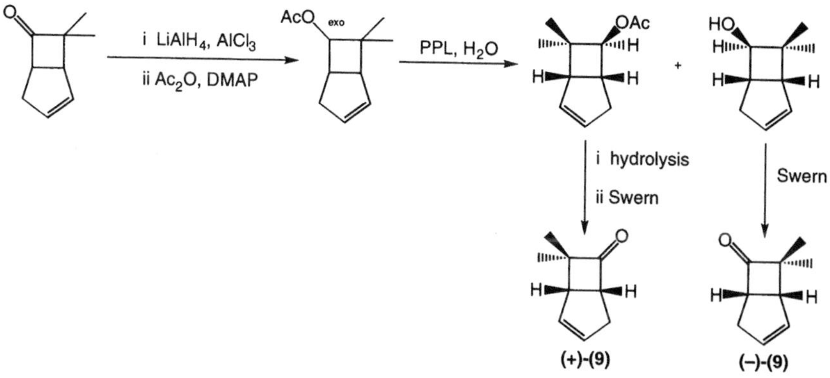

SCHEME 4.

The target molecule was then constructed from the two enantiomers (+)-(9) and (−)-(9) as shown in Scheme 5.

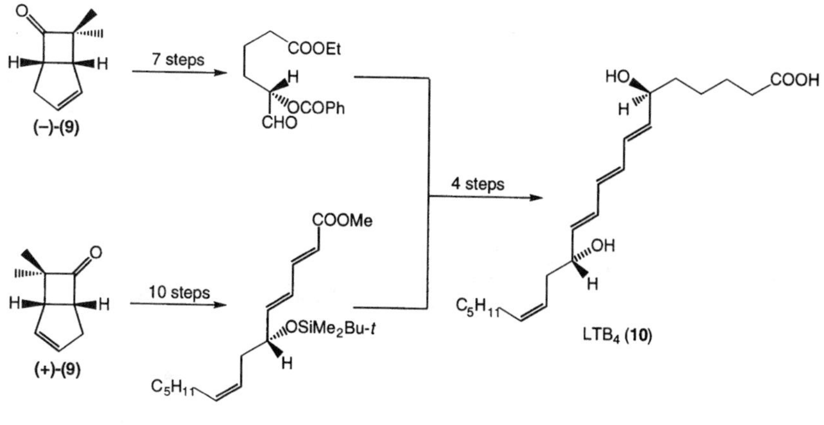

SCHEME 5.

17.4.2. The Bufuralol Story

Bufuralol (**11**) is a β-adrenoceptor agent (beta-blocker), developed by Roche [8,9]. As part of a continuing program on the synthesis of substrates and metabolites for Cytochrome P450 studies, we wished to prepare both optically active bufuralol and an important metabolite, the hydroxybufuralol (**12**). This brief account of our efforts illustrates the way in which classical resolution is giving way to asymmetric synthesis, whether mediated by synthetic chiral auxiliaries, conventional catalysts, or enzymes.

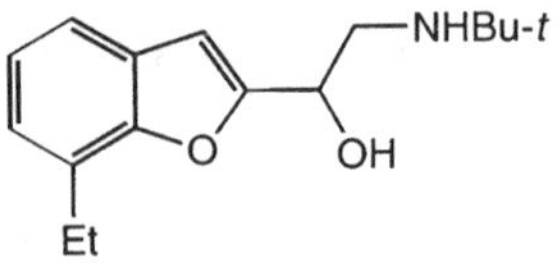

(hydrochloride)

11

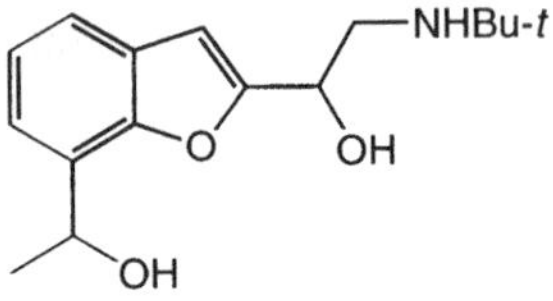

12

The published Roche synthesis of (±)-bufuralol required nine steps and gave an overall yield of approximately 1%. We were able to improve this somewhat, as outlined in Scheme 6.

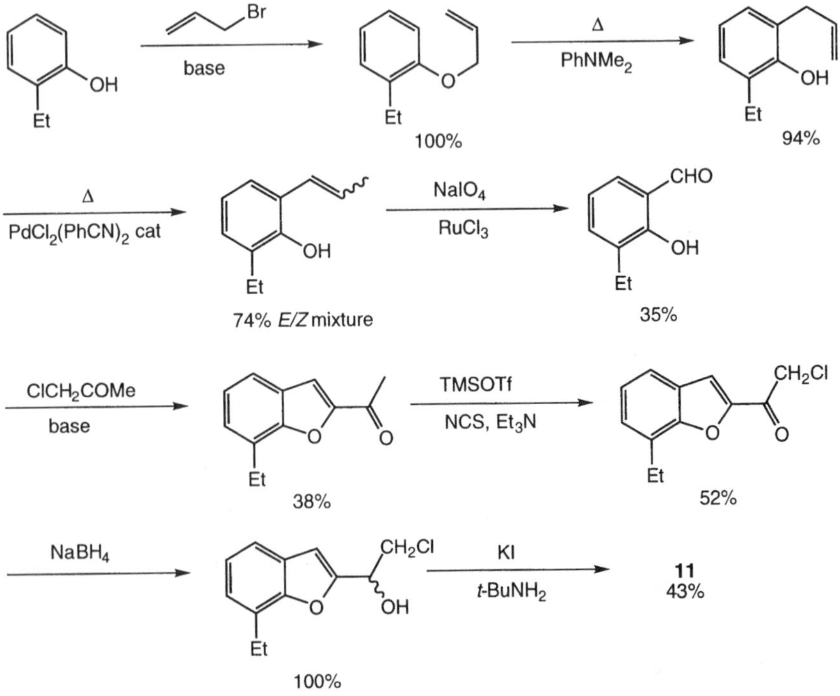

SCHEME 6.

Resolution might have been possible either by asymmetric reduction at the penultimate step, or by classical methods on the final product. However, the overall yield for the eight-step synthesis was still only 2%. Nevertheless, resolution was indeed accomplished at the penultimate stage using enantioselective transesterification and hydrolysis, both catalyzed by *Pseudomonas fluorescens* lipase, as shown in Scheme 7.

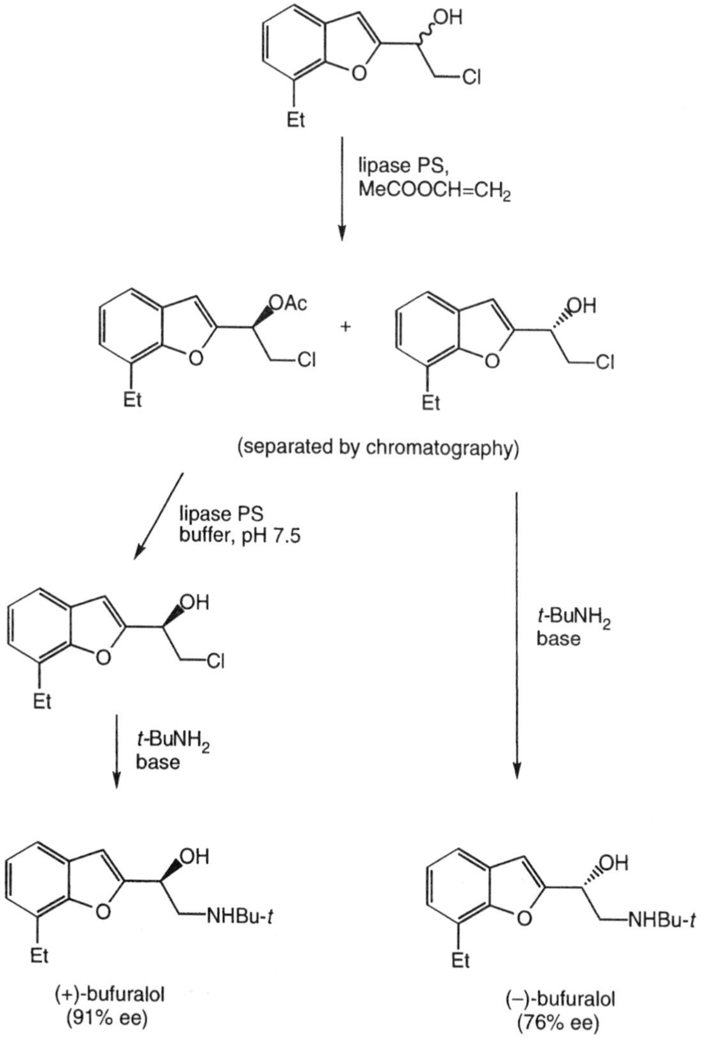

SCHEME 7.

The process development for the synthesis of hydroxybufuralol (**12**), as shown in Scheme 8, gave even better results; one step was saved, and the overall yield was 12% [10].

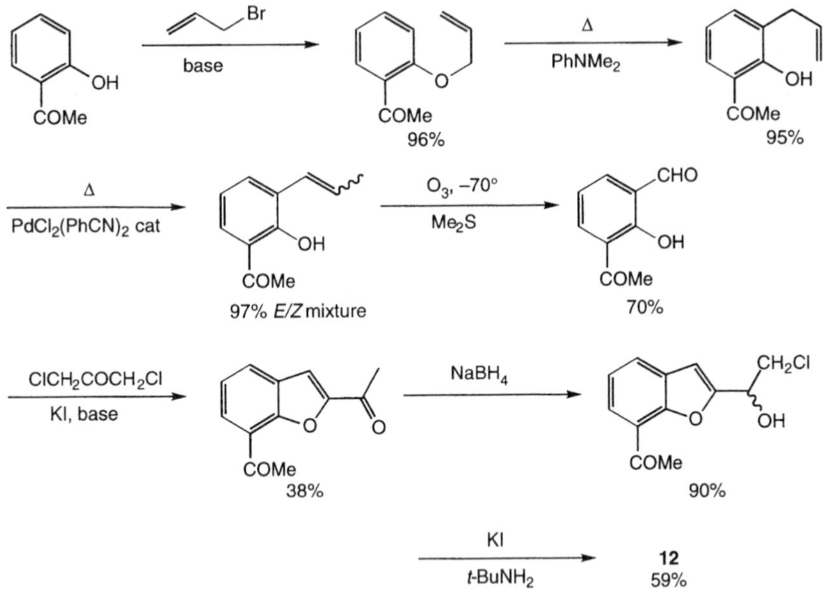

SCHEME 8.

ACKNOWLEDGEMENT

I thank Drs. Feodor Scheinmann and Andrew Stachulski for their help in the preparation of this chapter.

REFERENCES

1. Taylor, S. C. in *Opportunities in Biotransformations,* Coppang, L. G., Martin, R. E., Pickett, J. A., Bucke, C., Bunch, A. W., Eds; Elsevier: London, 1990, p. 170.
2. Ferguson, J. R., Lumbard, K. W., Scheinmann, F., Stachulski, A. V., Stjernlöf, P., Sundell, S. *Tetrahedron Lett.* 1995, *36,* 8867.
3. Stjernlöf, P., Trogen, L., Anderson, A. *Acta Chem. Scand.* 1989, *43,* 917.
4. Hart, T. W., Metcalfe, D. A., Scheinmann, F. *J. Chem. Soc. Perkin Trans. 1* 1978, 156.
5. Rokach, D., Zanboni, R., Lau, C., Guindon, Y. *Tetrahedron Lett.* 1981, *22,* 2759.

6. Cotterill, I. C., Jaouhari, R., Dorman, G., Roberts, S. M., Scheinmann, F., Wakefield, B. J. *J. Chem. Soc. Perkin Trans. 1* 1991, 2505.
7. Butt, S., Davies, H. G., Dawson, M. J., Lawrence, G. C., Leaver, J., Roberts, S. M., Turner, M. K., Wakefield, B. J., Wall, W. F., Winders, J. A. *Tetrahedron Lett.* 1985, *26,* 5077.
8. Fothergill, G. A., Francis, R. J., Hamilton, T. C., Osbond, J. M., Parkes, M. W. *Experientia* 1975, *31,* 1322.
9. Fothergill, G. A., Osbond, J. M., Wickens, J. C. *Arzneim. Forsch.* 1977, *27,* 978.
10. Collier, P.; Scheinmann, F.; Stachulski, A. V. *J. Chem. Research (S)* 1994, 312.

18

Asymmetric Catalysis: Development and Applications of the DuPHOS Ligands

MARK J. BURK

Chirotech Technology Limited, Chiroscience plc, Cambridge, England

18.1. INTRODUCTION

The importance of developing new effective methodologies for enantioselective synthesis is becoming manifestly evident. An increasing number of pharmaceutical, flavor and fragrance, and agrochemical products rely on the availability of enantiomerically pure intermediates that can serve as building blocks for further structural and/or stereochemical elaboration. Of the various techniques known for the preparation of specific optical isomers, asymmetric catalysis is emerging as one of the most efficient, versatile, and cost-effective methods for the preparation of a wide range of enantiomerically enriched compounds.

Asymmetric catalysis encompasses the use of both biocatalysts (e.g., enzymes) and chemical catalysts that possess an element of chirality (e.g., a transition metal complex bearing a chiral ligand). From a commercial perspective, the interest in asymmetric catalysis emanates from inherent economic and ecological benefits that are associated with the capacity to produce a large volume of valuable enantiomerically enriched material through the agency of a negligible quantity of a chiral catalyst. Asymmetric catalytic processes may involve kinetic resolution of a racemic substrate, or preferably, direct transformation of a prochiral substrate into the desired chiral molecule.

Although the practical advantages rendered by these methodologies have been demonstrated through numerous large-scale applications, the broad utility of asymmetric catalysis in the manufacture of chiral compounds has yet to be fully realized. It is important to recognize all of the attributes that a catalyst must embody to be acceptable for a robust large-scale manufacturing process. In addition to high enantioselectivities, high catalytic activities and catalytic efficiencies are imperative (see also Chapter 9).

The purpose of this chapter is to inspire future applications of asymmetric catalysis in industry. As a didactic example, we feature the design and development of a unique class of chiral diphosphine ligands that are useful in asymmetric chemocatalysis. We illustrate the availing nature of these ligands in asymmetric catalytic hydrogenation reactions. As will become apparent, effective ligand design is a quintessential component in the cultivation of viable asymmetric catalytic processes.

18.2. CHIRAL LIGANDS

The design and synthesis of new chiral ligands that coordinate to transition metals is a vital aspect of asymmetric catalysis research. A compilation of structural and electronic attributes that provide a blueprint for the generation of potent asymmetric catalysts is not available. The factors that are required for a catalyst to display a combination of high efficiency and high enantioselectivity remain nebulous and are often reaction specific. Rational or de novo design of effective chiral ligands continues as the exception rather than the rule.

Despite insufficient mechanistic detail, a growing number of useful chiral ligands that provide very high enantioselectivities are being identified. Of these ligands, chiral phosphines possess great potential because of the vast number of catalytic processes that use phosphine ligands. Indeed, some of the most successful and general ligands used in asymmetric catalysis are chiral phosphines. This fortune is limited, however, because many asymmetric transformations of interest are not yet possible due to the shortcomings of available catalysts. The need continues for alternative asymmetric ligands and/or catalysts that provide improved stereoinductive properties and catalytic efficiencies.

18.2.1. Design of New Phospholane Ligands

Our initial studies were dedicated to the design and synthesis of a new class of chiral phosphine ligands. Most of the successful chiral phosphine ligands available at the time possessed at least two aryl substituents on the phosphorus center. Aryl substituents render the phosphorus atoms, and thus the attached metal centers, relatively electron-poor. Because the electronic properties of phosphine ligands are well known to dramatically influence the reactivity and selectivity of transition metal centers, our first objective was to prepare new electron-rich (per-alkyl) chiral phosphines. At that time, the use of electron-rich chiral phosphines was largely unexplored.

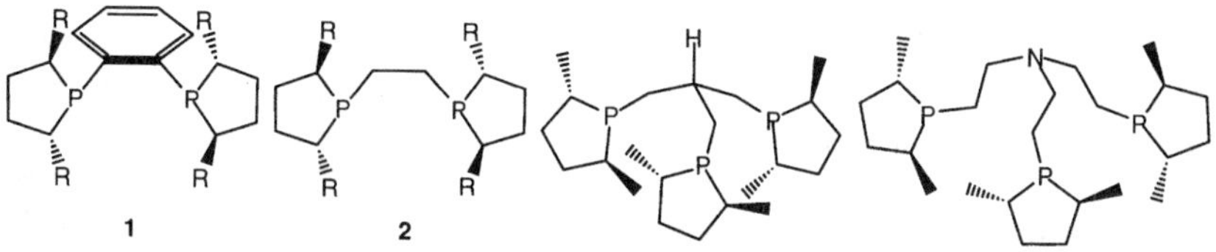

FIGURE 1 C_2-symmetric bis (phospholane) and C_3-symmetric tris (phospholand) ligands.

Conceptually, the ligands we sought were structurally homologous to other known C_2-symmetric *trans*-2,5-disubstituted five-membered heterocycles (i.e., B, N). The latter compounds were recognized to efficiently transfer their chirality with high levels of absolute stereocontrol [1]. These observations led us to synthesize the analogous phosphorus derivatives, *trans*-2,5-disubstituted phospholanes [2]. A series of chelating phospholane-based ligands followed, the most notable being the DuPHOS (**1**) and BPE (**2**) ligands (Figure 1) [3,4].

One of the true advantages that we envisaged our ligands might offer is a modular design that facilitates systematic variation of the steric environment imposed by these systems. The ability to manipulate the phospholane substituents of our ligands, we felt, would allow us to finely tune the steric environment of the ligands to accomodate the steric demands of different substrates of interest. It was anticipated that such "steric matching" would provide a facile means of optimizing rates and enantioselectivities in asymmetric catalytic transformations.

Many known asymmetric diphosphines are designed in such a way that the elements of chirality are confined to the backbone unit that links the two phosphorus atoms [5,6]. The incorporation of ligand chirality into the phospholane rings rather than the backbone tether provides the opportunity to use diverse backbone structures in an effort to control ligand conformational mobility. As will be evinced in the following section, these tenets were corroborated through ample experimentation, and the capacity to sterically and conformationally tailor a catalyst to a substrate provides the basis for the great versatility manifested by these ligand systems.

When designing new chiral ligands and catalysts, the effectiveness of the design must be verified by reference to certain criteria. New chiral phosphine ligands often are validated in model hydrogenation reactions through comparison with known benchmarks. Accordingly, we opened our catalytic screening program with asymmetric hydrogenation reactions. The great success we achieved in this singular enterprise has captivated our curiosity and guided our efforts through to the present. That many applications of the DuPHOS and BPE ligands remain to be uncovered both within and outside the compass of hydrogenation chemistry is evident, and numerous demonstrative examples are beginning to appear in the literature.

18.2.2. Manufacture of Chiral DuPHOS and BPE Ligands

Asymmetric catalysis is heralded for the potential advantages it can bestow in the production of chiral molecules. Ultimately, however, asymmetric catalysis is vindicated only through practical application. Commercialization of asymmetric catalytic processes relies on economical large-scale production of the requisite chiral ligands and/or catalysts. To realize the capabilities inherent in the DuPHOS/BPE technology, it was vital that we develop a practicable procedure for manufacture of these ligands.

The general route that we have optimized for fabrication of the DuPHOS/ BPE ligands is shown in Scheme 1 [demonstrated for (R,R)-Me-DuPHOS (**3**)]. In the original DuPHOS synthesis reported in the literature [7], access to the essential chiral 1,4-diols was provided by a Kolbe coupling procedure involving chiral β-hydroxy acids. Although expedient, this electrochemical method proved too capricious and impractical for large-scale ligand production. Hence, an alternative process for the synthesis of chiral 1,4-diols was required as a first step in the development of a viable ligand synthesis.

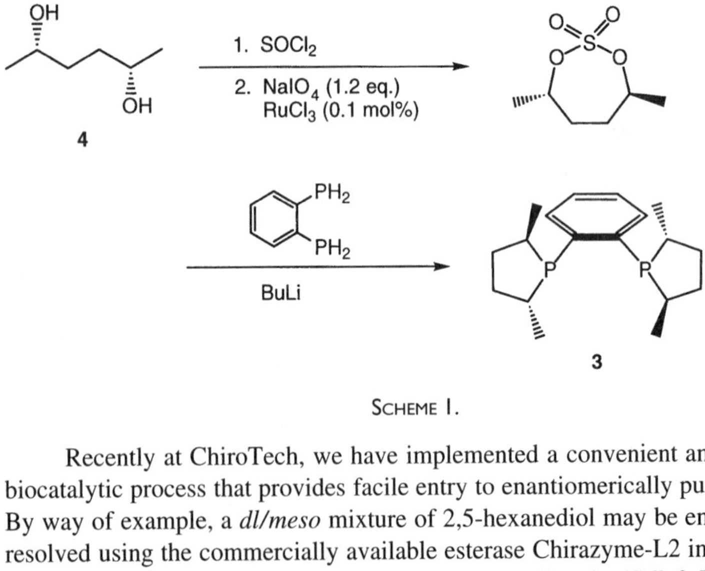

SCHEME 1.

Recently at ChiroTech, we have implemented a convenient and proficient biocatalytic process that provides facile entry to enantiomerically pure 1,4-diols. By way of example, a *dl/meso* mixture of 2,5-hexanediol may be enzymatically resolved using the commercially available esterase Chirazyme-L2 in vinyl butyrate. Upon completion, aqueous extraction yields directly (S,S)-2,5-hexanediol (**4**) in 17% yield. Further processing of the resultant mixture of butyrate esters allows isolation of antipodal (R,R)-2,5-hexanediol in enantiomerically and diastereomerically pure form in 42% overall yield. We have performed this chemistry on up to 200 kg of 2,5-hexanediol, thus providing ample material for production of the Me-DuPHOS and Me-BPE ligands. This development, combined with other process improvements, has rendered a suitable process for manufacture of multikilogram quantities of the DuPHOS and BPE ligands.

18.3. ASYMMETRIC CATALYTIC HYDROGENATION REACTIONS

Enantioselective hydrogenation reactions are probably the most thoroughly studied and well understood of any asymmetric catalytic processes yet developed (see Chapter 9) [8,9]. Asymmetric hydrogenation methods are characterized by convenience, cleanliness, high catalytic efficiency, versatility, and durability. Moreover, the highest enantioselectivities in asymmetric catalysis have been achieved predominantly in hydrogenation reactions. Asymmetric hydrogenation reactions are eminently amenable to large-scale production. Manifold prochiral unsaturated substrates are available, and thus asymmetric hydrogenation reactions potentially could provide direct and practical access to a panoply of chiral building blocks.

Yet despite these patent advantages, asymmetric hydrogenation processes have remained relatively underdeveloped and severely underused, especially in chemical manufacturing. A vast number of important substrate classes have yet to be hydrogenated with both high catalytic efficiencies and high enantioselectivities. Hence, at the commencement of the asymmetric hydrogenation program outlined in this chapter, it was apparent that further catalyst design and development were required before widespread application of this pivotal technology would transpire.

The contributions that the DuPHOS and BPE ligand series have made in the area of asymmetric hydrogenation chemistry are exemplified below. One purpose of this treatise is to underscore the broad scope of the DuPHOS and BPE catalyst systems for the generation of diverse collections of chiral compounds.

18.3.1. α-Enamides

Historically, the desire to develop practical routes to enantiomerically pure α-amino acid derivatives provided initial inspiration for the advent of asymmetric catalytic hydrogenation reactions almost 30 years ago [10,11]. Surprisingly, still prevailing in 1990 was the challenge to identify a truly general catalyst that would allow the enantioselective addition of hydrogen across the carbon-carbon double bond of a wide range of α-enamide substrates of generic structure **5**.

18.3.1.1. N-Acetyl α-Amino Acids

Preliminary investigations revealed that cationic rhodium complexes [(COD)Rh(-DuPHOS)]$^+$OTf$^-$ bearing the Et-DuPHOS or Pr-DuPHOS ligands [2,5-substituents on phospholanes (**1**) R = Et or Pr, respectively] were effective catalyst precursors for highly enantioselective hydrogenation of a broad range of N-acetyl α-enamide esters and acids (**5; R = Me**) (Scheme 2) [4].

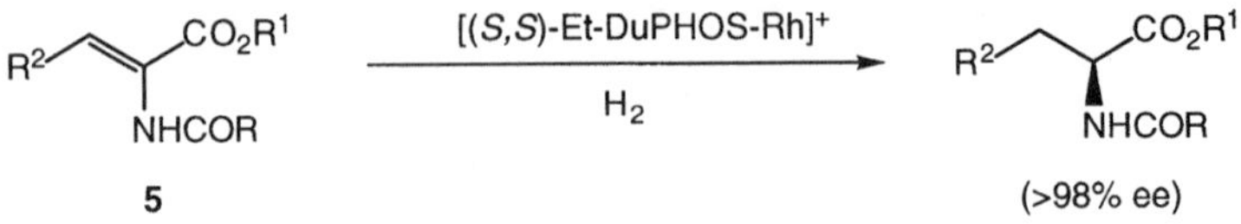

SCHEME 2.

The Et-DuPHOS-Rh catalyst was found to be extremely efficient, with substrate-to-catalyst (S/C) ratios up to 50,000/1 demonstrated. Moreover, these robust catalysts displayed tolerance to many different solvent types and organic functional groups. Hence, using this technology, we could gain ready access to a diverse array of unnatural α-amino acid derivatives [7,12]. Before this demonstration, no single asymmetric hydrogenation catalyst had been developed and shown to reduce such an expansive range of α-enamides to α-amino acid derivatives with very high enantioselectivities. However, although propitious, several issues concerning α-enamide hydrogenations remained to be addressed before a genuinely practicable technology could be claimed.

18.3.1.2. N-Cbz and N-Boc α-Amino Acids

Until very recently, most studies concerning asymmetric hydrogenation of α-enamides **5** have focused on N-acetyl or N-benzoyl protected substrates [13]. Application of the resultant amino acids in peptide synthesis requires subsequent removal of these protecting groups, which generally demands rather harsh acidic or basic conditions that commonly lead to varying degrees of racemization. Recent studies have revealed that hydrogenation of α-enamides possessing more easily removed N-protecting groups, such as N-Cbz or N-Boc, often leads to inferior results with most known catalysts [14,15].

We recently have found that the cationic Et-DuPHOS-Rh and Pr-DuPHOS-Rh catalysts allow the highly enantioselective hydrogenation of a wide range of N-Cbz and N-Boc-α-enamide esters and acids [12,16,17]. Moreover, these catalysts were found to reduce both (E)-and (Z)-N-Boc-β-alkyl-α-enamides (**5**; R = t-BuO, R^2 = alkyl) with high enantioselectivites. This finding was crucial as present routes to N-Boc-α-enamides generally afford $(E)/(Z)$-isomeric mixtures. Since N-Boc-α-enamides that possess (E)-β-aryl substituents were hydrogenated with lower enantioselectivities, we have developed a convenient route to a wide variety of isomerically pure (Z)-N-Boc-β-aryl-α-enamides **6** (Scheme 3) [17,18].

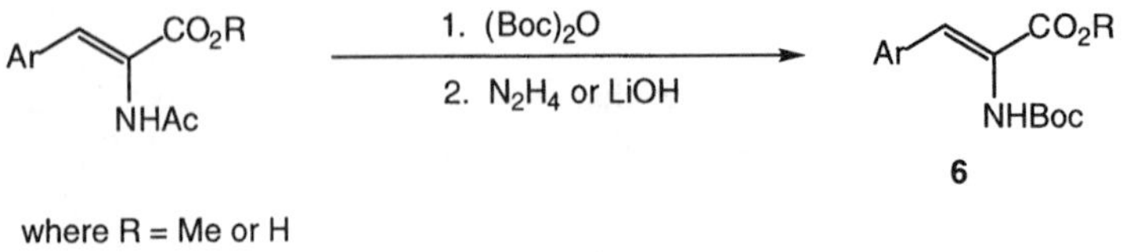

where R = Me or H

SCHEME 3.

Thus, treating readily available (*Z*)-*N*-acetyl α-enamides, obtained via Erlenmeyer condensation, with di-*tert*-butyldicarbonate followed directly by hydrazine or LiOH, affords the corresponding *N*-Boc α-enamide ester or acid, respectively. Hydrogenation of substrates **3** was found to proceed with enantioselectivities of > 98% ee by the use of the Et-DuPHOS-Rh catalyst. Alternatively, *N*-Boc amino acids may be obtained by initial hydrogenation of the *N*-acetyl α-enamides, followed by transmutation to the *N*-Boc derivative as previously outlined [18]. The mild nature of this method is evinced by the absence of detectable racemization in the *N*-Ac to *N*-Boc conversion performed after hydrogenation.

Schmid et al. at Hoffmann–La Roche recently reported similar results that bespeak the pre-eminence of the Me-DuPHOS-Rh catalyst for asymmetric hydrogenation of heterocyclic *N*-Cbz α-enamides [19]. Table 1 lists enantioselectivities achieved in the hydrogenation of a diverse range of *N*-Cbz and *N*-Boc α-enamides using the DuPHOS-Rh catalysts.

TABLE I DuPHOS-Rh-Catalyzed Hydrogenation of *N*-Cbz and *N*-Boc Enamides **5**[a]

5; R^2=	*N*-Protection	Ligand	% ee
H	Cbz	Pr-DuPHOS	>99
C_2H_5	Cbz	Pr-DuPHOS	>99
4-F-C_6H_4	Cbz	Pr-DuPHOS	>99
2-furanyl[b]	Cbz	Me-DuPHOS	>99
3-furanyl[b]	Cbz	Me-DuPHOS	99
2-thienyl[b]	Cbz	Me-DuPHOS	>99
3-thienyl[b]	Cbz	Me-DuPHOS	98
1 *H*-2-pyrrolyl[b]	Cbz	Me-DuPHOS	97
CH_3	Boc	Et-DuPHOS	>99
C_2H_5 (*Z*-enamide)	Boc	Pr-DuPHOS	>99
C_2H_5 (*E*-enamide)	Boc	Pr-DuPHOS	>99
C_3H_7	Boc	Pr-DuPHOS	>99
i-Pr	Boc	Pr-DuPHOS	95
t-Bu	Boc	Pr-DuPHOS	95
$C_6H_5CH_2$	Boc	Et-DuPHOS	97
C_6H_5	Boc	Et-DuPHOS	99
2-Br-C_6H_4	Boc	Pr-DuPHOS	>99
3-Br-C_6H_4	Boc	Pr-DuPHOS	>99
4-Br-C_6H_4	Boc	Pr-DuPHOS	>99
2-Naphthyl	Boc	Pr-DuPHOS	98
1-Naphthyl	Boc	Pr-DuPHOS	>99
2-Furanyl	Boc	Pr-DuPHOS	>99
2-Thienyl	Boc	Pr-DuPHOS	98

[a] All hydrogenations performed with cationic Rh catalyst precursors [(COD)Rh-DuPHOS]OTf and with α-enamide methyl ester substrates in MeOH, unless otherwise noted. All results involving *N*-Cbz substrates were extracted from references 7 and 19. All results involving *N*-Boc substrates derive from reference 17.

[b] α-Enamide *t*-butyl esters were used as substrates [19].

18.3.1.3.   Chemoselectivity and Regioselectivity

High levels of both chemoselectivity and regioselectivity are additional features manifested by the cationic DuPHOS-Rh catalysts in asymmetric hydrogenations. For instance, unsaturated functionality within an α-enamide substrate, such as aldehyde, ketone, imine, oxime, and nitro groups, typically are stable to reduction by these catalysts. Moreover, isolated carbon–carbon double and triple bonds are not reduced in the timeframe required for hydrogenation of the α-enamide moiety. The latter aspect has allowed the development of a convenient route to extremely valuable allylglycine derivatives 7, through highly regioselective and enantioselective hydrogenation of γ,δ-unsaturated α-enamides 8 (Scheme 4) [20,21]. The Et-DuPHOS-Rh catalysts were found paramount for this process, whereby a wide array of allylglycine derivatives 7 were obtained in > 98% ee and with < 1% overreduction to the fully saturated amino acid. High regioselectivity in this reaction likely derives from coordination of the amide carbonyl oxygen to the catalyst, thus directing hydrogenation to the α-enamide alkenyl unit.

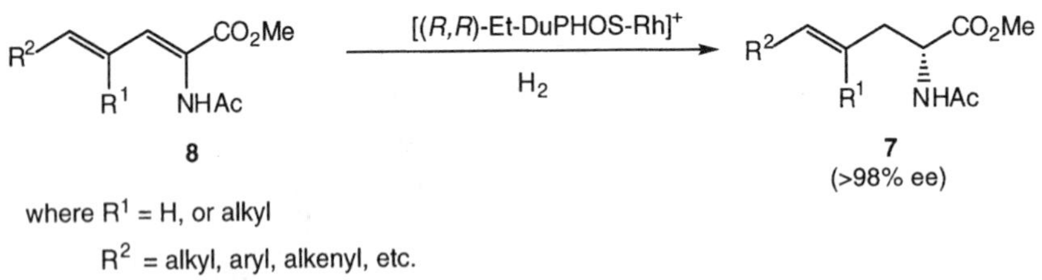

where R^1 = H, or alkyl

R^2 = alkyl, aryl, alkenyl, etc.

SCHEME 4.

Allylglycine derivatives of type 7 are found in nature and may serve as versatile synthetic intermediates by virtue of the remaining unsaturation in the molecule. A representative example of the synthetic utility of allylglycines 7 is shown in Scheme 5. Here, a concise asymmetric synthesis of the natural product

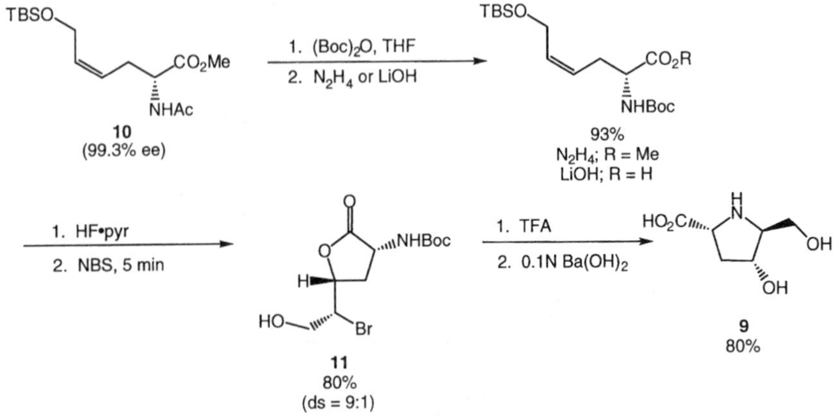

SCHEME 5.

(+)-bulgecinine (**9**) is achieved through conversion of the *N*-Ac allylglycine **10** into the *N*-Boc derivative as previously described [18], followed by bromolactonization to **11** and subsequent base-induced rearrangement to the desired product [21].

18.3.1.4. β-Branched α-Amino Acids

We have alluded to the fact that the cationic Et-DuPHOS-Rh catalysts were found capable of hydrogenating both (*E*)- and (*Z*)-isomeric α-enamides with high enantiomeric excesses. Deuteration studies eliminated substrate isomerization as a rationale for this uncommon behavior. This observation implied that these catalysts could tolerate a substituent in either the (*E*)-or (*Z*)-position of an α-enamide substrate. By corollary, we reasoned that our catalysts uniquely may allow the highly enantioselective hydrogenation of α-enamides possessing substituents in both the (*E*)- and (*Z*)-positions (i.e., β,β-disubstituted α-enamides **12**) (Scheme 6). Although the ability to hydrogenate β,β-disubstituted α-enamides with high ee's could provide practical access to an array of valuable, constrained β-branched α-amino acids, little success had been reported for enantioselective reduction of this substrate class [13,22].

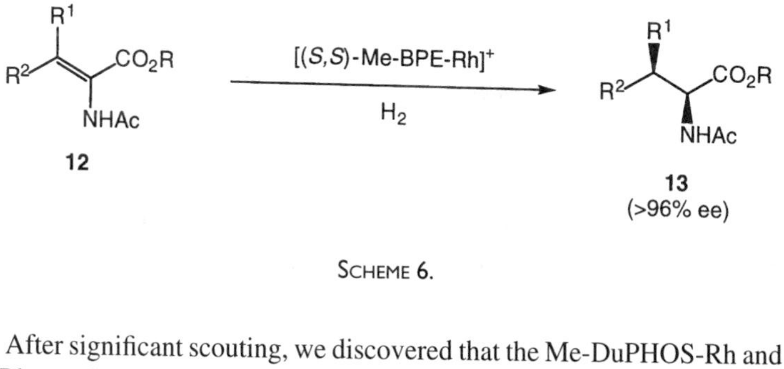

$$R^2 \overset{R^1}{=} \diagup CO_2R \quad \xrightarrow[H_2]{[(S,S)\text{-Me-BPE-Rh}]^+} \quad R^2 \overset{R^1}{\diagup} CO_2R$$

$$\underset{\text{NHAc}}{} \qquad \qquad \underset{\text{NHAc}}{}$$

12   **13**
                                                         (>96% ee)

SCHEME 6.

After significant scouting, we discovered that the Me-DuPHOS-Rh and Me-BPE-Rh catalysts provide a highly enantioselective route to a variety of β-branched amino acid derivatives **13** [16,17,23]. Hydrogenation of α-enamides **12** possessing dissimilar β-substituents allows the simultaneous generation of an additional new β-stereogenic center in the product. For instance, both *erythro* and *threo* diastereomers of the amino acid β-methylhomophenylalanine were produced in 96.2% ee and 95.8% ee, respectively, through hydrogenation of the corresponding (*E*)- and (*Z*)-enamides using (*R*,*R*)-Me-BPE-Rh. Other examples are shown in Figure 2.

The enantioselectivities we have achieved in the hydrogenation of β,β-disubstituted α-enamides are significantly higher than any previously reported for this challenging class of substrates. The sterically demanding nature of enamides **12** (tetrasubstituted olefins) has presented difficulties in the past with regard to both rates and enantioselectivities. These results lucidly illustrate the advantage

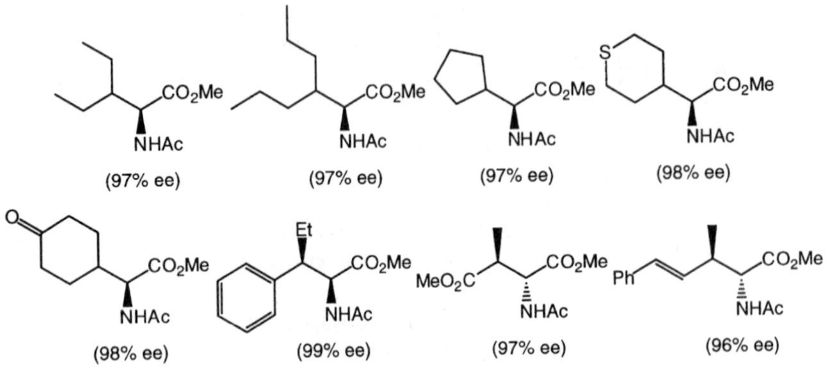

FIGURE 2 β-Branched amino acids via Me-BPE-Rh-catalyzed hydrogenation.

conferred by our ability to vary the steric environment imposed by the DuPHOS and BPE ligands. In the present case, the least encumbered Me-DuPHOS and Me-BPE ligands were found optimum with regard to enantioselectivity and rates in the hydrogenation of the sterically congested α-enamides **12**.

18.3.1.5. Amino Acids via Tandem Catalysis

Discovery and optimization of peptide and peptidomimetic therapeutics generally requires ready access to a diverse range of chiral building blocks. The DuPHOS-Rh catalysts provide one of the most convenient routes to novel amino acids, yet individual α-enamides must be prepared for each new amino acid needed. In an effort to enhance the efficiency of our hydrogenation methodology, we have developed a two-step tandem catalysis procedure for preparation of a wide range of aromatic amino acids and peptides (Scheme 7) [16,17,24].

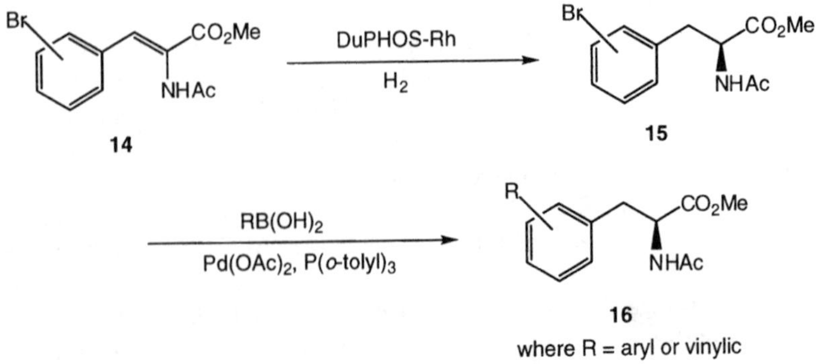

SCHEME 7.

Asymmetric hydrogenation of bromo-substituted aromatic α-enamides **14** affords the corresponding bromo-amino acid derivatives **15**, which subsequently is subjected to Pd-catalyzed cross-coupling with aryl and vinyl boronic acids. In addition to diverse phenylalanine derivatives **16**, a broad array of other novel aromatic and heterocyclic amino acids have been produced rapidly from a small number of bromo-functionalized intermediates [24]. This same two-step process may be applied to the production of many other classes of aromatic and heterocyclic chiral building blocks, such as arylalkylamines, amino alcohols, diamines, and directly on peptides as well.

18.3.2. β-Amino Alcohols

Optically active amino alcohols are important constituents of a multitude of biologically active natural products and routinely serve as intermediates for the synthesis of enantiomerically pure pharmaceuticals and agrochemicals. In an effort to gain access to such valuable chiral building blocks, we next focused our attention on the asymmetric hydrogenation of enamides of type **17**, which lack an α-carboxyl function. Although the significance of such a process is apparent, little success had been achieved in this area before our studies were conducted. We have found that hydrogenation of α-enamides **17** possessing hydroxymethyl substituents proceeds readily using the cationic Et-DuPHOS-Rh catalyst to afford a series of β-amino alcohol derivatives **18** directly and with very high enantioselectivities (> 95% ee) (Scheme 8) [25].

SCHEME 8.

18.3.3. α-1-Arylalkylamines

α-1-Arylalkylamines constitute another important class of compounds that have been used extensively as chiral auxiliaries, resolving agents, and building blocks for the synthesis of a wide variety of biologically active molecules. Generally, α-1-arylalkylamines have been challenging to prepare in enantiomerically pure form on commercial scales, and thus access is open to only a limited range of these valuable compounds. An asymmetric catalytic process that allows the preparation of α-1-arylalkylamines would have great merit and value.

Recently, we have found that our Me-DuPHOS-Rh and Me-BPE-Rh catalysts are highly effectual for the asymmetric hydrogenation of α-arylenamides **19** with enantioselectivities of > 95% ee (Scheme 9). Importantly, β-substituted enamides may be hydrogenated as *E/Z* mixtures with high enantioselectivities, thus providing simple entry to a diverse array of α-1-arylalkylamine derivatives **20** [26].

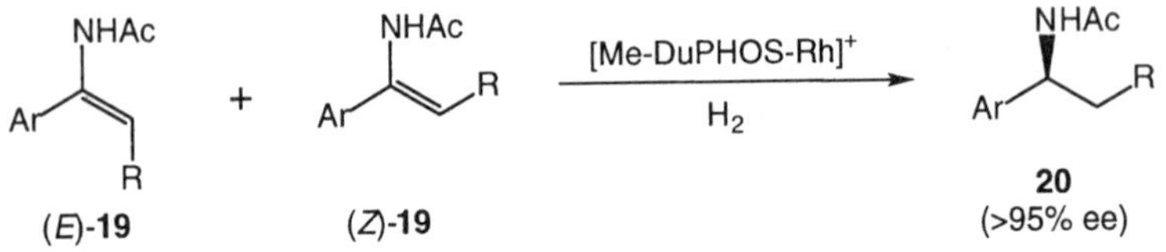

SCHEME 9.

This innovation has the verisimilitude of great practical worth. To date, attainment of high enantioselectivities in the hydrogenation of α-arylenamides is singularly associated with the Me-DuPHOS-Rh and Me-BPE-Rh catalysts. However, there remained a major obstacle to the establishment of a commercially practicable process for the production of α-1-arylalkylamines—no viable synthesis of α-arylenamides existed. In asymmetric catalysis, the actual catalytic step is only one aspect of the entire process. A muted yet crucial facet of industrial asymmetric catalysis deals with substrate synthesis. The great emphasis placed on the catalytic step often obscures the need for economical production of the substrate. Many potential asymmetric catalytic reactions have been rendered impractical because of inaccessibility of the requisite substrates.

We now have surmounted this hurdle through development of an efficacious route to α-arylenamides of type **19** directly from simple ketones [27]. The process involves two simple steps (Scheme 10): (1) conversion of an aromatic

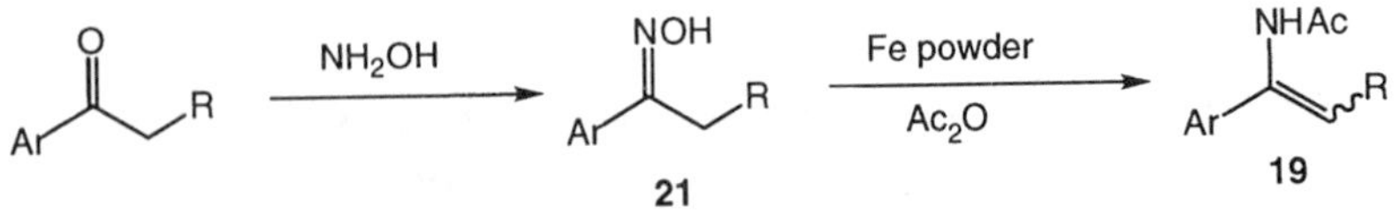

SCHEME 10.

ketone to its oxime derivative **21**; and (2) reduction of the oxime with iron metal in acetic anhydride to provide the desired α-arylenamide **19** as an *E/Z* mixture, where appropriate.

Overall, we have established a convenient and cost-effective three-step reductive amination process for the manufacture of multifarious α-1-arylalkylamines. Importantly, the new procedure outlined here also allows the preparation of previously unavailable enamides, including α-alkyl-substituted derivatives. For instance, this method has been used to convert pinacolone into enamide **22**, which upon hydrogenation with the (*S,S*)-Me-DuPHOS-Rh catalyst affords the corresponding *N*-acetyl amine **23** with very high enantiomeric excess (99% ee) (Scheme 11) [27].

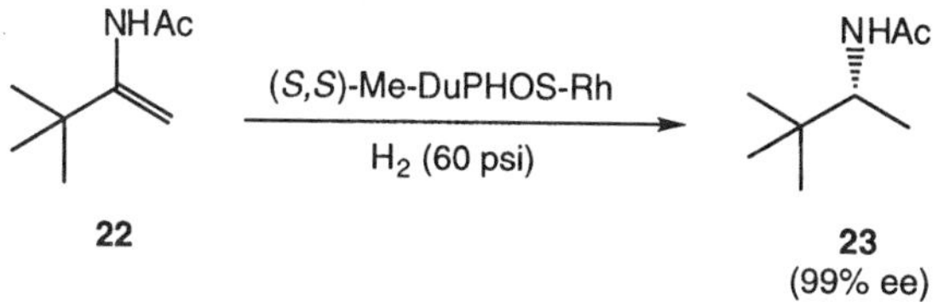

NHAc
(*S,S*)-Me-DuPHOS-Rh
⟶
H₂ (60 psi)

NHAc

22

23
(99% ee)

SCHEME 11.

18.3.4. 2-Alkylsuccinates

Enantiomerically pure 2-substituted succinic acid derivatives recently have attracted great interest because of their ability to serve as chiral building blocks and peptidomimetics in the design of pharmaceuticals, flavors and fragrances, and agrochemicals with improved properties. We recently have identified the most broadly effective catalysts for the asymmetric hydrogenation of itaconate derivatives **24,** thus providing facile access to a diverse collection of 2-alkylsuccinates **25** (Scheme 12). Interestingly, the itaconate substrates **24** may be prepared conveniently as crude *E/Z* isomeric mixtures via Stobbe condensation, and then used directly in the hydrogenation reaction without purification. As shown in Table 2, the Et-DuPHOS-Rh catalyst effects the asymmetric hydrogenation of a range of itaconates with exceedingly high enantioselectivites [28].

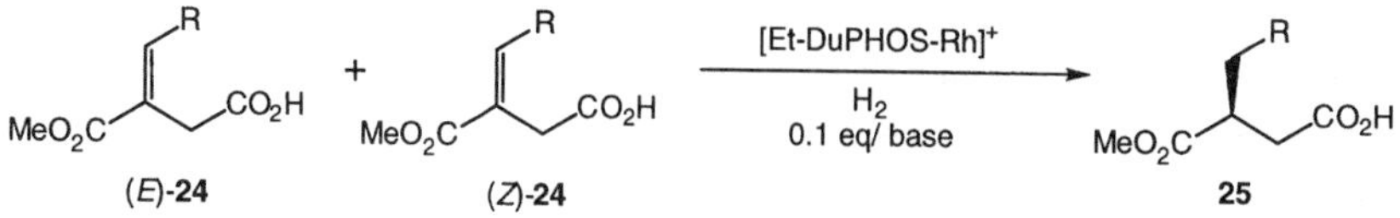

MeO₂C⟋⟍CO₂H + MeO₂C⟋⟍CO₂H $\xrightarrow[\text{0.1 eq/ base}]{\text{[Et-DuPHOS-Rh]}^{+} \quad \text{H}_2}$ MeO₂C⟋⟍CO₂H

(*E*)-**24**   (*Z*)-**24**   **25**

SCHEME 12.

TABLE 2   Asymmetric Hydrogenation
of Itaconates **24** Using Et-DuPHOS-
Rh Catalyst

Entry	R in 24	% ee
1	H	97
2	Et	99
3	n-Bu	98
4	CH_2CH_2Ph	99
5	i-Pr	99
6	c-Pr	99
7	Cy	98
8	t-Bu	99
9	Ph	97
10	1-Naphthyl	98
11	2-Naphthyl	98
12	2-Thienyl	99

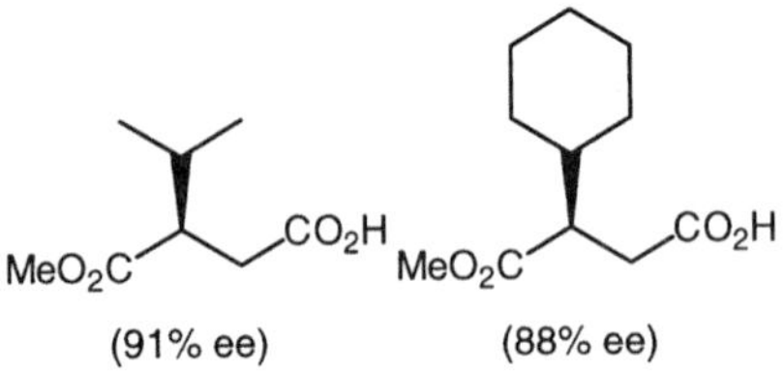

(91% ee)   (88% ee)

FIGURE 3   β-Branched succinates from (S,S)-Me-BPE-Rh-catalyzed hydrogenation.

Significantly, we also have found that β,β-disubstituted itaconates may be hydrogenated with high enantioselectivities using the cationic Me-BPE-Rh catalyst. Figure 3 depicts representative succinate derivatives we have prepared using this methodology [28]. The enantioselectivites listed are by far the highest yet achieved in the hydrogenation of β,β-disubstituted itaconates.

18.3.5. α-Hydroxy Esters and 1,2-Diols

Both α-hydroxy esters and 1,2-diols are widely used chiral synthetic intermediates. Development of a broadly effective and practicable catalyst for highly enantioselective hydrogenation of α-keto esters to α-hydroxy esters has not yet been reported [29,30]. We have observed similarly unsatisfactory results to date in the hydrogenation of α-keto esters with our cationic DuPHOS-Rh and BPE-Rh catalysts. An alternative approach to optically active α-hydroxy esters could be through enantioselective hydrogenation of the corresponding enol esters such as **26** (Scheme 13).

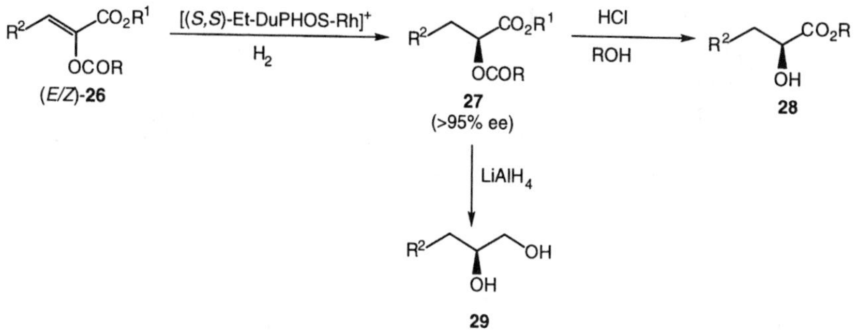

SCHEME 13.

We have found that the Et-DuPHOS-Rh catalysts are capable of effecting smooth and highly enantioselective hydrogenation of *E/Z* mixtures of enol esters **26** (R = Me, Ph, etc.) [31]. As delineated in Table 3, a wide substrate scope has been demonstrated with enantioselectivities of > 95% ee. Products **27** thus obtained may be directly converted to α-hydroxy esters **28** through mild acidic hydrolysis in alcohol, or alternatively **27** may be reduced to the corresponding 1,2-diols **29** by the use of reagents such as LiAlH$_4$, with no loss of enantiomeric purity.

TABLE 3 Et-DuPHOS-Rh-Catalyzed Hydrogenation of Enol Esters **26** (R, R^1 = Me)

Entry	R^2 in **26**	% ee
1	H	99
2	Me	96
3	*n*-Pr	98
4	C$_5$H$_{11}$	99
5	*i*-Pr	97
6	*i*-Bu	99
7	*c*-Pr	98
8	C$_6$H$_{11}$	95
9	C$_6$H$_5$	98
10	*m*-BrC$_6$H$_4$	97
11	2-Naphthyl	93
12	2-Thienyl	97

18.3.6. β-Hydroxy Esters

As previously mentioned, all attempts to use our cationic DuPHOS-Rh and BPE-Rh catalysts for enantioselective hydrogenation of ketones were unsuccessful. Because ruthenium catalysts were known to exhibit enhanced rates and enantioselectivities in the reduction of certain keto functionality, we initiated an investigation of Ru-DuPHOS and Ru-BPE catalysts. Despite the cursory nature of our studies to date, this inquiry has resulted in development of an efficient catalyst precursor, assigned the general structure i-Pr-BPE-RuBr$_2$, for the highly enantioselective hydrogenation of β-keto esters **30** to β-hydroxy esters **31** under mild conditions (Scheme 14)[16,17,32].

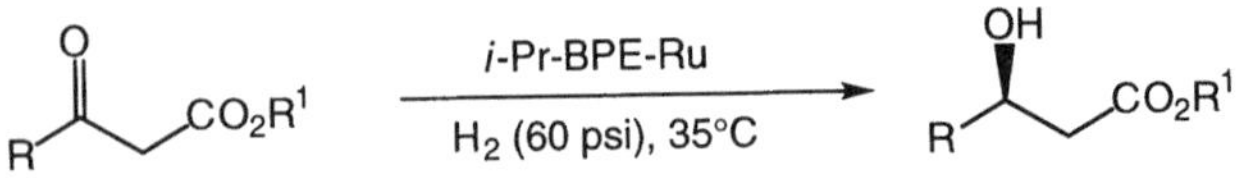

SCHEME 14.

Thus, at 60 psi H$_2$ and 35°C, we were able to actuate hydrogenation of a series of β-keto esters with > 98% ee (Table 4) [32]. Entry 8 implies that the hydrogenation process does not require the enolic form of substrate **30**. Dynamic

TABLE 4 Hydrogenation of β-keto Esters with (R,R)-i-Pr-BPE-Ru Catalyst

Entry	R in 30	% ee	Configuration
1	Me	99.3	S
2	Et	98.6	S
3	n-Pr	98.8	S
4	C$_{11}$H$_{23}$	98.7	S
5	CH$_2$OMe	95.5	R
6	i-Pr	99.0	R
7	C$_6$H$_{11}$	99.1	R
8	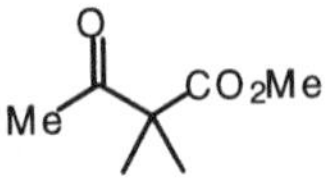	95.0	S
9	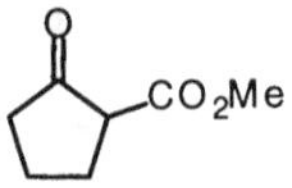	98.3 96.4	1S,2S 1R,2S
10	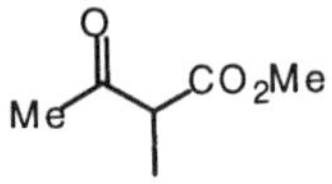	96.2	2R,3S

kinetic resolution was demonstrated in entry 9, where the *anti/syn* product ratio was 24/1. Lower diastereoselectivity (*syn/anti* = 1.4/1) was observed in entry 10. Given the importance of β-hydroxy esters as versatile chiral building blocks in organic synthesis and medicinal chemistry, further development of this β-keto ester hydrogenation process continues.

18.3.7. *N*-Acylhydrazones

In contrast to the high enantioselectivities and high catalytic activities observed in the hydrogenation of olefins, efficient catalysts for enantioselective hydrogenation of substrates containing a carbon-nitrogen double bond, such as imines, hydrazones, oximes, or oxime esters, has been more challenging. We have achieved preliminary results that suggest a possible new approach to the problem of enantioselective carbon-nitrogen double-bond hydrogenation [33].

Upon confronting the asymmetric carbon-nitrogen hydrogenation problem, we noted that, like α-enamides, *N*-acylhydrazones **32** possess an amide-like carbonyl oxygen that is similarly situated three atoms from the double bond to be reduced, and which could allow for chelation of the substrate to the catalytic Rh center.

Indeed, we quickly affirmed that the cationic Et-DuPHOS-Rh catalysts allowed smooth reduction of the *N*-benzoylhydrazones **32** in up to 97% ee (Scheme 15)[34]. Analogous rhodium catalysts based on known chiral diphosphines such as BDPP, BINAP, CHIRAPHOS, DIOP, and Ph β-Glup reduced *N*-benzoylhydrazones with low relative rates and low enantioselectivites (highest: 23% ee, 75% conversion over 48 hr with BINAP-Rh). This process provides simple access to α-arylhydrazines (**33**, R = aromatic or heteroaromatic ring) and α-hydrazino acids (**33**, R = CO$_2$Me) with high enantiomeric excesses.

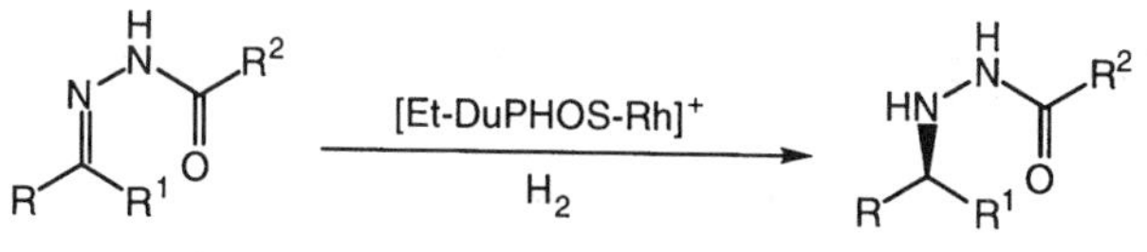

SCHEME 15.

18.3.8. Hydrogenations in Carbon Dioxide

Efficient asymmetric catalytic hydrogenation reactions intrinsically generate little waste; excess reagents typically are not required and generally few byproducts are formed. In chemical synthesis and manufacturing, solvents often constitute the preponderance of waste that must be disposed of or recovered. This scenario prevails in asymmetric hydrogenation reactions in which alcoholic solvents are the most commonly used. We envisaged that application of environmentally benign solvents such as supercritical carbon dioxide ($SC-CO_2$) may circumvent the use of hazardous solvents like methanol in this burgeoning area of chemical manufacturing. Indeed, performing asymmetric catalytic hydrogenation reactions in $SC-CO_2$ could be practically ideal; upon conclusion of the reaction, depressurization would provide directly the desired product together with a minor impurity ($< 0.1\%$ catalyst).

We recently have shown that this concept may be realized. A key feature of the cationic DuPHOS-Rh and BPE-Rh catalysts is their great tolerance to a broad range of solvents and organic functional groups. We have found that under the proper conditions, a wide array of both simple and β,β-disubstituted α-enamides 12 can be hydrogenated efficiently with these catalysts in $SC-CO_2$ (Scheme 6) [35].

More recently, we have found that virtually all substrate types highlighted in this chapter may be subjected to asymmetric hydrogenation in $SC-CO_2$ [36]. Enantioselectivites achieved in $SC-CO_2$ are very high and are comparable to those observed in conventional solvents. The only limiting feature unveiled thus far appears to be substrate solubility in the supercritical fluid (CO_2 is a very nonpolar solvent).

Interestingly, these reactions proceed even more effectively and afford products of higher enantioselectivity when the reactions are performed below the critical temperature (i.e., in liquid CO_2) [36]. Under the most suitable reaction conditions identified to date (1500 psi CO_2, 15 psi H_2, 22°C), this process can be operated routinely in standard high-pressure equipment to provide an efficient and clean route to a wide assortment of valuable compounds.

18.4. COMMERCIAL DEVELOPMENT AND APPLICATION OF ASYMMETRIC CATALYSIS

Asymmetric catalytic methodologies are well suited for the cost-effective manufacture of a diverse selection of enantiomerically pure compounds. Enantioselective hydrogenation processes are particularly attractive for large-scale production, and many possibilities can be envisaged (See Section 18.3). Presently, we are seeking to build on our preliminary observations and to establish these systems as robust, practicable technology. We and other investigators [37] recently have demonstrated the utility of the DuPHOS/BPE-based hydrogenation catalysts on medium scales in the pilot plant (up to 20-kg substrate). Several of these applications are likely to be operational in production units in the near future.

The following section delineates one recent application of the cationic DuPHOS/BPE rhodium catalysts aimed at the rapid generation of collections of chiral building blocks.

18.4.1. Chiral Libraries

With the establishment of an effective ligand synthesis, we sought commercial opportunities that would allow us to exploit this powerful class of catalysts. As previously illustrated, the DuPHOS-Rh hydrogenation catalysts are capable of furnishing many different classes of chiral compounds. Because drug discovery efforts often are limited by the diversity of readily available intermediates, we have embarked on a program to produce and supply novel libraries of chiral building blocks that may facilitate drug design and similar enterprises. Combinatorial chemistry programs may benefit distinctly from convenient access to such diversity libraries.

One chiral library in our *Chiro*Chem Collection consists of unnatural or nonproteinogenic amino acids. Synthetically, we are using the full gamut of our technology to generate these amino acid diversity sets. Asymmetric catalytic hydrogenation using the Et-DuPHOS-Rh catalyst obviously plays a vital role in this endeavor.

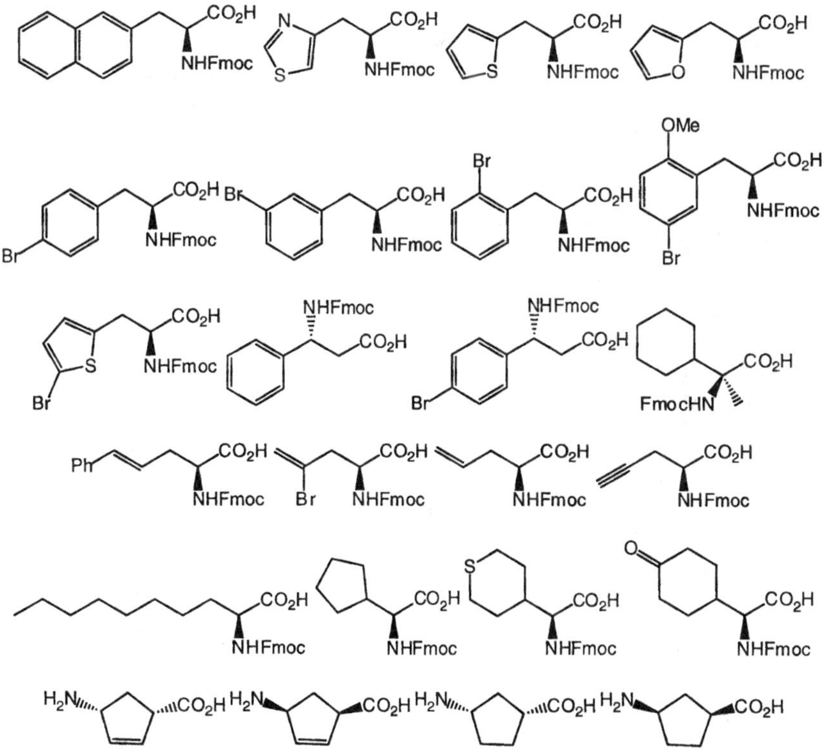

FIGURE 4 *Chiro*Chem collection of novel amino acids.

An example of a box-set of 25 amino acids that we now are marketing is shown in Figure 4. A range of aromatic, heterocyclic, β,β-disubstituted, γ,δ-unsaturated, and β-amino acids are represented in this sundry collection. Brominated derivatives are included to permit further functionalization through cross-coupling reactions such as those described in Section 18.3.1.5.

On the basis of our technological capabilities, many other libraries of chiral building blocks may be divined and are imminent.

18.5. SUMMARY

We have traced the development of an effective new class of chiral phosphine ligands and catalysts. The DuPHOS and BPE ligand series have been shown to be superior in a variety of asymmetric catalytic reactions. In particular, rhodium catalysts containing the DuPHOS and BPE ligands have been found uniquely capable of hydrogenating a range of substrate types with extremely high enantioselectivities and high catalytic efficiencies. A versatile ligand design principle has enabled systematic evaluation of steric effects in asymmetric catalytic reactions. This feature has allowed the optimization of rates and enantioselectivities in a growing list of useful asymmetric catalytic processes. The commercial potential of this technology presently is being realized.

REFERENCES

1. Whitesell, J. K. *Chem. Rev.* 1989, *89*, 1581.
2. Burk, M. J., Feaster, J. E., Harlow, R. L. *Organometallics* 1990, *9*, 2653.
3. Burk, M. J., Feaster, J. E., Harlow, R. L. *Tetrahedron: Asymmetry* 1991, *2*, 569.
4. Burk, M. J. *J. J. Am. Chem. Soc.* 1991, *113*, 8518.
5. Kagan, H. B. In *Asymmetric Synthesis;* Academic Press: New York, 1985; *Vol. 5;* p. 1.
6. Brunner, H., Zettlmeier, W. *Handbook of Enantioselective Catalysis with Transition Metal Compounds;* VCH: New York, 1993; *Vol. 2.*
7. Burk, M. J., Feaster, J. E., Nugent, W. A., Harlow, R. L. *J. Am. Chem. Soc.* 1993, *115*, 10125.
8. Ojima, I. In *Catalytic Asymmetric Synthesis;* VCH: New York, 1993; Chapter 1.
9. Noyori, R. In *Asymmetric Catalysis in Organic Synthesis;* Wiley: New York, 1993; Chapter 2.
10. Knowles, W. S., Sabacky, M. J. *J. Chem. Soc. Chem. Commun.* 1968, 1445.
11. Horner, L., Siegel, H., Büthe, H. *Angew. Chem. Int. Ed. Engl.* 1968, *7*, 942.
12. Burk, M. J., Bienewald, F. In *Transition Metals for Organic Synthesis and Fine Chemicals;* Bolm, C., Beller, M., Eds.; VCH: Wienheim, 1998, Chapter 1, pp. 13–25.

13. Koenig, K. E. In *Asymmetric Synthesis;* Morrison, J. D., Ed.; Academic: Orlando, 1985; *Vol. 5;* p. 71.
14. Kreuzfeld, H.-J., Di//bler, C., Krause, H. W., Facklam, C. *Tetrahedron: Asymmetry* 1993, *4*, 2047.
15. Krause, H. W., Kreuzfeld, H.-J., Schmidt, U., Dîbler, C., Michalik, M., Taudien, S., Fischer, C. *Chirality* 1996, *8*, 173.
16. Burk, M. J., Gross, M. F., Harper, T. G. P., Kalberg, C. S., Lee, J. R., Martinez, J. P. *Pure Appl. Chem.* 1996, *68*, 37.
17. Burk, M. J., Gross, M. F., Martinez, J. P., Straub, J. A. *J. Org. Chem.* submitted.
18. Burk, M. J., Allen, J. G. *J. Org. Chem.* 1997, *62*, 7054.
19. Masquelin, T., Broger, E., Mueller, K., Schmid, R., Obrecht, D. *Helv. Chim. Acta* 1994, *77*, 1395.
20. Burk, M. J., Allen, J. G., Kiesman, W. F., Stoffan, K. M. *Tetrahedron Lett.* 1997, *38*, 1309.
21. Burk, M. J., Allen, J. G., Kiesman, W. F. *J. Am. Chem. Soc.* 1998, *120*, 657.
22. Sawamura, M., Kuwano, R., Ito, Y. *J. Am. Chem. Soc.* 1995, *117*, 9602.
23. Burk, M. J., Gross, M. F., Martinez, J. P. *J. Am. Chem. Soc.* 1995, *117*, 9375.
24. Burk, M. J., Lee, J. R., Martinez, J. P. *J. Am. Chem. Soc.* 1994, *116*, 10847.
25. Burk, M. J., Lee, J. R. manuscript in preparation.
26. Burk, M. J., Wang, Y. M., Lee, J. R. *J. Am. Chem. Soc.* 1996, *118*, 5142.
27. Burk, M. J., Casy, G.; Johnson, N. B., *J. Org. Chem.,* 1998, in press.
28. Burk, M. J., Beinewald, F., Harris, M., Zanotti-Gerosa, A. *Angew. Chem. Int. Ed. Engl.,* 1998, *37*, 1931.
29. Spindler, F., Pittelkow, U., Blaser, U. *Chirality* 1991, *3*, 370.
30. Mashima, K., Kusano, K., Sato, N., Matsumura, Y., Nozaki, K., Kumobayashi, H., Sayo, N., Hori, Y., Ishizaki, T., Akutagawa, S., Takaya, H. *J. Org. Chem.* 1994, *59*, 3064.
31. Burk, M. J., Kalberg, C. S., Pizanno, A. *J. Am. Chem. Soc.,* 1998, *120*, 4345.
32. Burk, M. J., Harper, T. G. P., Kalberg, C. S. *J. Am. Chem. Soc.* 1995, *117*, 4423.
33. Burk, M. J., Feaster, J. E. *J. Am. Chem. Soc.* 1992, *114*, 6266.
34. Burk, M. J., Martinez, J. P., Feaster, J. E., Cosford, N. *Tetrahedron* 1994, *50*, 4399.
35. Burk, M. J., Feng, S., Gross, M. F., Tumas, W. *J. Am. Chem. Soc.* 1995, *117*, 8277.
36. Burk, M. J., Feng, S., Gross, M. F., Tumas, W. manuscript in preparation.
37. Scalone, M., Schmid, R., Broger, E., Burkart, W., Cereghetti, M., Crameri, Y., Foricher, J., Hennig, M., Kienzle, F., Montavon, F., Schoettel, G., Tesauro, D., Wang, S., Zell, R., Zutter, U., *ChiraTech '97 Proceedings,* 1997, Philadelphia.

Index

2164U90, 129
3TC (*see* Lamivudine)
4-Acetoxy azetidinone, synthesis, 153–154
Abzymes (*see* Catalytic antibodies)
Acetals, in Claisen rearrangement, 188
Acetates, with lipases, 104, 105
Acetophenone, reduction with oxazaborilidines, 214–215
Acetylation, enzymatic, 261
  D-glucose, 71
  sucrose, 71–72
Acrolein, in Diels-Alder reaction, 184
Acrylates, in Diels-Alder reaction, 182
N-Acryloylbornane-10,2-sultam, 182
3-Acyl-2-methylpropionic acid
  for captopril, 37
  resolution, 37
Acylases, for L-phenylalanine, 61–62
N-Acylhydrazones, asymmetric reductions, 354–355
N-Acyloxazolidinones, synthesis, 306–307
L-Alanine
  from aspartic acid, 270, 317
  for enalapril, 34
  by fermentation, 34

Alcohol dehydrogenases, 247
  mechanism, 247
  Prelog's rule, 247
Alcohols
  from allyl alcohols, 157
  to esters, with resolution, 264
  inversion, 103, 105
  from ketones, 247, 248, 250–253
    by asymmetric hydrosilylation, 170
    by chiral reduction, 213
  resolutions, 333
    with enzymes, 335
Aldehydes
  for 2,3-dihydroxy acids, 269
  to epoxides, 107
  for β-hydroxy amino acids, 256
Alder rule, 178
Alditols, synthesis, 231
Aldol reaction (*see also* Aldolases)
  alternative, 189, 192
  chiral auxiliaries, camphor, 93
  enzymatic, 256–257
  enzymatic alternative, 274
Aldolases, 268–270
  type I, 269
  type II, 269

Aldoses, synthesis, 231
Alkanes, from alkenes by asymmetric
    hydrosilylation, 170
Alkenes
    for alkanes, by asymmetric hydrosily-
        lation, 170
    for 1,2-amino alcohols, 234
    asymmetric hydrosilylation, 170
    for 1,2-diols, 227, 232–233
    for epoxides, 227, 236
    for halohydrins, 237
    Jacobsen epoxidation, 106
α-Alkoxy esters, in Claisen ester eno-
    late rearrangement, 189
Alkylations
    of 2-aminobutyric acid, 130
    chiral auxiliaries, camphor, 92
    of diketopiperazines, 308–310
    enantioselective, camphor, 92
N-Alkylglucamine, for resolution of na-
    proxen, 118–119
2-Alkylsuccinates, from itaconates, 351
Allyl alcohols
    for alcohols, 157
    from allyl ethers, by Wittig re-
        arrangement, 192
    from allyl sulfoxides, 194
    asymmetric hydrogenations, 157
    to 2,3-epoxy alcohols, 227–230
    by Evans rearrangement, 193–194
    kinetic resolution with Sharpless epox-
        idation, 229–230
    Sharpless epoxidation, 106
    by Wittig rearrangement, 192
Allyl amines
    asymmetric isomerizations of, 151–
        153
    to enamines, 151–153
Allyl ethers
    for allyl alcohols, 192
    Wittig rearrangement, 192
Allyl sulfoxides, from allyl alcohols,
    194
Allyl vinyl ethers, Claisen re-
    arrangement, 188

Allylic ethers
    for isoxazolines, 191
    with nitrile oxides, 191
Allylic substitutions (see Substitutions,
    allylic)
Alpine-Borane, 212
Amidases, 260
    for L-phenylalanine, by resolution,
        60–61
    resolutions, 60–61, 260
Amides, in Claisen rearrangement, 188
Amines
    from imines
        by asymmetric hydrogen transfer,
            169
        by asymmetric hydrosilylation, 170
    from ketones, by asymmetric hydroge-
        nation, 350–351
Amino acid dehydrogenase, 249–250
    alanine dehydrogenase, 249
    cofactor regeneration, 249
D-Amino acid transferase, 257
α-Amino acids (see also Amino acids
        and α,α-Disubstituted α-amino
        acids)
    alkylations, 92
    for β-amino acids, 271
    from amino esters, 265
    Claisen rearrangement, ester enolate,
        189
    from α-keto acids, 257–259
    for peptides, 266
    by transaminases, 257–259
Amino acids
    aromatic, biosynthetic pathway, 272
    by asymmetric hydrogenation, 147,
        162–164, 339–357
    from dehydromorpholinones, 304–
        306
    from diketopiperazines, 308–310
    from enamides, 147, 310–311, 343
    from α-keto acids, 249
    libraries, 356
    by oxazolidinones, 307
    from peptides, 259

[Amino acids]
synthesis, 301–315
by transaminases, 312–314
β-Amino acids
from α-amino acids, 271
Claisen rearrangement, ester enolate,
189
D-Amino acids (*see also under specific
compound names*)
with alanine dehydrogenase, 249
by hydantoinases, 135–138, 277
from α-keto acids, 249
in penicillin antiobiotics, 135
synthesis, 266
by transaminases, 313–314
L-Amino acids, by hydantoinases,
277
1,2-Amino alcohols
from alkenes, 234
from α-amino ketones, 165
by aminohydroxylation, 234
by asymmetric hydrogenation, 349
as chiral auxiliaries, 288
from cyclic sulfates, 235
from 1,2-diols, 235
from epoxides, with resolution, 237
as ligands in asymmetric hydrogen
transfers, 168–169
α-Amino amides, from amino esters,
265
α-Amino esters, for amino acids, 265
α-Amino ketones
for 1,2-amino alcohols, 165
asymmetric hydrogenations, 165
(R)-2-Amino-2-ethylhexanoic acid
from 2-aminobutyric acid, 130
resolution, 130–135
by Schöllkopf method, 130–131
synthesis, 129–130
3-Amino-2-hydroxy esters, from α,β-
unsaturated esters, 234
(S)-2-Amino-4-phenylbutanoic acid (*see*
L-Homophenylalanine)
Aminoacylases, resolutions, 62
D-2-Aminobutyrate, synthesis, 259

L-2-Aminobutyrate
synthesis, 259
by transaminases, 313
2-Aminobutyric acid, alkylation of,
130
7-Aminocephalosporanic acid
for cefaclor, 39
fermentation, 40
Aminohydroxylation, 234
Aminolysis, enzymatic resolution of
2-chloropropionic acid, 43
6-Aminopenicillanic acid
for amoxicillin, 36
synthesis, 36
Aminophosphines, as ligands, 158
Aminotransferases (*see* Transaminases
and also under specific enzymes)
Amoxicillin, synthesis, 36
Anatoxin-a, 330–331
Antibiotics (*see also under specific compounds and classes*)
containing D-amino acids, 135
macrolides, biosynthetic pathway,
274–276
by Wittig rearrangement, 192
Anticholesterol drug, 129
Aromatic aminotransferase, for L-phenylalanine, 53
Aromatic compounds, asymmetric oxidations, 238
Aryl alcohols, from aryl ketones, 168
Aryl ketones, for alcohols, 168
Arylalanines, by tandem catalysis, 348–
349
α-1-Arylalkylamines, 350
α-Arylglycinols, from styrene, 234
L-Ascorbic acid, 73
for R-solketal, 76
from D-sorbitol, 73
L-Asparagine, for L-aspartic acid, 317
Aspartame, 49, 317
synthesis, 40, 158, 256–257, 277
chemical, 40
enzymatic, 41
F-process, 40

[Aspartame]
Z-process, 40
TOSOH process, 41, 265
Aspartase (*see* L-Aspartate ammonia lyase)
Aspartate aminotransferase, 257
for L-phenylalanine, 53
L-Aspartate ammonia lyase, 268, 319
activation
chemical, 323
cloning, 320–321
by site-directed mutagenesis, 322–323
by truncation, 321–322
from *B. flavum*, 321
biocatalyst development, 319–323
from *E. coli*, 319–320
immobilization, 318
for L-aspartic acid, 317
thermostable, 323
Aspartic acid
equilibrium displacement, 258
with transaminases, 257
D-Aspartic acid
by aspartate racemase, 313
by resolution, 270, 312
with transaminases, 313–314
L-Aspartic acid
for l-alanine, 270
as chiral auxiliary, 308–310
commercial production, 317
by fermentation, 317
from fumaric acid, 318
from L-asparagine, 317
by L-aspartate ammonia lyase, 317
from maleic acid, 324–325
from maleic anhydride, 318
to oxaloacetate, 59
synthesis, 317–325
by transaminases, 58
with transaminases, 312–314
Asymmetric allylic alkyations, with ferrocenyldiphosphines, 160
Asymmetric catalysis, 144
biocatalysts, 339
chemical, 339

Asymmetric cyclopropanations, 171
Asymmetric dihydroxylations, 232–235
with cinchona alkaloid ligands, 232
with (DHQ)$_2$·PHAL ligands, 233
with (DHQD)$_2$·PHAL ligands, 233
kinetic resolutions, 234
mechanism, 233
for *o*-isopropoxy-*m*-methoxysytrene, 234
Asymmetric hydroborations, with ferrocenyldiphosphines, 160
Asymmetric hydrocarboxylation
for ibuprofen, 123
for naproxen, 119
Asymmetric hydrocyanation
for ibuprofen, 123
for naproxen, 119
Asymmetric hydroesterification, for naproxen, 119
Asymmetric hydroformylation, 169–170
for ibuprofen, 123
of 2-methoxy-6-vinylnaphthalene, 119
for naproxen, 119
Asymmetric hydrogenations, 143–146
acetoacetate, 77
of allyl alcohols, 157
to amino acids, 147
aryl ketones, by hydrogen transfer, 168–169
with BPPM, 164–165
with carbohydrate phosphinites, 162–163
with DeguPHOS, 163–164
with DuPHOS, 165, 343–357
with ferrocenyldiphosphines, 159–160
geraniol, 157
hetereogeneous, 165–168
homogeneous, 144
by hydrogen transfer, 168–169
diamine catalysts, 168–169
of α-hydroxy ketones, 156–157
for ibuprofen, 122–123, 125
of imines, 159
with Knowles' catalyst, 147–149, 310–311

[Asymmetric hydrogenations]
for Metolachlor, 159
for naproxen, 119
nerol, 157
with nickel and tartaric acid, 166–167
with platinum and chinchona alkaloids, 167–168
with PPM, 164–165
with Ru(BINAP) catalysts, 149–158
in supercritical carbon dioxide, 355–356
of α,β-unsaturated acids, 155–156
Asymmetric hydrosilylations, 170
of alkenes, 170
of imines, 170
of ketones, 170
with MeO-MOP, 170
Asymmetric isomerizations, 143
of allyl amines, 151–153
of N,N-diethylgeranylamine, 151
with Rh(BINAP) catalysts, 149–158
Asymmetric oxidations, 227–239 (see also Sharpless epoxidation and Asymmetric dihydroxylation)
Asymmetric reductions, 211–223 (see also Asymmetric hydrogenations and Reductions)
of ketones, 211–223
with Alpine-Borane, 212
with BINAL-H, 212
with Ipc₂BCl, 212
with organoboranes, 212
with oxazaborilidines, 211–223
with Alpine-Borane, 212
with BINAL-H, 212
with Ipc₂BCl, 212
with oxazaborilidines, 211–223
reaction parameters, 217–219
versatility, 219–220
with Yamagushi-Mosher reagent, 212
Asymmetric synthesis, definition, 1
Auxiliaries (see Chiral auxiliaries)
Avermectin, 274
2-Azetidinones, with oxazolidinones, 295
AZT, 74

Bacterial reductions, 253–255
Bao Gong Teng A, 296
Benazepril, 167–168
Benzazepin-2,3-dion, for benzazepin-3-hydroxy-2-one, 255
Benzazepin-3-hydroxy-2-one, benzazepin-2,3-dione, 255
2,3-Benzodiazepines, 250
Benzyl-(S)-(+)-hydroxybutyrate, from benzylacetoacetate, 252
Benzylacetoacetate, reduction, 252
Bicyclo[3.2.0]heptenenones, resolution, 333
BINAL-H, 212
BINAP, 144, 149–158 (see also Rh(BINAP) catalysts)
synthesis, 150
Biocatalysis, 2–3, 31, 245 (see also Enzymes and specific examples)
classification, 246
databases, 246
development, 319–323
hydrolases, 246
isomerases, 246
ligases, 246
lyases, 246
oxidoreductases, 246
screening, 276–278
transferases, 246
Biological catalysts (see Biocatalysts)
Biopol (see Poly-β-hydroxyalkanoates)
Biosynthetic pathways, common aromatic, 50
Biotin, synthesis, 162
Biotransformations, 2, 245–278
to L-phenylalanine, 49–64
Bisoxazolines
catalysis of Diels-Alder reaction, 181
as chiral auxiliaries, 288
3,4-Bisphosphinopyrrolines (see DeguPHOS)
BMS 181100, 212
BO-272, 253
Boc α-amino acids, by asymmetric hydrogenation, 344

Boc-Ala-anatoxin, 330
Boc-Pyrrolidine, for diphenylprolinol, 216
Boots-Hoechst-Celanese process, for ibuprofen, 123
BPE, 341
    in supercritical carbon dioxide, 355
    synthesis, 342
BPPM, 164
    from L-4-hydroxyproline, 164
β-Branched α-amino acids, by asymmetric hydrogenation, 347–348
Branched chain aminotransferase, 257
    for L-phenylalanine, 53
(S)-6-Bromo-β-tetralol, 251
6-Bromo-β-tetralone, reduction, 251
Bromocamphorsulfonic acid, for resolution, 135
Bromolactonization, with a chiral auxiliary, 297
Bucherer-Bergs reaction, 136
Bufuralol, 334–336
    racemate, 334
    resolution, 335
(+)-Bulgecinine, 346–347

Camphor, 86, 91
    as chiral auxiliary, 92–94
    aldol reaction, 93
    alkylations, 92
    Diels-Alder reaction, 94
    for steroid synthesis, 98
    by Diels-Alder reaction, 91
    from α-pinene, 91
d-Camphor, 91
l-Camphor, 91
Camphorsulfonic acid, for resolution, 135
Camphorsultams, as chiral auxiliaries, 288
Candida rugosa, hydroxylation with, 38
Captopril, 37
N-Carbamoyl amino acids, enzymatic hydrolysis, 277

D-N-Carbamylamino acids, for D-amino acids, 135
Carbohydrases, 259
Carbohydrate phosphinites, as ligands, 162–163
Carbohydrates, 69–79 (see also under specific carbohydrates and disaccharides)
    as building blocks, 69
    chiral pool, 70
    as chirons, 69
    derivatives, by Sharpless epoxidation, 231
    as food additives, 69
    in pharmaceuticals, 69
    in synthesis, 69–79
Carboxylic acids
    Claisen rearrangement, 188
    from esters, 259–260
    from nitriles, 259
Carnitine
    from epichlorohydrin, 42
    by microbial oxidation, 42
    synthesis, 42
Carveol, 88
Carvone, 86, 89
    from carveol, 88
dl-Carvone, 89–90
l-(−)-Carvone, 88–90
    from d-limonene, 90
    from (−)-α-pinene, 90
Catalysts (see Chiral catalysts and Biocatalysts)
Catalytic antibodies, 3
    for Diels-Alder reaction catalysis, 181
CBS method (see Oxazaborilidines)
N-Cbz α-amino acids, by asymmetric hydrogenation, 344
Cefaclor, 39
Chiral auxiliaries, 4, 30, 287–298, 303
    (see also under specific compounds)
    aldol reaction, 93
    amino acids, 303
    alkylations, 92

[Chiral auxiliaries]
 camphor, 92–94
 Diels-Alder reaction, 94, 182
 for 1,3-dipolar cycloadditions, 296
 menthol, 94–95
 N-acryloylbornane-10,2-sultam, 182
 oxazinones, 93
 pharmaceuticals, 289
 at scale, 287
 for steroid synthesis, 98
 terpenes, 83
Chiral boranes, 288
Chiral catalysts, 2, 3, 303
 amino acids, 303
 multiplication, 2
 transition metals, 6
Chiral compounds
 sources at scale, 13–27
 sourcing, 13–32
 synthesis by small company, 329–
  336
Chiral Dienes (*see* Dienes)
Chiral dienophiles
 in Diels-Alder reaction, 182
 type I, 182
 type II, 182
Chiral Lewis acids
 from bisoxazolines, for Diels-Alder reaction, 181
 from 1,2-diamines, for Diels-Alder reaction, 179–181
 for Diels-Alder reaction, 179–181
 from tartaric acid, for Diels-Alder reaction, 181
Chiral Lewis bases, catalysis of Diels-
  Alder reaction, 181
Chiral ligands, 2, 340
 design, 340
 terpenes, 83
Chiral multiplication, 2–3
Chiral pool, 2
 amino acids, 301, 314
 carbohydrates, 70
 leukotriene intermediates, 331
 terpenes, 83–100

Chiral reagents, 2, 303
 amino acids, 303
Chiral switch, ibuprofen, 122
Chiral synthesis, in agrochem, 43
Chiral templates, 4, 287, 303
 amino acids, 303
 dehydromorpholinones, 304–306
Chirald (*see* Yamagushi-Mosher
  reagent)
CHIRAPHOS, 144
Chirons (*see also* Chiral pool)
 carbohydrates, 69
 monoterpenes, 95–99
Chloramphenical, resolution of, 117
m-Chloro chloroacetophenone, reduction
  with oxazaborilidines, 221
p-Chloro chloroacetophenone, reduction
  with oxazaborilidines, 217–218
Chlorodiisopinocampheylborane, 212
Chloroketones, for epoxides by asymmetric reductions, 221
2-Chloropropionic acid
 enzymatic, 43
 aminolysis, 43
 dehalogenase, 44
 esterification, 43
 lipase, 44
 transesterification, 43
 for flamprop, 43
 for fluazifop-butyl, 43
 for phenoxypropionic acids, 43
 resolution, 329–330
 with (+)-α-phenylethylamine, 330
(*S*)-2-Chloropropionic acid, for phenoxy-
  propioic acids, 104
Chorismate mutase, for L-phenylalanine,
  52–55
Chorismic acid
 biosynthesis, 51–57
 for L-phenylalanine, 51–57
Chromatography, for resolutions, 115
Chromium salen, with epoxide, 237
trans-Chrysanthemic acid, 171
Chrysanthemic esters, resolution, 44
Chrysanthenone, 86

Chymotrypsin, 265
CI-1008, 291–292
Cinchona alkaloids
    as ligands
        asymmetric dihydroxylation, 232
        asymmetric hydrogenations, 167–168
Cinchonidine, for resolution of naproxen, 118
Cinmethylin, 98–99
Cinnamic acid
    for L-phenylalanine, 268
    with phenylalanine ammonia lyase, 57–58
(S)-Citramalic acid, from ketene, 195
Citronellal, 151
(R)-Citronellal, with 5-n-pentyl-1,3-cyclohexadione, 186
Claisen rearrangement, 187–190
    of acetals, 188
    of acids, 188
    allyl vinyl ethers, 188
    of amides, 188
    of enol ethers, 188
    ester enolates, 189–190
    of esters, 188
    Lewis acid catalysis, 189
    of orthoesters, 188
    of oxazolines, 188
    of thioesters, 188
Cobalt salen, with epoxides, 236–237
Confidential disclosure agreements, 28
Conglomerates, 116
Cope rearrangement, 188
    anionic oxy-Cope, 188
    of 1,5-hexadiene, 188
    oxy-Cope, 188
Corynantheine, derivatives, 191
D-CPP-ene, 160
Crotyl oxazolidinones, with cyclopentadiene, 179–181
Crystallization, for resolution, 44
Cyanides (see Nitriles)
Cyclic sulfates
    to 1,2-amino alcohols, 235
    from 1,2-diols, 234

[Cyclic sulfates]
    for DuPhos, 342
    with nucleophiles, 234–235
    reactions, 107, 234–235
    ring opening reactions, 234–235
    synthesis, 234
Cycloadditions
    dipolar, 191–192
    dipolar, 1,3-
        with chiral auxiliaries, 296
        with chiral templates, 305
    of nitrile oxides, 191
    photochemical, 194
    [2+2], 194–195
        imines, 194
        ketene, 195
        Reformatsky reagents, 194
Cycloalkenes, Claisen ester enolate rearrangement, 189
(R)-(+)-Cyclohex-3-enecarboxylic acid, 182
Cyclopentadiene
    with crotyl oxazolidinones, 179–181
    with dimenthyl fumarate, 185
    with dimethyl fumarate, 178
    with dimethyl maleate, 178
    with methacrolein, 179
(+)-α-Cyperone
    from (−)-3-caranone, 96
    from (+)-dihydrocarvone, 96
(+)-β-Cyperone, by Robinson annelation, 96

Darzen reaction, 36
Decahydroisoquinoline-3-carboxylic acid, N-t-butylamide, from L-phenylalanine, 314
Decarboxylases, 270
    resolutions of amino acids, 311–312
DeguPHOS
    as ligand, 163–164
    synthesis, 163–164
Dehalogenase, resolution, 44
Dehydrogenases (see also Oxidoreductases, Dehydrogenases, and under specific examples)

[Dehydrogenases]
alcohol dehydrogenases, 247
amino acid dehydrogenases, 249
cofactor regeneration, 248
cofactors, 247
lactate dehydrogenases, 248
Dehydromorpholinones
as chiral templates, 304–306
1,3-dipolar cycloadditions, 305
synthesis, 305
Deltamethrin, 44
Denopamine, 220
3-Deoxy-D-arabinoheptulosonate-
7-phosphate synthase, for
L-phenylalanine, 51–53
2-Deoxy-D-ribose, chiral pool, 332
Deoxyalditols synthesis, 231
6-Deoxyerythronolide B, 275
Desymmetrization, by allylic substitution, 109
1,2-Diamines
derivatives as chiral Diels-Alder catalysts, 179–181
from 1,2-diols, 234
as ligands for asymmetric hydrogen transfers, 168–169
Diels-Alder reaction, 177–187
Alder rule, 178
for camphor, 91
catalysis of, 179–181
by catalytic antibodies, 181
by chiral Lewis acids, 179–181
by chiral Lewis bases, 181
by 1,2-diamine derivatives, 179–181
by enzymes, 181
chiral auxiliaries, 94, 182
chiral dienes, 182–185
chiral dienophiles, 182
cooperative blocking groups, 184
of cyclopentadiene
with dimethyl fumarate, 178
with dimethyl maleate, 178
of dimethyl fumarate, with cyclopentadiene, 178
of dimethyl maleate, with cyclopentadiene, 178

[Diels-Alder reaction]
hetero, 187
for 3-hydroxy acids, 187
intramolecular, 186–187
hetero, 187
inverse electron demand, 186
Lewis acid catalysis, 186
for limonene, 86
regiochemistry, 178
solvent effects, 185–186
stereochemistry, 178
with 1-substituted dienes, 178
with 2-substituted dienes, 178
Dienes
1-substituted, in Diels-Alder reaction, 178
2-substituted, in Diels-Alder reaction, 178
chiral
with acrolein, 184
in Diels-Alder reaction, 182–185
in Diels-Alder reaction, 178
Diesters, to mono esters, 261
N,N-Diethylgeranylamine, isomerization of, 151
(+)-Dihydrocarvone, for (+)-α-cyperone, 96
Dihydrofurans, by Wittig rearrangement, 192
Dihydropinol, 99
Dihydropyrans, in Claisen ester enolate rearrangement, 189
Dihydroxy acetone, with aldolases, 269
2,3-Dihydroxy acids
from aldehydes, 269
by aldolases, 269
from dihydroxy acetone, 269
3,5-Dihydroxy ester, from β,δ-diketo ester, 254
Diisopinocampheylborane, 85
for (+)-b-pinene, 86
for purification of (+)-α-pinene, 85
β,δ-Diketo esters, to 3,5-dihydroxy ester, 254

β-Diketones
    asymmetric hydrogenations with Raney nickel and tartaric acid, 166
    for 1,3-diols, 250
    for β-hydroxy ketones, 166
    for β-hydroxy ketones, 250
Diketopiperazines
    for amino acids, 308–310
    as chiral auxiliaries, 288
Diltiazem
    resolution, 37
    synthesis, 36, 293, 294
Dimenthyl fumarate, with cyclopentadiene, 185
Dimethyl fumarate, with cyclopentadiene, 178
Dimethyl maleate, with cyclopentadiene, 178
(2S,3R)-4-Dimethylamino-1,2-diphenyl-3-methyl-2-butanol (see Chirald)
1,2-Diols
    from alkenes, 227, 232–233
    for 1,2-amino alcohols, 235
    from aromatic compounds, 238
    for cyclic sulfates, 107, 234
    for 1,2-diamines, 234
    from epoxides, 236, 259
    from α-keto esters, 351–352
    reactions, 234–235
1,3-Diols, from β-diketones, 250
1,4-Diols
    for DuPhos, 342
    resolution, 342
DIPAMP, 144 (see also Knowles' catalyst)
    synthesis, 144–147
Diphenyl oxazaborolidines, 214
Diphenylprolinol
    from Boc-pyrrolidine, 216
    in oxazaborilidines, 214, 216
    catalyst preparation, 215–216
    synthesis, 216
Dipolar cycloadditions (see Cycloadditions, dipolar
Disaccharides, 70–72 (see also under specific compounds)

Disparulene, 231
α,α-Disubstituted α-amino acids, 266
    from dehydromorphinones, 304–306
    resolution, 128–135
    by Schöllkopf method, 130–131
L-Dopa, 144
    resolution, 29
    synthesis, 147, 162, 310
L-DOPS, 155
Draflazine, 160
DuPhos, 165, 339–357
    asymmetric hydrogenations, 343–357
    in supercritical carbon dioxide, 355
    synthesis, 342
Dynamic kinetic resolutions (see Resolutions, kinetic, dynamic)

Enalapril, 34
Enamides
    for amino acids, 147, 311, 343
    asymmetric hydrogenations, 162–164
    synthesis, 311
Enamines, from allyl amines, 151–153
Endo rule (see Alder rule)
Ene reaction, 190–191
    for corynantheine derivatives, 191
    Lewis acids, 190
Enol ethers, in Claisen rearrangement, 188
Enzymes, 6, 245–278, (see also Biocatalysts, Biotransformations, Microbial, and specific enzymes and classes)
    acetylation
        D-glucose, 71
        sucrose, 71–72
    aldolases, 256–257
    amide bond formation, 40
    Aspartame synthesis, 41
    for cefaclor, 40
    coupled systems, 59
    coupling, 62
    databases, 246
    dehalogenases, 44
    in Diels-Alder reaction, 181
    esterases for resolutions, 128

[Enzymes]
hydrolases, 246
hydroxylations, 35, 38
immobilization, 318
isomerases, 246
isomerization in synthesis of naproxen, 120
ligases, 246
lipases for resolutions, 128, 130–135
lyases, 246
mutants, 3,4
oxidations, 238–239
D-sorbitol, 73
in synthesis of naproxen, 120
oxidoreductases, 246
resolutions, 3, 115, 128–138
D-hydantoinases, 135–138
dehalogenase, 44
ester hydrolysis, 44, 124
hydantoins, 62–63
lipases, 44
L-phenylalanine, 60–63
transaminases (see Transaminases)
transferases, 246
Epichlorohydrin
for carnitine, 42
equivalents, 105
Epoxidation (see Epoxides)
Epoxides (see also Sharpless epoxidation and Jacobsen epoxidation)
from aldehydes, 107
from alkenes, 227
Jacobsen epoxidation, 236
to 1,2-amino alcohols, 237
with azides, 236
from chloroketones, 221
for 1,2-diols, 126–127
to 1,2-diols, 236, 259
from 2,3-epoxy alcohols, 231
from limonene, 88
with nitrogen nucleophiles, 107
opening, 105–106
Payne rearrangement, 106
resolutions, 126–127, 236–237
with Jacobsen's catalysts, 126
from S-terpinene-4-ol, 98

[Epoxides]
with water, 126
resolution, 236
1,2-Epoxy alcohols
from 2,3-epoxy alcohols, 230
Payne rearrangement, 230, 332
with thiols, 230
2,3-Epoxy alcohols
from allyl alcohols, 227–230
for epoxides, 231
from 1,2-epoxy alcohols, 332
to 1,2-epoxy 3-alcohols, 230
with nucleophiles, 231
Payne rearrangement, 230
reactions, 230–232
with thiols, 230
to 1-t-butylthio 2,3-diols, 230
Epoxy esters, resolution, 37
Erythritol derivatives, 231
Erythromycin, 78
Erythromycin A, 274
D-Erythrose, 74
from D-fructose, 74
Erythrose-4-phosphate, for L-phenylalanine, 51–57
Essential oils, 84
Esterases, 260–264
resolutions, 128
Esters
from acids with enzymes, 261
from alcohols with resolution, 264, 333
Claisen rearrangement, 188
enzymatic resolution of 2-chloropropionic acid, 43
hydrolysis, 260
for ibuprofen, 124
to acids, 259, 260
Estrogens, 97
Ethers, by iodoetherification, 108
Ethyl 4-chloro-3-hydroxybutanoate, 253
Ethyl 4-chloroacetoacetate, reduction, 253
Ethyl L-lactate, for naproxen, 120
(R)-Etomoxir, 297
Evans rearrangement, 193–194

Fatty acid synthase, 274
Fermentation, 31
 L-alanine, 34
 L-aspartic acid, 317
 L-phenylalanine, 31, 49–64
 L-proline, 34
 lactic acid, 77
 Lovastatin, 39
 oxidation of D-glucose, 72
 macrolides, 78
 malic acid, 77
 Penicillin G, 36
 Penicillin V, 36
 Penicillium citrinum, 35
 Pravastatin, 35
 R,R-tartaric acid, 76
 Simvastatin, 35
Ferrocenyldiphosphines
 asymmetric allylic alkylations, 160
 asymmetric hydroborations, 160
 asymmetric hydrogenations, 158–162
  of enamides, 160
  of itaconates, 160
  of β-keto esters, 160
 as ligands, 158–162
 synthesis, 160
FK506, 274
α-Fluoro esters, in Claisen ester enolate
  rearrangement, 189
Fluoxetine
 synthesis, 220
 use of Yamagushi-Mosher reagent,
  212
Food additives, 69
Formate dehydrogenase, with phenylalanine dehydrogenase, 59–60
D-Fructose, 69
 to D-erythrose, 74
 from D-glucose, 271
 oxidative cleavage, 74
Fumaric acid
 fermentation, 324
 from glucose, 324
 for L-aspartic acid, 318
 by maleate isomerase, 324–325
 from maleic acid, 324–325

N-Fumaryl-L-phenylalanine methyl es-
  ters, for Aspartame, 277
Fungal reductions, 253–255
Furfuryl alcohols, kinetic resolutions, 230
Fusinopril, 262

Geraniol, asymmetric hydrogenation of,
 157
D-Gluconic acid, 69, 72–73
 from D-glucose, 72
D-Glucopyranoside, as ligand systems,
 162
D-Glucose, 69
 acetylation, enzymatic, 71
 for D-fructose, 271
 for D-gluconic acid, 72
 to D-sorbitol, 73
 for fumaric acid, 324
 oxidation, 72
  chemical, 73
  enzymatic, 72
 reduction, 73
 for Sucralose, 71
Glucose dehydrogenase, 249
Glucose isomerase, 271
D-Glutamic acid, 312
L-Glutamic acid, 50
Glyceraldehyde
 additions, 75
 derivatives, 75–76
Glyceric acid
 from D-mannitol, 75
 for *S*-solketal, 75
Glycidols, 231
Glycine allyic esters
 Claisen ester enolate rearrangement,
  190
 for γ,δ-unsaturated amino acids, 190
Glycosidases, 259
GR95168, 74

Halohydrins, 237
Halohydroxylations, 237–238
Halolactonization, 237–238
Herbicides
 cinmethylin, 98–99

[Herbicides]
  phenoxypropionic acids, 43
  phenylpropionic acids, 104
  synthesis, 43
Hetero Diels-Alder reaction (see Diels-
  Alder reaction, hetero)
1,5-Hexadiene, in Cope rearrangement,
  188
L-Hexoses, synthesis, 231–232
High fructose corn syrup, 271
High performance liquid chromatography, for resolution, 29
HIV Protease inhibitors, 106–107
HMG-CoA reductase inhibitor, 254
Homologations, in HIV protease inhibitor syntheses, 107
L-Homophenylalalnine
  synthesis, 250
  by transaminases, 313
L-Homoserine, from L-methionine,
  314
HPLC (see High performance liquid
  chromatography)
D-Hydantoinases, 135–138
  for D-amino acids, 136–138
  for D-p-hydroxyphenylglycine, 63,
    136–138
  for D-phenylglycine, 63, 136–138
  for D-valine, 136
  for L-phenylalanine, 63
Hydantoinases (see also D-Hydantoinases)
  dynamic resolutions, 136–138
  for L-phenylalanine, 63
Hydantoins
  Bucherer-Bergs reaction, 136
  for D-amino acids, 136–138
  for L-phenylalanine, 49, 62–63
  synthesis, 136
Hydration, of α-pinene, 90
Hydroborations, 7
  α-pinene, 85
  (−)-α-pinene, 90
Hydrogen transfer reactions (see Asymmetric hydrogenations, by hydrogen transfer)

Hydrogenations (see Asymmetric hydrogenations)
Hydrolases, 246, 259–268
  in carnitine synthesis, 42
Hydrolysis, of meso -diols, 261–262
β-Hydroxy α-amino acids, 256
α-Hydroxy acetone
  asymmetric hydrogenation, 156–157
  for propane 1,2-diol, 156–157
α-Hydroxy acids, 76–77 (see also under specific compounds)
  from α-keto acids, 249
β-Hydroxy acids, 77–78 (see also under
  specific compounds)
  by aldol reaction, 93
  by Diels-Alder reaction, 187
α-Hydroxy esters, from α-keto esters,
  351–352
β-Hydroxy esters
  dehydration, 274
  from β-keto esters, 154–155, 166,
    250–254, 274, 353–354
β-Hydroxy ketones
  from β-diketones, 166, 250
  from isoxazolines, 192
Hydroxy sulfones, from keto sulfone,
  252
Hydroxybufurfuralol, 334–336
3-Hydroxybutyric acid, 77
3R-Hydroxybutyric acid
  in Biopol, 77
  derivatives, 77–78
3S-Hydroxybutyric acid
  by reduction
    of acetoacetate, 77
  asymmetric hydrogenation, 77
  yeast, 77
Hydroxylations, enzymatic, 35, 38
D-p-Hydroxyphenylglycine
  for amoxicillin, 36
  in antibiotics, 135
  hydantoinase, 63
  by resolution, 63
  resolution
    by bromocamphorsulfonic acid,
      135

[D-*p*-Hydroxyphenylglycine]
  by salt formation, 135
  synthesis, 277–278
L-4-Hydroxyproline
  for BPPM, 164
  for PPM, 164

Ibuprofen, 156
  Boots-Hoechst-Celanese process, 123
  as chiral switch, 122
  resolution, 122–125
    enzymatic, 124
    with L-lysine, 123–124
    with *(S)*-α-methylbenzylamine, 124
  synthesis
    by asymmetric hydrocarboxylation,
      123
    by asymmetric hydrocyanation, 123
    by asymmetric hydroformylation,
      123
    by asymmetric hydrogenation,
      122–123, 125
    from isobutylbenzene, 123
(S)-(+)-Ibuprofen (*see* Ibuprofen)
Imines
  for amines
    by asymmetric hydrogen transfer,
      169
    by asymmetric hydrosilylation, 170
  asymmetric hydrogenations, 159
  asymmetric hydrosilylation, 170
  asymmetric reductions, by hydrogen
    transfer reactions, 169
  for β-lactams, 194–195
  with Reformatsky reagents, 194–195
Indinavir, 160, 237
Insecticides (*see under specific com-
    pounds*)
Intermediates, impurity profile, 28
Inversions, of alcohol, 103
Iodoetherification, 108
Iodolactonization, 108
Ipc$_2$BCl (*see* Chlorodiisopinocampheyl-
    borane)
Ipc$_2$BH (*see* Diisopinocampheylborane)

Ireland rearrangement (*see* Claisen re-
    arrangement, ester anolates)
Isobutylbenzene, for ibuprofen, 123
Isomalt
  from isomaltulose, 72
  synthesis, 72
Isomaltulose, 69
  hydrogenation, 72
  for isomalt, 72
  from sucrose, 72
Isomenthol, 89
Isomenthones, 89
Isomerases, 246, 270
  epimerases, 270
  mutases, 270, 271
  racemases, 270
Isomerization
  acid catalyzed, 85
  base catalyzed, 85
  enzymes, in synthesis of naproxen,
    120
  with KAPA, of (−)-β-pinene, 85
  of (+)-α-pinene, 86
  of (−)-β-pinene, 85
  transition metal catalyzed, 7
Isooxazoles, in pharmaceuticals, 290
Isopinocampheylborane
  as chiral auxiliaries, 288
Isoprene, for limonene, 86
o-Isopropoxy-*m*-methoxysytrene, by
    asymmetric dihydroxylation, 234
2,3-Isopropylidene-L-ribonic γ-lactone,
    238
Isoproterenol, 220
Isoxazolines
  from allylic ethers, 191
  for β-hydroxy ketones, 192
  from nitrile oxides, 191
Itaconates, for 2-alkylsuccinates, 351
Itsuno-Corey reduction (*see* Oxazabori-
    lidines)

Jacobsen epoxidation, 106, 236–237
  Diltiazem synthesis, 37
  resolutions, 126

KAPA, isomerizations with, 85
Ketene
for *(S)*-citramalic acid, 195
for *(S)*-malic acid, 195
α-Keto acids
for α-amino acids, 249, 257–259
for 1,2-diols, 351–352
for α-hydroxy acids, 249
for α-hydroxy esters, 351–352
β-Keto esters
asymmetric hydrogenations with Raney nickel and tartaric acid, 166
asymmetric reduction, 153
enzymatic, 274
for β-hydroxy esters, 166, 250–254, 353–354
Keto sulfones, reductions, 252
Ketones
for alcohols, 247, 248, 250–253
by asymmetric hydrosilylation, 170
for amines, 350–351
asymmetric hydrosilylation, 170
asymmetric reductions, 211–223
with Alpine-Borane, 212
with BINAL-H, 212
with Ipc₂BCl, 212
with oxazaborilidines, 211–233
with Yamagushi-Mosher reagent, 212
Knowles' catalyst *(see* [Rh(dipamp)-COD]BF₄)

L-699392, 221
L685,393, 252
Lactate dehydrogenase, 248
Lactic acid
by fermentation, 77
in herbicides, 77
R-Lactic acid, for phenoxypropionic acids, 104
S-Lactic acid, for phenoxypropionic acids, 104
Lactones, by iodolactonization, 108
D-Lactose, 69
Lamivudine, 74, 94
D-*tert*-Leucine, by resolution, 308

L-Leucine, by transaminases, 58
L-*tert*-Leucine, 60
by oxazolidinones, 307
synthesis, 250, 259, 264
by transaminases, 313
Leucine dehydrogenase, 60, 250
Leukotriene A₄, 331
Leukotriene B₄, 333
Leukotriene intermediates, 331–332
Levofloxacin, 156–157
Lewis acids
catalysis in intramolecular Diels-Alder raction, 187
Claisen rearrangement, 189
ene reaction, 190
Ligands, 144 *(see also under specific compounds)*
Ligases, 246
Limonene, 86–88
epoxidation, 88
from isoprene, 86
synthesis, 86
d-Limonene, 86–88
for *l*-carvone, 90
l-Limonene, 86–88
Limonene epoxide, 88
for carveol, 88
from limonene, 88
Lipases, 260–264
Candida cylindracea 44
resolutions, 44, 128
acetates, 104, 105
of 2-amino-2-ethylhexanoic acid, 130–135
bufuralol, 335
epoxy esters, 37
ester formation, 333
Lovastatin, 39
Serratia marcescens, 37
Lisinopril, 36
Lovastatin
fermentation, 39
synthesis, 39
Lustral *(see* Sertraline)
LY-300164, 250–251
LY309887, 294–295

Lyases, 246, 268–270
L-Lysine
  for lisinopril, 36
  resolution of ibuprofen, 123–124

Macrolides, 78 (see also under specific
    compounds)
Maleic acid
  for fumaric acid, 324–325
  with maleate isomerase, 324–325
Maleic anhydride, for L-aspartic acid,
    318
Maleic isomerase, 324–325
Malic acid, 77
  by fermentation, 77
  (S)-Malic acid, from ketene, 195
D-Maltose, 69
Manganese salen (see Jacobsen epoxida-
    tion)
D-Mannitol, 69
  for glyceric acid, 75
  oxidative cleavage, 75
  for S-solketal, 75
Me-CBS (see Methyl-oxazaborilidine)
p-Menthanes, 89
Menthol, 89
  as chiral auxiliary, 94–95
  inversion of hydroxy, 105
  synthesis, 40, 151–153
(−)-Menthol. as chiral auxiliary, 94
l-Menthol, oxidation, 89
Menthone, 89
l-Menthone, 89
  from l-menthol, 89
Menthones, 89
Menthoxyaluminum complexes, as
    Diels-Alder catalysts, 179
MeO-MOP, 170
Metabolic pathway engineering, 272–
    276
Methacrolein, with cyclopentadiene, 179
L-Methionine, for l-homoserine, 314
2-Methoxy-6-vinylnaphthalene, asym-
    metric hydroformylation of, 119
Methyl (S)-lactate, as chiral auxiliary,
    296

α-Methyl-L-dopa, resolution of, 117
Methyl-oxazaborilidine, 216
α-Methylbenzylamine (see Phenylethyl-
    amine)
cis-Methylnopinone, for Robinson annel-
    ation, 97
Metolachlor, 43, 159
Michael addition, with chiral auxiliaries,
    297
MK-0417, 221
MK-0499, 221, 251
Mono esters, from diesters, 261
Monosaccharides, 72–75
  from polysaccharides, 259
Monoterpenes, 83–100 (see also under
    specific compounds)
  chiral pool, 83–100
  as chirons, 95–99
  reactions, 92–99
  in Robinson annelation, 96

Naltiazem, 293
Naproxen, 155
  by isomerization, 120
  by oxidation, 120
  resolution, 29, 118–122
    by cinchonidine salt, 118
    by N-alkylglucamine salt, 118–119
    by salt formation, 121
  synthesis, 155–156
    by asymmetric hydrocarboxylation,
      119
    by asymmetric hydrocyanation, 119
    by asymmetric hydroesterification,
      119
    by asymmetric hydroformylation,
      119
    by asymmetric hydrogenation, 119
    from ethyl L-lactate, 120
    Zambon process, 119, 290–291
(S)-Naproxen (see Naproxen)
Neoisomenthol, 89
Neomenthol, 89
Nerol asymmetric hydrogenation of, 157
Nitrilases, 259, 260
Nitrile hydratases, 260

Nitrile oxides
  with allylic ethers, 191
  cycloadditions, 191
  for isoxazolines, 191
Nitriles, for acids, 259
Norephedrine, oxazolidinone from, 291
Nucleosides, 74 (*see also under specific compounds*)
  from D-ribose, 74

Olestra (*see* sucrose, esters)
Organoboranes, asymmetric reductions of ketones, 212
Orthoesters, in Claisen rearrangement, 188
Osmium tetroxide, for asymmetric dihydroxylations, 232
Oxaloacetic acid, from L-aspartic acid, 59
Oxazaborilidines, 211–223
  with acetophenone, 214–215
  with borane reducing agents, 217
  catalyst preparation, 215–216
  as chiral auxiliaries, 288
  containing diphenylprolinol, 214, 216
  with *m*-chloro chloroacetophenone, 221
  with *p*-chloro chloroacetophenone, 217–218
  with phenoxyphenylvinyl methyl ketone, 218–219
  reductions
    for denopamine, 220
    for Fluoxetine, 220
    for isoproterenol, 220
    reaction parameters, 217–219
    versatility, 219–220
Oxazinones, as chiral auxiliaries, 93, 288
Oxazoles, in pharmaceuticals, 290
Oxazolidinones
  for amino acids, 307
  as chiral auxiliaries, 288
  for CI-1008, 291–292
  from D-phenylalanine, 307
  for LY309887, 294

[Oxazolidinones]
  for matrix metalloproteinase inhibitors, 297
  for *N*-acyloxazolidinones, 307
  from norephedrine, 291
  in pharmaceuticals, 289
  for renin inhibitors, 296
  synthesis, 306
  for (+) SCH 54016, 295
Oxazolines
  as chiral auxiliaries, 288
  Claisen rearrangement, 188
Oxidations (*see also* Oxidative cleavage)
  enzymatic, 72
  D-glucose, 72
  D-sorbitol, 73
    in synthesis of naproxen, 120
  fermentation, 72
  l-menthol, 89
  microbial, 42
  transition metal catalyzed, 6
Oxidative cleavage
  D-fructose, 74
  D-mannitol, 75
Oxidoreductases, 246, 247
  dehydrogenases, 247–255
Oxo process (*see* Asymmetric hydroformylation)
Oxy-Cope rearrangement (*see* Cope rearrangement, oxy-Cope)

Papain, 265
Payne rearrangement, 106, 230–232, 332
Penicillin G acylase, 260
5-*n*-Pentyl-1,3-cyclohexadione, with (*R*)-citronellal, 186
Peptides
  cleavage, 259
  formation, 265
  from amino acids, 266
Pericyclic reactions, 7, 177–195
Pharmaceuticals, 289
  carbohydrates, 69
  top-sellers, 34

Phenoxyphenylvinyl methyl ketone, reduction with oxazaborilidines,
218–219
Phenoxypropionic acids (*see also under
specific compound names*)
from 2-chloropropioic acid, 104
from lactic acid, 104
synthesis, 43, 104
D-Phenylalanine
oxazolidinone from, 295, 307
by oxazolidinones, 307
by resolution, 63
L-Phenylalanine
for Aspartame, 49
by asymmetric hydrogenation, 147,
158, 163
biosynthetic pathway, 50–57
by biotransformations, 49–64
by chemoenzymatic methods, 49–64
from chorismic acid, 51–57
from cinnamic acid, 57–58, 268
for decahydroisoquinoline-3-carbox-
ylic acid, N-t-butylamide, 314
from erythrose-4-phosphate, 51–57
by fermentation, 31, 49–64
by hydantoins, 62–63
oxazolidinone from, 297, 306
by oxazolidinones, 307
pathway engineering, 272
by phenylalanine ammonia lyase, 57–
58
by phenylalanine dehydrogenase, 59–
60
from phenylpyruvate, 53, 58–59, 250
from phosphoenolpyruvate, 51–57
production, 49–64
resolution, enzymatic, 60–63
role of chorismate mutase, 52
role of prephenate dehydratase, 51
role of 3-deoxy-D-arabinoheptuloso-
nate-7-phosphate synthase, 51
secretion, 56
synthesis, 250
for tetrahydroisoquinoline-3-carbox-
ylic acid, 314
by transaminase enzymes, 53

Phenylalanine ammonia lyase, 57, 268
cinnamic acid, 57
for L-phenylalanine, 57–58
Phenylalanine dehydrogenase, 250
with formate dehydrogenase, 59–60
for L-phenylalanine, 59–60
(*1R,2S*)-2-Phenylcyclohexanol, as chiral
auxiliary, 293
(+)-α-Phenylethylamine
for resolutions
2-chloropropionic acid, 330
ibuprofen, 124
D-Phenylglycine
in antibiotics, 135
for cefaclor, 40
hydantoinase, 63
oxazolidinone from, 295
resolution, 63
by camphorsulfonic acid, 135
by salt formation, 135
synthesis, 277–278
Phenylpyruvate
for L-phenylalanine, 53, 58–59, 250
transamination, 53, 58–59
α-Phenylthio esters, Claisen ester enolate rearrangement, 189
Phosphinothricin, 259
Phosphoenolpyruvate, for L-phenylalanine, 51–57
Photooxygenation, α-pinene, 86
Pig liver esterase, 260
for resolution of, 2-amino-2-ethylhexa-
noic acid, 130
β-3-Pinanyl-9-borabicyclo[3.3.1]nonane
(*see* Alpine-Borane)
Pinenes, for other terpenes, 86
α-Pinene, 84–85
for camphor, 91
for l-carvone, 90
hydration, 90
hydroboration, 85
photooxygenation, 86
for α-terpineol, 90
for *trans*-verbenyl acetate, 86
Wagner-Meerwein rearrangement,
91

(+)-α-Pinene, 84
   isomerization, 86
   for (+)-β-pinene, 86
   purification, 85
(−)-α-Pinene, 84
   hydroboration, 90
   for l-(−)-carvone, 90
   from (−)-β-pinene, 85
β-Pinene, 85–86
(+)-β-Pinene, from (+)-α-pinene, 86
(−)-β-Pinene, 85–86
   isomerization, 85
   for (−)-α-pinene, 85
α-Pinene epoxide, 86
Pinocampheols, 86
Pinocamphones, 86
trans-Pinocarveol, 86
Piperazinediones (see Diketopiperazines)
Platinit (see Isomalt)
Platinum, asymmetric hydrogenations
   with cinchona alkaloid ligands, 167–168
PLE (see Pig liver esterase)
PNNP, 158
Poly-β-hydroxyalkanoates, 77, 272, 273
Poly-β-hydroxybutyrate, 273
Poly-β-hydroxyvalerate, 273
Polyketides (see Antibiotics, macrolides)
Polysaccharides, cleavage, 259
Potassium 3-aminopropylamide (see KAPA)
PPM, 164
   from l-4-hydroxyproline, 164
Pravastatin, 35
Prelog's rule, 247
Prephenate dehydratase, for l-phenylalanine, 51–55
D-Proline, 125–126
   from l-proline, 125
   resolution
   by resolution, 308
      with D-tartaric acid, 125
   by salt formation, 125

L-Proline
   for captopril, 37
   as chiral auxiliary, 297
   for D-proline, 125
   for enalapril, 34
   by fermentation, 34
   for lisinopril, 36
   racemization, 125
Proline derivatives, by chiral templates, 305–306
Propane-1,2-diol, from hydroxy acetone, 156–157
1,2,3-Propanetriol (see Solketal)
(S)-Propranolol, 237
Prostaglandins, 222
Proteases, 259, 264–268
   for amino acid coupling, 266
Prozac (see Fluoxetine)
Pyrethroids
   by asymmetric cyclopropanations, 171
   crysanthemic esters, 44
   deltamethrin, 44
   Sumitomo synthesis, 103
   synthesis, 44, 103

Racemates, 116–117
Racemic mixtures, 116
Racemic switch (see Chiral switch)
RAMP, 288
Raney nickel
   asymmetric hydrogenations with, 166–167
   with β-diketones, 166
   with β-keto esters, 166
Reductions, 6, 7 (see also Asymmetric hydrogenations)
   bacterial, 253–255
   fungal, 253–255
   transition metal catalyzed, 6
   yeast, 77, 250–253
Reductive aminations, 249
   phenylpyruvate, 59–60
Reformatsky reagents
   with imines, 194–195
   for β-lactams, 194–195

Reichstein-Grussner process, 73
Resolutions, 5, 115–138, 303
  acetates, by lipases, 104, 105
  3-acyl-2-methylpropionic acid, 37
  alcohols, 264
  allylic acetates, by substitution, 108–109
  amide, 42
  amino acids, 302, 307–308
  aspartic acid, with decarboxylases, 270
  bicyclo[3.2.0]heptenones, 333
  bufuralol, 335
  2-chloropropionic acid, 329–330
  chromatographic, 115
  crystallization, 29, 44
    Diltiazem, 37
  decarboxylases, 311–312
  diastereoisomers, 115–126
  Diltiazem, 37
  1,4-diols, 342
  α,α-disubstituted α-amino acids, 128–135
  dynamic, 115
    with hydantoinases, 136–138
  enzymatic, 3, 115, 128–138
    with amidases, 260
    amide, 42
    2-amino-2-ethylhexanoic acid, 130–135
    D-hydantoinases, 135–138
    ester hydrolysis, 124
    with esterases, 128
    with lipases, 128
  L-Dopa, 29
  by salt formation, 44, 116, 123–124
  epoxides, 126–127, 236–237
  ester formation with enzymes, 333
  high performance liquid chromatography, 29
  hydantoinase, 63
  hydantoins, 62–63
  ibuprofen, 122–125
  in carnitine synthesis, 42
  kinetic, 29, 115
    allyl acetates, 108

[Resolutions]
  of allyl alcohols, by Sharpless epoxidation, 229–230
  asymmetric dihydroxylations, 234
  dynamic, 263–264
  asymmetric hydrogenations, 153
    with Ru(BINAP) catalysts, 153
    α-substituted β-keto esters, 153–154
  D-p-hydroxyphenylglycine, 135–138
  D-phenylglycine, 135–138
  Naproxen, 29, 118–122
  Sertaline, 39
    by Sharpless epoxiation, 229–230
  simulated moving bed, 30
  S,S-tartaric acid, 76
Resolving agents, terpenes, 83
Rh(BINAP) catalysts, isomerizations with, 151–153
[Rh(dipamp)COD]BF$_4$ 147–149
  (see also Dipamp)
D-Ribose, for nucleosides, 74
Robinson annelation, 96
  of (−)-3-caranone, 96
  of cis-methylnopinone, 97
  for (+)-β-cyperone, 96
  of (+)-dihydrocarvone, 96
Ru(BINAP) catalysts
  with allyl alcohols, 157
  dynamic resolutions, 153
  for β-keto esters, 153
Ru(diamine) catalysts
  with aryl ketones, 168–169
  hydrogen transfer reactions, 168–169
  with imines, 169

Salen complexes (see Jacobsen epoxidation and under specific metals)
SAMP, 288
Scalemic, 84
SCH 51048, 261–262
(+) SCH 54016, 295
Schöllkopf method, for α-amino acids, 130–131
Serine hydroxymethyltransferase (see Threonine aldolase)

Sertaline, 39, 222
Sharpless epoxidation, 106, 227–230
    catalytic method, 228
    kinetic resolutions, 229–230
    stereochemical outcome, 228
    structures of intermediates, 228–229
[2,3]-Sigmatropic rearrangements, 192–
        194 (see also Wittig
        rearrangement and Evans
        rearrangement)
    anionic, 192
    neutral, 192
Simulated moving bed, for resolutions, 30
Simvastatin, 35
SMB (see Simulated moving bed)
S_N2 reactions (see Substitution reac-
        tions, S_N2)
SNZ-ENA-713, 222
R-Solketal, from L-ascorbic acid, 76
S-Solketal, 75
    from D-mannitol, 75
    from glyceric acid, 75
D-Sorbitol, 69, 73
    from D-glucose, 73
    enzymatic oxidation, 73
    to L-ascorbic acid, 73
L-Sorbose, 69
Sourcing, 11–32 (see also Suppliers)
    amino acids, 13–27, 29
    chiral compounds, 11–32
    costs, 31
    development times, 31
SQ31,765, 255
Steroids (see also under specific com-
        pounds)
    synthesis, 97–98
    by Wittig rearrangement, 192
Strecker reaction, 61
Styrene, for α-arylglycinols, 234
α-Substituted β-keto esters
    asymmetric reductions, 153
    for β-hydroxy esters, 154, 155
α-Substituted carboxylic acids, from
        α,β-unsaturated acids, 155–156
Substitution reactions, 7, 103–110
    allyl acetates, resolution, 108–109

[Substitution reactions]
    allylic, 108–109
    S_N2, 7, 103–105
Subtilisin, 265
Sucralose
    from D-glucose, 71
    as food additive, 70
    from sucrose, 70–71
    synthesis, 70–72
    chemical, 70–71
Sucrose, 69, 70
    acetylation, enzymatic, 71–72
    esters, 70
    as food additive, 70
    to isomaltulose, 72
    in synthesis of Sucralose, 70
Suppliers
    chiral center creators, 12
    chiral center elaborators, 12
    chiral compounds, 12
    confidential disclosure agreements, 28
    technology, 28

Tandem catalysis, 348
Tartaric acid, 76
    as chiral auxiliary, 290–291
    derivatives as catalysts for Diels-
        Alder reaction, 181
    esters as ligand in Sharpless epoxida-
        tion, 227–230
    as ligand, 76
D-Tartaric acid, resolution of D-proline, 125
R,R-Tartaric acid, 76
    by fermentation, 76
    as ligand for asymmetric hydrogena-
        tions, 166–167
S,S-Tartaric acid, 76
    resolution, 76
Templates (see Chiral templates)
Terpenes, 83–100 (see also under spe-
        cific compounds)
    as auxiliaries, 83
    isolation, 84
    as ligands, 83
    resolving agents, 83
    scalemic, 84

Terpineol, 86, 89, 90–91
α-Terpineol, 88, 90
  from α-pinene, 90
β-Terpineol, 88, 90
S-Terpinene-4-ol
  for cinmethylin, 98
  for dihydropinol, 99
1,2,3,4-Tetrahydroisoquinoline-3-
  carboxylic acid, from L-phenylalanine, 314
Tetrahydrolipostatin, 167
Tetronic acids, Claisen ester enolate rearrangement, 189
Thermolysin, in Aspartame synthesis,
  256
L-Thienylalanine, 259
Thioesters, Claisen rearrangement, 188
L-Threitol, 231
Threonine aldolase, 256–257
TOSOH process, for Aspartame, 265
Transaminases, 257–259 (see also under specific examples)
  for amino acids, 312–314
  coupled enzyme systems, 313
  for D-amino acids, 313–314
  for L-aspartate, 58–59
  for L-leucine, 58–59
  for L-phenylalanine, 58–59
  for L-tyrosine, 58–59
  for L-valine, 58–59
Transesterification, for enzymatic resolution of 2-chloropropionic acid,
  43
Transferases, 246
Transition metals
  catalyzed reactions, 143–173
  ligands, 144
4,1′,6′-trichloro-4,1′6′-trideoxy-galacto-
  sucrose (see Sucralose)
Trypsin, 265
L-Tryptophan, pathway engineering, 272
L-Tyrosine
  biosynthesis, 51
  by transaminases, 58
Tyrosine aminotransferase, 257

α,β-Unsaturated acids, for asymmetric
  hydrogenations, 155–156
γ,δ-Unsaturated amino acids
  by asymmetric hydrogenation, 345
  from glycine allylic esters, 190
α,β-Unsaturated esters, for, 3-amino-2-
  hydroxy esters, 234
D-Valine, by hydantoinases, 136
L-Valine
  oxazolidinone from, 294
  by transaminases, 58
cis-Verbenol, 86
trans-Verbenol, 86
Verbenone, 86
trans-Verbenyl acetate
  from α-pinene, 86
  for trans-verbenol, 86
Vitamin B_T (see Carnitine)
Vitamin C (see L-Ascorbic acid)

Wagner-Meerwein rearrangement, of
  α-pinene, 91
Wittig rearrangement, 192–193
  allyl ethers, 192
  for antibiotics, 192
  for dihydrofurans, 192
  for steroids, 192

D-Xylose, 69
D-Xylose isomerase (see Glucose isomerase)

Yamagushi-Mosher reagent, 212
Yeast reductions, 247, 250–253
  of acetoacetate, 77
  of benzulacetoacetate, 252
  of 6-bromo-β-tetralone, 251
  of β-diketones, 250
  of ethyl 4-chloroacetoacetate, 253
  of β-keto ester, 250–253
  of keto sulfones, 252
  of ketones, 250–253

Zambon process, for naproxen, 119,
  290–291